# A TEXTBOOK OF
# DISCRETE MATHEMATICS

# A TEXTBOOK OF
# DISCRETE MATHEMATICS

## Second Edition


**Harish Mittal**

M.Tech

*Principal and Head*
Department of CSE
BM Institute of
Engineering and Technology
Sonepat, Haryana

**Vinay Kumar Goyal**

M.Tech, Ph.D.

*Vertical Head – IT*
iNurture Education
Solutions Private Limited
Chandigarh

**Deepak Kumar Goyal**

M.Tech.

*Principal and Professor*
Dept. of CSE
Vaish College of Engineering
Rohtak, Haryana

**I.K. International Pvt. Ltd.**

NEW DELHI

*Published by*
I.K. International Pvt. Ltd.
4435-36/7, Ansari Road, Daryaganj
New Delhi-110 002 (India)
E-mail: info@ikinternational.com
Website: www.ikbooks.com

ISBN: 978-93-90620-25-8



Published by Krishan Makhijani for I.K. International Pvt. Ltd., 4435-36/7, Ansari Road, Daryaganj, New Delhi-110 002 and Printed by Rekha Printers Pvt. Ltd., Okhla Industrial Area, Phase II, New Delhi-110 020.

*Dedicated to our Parents*

# Preface

Discrete mathematics is the part of mathematics devoted to the study of discrete objects, here discrete means consisting of distinct or unconnected elements. The kind of problems solved using discrete Mathematics include

- How many ways are there to choose a valid password on a computer system?
- What is the probability of winning a lottery?
- What is the shortest path between two cities using a transportation system?
- How can it be proved that a sorting algorithm correctly sorts a list?
- How many valid Internet addresses are there?

You will learn the discrete structures and techniques needed to solve problems such as these. More generally, discrete mathematics is used whenever objects are counted, when relationships between finite (or countable) sets are studied, and when processes involving a finite number of steps are analyzed. A key reason for the growth in the importance of discrete Mathematics is that information is stored and manipulated by computing machines in a discrete fashion.

There are several important reasons for studying discrete Mathematics. First, through this course you can develop your mathematical maturity, that is, your ability to understand and create mathematical arguments. Discrete Mathematics provides the mathematical foundations for many computer science courses, including data structures, algorithms, database theory, automata theory, computer security, and operating systems. Students find these courses much difficult when they have not had the appropriate mathematical foundations from discrete math.

Courses based on the material studied in discrete mathematics include logic, set theory, number theory, combinatories, graph theory and probability theory. Also, discrete Mathematics contains the necessary mathematical background for solving problems in operations research including many discrete optimization techniques, chemistry, engineering, biology, and so on.

There are plenty of exercises in this text similar to those addressed in the examples, a large percentage of the exercises require original thought. The material discussed in the text provides the tools needed to solve these exercises. You will learn the most by actively working exercises. I suggest that you solve as many as you possibly can. This book covers syllabi of BE/B.Tech. (CSE/IT), MCA, BCA and B.Sc.

**HARISH MITTAL**
**VINAY KUMAR GOYAL**
**DEEPAK KUMAR GOYAL**

# Acknowledgements

I owe my deep concern to many individuals whose encouragement, guidance and suggestions have resulted in a better manuscript:

| | |
|---|---|
| Prof. (Dr.) J. P. Mittal | Principal & Head, Department of Mathematics, Satpriya Institute of Engineering & Technology, Rohtak |
| Prof. (Dr.) Dharmender Kumar | Dean (FET), Chairman, Department of CSE, Guru Jambheshwar University of Science and Technology, Hisar |
| Dr. P. K. Bhatia | Reader, Department of CSE, Guru Jambheshwar University of Science and Technology, Hisar |
| Mr. Mukesh Singhal | AP, Department of IT, Vaish College of Engineering, Rohtak |
| Mr. Suresh Ahalawat | AP & Head, Department of CSE, Satpriya Institute of Engineering & Technology, Rohtak |
| Mr. Manoj Gupta | M.Tech., IIT Roorkee |

Additionally, I thank the management, administration and staff of Satpriya Institute of Engineering & Technology, Rohtak, for their valuable suggestions and whole hearted cooperation. My special thanks to students of Vaish College of Engineering, Rohtak, who provided me valuable feedback and critical appraisal.

It was a pleasure working with IK International publishers. I specially thank Mr. Anand for his encouragement, support and assistance. Finally, I pay my gratitude to my family members for their love and patience during writing this book.

**HARISH MITTAL**

I owe my deep concern to many individuals whose encouragement, guidance and suggestions have resulted in a better manuscript:

| | |
|---|---|
| Dr. Kuldeep Bansal | Chairman, Department of Mathematics, GJUS&T, Hisar |
| Prof. Rajesh Gargi | Head (CSE) JIET, Jind |
| Er. Ravinder Ahuja | Lecturer, JIET, Jind |

Additionally I thank to management and administration and staff of JIET, Jind for their valuable suggestions and whole hearted cooperation. My special thanks to students of JIET, Jind who provided me valuable feedback and critical appraisal.

It was a pleasure working with IK International publishers. I specially thank Mr. Anand for his encouragement, support and assistance. Finally, I pay my gratitude to my family members for their love and patience during writing this book.

**VINAY KUMAR GOYAL**

I owe my deep concern to many individuals whose encouragements, guidance and suggestions have resulted in a better manuscript, especially to my family members for their love and patience during writing this book. Additionally I thank the management, administration and staff of Vaish College of Engineering, Rohtak for their valuable suggestions and whole hearted cooperation. My special thanks to students of Vaish College of Engineering, Rohtak who provide me valuable feedback and critical appraisal.

**DEEPAK KUMAR GOYAL**

# Contents

## SECTION 2

## SECTION 3

**SECTION 4**

**10. GRAPH THEORY** 309

## SECTION 5

# Section 1

# 1

# Sets

---

## 1.1 SETS

### 1.1.1 The Concept of a Set

Sets are used to group objects having similar properties. Set theory is a means to study such groups in an organised way. The German mathematician George Cantor (1845–1918) is considered the founder of the set theory. It has a great contribution in the study of different branches of Mathematics.

**Definition**: A set is a collection of well defined and distinct objects. The objects of a set are called elements or points. If $a$ is an element of a set S, then $a \in$ S otherwise $a \notin$ S. The set whose elements are $a$, $b$, $c$ is denoted as $\{a, b, c\}$. We usually give upper case letter as name to a set. For example, we write S $= \{a, b, c\}$. Elements are denoted by lower case letters.

### 1.1.2 Representation of Sets

Sets can be represented in two ways:

    (i) Tabular or Roster Method
    (ii) Builder Method or Rule Method

**(i) Tabular or Roster Method:** A set is represented by listing the members within $\{\}$. Elements are separated from one another by comma ','.

    e.g., $\{1, 2, 3\}$.

**(ii) Builder Method or Rule Method:** A set is represented by the property which the elements satisfy.

    e.g., set of all rational numbers (Q) $= \{x: x = a/b, a, b \in \text{I}, b \neq 0\}$

- When the number of elements in a set is small, we use listing method, but when the number of elements in the set is large or infinite, we use the set builder form.

Some **examples** of sets are:

- Set of all natural numbers (N) = {1, 2, 3...}.
- Set of all integers (Z or I) = {..., −3, −2, −1, 0, 1, 2 ...}.
- Set of all rational numbers (Q) = {$x \mid x = a/b, a, b \in$ I, $b \neq 0$}.
- Set of all real numbers = {$x \mid x$ is either rational or irrational}.
- $R^+$ is set of all +ve real numbers.
- $I_0$ is set of all integers excluding 0.
- $I_+$ is set of all +ve integers.
- $N_k$ is set of all natural numbers $\leq k$ = {1, 2, ... $k$}.
- The set of vowels of English alphabet = {$a, e, i, o, u$}.
- The set of even numbers between 1 and 15 {2, 6, 8, 10, 12, 14}.

**Example 1:**  Write the set of the letters in the word INDIA.

*Solution*

As no element can be repeated in a set, the set is {I, N, D, A}

**Example 2:**  Write the following sets into the tabular form

(i)  A = {$x: 0 < x < 6, x$ is the set of integers}.
(ii)  A = {$x: x^2 - 1 = 0, x$ is a natural number}.

*Solution*

(i)  Integers satisfying $0 < x < 6$ are 1, 2, 3, 4, 5

∴   A = {1, 2, 3, 4, 5}

(ii)  $x^2 - 1 = 0$

$(x - 1)(x + 1) = 0$

∴   $x = 1, -1$

But x is a natural number. Therefore discarding $x = -1$

Hence A = {1}

## Some Basic Definitions

### 1.1.3 Empty Set or Null Set or Void Set

A set having no element is called empty or null or void set and is denoted by $\phi$ or {}.

e.g.,   A = {$x: x^2 + 1 = 0, x$ is a real number}.

### 1.1.4 Singleton Set or Unit Set

A set having only one element.

e.g., $\{0\}, \{x\}, \{\phi\}$.

### 1.1.5 Subset

The set A is subset of set B if each element of A is also an element of B, we write $A \subseteq B$. e.g., If $A = \{a, b, c\}$ and $B = \{a, b, c, e\}$. Then clearly $A \subseteq B$. If A is not a subset of B, we denote $A \nsubseteq B$.

- For every set A, $A \subseteq A$, i.e., A is itself a subset of A.
- The empty set $\phi$ is always a subset of every set.

**Proper Subset:** If A is a subset of B and $A \neq B$, then A is a proper subset of B, we write $A \subset B$ e.g., if $A = \{1, 2, 3\}$ and $B = \{1, 2, 3, 4\}$, then $A \subset B$.

**Super Set:** If A is a subset of B then B is a superset of A, we write $B \supseteq A$.

### 1.1.6 Equal Sets

Two sets A and B are equal $(A = B)$, only if both contain same elements. Two sets are equal if and only if $A \subseteq B$ and $B \subseteq A$.

e.g., If $A = \{1, 2\}$, $B = \{1, 2\}$, then $A = B$

### 1.1.7 Equivalent Sets

Two finite sets A and B are equivalent if they have the same number of elements, we write $A \sim B$. Two sets are called equivalent sets, if and only if there is one to one correspondence between their elements.

e.g., If $A = \{a, b, c\}$ and $B = \{1, 2, 3\}$, then $A \sim B$.

*Note:* If two sets are equal, they are equivalent but two equivalent sets are not necessarily equal.

### 1.1.8 Comparable Sets

Two sets, A and B, are said to be comparable if either $A \subseteq B$ or $B \subseteq A$. For example,

(i) If $A = \{1, 2, 3\}$, $B = \{1, 2, 3, 4\}$. Then $A \subseteq B$ so that A and B are comparable.
(ii) If $A = \{a, b\}$ and $B = \{a\}$. Then $B \subseteq A$, so that A and B are comparable.

### 1.1.9 Universal Set

In any application of set theory all sets under consideration will likely be sub-sets of a finite set called "universal set", denoted by U. Different universal sets are used in different contexts. The choice of the universal set is not unique e.g., in the study of different set of letters of English alphabet, the universal set is the set of all letters of English alphabet.

**Theorem 1** The empty set is the subset of every set.

*Proof*

Let S be any set. To show that $\phi \subseteq S$, we must show that $\forall\, x\, (x \in \phi \rightarrow x \in S)$ is true. Since the empty set $\phi$ contains no elements, it follows that $x \in \phi$ is always false. It follows that the implication $x \in \phi \rightarrow x \in S$ is always true, since its hypothesis is always false (an implication with a false hypothesis is true) i.e. $\forall\, x\, (x \in \phi \rightarrow x \in S)$ is true, so $\phi \subseteq S$.

**Theorem 2** Every set is the subset of itself.

*Proof*

Let S be any set. Each element of S is an element of the set S. By definition of the subset, $S \subseteq S$

**Theorem 3** If $A \subseteq B$ and $B \subseteq A$, then $A = B$.

*Proof*

$A \subseteq B$

$\therefore \quad x \in A$ and $x \in B$ $\hspace{6cm}$ (1)

Also $B \subseteq A$.

$\therefore \quad x \in B$ and $x \in A$ $\hspace{6cm}$ (2)

From (1) and (2), $\quad x \in A$ and $x \in B$.

$\therefore \quad A = B$

**Theorem 4** If $A \subseteq B$ and $B \subseteq C$, then $A \subseteq C$.

*Proof*

$A \subseteq B$

$\therefore \quad x \in A$ and $x \in B$ $\hspace{6cm}$ (1)

$B \subseteq C$

$\therefore \quad x \in B$ and $x \in C$ $\hspace{6cm}$ (2)

From (1) and (2), it is clear

$x \in A$ and $x \in C$

$\therefore \quad A \subseteq C$

**Theorem 5** If a set S contains $n$ elements, then $P(S)$ contains $2^n$ elements.

***Proof***

Let $a_1, a_2, a_3 \dots a_n$, be $n$ elements of the set S.

Number of subsets of S having one element each $= {}^nC_1$

Number of subsets of S having two elements each $= {}^nC_2$

Number of subsets of S having three elements each $= {}^nC_3$

Number of subsets of S having $n$ elements each $= {}^nC_n$

Also, $\phi$ is the subset of S, its number is $= {}^nC_0$

$\therefore \quad$ Total number of subsets of $A = P(S) = {}^nC_0 + {}^nC_1 + {}^nC_2 \dots + {}^nC_n$

$$= (1 + 1)^n \quad \text{[by Binomial theorem]}$$

$$= 2^n$$

### 1.1.10 Venn-Diagram

Venn diagram is a graphical representation of a set. The universal set, U, is represented by a rectangle, and a subset A of U by interior of a circle or some other simple closed curve.

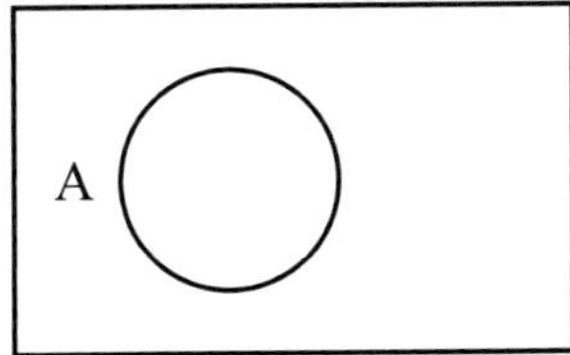
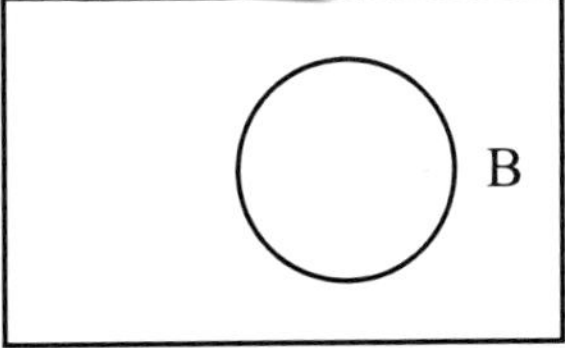

- The English mathematician John Venn introduced the use of these diagrams.
- Venn diagrams are used to indicate the relationship between sets.

## 1.2 OPERATIONS ON SETS

### 1.2.1 Union of Sets

The union of two sets A and B, written as $A \cup B$, where $A \cup B = \{x \mid x \in A \vee x \in B\}$.

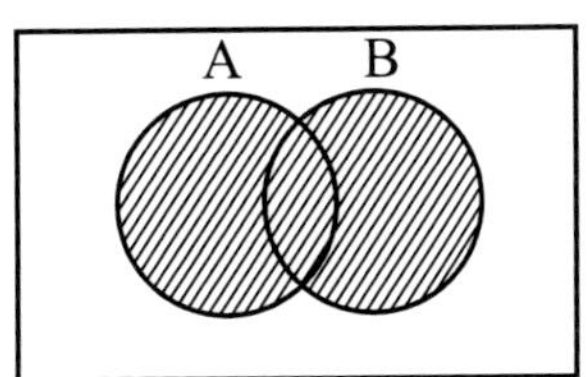

$A \cup B$ (Shaded)

### 1.2.2 Intersection of Sets

The intersection of two sets A and B, written as $A \cap B$, where $A \cap B = \{x \mid x \in A \ \forall \ x \in B\}$.

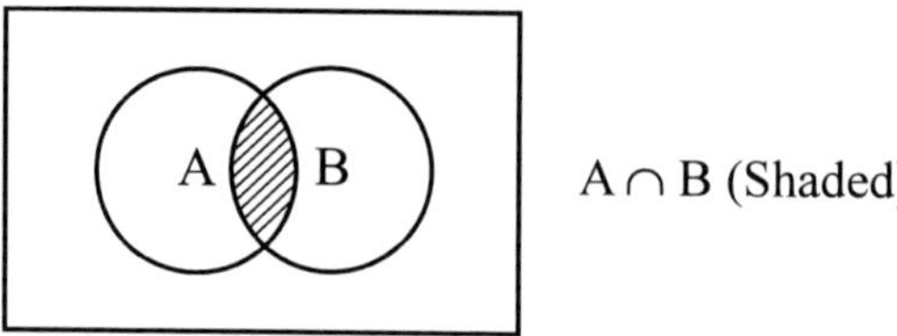

$A \cap B$ (Shaded)

### 1.2.3 Disjoint Sets

Two sets A and B are called disjoint if they have no common elements. We write $A \cap B = \phi$ e.g., the set of all even integers and the set of all odd integers are disjoint sets.

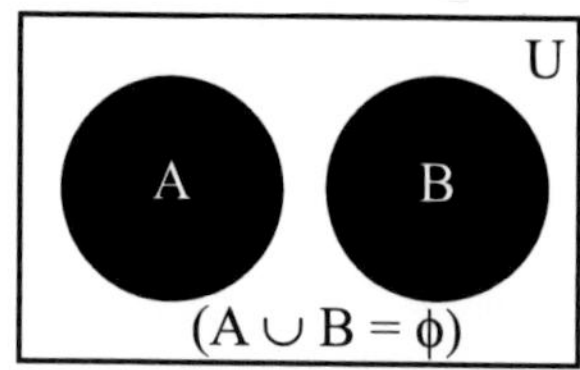


### 1.2.4 Complement of a Set

Complement of a set A, written as $A'$ or $A^c$, where, $A' = U - A$, $U' = \phi$

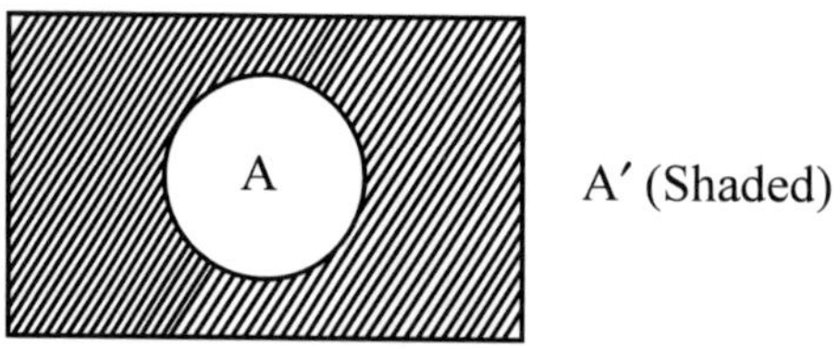


$A'$ (Shaded)

e.g.,   (i)  If $U = \{1, 2, 3, 4, 5, 6\}$ and $A = \{1, 2, 4\}$
  Then $A' = \{3, 5, 6\}$

  (ii)  If U is the set of all letters of English alphabet and A the set of vowels then $A'$ is the set of letters of the English alphabet other than the vowels.

### 1.2.5 Difference of Sets

Difference of two sets A and B, written as $A - B$, where $A - B = \{x \mid x \in A \text{ and } x \notin B\}$.

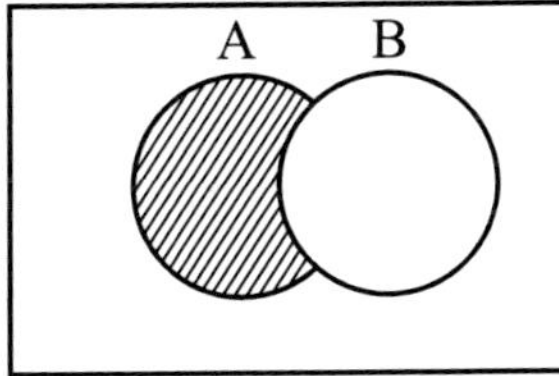

A – B (Shaded)

## 1.2.6 Symmetric Difference of Sets

Symmetric Difference of two sets A and B, written as $(A \oplus B)$, contains elements which are in A or B but not in both.

$$(A \oplus B) = (A - B) \cup (B - A)$$

- $\{a, b\} \oplus \{b, c\} = \{a, c\}$.
- $\{a, b\} \oplus \{a, b\} = \phi$.

**Theorem 6** (On Symmetric difference) If A and B are any two sets, then $A \oplus B = (A \cup B) - (A \cap B)$.

***Proof***

Let $x \in A \oplus B$.

Then $x \in (A - B) \cup (B - A)$

$\Leftrightarrow (x \in A \text{ and } x \notin B) \text{ or } (x \in B \text{ and } x \notin A)$
$\Leftrightarrow (x \in A \text{ and } x \notin B) \text{ or } x \in B \text{ and } [x \in A \text{ and } x \notin B] \text{ or } x \in A$
$\Leftrightarrow [x \in B \text{ or } (x \in A \text{ and } x \notin B)] \text{ and } [x \notin A \text{ or } (x \in A \text{ and } x \in B)]$
$\Leftrightarrow [(x \in B \text{ or } x \in A) \text{ and } (x \in B \text{ or } x \notin B)] \text{ and } [(x \notin A \text{ or } x \in A) \text{ and } (x \notin A \text{ or } x \notin B)]$
$\Leftrightarrow [x \in A \cup B \text{ and } x \in U] \text{ and } [x \in U \text{ and } x \notin A \cap B)]$
$\Leftrightarrow x \in (A \cup B) \text{ and } x \notin (A \cap B)$
$\Leftrightarrow x \in [(A \cup B) - (A \cap B)]$

i.e. $A \oplus B = (A \cup B) - (A \cap B)$

**Example 3:**

Let
$\quad U = \{1, 2, 3, 4, 5, 6, 7, 8, 9, 10\}$
$\quad A = \{1, 2, 3, 4\}$
$\quad B = \{5, 6, 7, 8\}$
$\quad C = \{9, 10\}$
$\quad D = \{1, 3, 5, 7, 9\}$

Find the following subsets of U:

$\quad$ (i) $A \cup B$ $\quad$ (ii) $A \oplus B$ $\quad$ (iii) $C - D$ $\quad$ (iv) $A \cap \phi$ $\quad$ (v) $(C \cup D)'$

*Solution*

(i) $A \cup B = \{x : x \in A \text{ or } x \in B\}$
$= \{1, 2, 3, 4, 5, 6, 7, 8\}$

(ii) $A \oplus B = \{x : x \in A \text{ or } x \in B, x \notin A \cap B)$
$= \{1, 2, 3, 4, 5, 6, 7, 8\}$

(iii) $C - D = \{x : x \in C, x \notin D\}$
$= \{10\}$

(iv) $A \cap \phi = \{x : x \in A, x \in \phi\}$
$= \phi$

(v) $(C \cup D)'$

$(C \cup D) = \{1, 3, 5, 7, 9, 10\}$

$(C \cup D)' = U - (C \cup D) = \{2, 4, 6, 8\}$

**Example 4:** Are the sets $\phi$, $\{\phi\}$, $\{0\}$ different?

*Solution*

Set $\phi$ contains no element. The set $\{\phi\}$ contains one element $\phi$.
The set $\{0\}$ contains one element 0. All the three sets are different.

**Example 5:** Are the following sets equal?
$S_1 = \{1, 2, 2, 3\}$, $S_2 = \{1, 2, 3\}$

*Solution*

We have $S_1 = \{1, 2, 2, 3\} = \{1, 2, 3\} = S_2$

**Example 6:** Find the number of proper subsets of the set of letters of the HINDUSTAN.

*Solution*

The set of letters of the word HINDUSTAN is $\{H, I, N, D, U, S, T, A\}$.

Because there are 8 elements in the set, therefore the total number of subsets is $= 2^8$. Since the set consisting of all the 8 elements is not a proper set. Therefore the required number of proper subsets

$= 2^8 - 1$

$= 256 - 1 = 255$

**Example 7:** Prove that $A \subset \phi \Rightarrow A = \phi$.

*Solution*

Let $A \subset \phi$.

Since $\phi$ is a subset of every set.

Therefore, in particular $\phi$ is a subset of A, i.e. $\phi \subset A$.

Now $\phi \subset A$ and $A \subset \phi \Rightarrow A = \phi$.

**Example 8:** Is a set A comparable with itself?

### Solution

Since $A \subset A$ is true for every set A, by definition, this declares that A is comparable with itself.

### 1.2.7 Fundamental Products

**Fundamental Product:** A fundamental product of sets $A_1$, $A_2$, ... $A_n$ is an expression of the form $A_1^* \cap A_2^* \cap A_3^* \cap \ldots \ldots \cap A_n^*$, where $A_i^*$ is either $A_i$ or $A_i^c$. For $n$ sets we may have $2^n$ fundamental products.

**Example 9:** List the fundamental products of three sets A, B and C.

There are $2^3 = 8$ fundamental products as follows

$$P1 = A \cap B \cap C \qquad P2 = A \cap B \cap C^c \qquad P3 = A \cap B^c \cap C$$
$$P4 = A \cap B^c \cap C^c \qquad P5 = A^c \cap B \cap C \qquad P6 = A^c \cap B \cap C^c$$
$$P7 = A^c \cap B^c \cap C \qquad P8 = A^c \cap B^c \cap C^c$$

Each of these fundamental products can be represented by the Venn diagram as follows

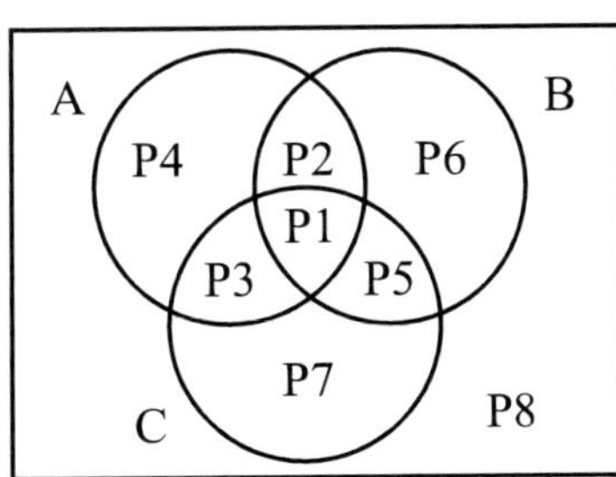


**Example 10:** Prove that for any three sets A, B and C.

(a) $A \cap (B - C) = (A \cap B) - (A \cap C)$
(b) $A - (B - C) = (A - B) \cup (A \cap C)$

***Proof***

**(a)** $A \cap (B - C) = (A \cap B) - (A \cap C)$

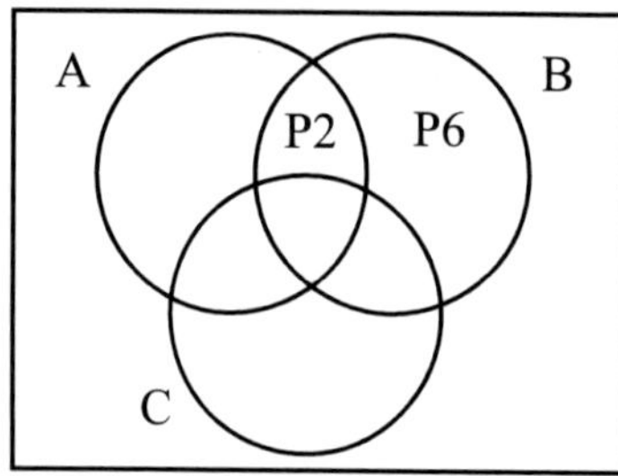

Fig. 1

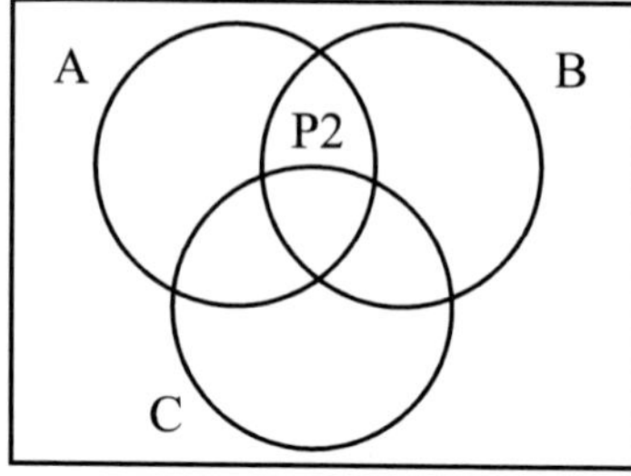

Fig. 2

$B - C$ represented by area P2 and P6 (Fig. 1), P2 represents $A \cap (B - C)$ (Fig. 2).

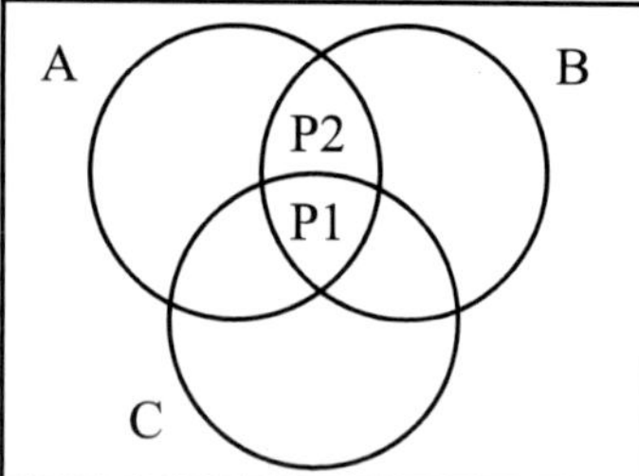

Fig. 3

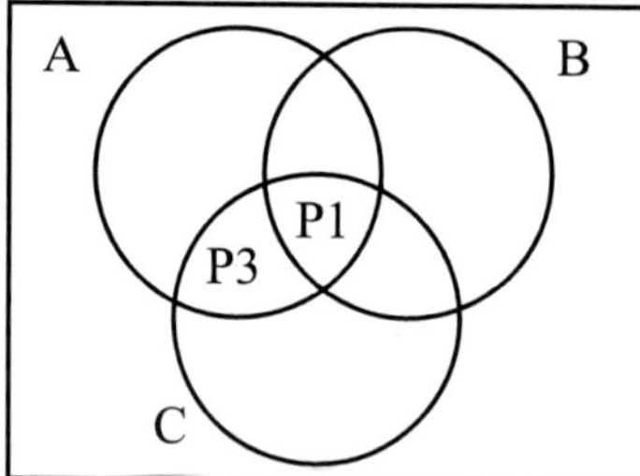

Fig. 4

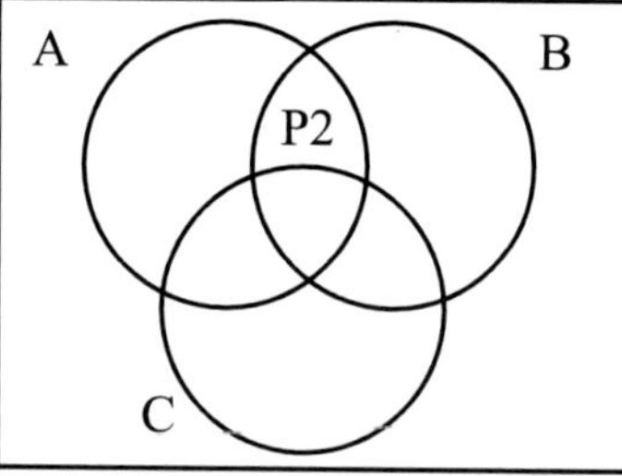

Fig. 5

$A \cap B$ represented by area P2 and P1 (Fig. 3), $A \cap C$ represented by area P1 and P3 (Fig. 4). $A \cap B - A \cap C$ represented by area P2 (Fig. 5).

Area represented by Venn diagram for $A \cap (B - C)$ and $(A \cap B) - (A \cap C)$ is same, hence

$$A \cap (B - C) = (A \cap B) - (A \cap C)$$

**(b)** $A - (B - C) = (A - B) \cup (A \cap C)$

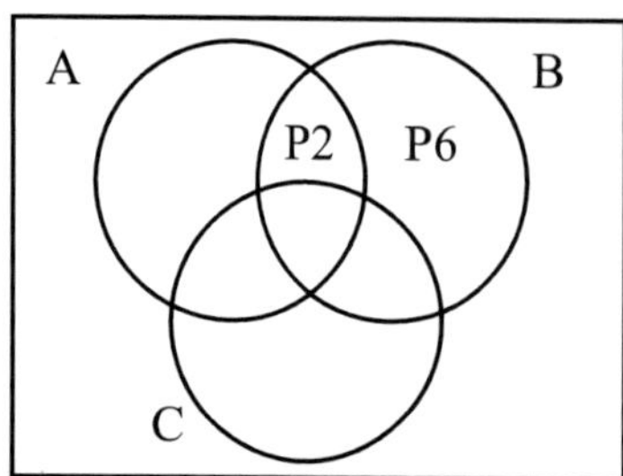

Fig. 1

Fig. 2

B – C represented by area P2 and P6 (Fig. 1), A – (B – C) represented by area P1, P3 and P4 (Fig. 2).

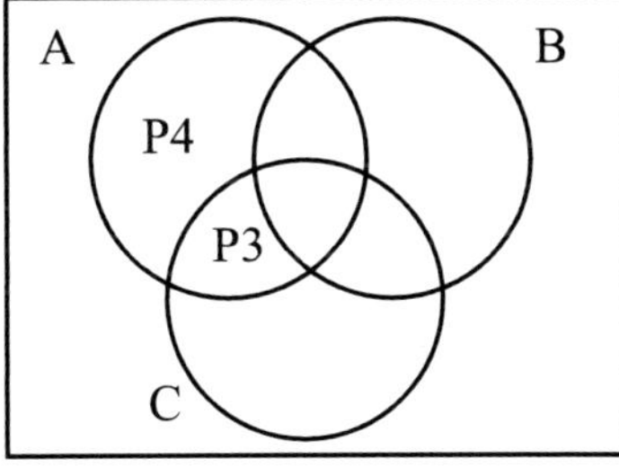

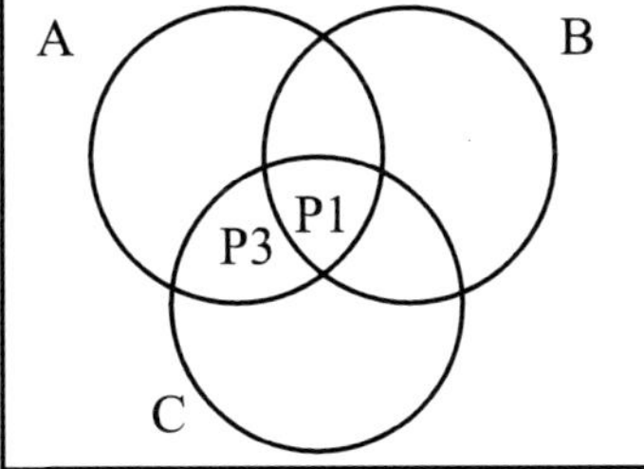

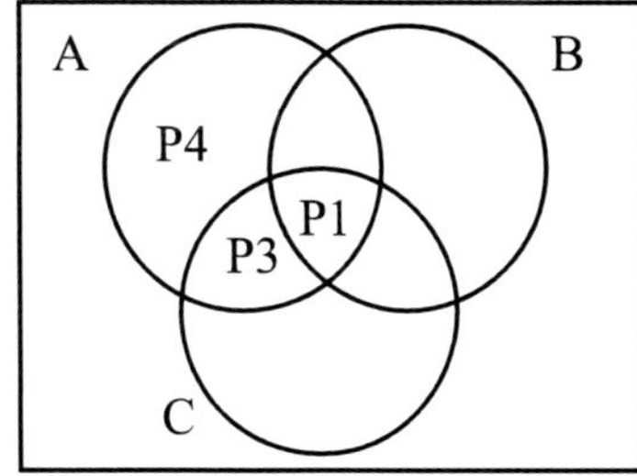

|  Fig. 3  |  Fig. 4  |  Fig. 5  |

A – B represented by area P3 and P4 (Fig.3), A – C represented by area P1, P3 (Fig. 4), (A – B) $\cup$ (A – C) represented by area P1, P3 and P4 (Fig. 5).

Area represented by Venn diagram for A – (B – C) and (A – B) $\cup$ (A – C) is same, hence

$$A - (B - C) = (A - B) \cup (A - C)$$

## Exercise 1.1

1. Is the set A = {x: x + 1 = 1} null?

2. Write down all the subsets of the set {1, 2, 3}.

3. How many subsets of the letters of the word ALLAHABAD will be formed?

4. Are the following sets equal?

   (i) A = {$x$: $x$ is a letter in the word WOLF}.

   (ii) B = {$x$: $x$ is a letter in the word FOLLOW}.

5. If A = {1, 2, 3, 4}, B = {2, 3, 4, 5} & C = {4, 5, 6, 7}, find A – (B – C).

6. If A = {1, 3, 6, 10, 15, 21}, & B = {15, 3, 6}, find (A – B) $\cap$ (B – A).

7. If X = {1, 2, 3, 4, 5} & Y = {1, 3, 5, 7, 9}, find the values of X $\cap$ Y and (X – Y) $\cup$ (Y – X).

8. If A = {1, 2, 3, 4, 5} & B = {1, 3, 5, 7, 9}, find the symmetric difference of A & B.

9. If A = {a, b, c, d}, & B = {e, f, c, d}, find A $\oplus$ B.

10. Let A = {1, 2, 3, 4, 5} and B = {0, 3, 6}. Find  a) A $\cup$ B  b) A $\cap$ B  c) A – B  d) B – A

**Answers to Selected Problems**

1. No

2. {1}, {2}, {3}, {1, 2}, {1, 3}, {2, 3}, {1, 2, 3}, $\phi$

3. {A, L, H, B, D}, No. of subsets $2^5 = 32$

4. Yes

6. $\phi$

7. {1, 3, 5}, {2, 4, 7, 9}

8. {2, 4, 7, 9}

10. a. {0, 1, 2, 3, 4, 5, 6}
    b. {3}
    c. {1, 2, 3, 4, 5}
    d. {0, 6}

## 1.3 ALGEBRA OF SETS

The operations union, intersection and complement of sets satisfy various laws or set identities.

### 1.3.1 Laws of the Algebra of Sets or Set Identities

| | |
|---|---|
| Identity laws | 1a. $A \cup \phi = A$ <br> 1b. $A \cap U = A$ |
| Identity laws | *Domination laws* <br> 2a. $A \cup U = U$ <br> 2b. $A \cap \phi = \phi$ |
| | *Idempotent laws* <br> 3a. $A \cup A = A$ <br> 3b. $A \cap A = A$ |
| Complementation law or Involution law | 4. $(A')' = A$ |
| Commutative laws | 5a. $A \cup B = B \cup A$ <br> 5b. $B \cup A = A \cup B$ |
| Associative laws | 6a. $(A \cup B) \cup C = A \cup (B \cup C)$ <br> 6b. $(A \cap B) \cap C = A \cap (B \cap C)$ |
| Distributive laws | 7a. $A \cap (B \cup C) = (A \cap B) \cup (A \cap C)$ <br> 7b. $A \cup (B \cap C) = (A \cup B) \cap (A \cup C)$ |
| De Morgan's laws | 8a. $(A \cup B)' = A' \cap B'$ <br> 8b. $(A \cap B)' = A' \cup B'$ |
| Absorption laws | 9a. $A \cup (A \cap B) = A$ <br> 9b. $A \cap (A \cup B) = A$ |
| Complement laws | 10a. $A \cup A' = U$ <br> 10b. $A \cap A' = \phi$ <br> 11a. $U' = \phi$ <br> 11b. $\phi' = U$ |

**Example 11:** Prove the following laws:

1. Idempotent law: $A \cup A = A$
2. Commutative law: $A \cup B = B \cup A$
3. Associative law: $(A \cup B) \cup C = A \cup (B \cup C)$
4. Identity law: (a) $A \cup \phi = A$,

   (b) $A \cup U = U$

***Proof***

1. To prove $A \cup A = A$

   Let $x \in A \cup A$
   $\Rightarrow x \in A$ or $x \in A$
   $\Rightarrow x \in A$
   i.e., $A \cup A \subseteq A$    (1)

   Let $y \in A$. Then
   $\Rightarrow y \in A \rightarrow y \in A$ or $y \in A$
   $\Rightarrow y \in A \cup A$
   i.e., $A \subseteq A \cup A$    (2)

   From (1) & (2), we have

   $$\mathbf{A \cup A = A}$$

2. To prove $A \cup B = B \cup A$

   Let $x \in A \cup B$
   $\Rightarrow x \in A$ or $x \in B$
   $\Rightarrow x \in B$ or $x \in A$
   $\Rightarrow x \in B \cup A$
   i.e., $A \cup B \subseteq B \cup A$    (1)

   Let $y \in B \cup A \rightarrow y \in B$ or $y \in A$
   $\Rightarrow y \in A$ or $y \in B$
   $\Rightarrow y \in A \cup B$
   i.e., $B \cup A \subseteq A \cup B$    (2)

   From (1) & (2), we have

   $$\mathbf{A \cup B = B \cup A}$$

3. To prove $(A \cup B) \cup C = A \cup (B \cup C)$

   Let $x \in (A \cup B) \cup C$
   $\Rightarrow x \in (A \cup B)$ or $x \in C$
   $\Rightarrow (x \in A$ or $x \in B)$ or $x \in C$
   $\Rightarrow x \in A$ or $(x \in B$ or $x \in C)$
   $\Rightarrow x \in A$ or $(x \in B \cup C)$

   i.e., $x \in A \cup (B \cup C)$    (1)

   Let $y \in A \cup (B \cup C)$
   $\Rightarrow y \in A$ or $(y \in B \cup C)$
   $\Rightarrow y \in A$ or $(y \in B$ or $y \in C)$

$\Rightarrow (y \in A \text{ or } y \in B) \text{ or } (y \in C)$
$\Rightarrow (y \in A \cup B) \text{ or } (y \in C)$
$\Rightarrow y \in (A \cup B) \cup C$

i.e., $A \cup (B \cup C) \subseteq (A \cup B) \cup C$         (2)

From (1) & (2), we have

$$(A \cup B) \cup C = A \cup (B \cup C)$$

4.  To prove $A \cup \phi = A$ and $A \cup U = U$

Let $x \in A \cup \phi$
$\Rightarrow x \in A \text{ or } x \in \phi$
$\Rightarrow x \in A$

i.e., $A \cup \phi \subseteq A$         (1)

But $A \subseteq A \cup \phi$         (2)

From (1) & (2), we have

$$A \cup \phi = A$$

Again let $y \in A \cup U$
$\Rightarrow y \in A \text{ or } y \in U$
$\Rightarrow y \in U$

i.e., $A \cup U \subseteq U$         (1)

But $U \subseteq A \cup U$         (2)

From (1) and (2), we have

$$A \cup U = U$$

**Example 12:**  Prove the following laws:

1.  Idempotent law: $A \cap A = A$
2.  Commutative law: $A \cap B = B \cap A$
3.  Associative law: $(A \cap B) \cap C = A \cap (B \cap C)$
4.  Identity law: $A \cap U = A, A \cap \phi = \phi$
5.  Distributive law: $A \cup (B \cap C) = (A \cup B) \cap (A \cup C)$
    $$A \cap (B \cup C) = (A \cap B) \cup (A \cap C)$$

***Proof***

1.  To prove $A \cap A = A$

Let $x \in A \cap A$
$\Rightarrow x \in A \text{ and } x \in A$
$\Rightarrow x \in A$

i.e. $A \cap A \subseteq A$ (1)

Let $y \in A$

$\Rightarrow y \in A$

$\Rightarrow y \in A$ and $y \in A$

$\Rightarrow y \in A \cap A$

i.e., $A \subseteq A \cap A$ (2)

From (1) and (2), we have

$$A \cap A = A$$

2. To prove $A \cap B = B \cap A$

Let $x \in A \cap B$

$\Rightarrow x \in A$ and $x \in B$

$\Rightarrow x \in B$ and $x \in A$

$\Rightarrow x \in B \cap A$

i.e., $A \cap B \subseteq B \cap A$ (1)

Let $y \in B \cap A$

$\Rightarrow y \in B$ and $y \in A$

$\Rightarrow y \in A$ and $y \in B$

$\Rightarrow y \in A \cap B$

i.e. $B \cap A \subseteq A \cap B$ (2)

From (1) and (2), we have

$$A \cap B = B \cap A$$

3. To prove $(A \cap B) \cap C = A \cap (B \cap C)$

Let $x \in (A \cap B) \cap C$

$\Rightarrow x \in (A \cap B)$ and $x \in C$

$\Rightarrow (x \in A$ and $x \in B)$ and $x \in C$

$\Rightarrow x \in A$ and $(x \in B$ and $x \in C)$

$\Rightarrow x \in A$ and $(x \in B \cap C)$

$\Rightarrow x \in A \cap (B \cap C)$

i.e., $(A \cap B) \cap C \subseteq A \cap (B \cap C)$ (1)

Let $y \in A \cap (B \cap C)$

$\Rightarrow y \in A$ and $y \in (B \cap C)$

$\Rightarrow y \in A$ and $(y \in B$ and $y \in C)$

$\Rightarrow (y \in A$ and $y \in B)$ and $y \in C$

$\Rightarrow y \in (A \cap B) \cap C$

$$\text{i.e. } A \cap (B \cap C) \subseteq (A \cap B) \cap C \tag{2}$$

From (1) and (2), we have

$$\mathbf{(A \cap B) \cap C = A \cap (B \cap C)}$$

4. To prove $A \cap U = A$, $A \cap \phi = \phi$

Let $x \in A \cap U$
$\Rightarrow x \in A$ and $x \in U$
$\Rightarrow x \in A$   [Because $A \subset U$]

$$\text{i.e., } A \cap U \subseteq A \tag{1}$$

Again, let $y \in A$
$\Rightarrow y \in A$ and $y \in U$   [Because $A \subset U$]

$$\text{i.e. } A \subseteq A \cap U \tag{2}$$

From (1) and (2), we have

$$\mathbf{A \cap U = A}$$

Again to prove $A \cap \phi = \phi$

The set $A \cap \phi$ is the set of all those elements, which are common to both A and the empty set $\phi$. But $\phi$ contains no elements. Therefore the set $A \cap \phi$ contains no elements, Thus $A \cap \phi = \phi$.

**Example 13:** Prove the following laws:

1. $A \cup A' = U$
2. $A \cap A' = \phi$
3. $(A')' = A$

***Proof 1***

To prove $A \cup A' = U$

Since every set is a subset of a universal set,
therefore the set $A \cup A' \subseteq U$ $\hspace{2cm}$ (1)

Again, let $x \in U$, then

$x \in U \Rightarrow x \in A$ or $x \notin A'$.
$\hspace{1.5cm} \Rightarrow x \in (A \cup A')$
i.e., $U \subseteq A \cup A'$ $\hspace{3cm}$ (2)

From (1) & (2), we have

$$\mathbf{A \cup A' = U}$$

### Proof 2

To prove $A \cap A' = \phi$

Since the null set $\phi$ is a subset of every set, therefore it follows that

$$\phi \subseteq A \cap A' \tag{1}$$

Again, let $x \in A \cap A'$
$$\Rightarrow x \in A \text{ and } x \in A'$$
$$\Rightarrow x \in A \text{ and } x \notin A$$
$$\Rightarrow x \in \phi$$

i.e. $A \cap A' = \phi \tag{2}$

From (1) and (2), we have

$$\mathbf{A \cap A' = \phi}$$

### Proof 3

To prove $(A')' = A$

Let $x \in (A')'$

Then $x \in (A')' \Rightarrow x \notin A'$
$$\Rightarrow x \in A$$

i.e. $(A')' \subseteq A \tag{1}$

Again, let $x \in A$

Then $x \in A \Rightarrow x \notin A'$
$$\Rightarrow x \in (A')'$$

i.e. $A \subseteq (A')' \tag{2}$

From (1) and (2), we have

$$\mathbf{(A')' = A}$$

**Example 14:** Prove De Morgan laws
1) $(A \cup B)' = A' \cap B'$     2) $(A \cap B)' = A' \cup B'$

### Proof 1

$$
\begin{aligned}
(A \cup B)' &= \{x: x \notin A \cup B\} \\
&= \{x: \neg\, x' \in (A \cup B)\} \\
&= \{x: \neg\, (x \in A \vee x \in B) \\
&= \{x: x \notin A \wedge x \notin B\} \\
&= \{x: x \in A' \wedge x \in B'\} \\
&= \{x: x \in A' \cap B'\} \\
&= A' \cap B'
\end{aligned}
$$

### *Proof 2*

$$
\begin{aligned}
(A \cap B)' &= \{x : x \notin A \cap B\} \\
&= \{x : \neg\, x \in (A \cap B)\} \\
&= \{x : \neg(x \in A \wedge x \in B) \\
&= \{x : x \notin A \vee x \notin B\} \\
&= \{x : x \in A' \wedge x \in B'\} \\
&= \{x : x \in A' \cup B'\} \\
&= A' \cup B'
\end{aligned}
$$

**Distributive laws:**  
1) $A \cup (B \cap C) = (A \cup B) \cap (A \cup C)$  
2) $A \cap (B \cup C) = (A \cap B) \cup (A \cap C)$

### *Proof 1*

Proof by Membership Table

| A | B | C | B∩C | A∪(B∩C) | (A∪B) | (A∪C) | (A∪B)∩(A∪C) |
|---|---|---|---|---|---|---|---|
| 1 | 1 | 1 | 1 | 1 | 1 | 1 | 1 |
| 1 | 1 | 0 | 0 | 1 | 1 | 1 | 1 |
| 1 | 0 | 1 | 0 | 1 | 1 | 1 | 1 |
| 1 | 0 | 0 | 0 | 1 | 1 | 1 | 1 |
| 0 | 1 | 1 | 1 | 1 | 1 | 1 | 1 |
| 0 | 1 | 0 | 0 | 0 | 1 | 0 | 0 |
| 0 | 0 | 1 | 0 | 0 | 0 | 1 | 0 |
| 0 | 0 | 0 | 0 | 0 | 0 | 0 | 0 |

Since the columns for $A \cup (B \cap C)$ and $(A \cup B) \cap (A \cup C)$ are the same, the identity is valid.

### *Proof 2*

Proof by Membership Table

| A | B | C | B∪C | A∩(B∪C) | (A∩B) | (A∩C) | (A∩B)∪(A∩C) |
|---|---|---|---|---|---|---|---|
| 1 | 1 | 1 | 1 | 1 | 1 | 1 | 1 |
| 1 | 1 | 0 | 1 | 1 | 1 | 0 | 1 |
| 1 | 0 | 1 | 1 | 1 | 0 | 1 | 1 |
| 1 | 0 | 0 | 0 | 0 | 0 | 0 | 0 |
| 0 | 1 | 1 | 1 | 0 | 0 | 0 | 0 |
| 0 | 1 | 0 | 1 | 0 | 0 | 0 | 0 |
| 0 | 0 | 1 | 1 | 0 | 0 | 0 | 0 |
| 0 | 0 | 0 | 0 | 0 | 0 | 0 | 0 |

Since the columns for $A \cap (B \cup C)$ and $(A \cap B) \cup (A \cap C)$ are the same, the identity is valid.

***Alternate Proof***

(i)  To prove $A \cup (B \cap C) = (A \cup B) \cap (A \cup C)$

Let $x \in A \cup (B \cap C)$
$\Rightarrow x \in A$ or $(x \in B \cap C)$
$\Rightarrow x \in A$ or $(x \in B$ and $x \in C)$
$\Rightarrow (x \in A$ or $x \in B)$ and $(x \in A$ or $x \in C)$
$\Rightarrow (x \in A \cup B)$ and $(x \in A \cup C)$
$\Rightarrow x \in (A \cup B) \cap (A \cup C)$

i.e., $A \cup (B \cap C) \subseteq (A \cup B) \cap (A \cup C)$      (1)

Again, let $y \in (A \cup B) \cap (A \cup C)$
$\Rightarrow (y \in A \cup B)$ and $(y \in A \cup C)$
$\Rightarrow (y \in A$ or $y \in B)$ and $(y \in A$ or $y \in C)$
$\Rightarrow y \in A$ or $(y \in B$ and $y \in C)$
$\Rightarrow y \in A$ or $(y \in B \cap C)$
$\Rightarrow y \in A \cup (B \cap C)$

i.e., $(A \cup B) \cap (A \cup C) \subseteq A \cup (B \cap C)$      (2)

From (1) and (2), we have

$$\mathbf{A \cup (B \cap C) = (A \cup B) \cap (A \cup C)}$$

(ii) To prove $A \cap (B \cup C) = (A \cap B) \cup (A \cap C)$

Let $x \in A \cap (B \cup C)$
$\Rightarrow x \in A$ and $(x \in B \cup C)$
$\Rightarrow x \in A$ and $(x \in B$ or $x \in C)$
$\Rightarrow (x \in A$ and $x \in B)$ or $(x \in A$ and $x \in C)$
$\Rightarrow (x \in A \cap B)$ or $(x \in A \cap C)$
$\Rightarrow x \in (A \cap B) \cup (A \cap C)$

i.e.,  $A \cap (B \cup C) \subseteq (A \cap B) \cup (A \cap C)$      (1)

Similarly, let $y \in (A \cap B) \cup (A \cap C)$
$\Rightarrow y \in (A \cap B)$ or $y \in (A \cap C)$
$\Rightarrow (y \in A$ and $y \in B)$ or $(y \in A$ and $y \in C)$
$\Rightarrow y \in A$ and $(y \in B$ or $y \in C)$
$\Rightarrow y \in A$ and $(y \in B \cup C)$
$\Rightarrow y \in A \cap (B \cup C)$

i.e.,  $(A \cap B) \cup (A \cap C) \subseteq A \cap (B \cup C)$      (2)

From (1) and (2), we have

$$\mathbf{A \cap (B \cup C) = (A \cap B) \cup (A \cap C)}$$

### 1.3.2 Duality

The dual of an equation involving sets is obtained by interchanging $\cup$ and $\cap$ and also U and $\phi$ in it. Each law of algebra imply its own dual.

### PRINCIPLE OF DUALITY

For any given theorem and its proof, the dual of the theorem can be proved in the same way by using the dual of each step in the proof of the original theorem, each law of algebra imply its own dual. Thus the principle of duality applies to the algebra of sets.

### SUCCESSOR OF SET

**The successor** of the set $A$ is the set $A \cup \{A\}$ and is denoted by $A^+$.

**Example 16:** Find the successors of the set $A = \{1, 2, 3, 4\}$

*Solution*

$A = \{1, 2, 3, 4\}$
$\{A\} = \{\{1, 2, 3, 4\}\}$
$A^+ = A \cup \{A\} = \{1, 2, 3, 4, \{1, 2, 3, 4\}\}$

**Example 17:** Find the successors of the set $\phi$ and $\phi^+$.

*Solution*

Successor of $\phi = \phi^+ = \phi \cup \{\phi\} = \{\phi\}$
Successor of $\phi^+ = \{\phi\} \cup \{\{\phi\}\}$
$\qquad\qquad = \{\phi, \{\phi\}\}$

### 1.3.3 Finite and Infinite Sets

*Finite Set:* A set is said to be a finite set if in counting its different elements, the counting process comes to an end. Thus a set with finite number of elements is a finite set, e.g.,

   (i) The set of vowels = {a, e, i, o, u}
  (ii) The set of people living in INDIA.

*Infinite Set:* A set, which is neither a null set nor a finite set is called an infinite set. The counting process can never come to an end in counting the elements of this set, e.g., N, the set of natural numbers = $\{1, 2, 3, 4...\}$.

### 1.3.4 Classes of Sets or Family of Sets or Set of Sets

A set whose elements are sets is called a class of sets or family of sets or set of sets, e.g., $\{\{1\},\{1,2\},\{4,5,6\}\}$.

Given a set S, we might wish to talk about some of its subsets. Thus we would be considering a set of sets. Whenever such a situation occurs, to avoid confusion we will speak of a class of sets or collection of sets rather than a set of sets. If we wish to consider some of the sets in a given class of sets, then we speak of a subclass or subcollection. **For example,** suppose S = {1, 2, 3, 4}. Let A be the class of subsets of S which contain exactly three elements of S. Then A = [{1, 2, 3}, {1, 2, 4}, {1, 3, 4}, {2, 3, 4}]. The elements of A are the sets {1, 2, 3}, {1, 2, 4}, {1, 3, 4}, and {2, 3, 4}. Let B be the class of subsets of S which contain 2 and two other elements of S. Then B = [{1, 2, 3}, {1, 2, 4}, {2, 3, 4}].

The elements of B are the sets {1, 2, 3}, {1, 2, 4}, and {2, 3, 4}. Thus B is a subclass of A, since every element of B is also an element of A. (To avoid confusion, we will sometimes enclose the sets of a class in brackets instead of braces.)

### 1.3.5 Power Set

The set of all possible subsets of a given set A is called the power set A. The power set of A is denoted by P (A). If the number of elements in a set is $n$, then the number of elements in power set is $2^n$ e.g., if A = {$a, b, c$}. Then its subsets are $\phi$, {$a$}, {$b$}, {$c$}, {$a, b$}, {$a, c$}, {$b, c$}, {$a, b, c$}.

$$P (A) = \{\phi, \{a\}, \{b\}, \{c\}, \{a, b\}, \{a, c\}, \{b, c\}, \{a, b, c\}\}$$

### 1.3.6 Multiset

A set is a collection of well defined and distinct objects. But when we make the collection of names of students in a class, it may contain a name more than once, as we may have two or more students having the same name.

***Definition***: A multiset is the collection of objects that are not necessarily distinct. The multiplicity of an element in a multiset is defined as the number of times the element appears in the multiset. The notation {$m_1.a_1, m_2.a_2, \ldots\ldots, m_r.a_r$} denotes the multiset with element $a_1$ occurring $m_1$ times, element $a_2$ occurs $m_2$ times, and so on. The numbers $m_i = 1, 2, \ldots, r$ are called the multiplicities of the elements $a_i,\ i = 1, 2, \ldots.., r$.

- {$3.a, 1.b, 2.c$}, {$1.a, 1.b, 1.c$}, and {} are examples of multisets.
- Sets are special cases of multisets in which the multiplicity of an element is 0 or 1.

### OPERATIONS ON MULTISETS

Let P and Q be the multisets, the **union** of the multisets P and Q is the multiset where the multiplicity of an element is the maximum of its multiplicities in P and Q. The **intersection** of P and Q is the multiset where the multiplicity of an element is the minimum of its multiplicities in P and Q. The **difference** of P and Q is the multiset where the multiplicity of an element is the multiplicity of the element in P less its multiplicity in Q unless this difference is negative, in which case the multiplicity is O. The **sum** of P and Q is the multiset where the multiplicity of an element is the sum of multiplicities in P and Q. The union, intersection, and difference of

P and Q are denoted by $P \cup Q$; $P \cap Q$; $P - Q$, respectively (Where these operations should not be confused with the analogous operations for sets.) The sum of P and Q is denoted by $P + Q$.

**Example 17:** Let A and B be the multisets $\{3.a, 1.b, 2.c\}$, $\{1.a, 2.b, 4.c\}$, respectively. Find $A \cup B$ (ii) $A \cap B$ (iii) $A - B$ (iv) $A + B$.

### *Solution*

   (i)  $A \cup B = \{3.a, 2.b, 4.c\}$
   (ii)  $A \cap B = \{1.a, 1.b, 2.c\}$
   (iii)  $A - B = \{2.a, 0.b, 0.c\}$
   (iv)  $A + B = \{4.a, 3.b, 6.c\}$

### 1.3.7 Fuzzy Sets

**Fuzzy sets** are used in artificial intelligence. Each element in the universal set U has a **degree of membership,** which is a real number between 0 and 1 (including 0 and 1), in a fuzzy set S. The fuzzy set S is denoted by listing the elements with their degrees of membership (elements with 0 degree of membership are not listed) e.g.

   (i)  S (of famous persons) = {0.6 Rama, 0.9 Krishna, 0.4 Ankur, 0.1 Mohan, 0.5 Sohan}, where Rama, Krishna, Ankur, Mohan, and Sohan have 0.6, 0.9, 0.4, 0.1, and 0.5 degree of membership respectively in F (so that Krishna is the most famous and Mohan is the least famous of these people).
   (ii)  T (of rich people) = {0.4 Rama, 0.8 Krishna, 0.2 Ankur, 0.9 Mohan, 0.7 Sohan}.

### OPERATIONS ON FUZZY SETS

The **union** of two fuzzy sets S and T is the fuzzy set where the degree of membership of an element is the maximum of degrees of membership of this element in S and T. The **intersection** of S and T is the fuzzy set where the degree of membership of an element is the minimum of degrees of membership of this element in S and T. The **complement** of a fuzzy set S is the set S′, with the degree of membership of an element in S′ equal to 1 minus the degree of membership of this element in S.

**Example 18:** Given two fuzzy sets
S (of famous persons) = {0.6 Rama, 0.9 Krishna, 0.4 Ankur, 0.1 Mohan, 0.5 Sohan},
T (of rich people) = {0.4 Rama, 0.8 Krishna, 0.2 Ankur, 0.9 Mohan, 0.7 Sohan}.
Find $S \cup T$, $S \cap T$, and S′.

### *Solution*

S (of famous persons) = {0.6 Rama, 0.9 Krishna, 0.4 Ankur, 0.1 Mohan, 0.5 Sohan},
   T (of rich people) = {0.4 Rama, 0.8 Krishna, 0.2 Ankur, 0.9 Mohan, 0.7 Sohan}.

S $\cup$ T(set of rich or famous people) = {0.6 Rama, 0.9 Krishna, 0.4 Ankur, 0.9 Mohan, 0.7 Sohan}.

S $\cap$ T (set of rich and famous people) = {0.4 Rama, 0.8 Krishna, 0.2 Ankur, 0.1 Mohan, 0.5 Sohan}.

S$'$ = {0.4 Rama, 0.1 Krishna, 0.6 Ankur, 0.9 Mohan, 0.5 Sohan}.

### 1.3.8 Cartesian Product

**Cartesian Product:** Cartesian product of two sets A and B is defined as A $\times$ B = {$(a, b) \mid a \in$ A, $b \in$ B}.

- A $\times$ B $\neq$ B $\times$ A
- A $\times$ B = $\phi$ if one or both sets are empty
- A $\times$ (B $\cup$ C) = (A $\times$ B) $\cup$ (A $\times$ C)

**Example 19:** If A = {$a, b, c, d$} and B = {1, 2}. Find A $\times$ B and B $\times$ A.

$$A \times B = \{(a, 1), (a, 2), (b, 1), (b, 2), (c, 1), (c, 2),(d, 1),(d, 2)\}$$
and   $$B \times A = \{(1, a), (1, b), (1, c), (1, d), (2, a), (2, b), (2, c),(2, d)\}.$$

We observe that A contains 4 elements, B contains 2 elements. A $\times$ B and B $\times$ A contains 8 elements.

Here $(1, a) \in$ A $\times$ B but $(1, a) \notin$ B $\times$ A

Thus         A $\times$ B $\neq$ B $\times$ A.

**Example 20:** Prove that for any three sets A, B and C

$$A \times (B \cap C) = (A \times B) \cap (A \times C)$$

***Proof***

$A \times (B \cap C)$
$\Rightarrow \{(x, y): x \in$ A and $y \in$ (B $\cap$ C)$\}$
$\Rightarrow \{(x, y): x \in$ A and $(y \in$ B, $y \in$ C)$\}$
$\Rightarrow \{(x, y): x \in$ A and $y \in$ B, $x \in$ A and $y \in$ C$\}$
$\Rightarrow \{(x, y): (x, y) \in$ (A $\times$ B), $(x, y) \in$ (A $\times$ C)$\}$

i.e.   (A $\times$ B) $\cap$ (A $\times$ C) $\subseteq$ A $\times$ (B $\cap$ C)         (1)

also (A $\times$ B) $\cap$ (A $\times$ C)

$\Rightarrow \{(x, y): (x, y) \in$ (A $\times$ B), $(x, y) \in$ (A $\times$ C)$\}$
$\Rightarrow \{(x, y): x \in$ A and $y \in$ B, $x \in$ A and $y \in$ C$\}$
$\Rightarrow \{(x, y): x \in$ A and $(y \in$ B, $y \in$ C)$\}$
$\Rightarrow \{(x, y): x \in$ A and $y \in$ (B $\cap$ C)$\}$

i.e.   A $\times$ (B $\cap$ C) $\subseteq$ (A $\times$ B) $\cap$ (A $\times$ C)         (2)

Hence from (1) and (2)

$$A \times (B \cap C) = (A \times B) \cap (A \times C)$$

**Example 21:** If $A = \{1, 2\}$, $B = \{2, 3\}$, $C = \{2, 3, 4\}$. Show that

(i) $A \times (B \cup C) = (A \times B) \cup (A \times C)$

(ii) $A \times (B \cap C) = (A \times B) \cap (A \times C)$

*Solution*

$A \times B = \{1, 2\} \times \{2, 3\} = \{(1, 2), (1, 3), (2, 2), (2, 3)\}$

$A \times C = \{1, 2\} \times \{2, 3, 4\} = \{(1, 2), (1, 3), (1, 4), (2, 2), (2, 3), (2, 4)\}$

(i) $(A \times B) \cup (A \times C) = \{(1, 2), (1, 3), (2, 2), (2, 3), (1, 4), (2, 4)\}$   (1)

$B \cup C = \{2, 3, 4\}$

$\therefore \quad A \times (B \cup C) = \{(1, 2), (1, 3), (2, 2), (2, 3), (1, 4), (2, 4)\}$   (2)

From (1) and (2),

$$A \times (B \cup C) = (A \times B) \cup (A \times C)$$

(ii) $(A \times B) \cap (A \times C) = \{(1, 2), (1, 3), (2, 2), (2, 3)\}$   (3)

$B \cap C = \{2, 3\}$

$A \times (B \cap C) = \{(1, 2), (1, 3), (2, 2), (2, 3)\}$   (4)

From (3) and (4),

$$A \times (B \cap C) = (A \times B) \cap (A \times C)$$

**Example 22:** If A, B, C are three sets such that $A \subseteq B$. Show $(A \times C) \subseteq (B \times C)$.

*Solution*

Let $(x, y) \in (A \times C)$

$\Rightarrow x \in A$ and $y \in C$

$\Rightarrow x \in B$ and $y \in C;\ (A \subseteq B)$

$\Rightarrow (x, y) \in B \times C$

$\Rightarrow (A \times C) \subseteq (B \times C)$

**Example 23:** How many elements in $A \times B$ and $B \times A$ are common, if $n$ elements are common to A and B?

*Solution*

Let C be the set common to both A and B, then $C \subseteq A$ and $C \subseteq B$.

Now $(x, y) \in C \times C$

$$\Leftrightarrow x \in C \text{ and } y \in C$$
$$\Leftrightarrow (x \in C \text{ and } y \in C) \text{ and } (x \in C \text{ and } y \in C)$$
$$\Leftrightarrow (x \in A \text{ and } y \in B) \text{ and } (x \in B \text{ and } y \in A)$$
$$\Leftrightarrow (x, y) \in A \times B \text{ and } (x, y) \in B \times A$$
$$\Leftrightarrow (x, y) \in (A \times B) \cap (B \times A)$$

But C has $n$ elements.

$\therefore$　　$C \times C$ has $n^2$ elements

$\therefore$　　$n^2$ elements are common.

## Exercise 1.2

1. If $A \subseteq B$, $B \subseteq C$ and $C \subseteq A$, show that $B = A$.

2. If $A = A \cup B$, show $B = A \cap B$.

3. Show that if A, B, and C are sets, then $(A \cap B \cap C)' = A' \cup B' \cup C'$

   (a) By showing each side is a subset of the other side.
   (b) Using a membership table.

4. Show that if A and B are sets, then $A - B = A \cap B'$.

5. Let $A = \{0, 2, 4, 6, 8, 10\}$, $B = \{0, 1, 2, 3, 4, 5, 6\}$, and $C = \{4, 5, 6, 7, 8, 9, 10\}$. Find

   (a) $A \cap B \cap C$
   (b) $A \cup B \cup C$
   (c) $(A \cup B) \cap C$
   (d) $(A \cap B) \cup C$

6. What can you say about the sets A and B if we know that

   (a) $A \cup B = A$
   (b) $A \cap B = A$
   (c) $A - B = A$
   (d) $A \cap B = B \cap A$
   (e) $A - B = B - A$

7. Show that if A is a subset of a universal set U, then

   (a) $A \oplus A = \phi$
   (b) $A \oplus 0 = A$
   (c) $A \oplus U = A$
   (d) $A \oplus A = U$

8. What can you say about the sets A and B if $A \oplus B = A$?

9. Suppose that A, B, and C are sets such that $A \oplus C = B \oplus C$. Must it be the case that $A = B$?

10. If A, B, C, and D are sets, does it follow that $(A \oplus B) \oplus (C \oplus D) = (A \oplus D) \oplus (B \oplus C)$?

11. If A, B are subsets of a set $S_1$ and A′, B′ are the complements of A and B respectively. Prove that $A \subseteq B \Rightarrow B' \subseteq A'$.

12. Prove that for any two sets A & B,

    (i) $(A - B) \cup (B - A) = (A \cup B) - (A \cap B)$.
    (ii) $(A - B) = A - (B \cup C)$
    (iii) $(A - B) - C = (A - C) - B$

13. If $A = \{1, 2, 3, 4\}$, $B = \{2, 3, 5, 6\}$ and $C = \{4, 5, 6, 7\}$. Then verify that

    (i) $A \cup (B \cap C) = (A \cup B) \cap (A \cup C)$
    (ii) $A \cap (B \cup C) = (A \cap B) \cup (A \cap C)$

14. Suppose that A is the multiset that has as its elements the types of computer equipment needed by one department of a university where the multiplicities are the number of pieces of each

type needed, and B is the analogous multiset for a second department of the university. For instance, A could be the multiset {107 personal computers, 44 routers, 6 servers} and B could be the multiset {14 personal computers, 6 routers, 2 mainframes}.

(a) What combination of A and B represents the equipment the university should buy assuming both departments use the same equipment?

(b) What combination of A and B represents the equipment that will be used by both departments if both departments use the same equipment?

(c) What combination of A and B represents the equipment that the second department uses, but the first department does not, if both departments use the same equipment?

(d) What combination of A and B represents the equipment that the university should purchase if the departments do not share equipment?

15. Find the successors of the set
$A = \{1, 2, 3, 4, 5\}$

16. Determine the power set of the following

(i) {a}   (ii) {{a}}   (iii) {$\phi$, {a}}

17. Let N denote the set of all natural numbers. Let S denote the set of all finite subsets of N. What is the cardinality of S? Justify your answer.

**Answers to Selected Problems**

5. a. {4, 6}
   b. {0, 1, 2, 3, 4, 5, 6, 7, 8, 9, 10}
   c. {4, 5, 6, 8, 10}
   d. {0, 2, 4, 5, 6, 7, 8, 9, 10}

15. $A^+ = \{1, 2, 3, 4, 5, \{1, 2, 3, 4, 5\}\}$

16. (i) $p(A) = \{\phi, \{a\}\}$
    (ii) $p(A) = \{\phi, \{\{a\}\}$
    (iii) $p(A) = \{\phi, \{\phi\}, \{\{a\}\}, \{\phi, \{a\}\}\}$

# 2

# Relations and Functions

## 2.1 INTRODUCTION

The word relation implies an association of two objects, two persons, etc., according to some property possessed by them, e.g.

(i) A is the son of B. The relation R here is "is the son of".
(ii) Delhi is the capital of INDIA. The relation R here is "is the capital of".

In Mathematics we can study relationship between a real number and a number which is smaller than it, between a program and a variable it uses. Relationship between elements of sets is represented using a structure called relation.

## 2.2 RELATION OR BINARY RELATION

Let A and B be two non-empty sets. A binary relation or simply a relation from A to B is a subset of $A \times B$. Given $x \in A$ and $y \in B$, we write $x \, R \, y$ if $(x, y) \in R$ and $x \, \cancel{R} \, y$ if $(x, y) \notin R$. If R is relation from A to A, then R is said to be relation on A.

A binary relation on a set A is a subset of $A \times A$. This definition needs some explanation, since the connection between it and the informal idea of a relation might not be obvious.

*Note:* The word 'binary' refers to the fact that the relation is between two elements of A. As this is the only kind of relation we will be studying, we will omit the word 'binary' in what follows.

The easiest way to do this is by means of an example.

Let $A = \{1, 3, 5\}$ be a set. Then $A \times A = \{(1, 1), (1, 3), (1, 5), (3, 1), (3, 3), (3, 5), (5, 1), (5, 3), (5, 5)\}$.

Let $x$ be the first component of the ordered pairs, and $y$ the second component. We may have the following three kinds of relations:

1. $x < y$
2. $x = y$
3. $x > y$

1. If $x < y$. The relation $x\,R_1\,y$ means to select from the set $A \times A$ the set of those ordered pairs in which the first component is less than the second.

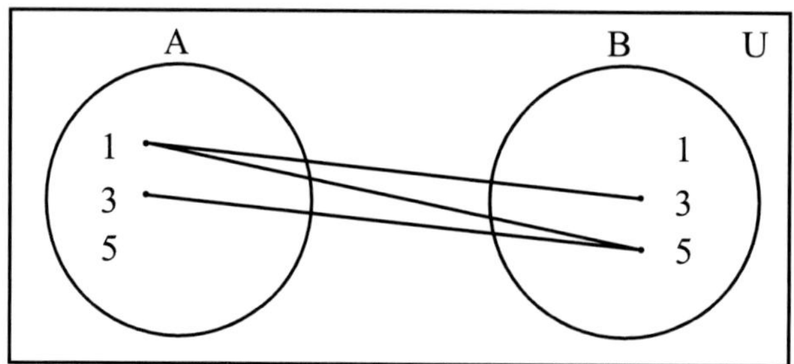



These ordered pairs are: $(1, 3)$, $(1, 5)$, $(3, 5)$.

The set of these ordered pairs is $R_1 = \{(1, 3), (1, 5), (3, 5)\}$, a subset of $A \times A$ and a relation on A.

2. If $x = y$. The relation $x\,R_2\,y$ means the ordered pairs in which the first component is equal to the second one.

These are $(1, 1)$, $(3, 3)$, $(5, 5)$

$\therefore \quad R_2 = \{(1, 1), (3, 3), (5, 5)\}$, a subset of $A \times A$ and a relation on A.

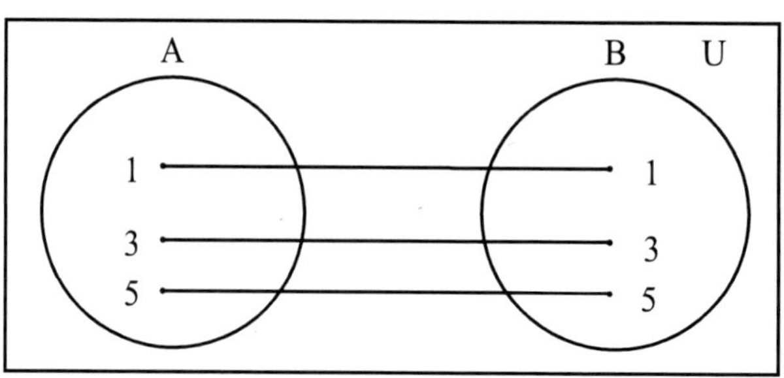

3. If $x > y$. The relation $x\,R_3\,y$ means the ordered pairs in which the first element is greater than the second. These are $(3, 1), (5, 1),$ and $(5, 3)$.

$\therefore$   $R_3 = \{(3, 1), (5, 1), (5, 3)\}$, a subset of $A \times A$ and so is a relation of A.

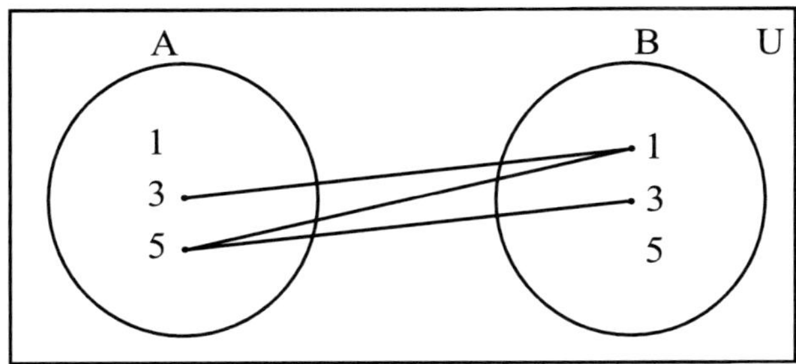


Hence, in all the above cases, we have

$$R_1 \subseteq A \times A,$$
$$R_2 \subseteq A \times A,$$
and
$$R_3 \subseteq A \times A.$$

Thus a relation R on any set is any subset of the Cartesian product $A \times A$.

## 2.3 DOMAIN AND RANGE OF RELATION

For a relation R, domain is a set of elements in A which are related to some element in B. Similarly range of R is a set of elements in B which are related to some element in A.

**Example 1:**  If $A = \{a, b, c\}, B = \{x, y, z\}$; and $R = \{(a, x), (a, z), (b, y), (c, x)\}$. Find the domain and range of R.

*Solution*

Domain of R = the set of first components of the ordered pairs in R = $d(R) = \{a, b, c\}$
Range of R = the set of second components of the ordered pairs in R = $r(R) = \{x, y, z\}$.

**Example 2:**  Find the number of relations from $A = \{a, b, c\}$ to $B = \{x, y, z\}$.

*Solution*

There are $3 \times 3 = 9$ elements in $A \times B$ and hence there are $m = 2^9 = 512$ subsets of $A \times B$. Thus there are $m = 512$ relations from $A \times B$

## 2.4 COMPLEMENT OF RELATION

Let R be a binary relation from a set A to a set B. The complement, $R'$ or $\bar{R}$, of relation R is given by $R' = \{(a, b) : (a, b) \notin R \text{ and } (a, b) \in A \times B\}$, where $a \in A, b \in B$.

## 2.5 INVERSE RELATION

Let R be a relation from a set A to a set B. Then $R^{-1} = \{(y, x): (x, y) \in R\}$ is called the inverse relation of $R = \{(x, y): x \in A, y \in B\}$. Clearly, domain of $R^{-1}$ = range R, and range of $R^{-1}$ = domain of R.

**Example 3:** If $A = \{a, b, c\}$, $B = \{x, y, z\}$ and $R = \{(a, x), (b, z), (c, x), (c, y)\}$. Find $R^{-1}$.

*Solution*

$R^{-1} = \{(x, a), (z, b), (x, c), (y, c)\}$.

**Example 4:** Prove that the inverse of an inverse relation is the relation itself.

*Solution*

Let R be a relation in $A \times B$.

$R = \{(x, y) : x \in A, y \in B\}$.
$R^{-1} = \{(y, x) : y \in B, x \in A\}$.

$\therefore$  The inverse of $R^{-1}$ is $= (R^{-1})^{-1} = \{(x, y) : x \in A, y \in B\} = R$

## 2.6 REPRESENTATION OF RELATIONS

Relations can be represented by

  (i) Table
  (ii) Arrow diagram
  (iii) Directed graph
  (iv) Matrix

**(i) Representation by Table:** Let, $m$ and $n$ be number of elements in set A and B respectively, then a relation R from A to B can be represented by a table M with $m$ rows and $n$ columns, having $\sqrt{}$ for the boxes which represent relation of element from set A to B.

  Let  $A = \{1, 2, 3\}$  and  $B = \{5, 6\}$
      $R = \{(1, 5), (1, 6), (2, 6), (3, 6)\}$

The relation R can be represented by a table with 3 rows and 2 columns as follows

| R | 5 | 6 |
|---|---|---|
| 1 | √ | √ |
| 2 |   | √ |
| 3 |   | √ |

**(ii) Representation by Arrow Diagram:** The elements of the set A and elements of set B are written in two disjoint ellipses. If element $a \in$ A is related to element $b \in$ B, then we draw an arrow from $a$ to $b$.

Let  A = {1, 2, 3}  and  B = {5, 6}
R = {(1, 5), (1, 6), (2, 6), (3, 6)}

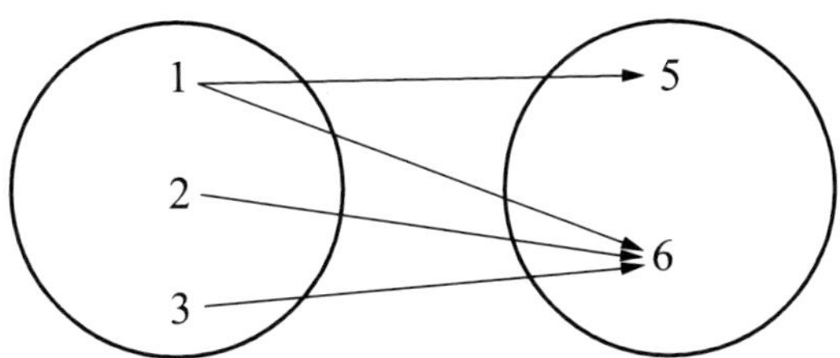

**(iii) Representation by Directed Graph:** Directed graph is used for a relation R on a set A. It is not defined for a relation from one set to another set. Write down the elements of A, and then draw an arrow from an element $a$ to an element $b$ whenever $(a, b) \in$ R.

Let A = {1, 2, 3, 4} and R the relation on A defined by "$a$ divides $b$", written $a|b$. R = {(1, 1), (1, 2), (1, 3), (1, 4), (2, 2), (2, 4), (3, 3), (4, 4)}

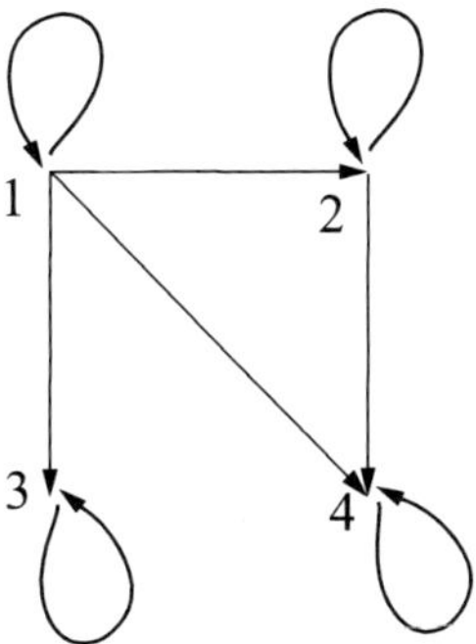

**(iv) Representation by Matrix:** Let, $m$ and $n$ be number of elements in set A and B respectively, then a relation R from A to B can be represented by a $m \times n$ matrix M = [ $M_{ij}$ ], where

$$M_{ij} = \begin{cases} 0 \text{ if } \left(a_i, b_j\right) \notin R \\ 1 \text{ if } \left(a_i, b_j\right) \in R \end{cases}$$

Let   A = {1, 2, 3} and B = {5, 6}
      R = {(1, 5), (1, 6), (2, 6), (3, 6)}

$$\begin{array}{c c} & \begin{array}{cc} 5 & 6 \end{array} \\ \begin{array}{c} 1 \\ 2 \\ 3 \end{array} & \left[\begin{array}{cc} 1 & 1 \\ 0 & 1 \\ 0 & 1 \end{array}\right] \end{array}$$

## 2.7 COMPOSITION OF RELATIONS

Let A, B, C be three sets and a relation $R_1$ from the set A to B be $a\ R_1\ b$ where $a \in A$ and $b \in B$, and the relation $R_2$ from B to C be b $R_2$ c where $b \in B$ and $c \in C$. Then the composition of the relation $R_1$ and $R_2$ denoted by $R_1$ o $R_2$ or $R_1 R_2$ is a relation from A to C defined by

$R_1$ o $R_2$ = {$(a, c)$: b $\in$ B is an element such that $aRb$ and $bRc$}, where $a \in A$ and $c \in C$.

Computing the composite of two relations requires that we find elements that are the second element of ordered pairs in first relation and the first element of ordered pairs in the second relation.

**Example 5:** Let A = {1, 3, 5}, Let R be the relation such that $xRy$, iff $y = x + 2$ and S be the relation such that $xRy$ iff $x < y$.

(a)  Find RS.
(b)  Find SR.
(c)  Illustrate RS and SR via a diagram.
(d)  Is the relation (set) RS equal to the relation SR? Why?

*Solution*

R = {(1, 3), (3, 5)}
S = {(1, 3), (1, 5), (3, 5)},

(a)  RS = {(1, 5)}

(b)  SR = {(1, 5)}

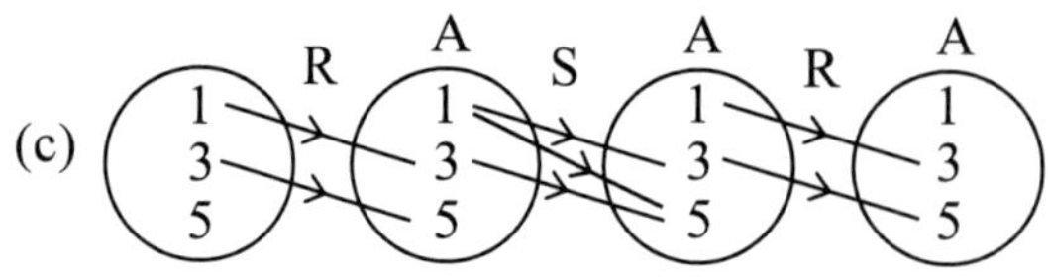

(c)

(d)  ∴ The relations RS = SR

Relations are same as the elements of the set.

## 2.8 TYPES OF RELATIONS OR PROPERTIES OF RELATION

There are several properties that are used to classify relations on a set.

    (i) **Reflexive:** R is reflexive if $a$ R $a$ $\forall$ $a \in$ A.

    (ii) **Symmetric:** R is symmetric if $a$ R $b \Rightarrow b$ R $a$.

    (iii) **Asymmetric:** If R is not symmetric, it is called Asymmetric relation.

    (iv) **Anti-symmetric:** R is anti-symmetric if $a$ R $b$ and $b$ R $a \Rightarrow a = b$; i.e. if $(a, b) \in$ R then $(b, a) \notin$ R unless $a = b$

    (v) **Transitive:** R is transitive if $a$ R $b$ and $b$ R $c \Rightarrow a$ R $c$. $\forall$ $a, b, c$.

These properties are only defined for a relation on a set.

**Example 6:** If A = $\{a, b, c, d\}$ and R = $\{(a, a), (b, b), (c, c), (d, d)\}$. Prove that R is reflexive.

*Solution*

$a \in$ A, $(a, a) \in$ R,
$b \in$ A, $(b, b) \in$ R,
$c \in$ A, $(c, c) \in$ R,
$d \in$ A, $(d, d) \in$ R.

$\therefore$    $a$ R $a$ $\forall$ $a \in$ A, hence R is reflexive.

**Example 7:** Prove that the relation R = $\{(1, 1),(2, 1), (2, 2),(2, 3),(2, 4),(3, 1),(3, 2),(3, 3),(3, 4)\}$ on set A = $\{1, 2, 3, 4\}$ is neither reflexive, nor symmetric or anti-symmetric but transitive.

*Solution*

A = $\{1, 2, 3, 4\}$
$(4, 4) \notin$ A, hence R is not reflexive.
$(2, 1) \in$ R and $(1, 2) \notin$ R, hence R is not symmetric.
$(2, 3) \in$ R and $(3, 2) \in$ R, but $2 \neq 3$, hence R is not anti-symmetric.
$(a, b) \in$ R and $(b, c) \in$ R imply $(a, c) \in$ R $\forall$ $a, b, c \in$ A, hence transitive.

### 2.8.1 Universal Relation

A relation R in a set A is called universal relation if R = A $\times$ A. e.g.: If A = $\{x, y, z\}$. Then R = $\{(x, x), (x, y), (x, z), (y, x), (y, y), (y, z), (z, x), (z, y), (z, z)\}$ is a universal relation in A.

### 2.8.2 Identity Relation

A relation in a set A is called identity relation if R = $\{(x, x): x \in$ A$\}$ e.g.: If A $\{x, y, z\}$. Then R = $\{(x, x), (y, y), (z, z)\}$ is an identity relation in A.

### 2.8.3 Equivalence Relation

A relation R on a set A is equivalence relation if it is reflexive, symmetric and transitive. Two elements that are related by equivalence relation are called equivalent.

### 2.8.4 Equivalence Class

Equivalence class of an element $a \in A$, denoted by $[a]$, having relation R on A, is the set of elements of A to which a is related i.e., $[a] = \{x: (a, x) \in R\}$.

**Example 8:** Let R be the relation on the set N of positive integers defined by R= $\{(a, b): a + b$ is even$\}$. Prove that R is an equivalent relation.

### *Solution*

R is reflexive: $\forall\, a \in N$, $a + a$ is even, hence $a\,R\,a\,\forall\, a$.
R is symmetric: If $a + b$ is even then $b + a$ is even, hence $a\,R\,b \Rightarrow b\,R\,a$.
R is transitive: Let $a, b, c$ have the same parity i.e., these are either odd or even. We see if $a\,R\,b$ and $b\,R\,c$ then $a\,R\,c$. Hence R is equivalence relation.

**Example 9:** Let R be the relation on the set Z of integers defined by $R = \{(x, y): x \in Z, y \in Z,$ $(x - y)$ is divisible by 6$\}$. Prove that R is an equivalent relation.

### *Solution*

R is an equivalence relation if it is reflexive, symmetric and transitive.
(a) **Reflexivity:** $a - a = 0$ is divisible by 6, hence $a\,R\,a$ and R is reflexive.
(b) **Symmetry:** $a - b = -(b - a)$, if $a - b$ is divisible by 6, $(b - a)$ is also divisible by 6, hence $a\,R\,b = b\,R\,a$ and R is symmetric.
(c) **Transitivity:** If $a - b$ and $b - c$ are divisible by 6, let $a - b = 6m$, $m \in Z$ and $b - c = 6n$, $n \in Z$ so that $(m + n) \in Z$ and $(a - c) = 6(m + n)$, divisible by 6, accordingly if $a\,R\,b$ and $b\,R\,c$ then $a\,R\,c$, so that R is transitive.

**Example 10:** If A is a set of all integers. Then prove that $R = \{(a, b): a, b \in A, (a - b)$ is an even integer$\}$ is an equivalence relation on A.

### *Solution*

$0 = a - a$ is an even integer. So $(a, a) \in R$

If $(a, b) \in R$, $a - b$ is an even integer. This implies $b - a = -(a - b)$ is also an even integer which in turn implies that $(b, a) \in R$

If $(a, b) \in R$, $(b, c) \in R$. Then $a - c = (a - b) + (b - c)$

$\Rightarrow a - c$ is an even integer (because the terms on RHS are even)

$\Rightarrow (a, c) \in R$

Thus the relation is reflexive, symmetric and transitive, and is equivalence relation

**Example 11:** In a set $S = \{a, b, c, d\}$ of four men, $a$ is younger to the other three, $b$ is younger to $c$ and $d$ only and $c$ is younger to $d$ only. Is the relation "is younger to" an equivalence relation? Give reasons for your answer.

### Solution

Define the relation '$<$' on S as follows:

$a < b$ if $a$ is younger to $b$.

$\therefore$ In the given problem,

$$a < b, a < c \text{ and } a < d$$
$$b < c, b < d$$
$$c < d$$

Combining these equations, we have $a < b < c < d$.

It is clear that the given relation is neither reflexive nor symmetric. Hence, it is not equivalence relation.

**Example 12:** If N is a set of natural numbers, show that the relation R defined by $(a, b) R (c, d) \Rightarrow a + d = b + c$; $(a, b), (c, d) \in N \times N$, is an equivalence relation.

### Solution

(i) The relation R is reflexive, because
$$a + b = b + a, a, b \in N$$
$$\therefore \quad (a, b) R (a, b)$$

(ii) The relation R is symmetric, because
$$(a, b), (c, d) \in N \times N$$
$$(a, b) R (c, d) \Rightarrow a + d = b + c$$
$$\Rightarrow b + c = a + d$$
$$\Rightarrow c + b = d + a$$
$$\Rightarrow (c, d) R (a, b)$$

(iii) R is transitive because $(a, b), (c, d), (e, f) \in N \times N$
$$(a, b) R (c, d) \Rightarrow a + d = b + c,$$
$$\text{and } (c, d) R (e, f) \Rightarrow c + f = d + e$$

$$\therefore \quad (a, b) \, R \, (c, d) \text{ and } (c, d) \, R \, (e, f)$$
$$\Rightarrow a + d = b + c \text{ and } c + f = d + e$$
$$\Rightarrow a + d + c + f = b + c + d + e$$
$$\Rightarrow a + f + (c + d) = b + e + (c + d)$$
$$\Rightarrow a + f = b + e$$
$$\Rightarrow a + f = e + b$$
$$\Rightarrow (a, b) \, R \, (e, f)$$

$$\therefore \quad (a, b) \, R \, (c, d) \quad \text{and} \quad (c, d) \, R \, (e, f)$$
$$\Rightarrow (a, b) \, R \, (e, f)$$

i.e., R is transitive. Hence R, is an equivalence relation.

**Example 13:** Let S be a set of triangles in a plane, and define R as the set $R = \{(a, b) \mid a, b \in S, a \text{ is congruent to } b\}$. Then show that R is an equivalence relation.

*Note:* Two triangles are congruent if the area of one triangle is equal to the area of the other.

***Solution***

If $a$ and $b$ are triangles in a plane, then $(a, b) \in R$, iff, $a$ is congruent to $b$, i.e., area of triangle $a$ = area of the triangle $b$ or $\Delta a = \Delta b$.

Here

(i)  Since $\Delta a = \Delta a \; a \in S$, for $a$ is congruent to $a$, i.e., $(a, a) \in R$

Hence R is reflexive.

(ii) $(a, b) \in R \Rightarrow a$ is congruent to $b$.
$$\Rightarrow \Delta a = \Delta b$$
$$\Rightarrow \Delta b = \Delta a$$
$$\Rightarrow b \text{ is congruent to } a.$$
$$\Rightarrow (b, a) \in R$$

Hence R is symmetric.

(iii) $(a, b) \in R, (b, c) \in R$

$\Rightarrow a$ is congruent to $b$ and $b$ is congruent to $c$.
$$\Rightarrow \Delta a = \Delta b \text{ and } \Rightarrow \Delta b = \Delta c$$
$$\Rightarrow \Delta a = \Delta c$$
$$\Rightarrow a \text{ is congruent to } c.$$
$$\Rightarrow (a, c) \in R.$$

Hence R is transitive; Thus R is an equivalence relation.

**Example 14:** Define congruent modulo $m$. Show that the relation of congruence modulo $m$, $a \equiv b \pmod{m}$, in the set Z is an equivalence relation. That is the relation R $= \{(a, b): a - b = km$ for some fixed integer $m$ and $a, b, k \in$ Z$\}$ is an equivalence relation.

### *Solution*

Let $m$ be a fixed integer. Two integers are said to be congruent modulo $m$, written $a \equiv b \pmod{m}$ if $m$ divides $a - b$. That is, a $\equiv$ b $\pmod{m}$ if $a - b =$ km for some integer $k$.

To prove the relation R is an equivalence relation we verify the following properties:

(i) Since a $-$ a $= 0.$ $m$, a $\in$ Z, (a, a) $\in$ R.

Hence R is reflexive.

(ii) $(a, b) \in$ R $\Rightarrow a - b = km$ for some integer $k$,

$\Rightarrow b - a = (-k)\, m$, for some integer $-k$

$\Rightarrow (b, a) \in$ R

Hence R is symmetric.

(iii) $(a, b) \in$ R, $(b, c) \in$ R

$\Rightarrow a - b = km$, $b - c = lm$, for some integers $k$ and l.

$\Rightarrow (a - b) + (b - c) = km + lm$

$\Rightarrow (a - c) = (k + l)\, m$, for the integer $k + l$. ·

$\Rightarrow (a, c) \in$ R.

Hence R is transitive.

This completes the proof that R on Z is an equivalence relation.

**Example 15:** Define circular relation, show that a relation is reflexive and circular iff it is reflexive, symmetric and transitive.

### *Solution*

**Circular Relation:** A relation R is called circular if $(a, b) \in$ R and $(b, c) \in$ R $\Rightarrow \square (c, a) \in$ R.

**Necessary Part:** Let the relation R be reflexive and circular. Then we shall prove that R is reflexive, symmetric and transitive.

$(a, b) \in$ R, $(b, c) \in$ R $\Rightarrow (c, a) \in$ R, since R is circular and $(a, a) \in$ R, since R is reflexive.

We have $(c, a) \in$ R, $(a, a) \in$ R $\Rightarrow (a, c) \in$R, since R is circular.

This shows $(a, c) \in$R and $(c, a) \in$R. Hence R is symmetric.

$$(a, b) \in R, (b, c) \in R \Rightarrow (c, a) \in R, \text{ since R is circular.}$$
$$\Rightarrow (a, c) \in R, \text{ since R is symmetric.}$$
$$\Rightarrow R \text{ is transitive.}$$

It is given that R is reflexive which completes the proof of necessary part.

**Sufficient Part:** If R is reflexive, symmetric and transitive. Then we have to show that R is reflexive and circular.

$$(a, b) \in R, (b, c) \in R \Rightarrow (a, c) \in R, \text{ since R is transitive.}$$
$$\Rightarrow (c, a) \in R, \text{ since R is symmetric.}$$
$$\Rightarrow R \text{ is circular.}$$

$$(a, c) \in R, (c, a) \in R \Rightarrow (a, a) \in R, \text{ since R is transitive.}$$
$$\Rightarrow R \text{ is reflexive.}$$

This completes the proof of sufficient part.

**Example 16:** On the set of integers $Z$, the relation $a R b$ iff $a - b$ is multiple of 5, is an equivalence relation. Find the equivalence classes.

***Solution***

  (i) The relation $a R b$ iff $a - b$ is multiple of 5, is reflexive because for $a \in Z$, $a - a = 0$, multiple of 5.

  $$\therefore \quad a R a$$

 (ii) The relation $a R b$ iff $a - b$ is a multiple of 5, is symmetric because for $a, b \in Z$ if $a - b$ is a multiple of 5, then $b - a$ is also a multiple of 5.

  $$\therefore \quad a R b \Rightarrow b R a, a, b \in Z.$$

(iii) The relation $a R b$ is transitive because for $a, b, c, \in Z$, $a - b$ is a multiple of 5, and $b - c$ is a multiple of 5.

  $$\therefore (a - b) + (b - c) \text{ is a multiple of 5.}$$

  $$\Rightarrow a - c \text{ is a multiple of 5.}$$

  $$\therefore \quad a R b \text{ and } b R c \Rightarrow a R c.$$

Now the relation $a R b$ if $a - b$ is a multiple of 5 means $a \sim b$ iff $a - b = 5k$, where $k$ is an integer, which again means that if we divide the integer $a$ by 5, the remainder is $b$. Clearly that remainder may be 0, 1, 2, 3, or 4.

The equivalence class $= \{x : x \in Z, x R a\} = \{x : x \in Z, x - a = 5k\}$

$$= \{x : x \in Z, x - 0 = 5k\} = \{x : x \in Z, x = 5k\}, = \{0, \pm 5, \pm 10, ...\}$$
$$k = 0, \pm 1, \pm 2, ...$$

$$= \{x : x \in Z, x - 1 = 5k\} = \{x : x \in Z, x = 5k + 1\}$$

$$= \{1, 1 \pm 5, 1 \pm 10, 1 \pm 15, ...\} - \{..., -9, -4, 1, 6, 11, ...\}$$

$$= \{x : x \in Z, x - 2 = 5k\} = \{x : x \in Z, x = 2 + 5k\}$$

$$= \{..., -8, -3, +2, 7, 12, 17, ...\}$$

$$= \{x : x \in Z, x - 3 = 5k\}$$

$$= \{x : x \in Z, x = 3 + 5k\} = \{3, 3 \pm 5, 3 \pm 10, ...\}$$

$$= \{..., -7, -2, 3, 8, 13, ...\}$$
$$= \{x: x \in Z, x - 4 = 5k\}$$
$$= \{x: x \in Z, x = 4 + 5k\} = \{4, 4 \pm 5, 4 \pm 10, ...\}$$
$$= \{... -11, -6, -1, 4, 9, 14, ...\}$$

### 2.8.5 Partition

If A is a non-empty set. A partition of set A is a collection $P = \{A_i\}$ of non-empty subsets of A such that:

(i) Each element $a$ in A belongs to one of the subset $A_i$ i.e., the union of $A_i$ is equal to the set A.

(ii) The sets of P are mutually disjoint i.e., the partitions divide the elements of the set A into disjoint sets or $A_i \cap A_j = \phi$.

The subsets in a partition are called cells or blocks. The Figure shows, Venn diagram of a partition of the rectangular set of points into five cells

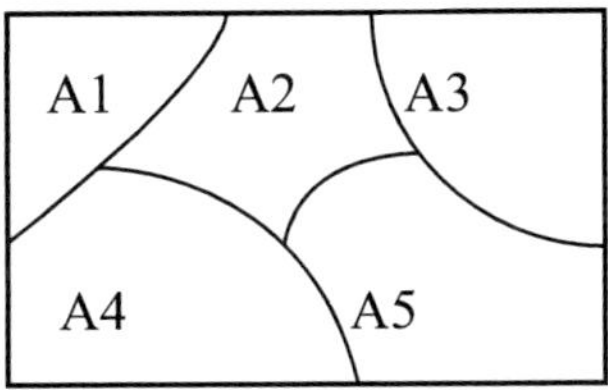


**Example 17:** Let A= $\{1, 2, 3\}$

There are 5 partitions of A

(i) $\{1, 2, 3\}$
(ii) $\{\{1\}, \{2, 3\}\}$
(iii) $\{\{2\}, \{1, 3\}\}$
(iv) $\{\{3\}, \{1, 2\}\}$
(v) $\{\{1\}, \{2\}, \{3\}\}$

Each partition of A contains either 1, 2, or 3 distinct sets.

**Example 18:** Let A be the set of non-zero integers and let R be the relation on $A \times A$ defined by $(a, b) \, R \, (c, d) \Leftrightarrow ad = bc$

(i) Show that R is an equivalence relation

(ii) Find equivalence class of $(1, 2)$

***Solution***

$A = \{a : a \in I_0\}$

(i) R is an equivalence relation if it is reflexive, symmetric and transitive.

    (a) **Reflexivity:** $(a, b)$ R $(a, b) \Leftrightarrow ab = ba$, for $\forall\ a, b \in I_0$; hence R is reflexive.

    (b) **Symmetry:** Suppose $(a, b)$ R $(c, d) \Leftrightarrow ad = bc$. Accordingly, $cb = da$ and hence $(c, d)$ R $(a, b)$. Thus R is symmetric.

    (c) **Transitivity:** Suppose $(a, b)$ R $(c, d)$ and $(c, d)$ R $(e, f)$. Then $ad$ R $bc$ and $cf$ R $de$. Multiplying corresponding terms of equations gives $(ad)(cf) = (bc)(de)$. Cancelling $c \neq 0$ and $d \neq 0$ from both sides, we get $af = be$, hence $(a, b)$ R $(e, f)$. Thus R is transitive.

    Accordingly R is an equivalence relation.

(ii) Equivalence class of an element $a \in A$, denoted by $[a]$, is the set of elements of A to which $a$ is related i.e., $[a] = \{x: (a, x) \in R\}$

    As $(a, b)$ R $(c, d) \Leftrightarrow ad = bc$, for $\forall\ a, b, c, d \in I_0$

    Accordingly $[(1, 2)] = \{(k, 2k): k \in I_0)\}$

## 2.8.6 Relation Between Equivalence Relations and Partitions

Let R be an equivalence relation on a set S. Then the equivalence classes of R form a partition of S. Conversely given a partition $\{A_i: i \in I\}$ of the set S, there is an equivalence relation R that has the sets $A_i$, $i \in I$, as its equivalence classes.

If R be an equivalence relation on a nonempty set A, then

    (a) equivalence class $[a]$ is the set of elements of A to which $a$ is related, i.e. $[a] = \{x: (a, x) \in R\}$

    (b) A/R is the collection of equivalence classes, i.e. A/R $= \{[a] : a \in A\}$

    A/R is partition of A i.e.

        (i) $a \in [a]$, for every $a \in A$

        (ii) $[a] = [b]$ iff $(a,b) \in R$

        (iii) if $[a] \neq [b]$ then $[a]$ and $[b]$ are disjoint.

## 2.8.7 Closures of Relations

In this section we wish to consider the situation of constructing a new relation $R^+$ from previously known relation R such that $R^+$ contains R. This relation is called the closure of R.

In general let R be a relation on a set A. R may or may not have some property say $p$, such as reflexivity, symmetry or transitivity. If there is a relation $R^+$ with property $p$ containing R such that $R^+$ is a subset of every relation with property $p$ containing R, then $R^+$ is called closure of R with property $p$. Here we will show how reflexive, symmetric and transitive closures of relations can be found.

**Reflexive Closure**

Let R be the relation on a set A = $\{a, b, c\}$ such that R = $\{(a, a), (a, b), (b, a), (a, c)\}$. R is not a reflexive relation. Now we produce a reflexive relation containing R that is as small as possible. This can be done by adding $(b, b)$ and (c, c) to R and then the resultant relation $R^+$ = $\{(a, a),$ $(b, b), (c, c), (a, b), (b, a), (a, c)\}$. Now $R^+$ is reflexive containing R and $R^+$ is contained within every reflexive relation that contains R, it is called reflexive closure of R.

Therefore from the above example it is clear that for a given relation R on set A, the reflexive closure of R can be formed by adding to R all pairs of the form $(a, a)$ with a $\in$ A, not already in R. That is reflexive closure of R equals R $\cup$ $\Delta$, where $\Delta$ = $\{(a, a) \mid a \in A\}$.

**Example 19:** What is the reflexive closure of the relation R = $\{(X, Y) \mid X \subset Y\}$.

*Solution*

By the above discussion the reflexive closure of R is

$$R \cup \Delta = \{(X, Y) \mid (X \subset Y) \cup (X, X)\} = \{(X, Y) \mid (X \subseteq Y)\}$$

**Example 20:** Find reflexive closure of the relation R = $\{(a, b) \mid a > b\}$ on the set of integers.

*Solution*

$$R^+ = R \cup \Delta = \{(a, b) \mid a > b\} \cup \{(a, a) \mid (a \in Z\}$$
$$= \{(a, b) \mid a \geq b\}$$

**Symmetric Closure**

The symmetric closure of a relation R can be constructed by adding all ordered pairs of the form $(b, a)$ whenever $(a, b)$ in the relation that are not already present in R. By doing this we can produce a relation that is symmetric, contains R and that is contained in any symmetric relation that contains R. Therefore the symmetric closure, can be constructed as

$$R^+ = R \cup R^{-1} \text{ where } R^{-1} = \{(b, a) \mid (a, b) \in R\}$$

**Example 21.** Let R = $\{(1, 1), (1, 2), (2, 2), (2, 3), (2,1), (3, 1), (3, 2)\}$. What is the Symmetric closure of R?

*Solution*

Now R is not a symmetric relation because the order pair $(3, 1.) \in$ R but $(1, 3) \notin$ R. Therefore we can find symmetric closure of R by adding the ordered pair $(1, 3)$ to R and the resultant relation $R^+$ will be symmetric closure of R and $R^+$ = R U $R^{-1}$

$$R^{-1} = \{(1, 1), (2, 1), (2, 2), (3, 2), (1, 2), (1, 3), (2, 3)\}$$
$$R \cup R^{-1} = \{(1, 1) (1, 2), (2, 2), (2, 3), (2, 1), (3, 1), (3, 2), (1, 3)\}$$

**Example 22:** What is the symmetric closure of the relation R = $\{(a, b) \mid a < b\}$: on the set of positive integers.

$$R^+ = R \cup R^{-1} = \{(a, b) \mid a < b\} \cup \{(b, a) \mid a < b\} = \{(a, b) \mid a \neq b\}$$

## Transitive Closure

Consider a telephone network in which the main office $x$ is connected to and can communicate to, individuals $y$ and $z$. Both $y$ and $z$ can communicate to another person, $t$ ; however, the main office cannot communicate with $t$. Assume that communication is only one way. This program can be described by the relation R = $\{(x, y), (x, z), (y, t), (z, t)\}$. Now suppose we want to change the program such that the main office $x$ can communicate with person $t$ and still maintain the previous system, i.e., most economical system. This can be done by finding the smallest relation $R^+$ which contains R as a subset and which is transitive,

Let, $R^+ = \{(x, y), (x, z), (y, t), (z, t), (x, t)\}$.

Therefore, we can define the transitive closure as follows, let A be a set and R be a relation, on a set A. The transitive closure of R denoted by $R^+$ is the smallest relation which contains R as subset and which is transitive.

**Example 23:** Let A = $\{1, 2, 3, 4\}$. Consider the successor relation on A, i.e. R = $\{(1, 2), (2, 3), (3, 4)\}$. Find transitive closure of R.

### *Solution*

Now we analyse that the given relation R is not a transitive relation because the order pairs (1,3) and (2,4) are not there. We can find the transitive relation R by adding these two ordered pairs in R. $R \cup \{(1, 3), (2, 4)\} = \{(1, 2), (2, 3), (3, 4), (1, 3), (2, 4)\}$ Is this relation is transitive? Again by inspection (1,4) is not there as $1R^2$ and $2R^4$ then 1 should relate to 4 under the relation R.

$$\therefore \quad R^+ = \{(1, 2), (2, 3), (3, 4), (1, 3), (2, 4), (1, 4)\}$$

It is difficult to find a transitive closure by this procedure. From the above example it is clear that in the first step. If $(a, b) \in R$ and $(b, c) \in R$ then $(a, c) \in R^+$. This condition is exactly the membership requirement for the pair $(a, c)$, to be in the composition $R R = R^2$. So every element in $R^2$ must be an element in $R^+$. Therefore, $R^+$ contains at least $R \cup R^2$. In particular for the above example

$$R = \{(1, 2), (2, 3), (3, 4)\}$$
$$R^2 = \{(1, 3), (2, 4)\}$$

and we have

$$R \cup R^2 = \{(1, 2), (2, 3), (3, 4), (1, 3), (2, 4)\}$$

and still we require the pair (1, 4). As $(1, 3) \in S^2$ and $(3, 4) \in S$ and the composition $S^2 S = S^3$ produces (1, 4). This implies $R^3 \subseteq R^+$. This process must be continued until the resulting relation is transitive. Here $R^+ = R \cup R^2 \cup R^3$.

**Theorem 1** If R is a relation on a set A and if dim A = $n$, then transitive closure $R^+$ of R is $R^+$ = $R \cup R^2 \cup R^3 \cup$ .... $\cup R^n$.

Let's now consider the matrix analogue of the transitive closure.

**Theorem 2** Let $M_R{}^+$ be the matrix of $R^+$. The transitive closure of R, which is a relation on a set of $n$ elements. Then $M_R{}^+ = M_R + M_R{}^2 + ..... + M_R{}^n$, where addition is done using Boolean arithmetic.

**Example 24.** Consider the relation R = {(1, 4), (2, 1), (2, 2), (2, 3), (3, 2), (4, 3), (4, 5), (5, 1)} on the set A = {1, 2, 3, 4, 5}. The matrix $M_R$ of the relation R is

$$M_R = \begin{bmatrix} 0 & 0 & 0 & 1 & 0 \\ 1 & 1 & 1 & 0 & 0 \\ 0 & 1 & 0 & 0 & 0 \\ 0 & 0 & 1 & 0 & 1 \\ 1 & 0 & 0 & 0 & 0 \end{bmatrix}$$

Recall $R_2$, $R_3$, ... can be determined through computing the matrices $MR^2$, $MR^3$, Here

$$M_R{}^2 = M_R \times M_R$$

$$= \begin{bmatrix} 0 & 0 & 0 & 1 & 0 \\ 1 & 1 & 1 & 0 & 0 \\ 0 & 1 & 0 & 0 & 0 \\ 0 & 0 & 1 & 0 & 1 \\ 1 & 0 & 0 & 0 & 0 \end{bmatrix} \times \begin{bmatrix} 0 & 0 & 0 & 1 & 0 \\ 1 & 1 & 1 & 0 & 0 \\ 0 & 1 & 0 & 0 & 0 \\ 0 & 0 & 1 & 0 & 1 \\ 1 & 0 & 0 & 0 & 0 \end{bmatrix} = \begin{bmatrix} 0 & 0 & 1 & 0 & 1 \\ 1 & 1 & 1 & 0 & 0 \\ 1 & 1 & 1 & 0 & 0 \\ 1 & 1 & 0 & 0 & 0 \\ 0 & 0 & 0 & 1 & 0 \end{bmatrix}$$

$$M_R{}^3 = M_R{}^2 \times M_R$$

$$= \begin{bmatrix} 0 & 0 & 1 & 0 & 1 \\ 1 & 1 & 1 & 0 & 0 \\ 1 & 1 & 1 & 0 & 0 \\ 1 & 1 & 0 & 0 & 0 \\ 0 & 0 & 0 & 1 & 0 \end{bmatrix} \times \begin{bmatrix} 0 & 0 & 0 & 1 & 0 \\ 1 & 1 & 1 & 0 & 0 \\ 0 & 1 & 0 & 0 & 0 \\ 0 & 0 & 1 & 0 & 1 \\ 1 & 0 & 0 & 0 & 0 \end{bmatrix} = \begin{bmatrix} 1 & 1 & 0 & 0 & 0 \\ 1 & 1 & 1 & 0 & 0 \\ 1 & 1 & 1 & 1 & 0 \\ 1 & 1 & 1 & 1 & 0 \\ 0 & 0 & 1 & 0 & 1 \end{bmatrix}$$

$$M_R^4 = M_R^3 \times M_R$$

$$= \begin{bmatrix} 1 & 1 & 0 & 0 & 0 \\ 1 & 1 & 1 & 0 & 0 \\ 1 & 1 & 1 & 1 & 0 \\ 1 & 1 & 1 & 1 & 0 \\ 0 & 0 & 1 & 0 & 1 \end{bmatrix} \times \begin{bmatrix} 0 & 0 & 0 & 1 & 0 \\ 1 & 1 & 1 & 0 & 0 \\ 0 & 1 & 0 & 0 & 0 \\ 0 & 0 & 1 & 0 & 1 \\ 1 & 0 & 0 & 0 & 0 \end{bmatrix} = \begin{bmatrix} 1 & 1 & 1 & 1 & 0 \\ 1 & 1 & 1 & 1 & 0 \\ 1 & 1 & 1 & 1 & 1 \\ 1 & 1 & 1 & 1 & 1 \\ 1 & 1 & 0 & 0 & 0 \end{bmatrix}$$

$$M_R^5 = M_R^4 \times M_R$$

$$= \begin{bmatrix} 1 & 1 & 1 & 1 & 0 \\ 1 & 1 & 1 & 1 & 0 \\ 1 & 1 & 1 & 1 & 1 \\ 1 & 1 & 1 & 1 & 1 \\ 1 & 1 & 0 & 0 & 0 \end{bmatrix} \times \begin{bmatrix} 0 & 0 & 0 & 1 & 0 \\ 1 & 1 & 1 & 0 & 0 \\ 0 & 1 & 0 & 0 & 0 \\ 0 & 0 & 1 & 0 & 1 \\ 1 & 0 & 0 & 0 & 0 \end{bmatrix} = \begin{bmatrix} 1 & 1 & 1 & 1 & 1 \\ 1 & 1 & 1 & 1 & 1 \\ 1 & 1 & 1 & 1 & 1 \\ 1 & 1 & 1 & 1 & 1 \\ 1 & 1 & 1 & 1 & 0 \end{bmatrix}$$

$$\therefore \quad M_R^+ = M_R + M_R^2 + M_R^3 + M_R^4 + M_R^5$$

Using Boolean Addition.

### Exercise 2.1

1. Which of the following relations in the set of real numbers are equivalence relations?
   (i) $aRb$ if $|a| = |b|$
   (ii) $aRb$ if $|a| \geq |b|$
   (iii) $aRb$ if $a - b \geq 0$

2. Prove that a relation R on a set A is symmetric if $R^{-1} = R$.

3. Give an example of a relation that is reflexive but neither symmetric nor transitive.

4. Show that the relation 'is perpendicular to' over the set of all straight lines in the plane is symmetric but neither reflexive nor transitive.

5. Let S and T be sets with $m$ and $n$ element respectively. How many elements has $S \times T$? How many relations are there in $S \times T$?

6. If R and S are equivalence relations in the set X, prove that $R \cap S$ is an equivalence relation.

7. Show that the relation of congruence modulo $m$ has $m$ distinct equivalence classes.

8. Show that a partition of a set S determines an equivalence relation in S.

9. Let $S = \{n: n \in N \text{ and } n > 1\}$. If $a, b \in S$ define $a \sim b$ to mean that $a$ and $b$ have the same number of positive prime factors (distinct or identical). Show that $\sim$ is an equivalence relation.

10. Prove that in the set N × N, the relation R defined by $(a, b) \, R \, (c, d) \Leftrightarrow ad = bc$ is an equivalence relation.

11. In the set of integers a relation R is defined as follows: $aRb$ iff $3a + 4b$ is divisible by 7. Prove that R is an equivalence relation.

12. Let P be a set of all people. Let R be the binary relation on R i.e., $aRb$ if $a$ is the brother of $b$. Is R reflexive, symmetric, anti-symmetric, transitive, an equivalence relation?

13. Let R be a binary relation on the set of all strings of 0's and 1's such that R = $\{(a, b) \mid a$ and $b$ are strings that have same number of 0's$\}$. Is R reflexive?

Symmetric?        Anti-symmetric?   Transitive?

14. Let S= {1, 2, 3, 4} and A= S × S. Define the relation R on A as follows: we say that R $(a', b')$ iff $ab' = a'b$. Show that R is an equivalence relation. Compute A/R also.

**Answers to Selected Problems**

1. (i) Yes
   (ii) No
   (iii) No

5. No. of elements in S × T = m × n

   No. of relations = $2^{m \times n}$

12. R is reflexive, symmetric, transitive and equivalence relation

13. R is reflexive, symmetric and transitive.

## 2.9 FUNCTIONS / MAPPINGS

If to each element of A there is assigned a unique element of B, the collection of such assignments is called a function, (or mapping or map) from A into B.

f: A → B is a function $f$ from A into B. A is called **domain** of $f$, B is called **Co-domain** of $f$. If $f(a) = b$ then $b$ is called image of $a$ and $a$ is called pre-image of $b$. **Range** of $f$ is set of all images under $f$ and is denoted by $f(A)$

$R_f$ = range $f = f(A) = \{b \in f(a)$ for all $a \in A\}$
$R_f = f(A) \subseteq B$

*Note:* If $(x, y_1) \in f$ and $(x, y_2) \in f$ then $y_1 = y_2$

## 2.10 TYPES OF FUNCTIONS

**Many-One function:** The function $f: A \to B$ is many-one if two or more elements in A have same image under $f$.

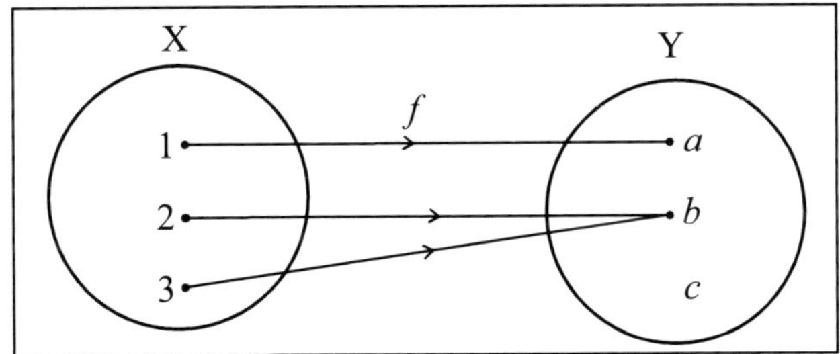

**One-to-one function:** A function $f: X \rightarrow Y$ is one to one (Injective or 1:1) if different elements in domain X have distinct images. i.e. $f(x) = f(x')$ implies $x = x'$.

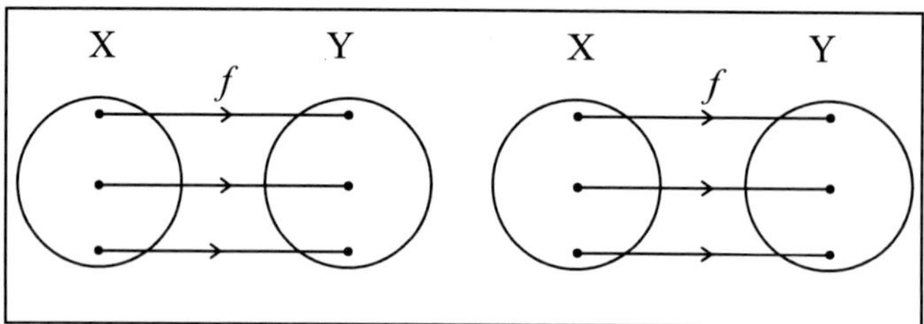

**Into function:** The function $f: A \rightarrow B$ is into function, if range of $f$ is a proper subset of B. $f(A) \subset B$ and $f(A) \neq B$. In this case there is at least one element of B which is not an image of any element of A.

**Example 25:** If $X = \{1, 2, 3\}$ and $Y = \{a, b, c\}$. Prove that $f = \{(1, a), (2, b), (3, b)\}$ is an into mapping.

### *Solution*

In other words, the mapping $f: X \rightarrow Y$ is an into mapping if the range of $f$ is the proper subset of the co-domain, i.e., $\{f(x)\} \subset Y, x \in X$.

Here, the element of $c$ of Y is not the $f$-image of any element of X i.e., $\{f(x)\} \subset Y, x \in X$. Hence it is into function.

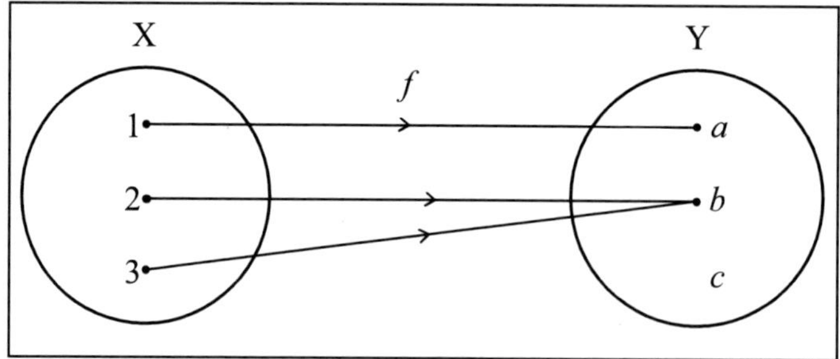

**Onto function:** A function $f: A \rightarrow B$ is onto or surjective if each element of B is image of some element of A.

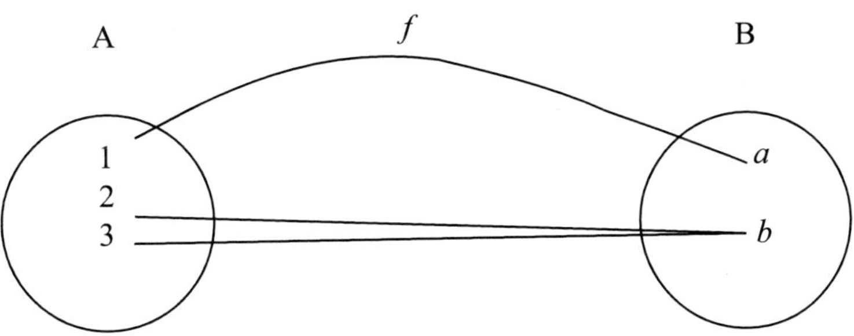

**Onto function (surjective)**

**Example 26:** If X = {1, 2, 3, 4,}, Y = {a, b, c}. Prove that $f$ = {(1, a), (2, b), (3, b), (4, c)} is onto function.

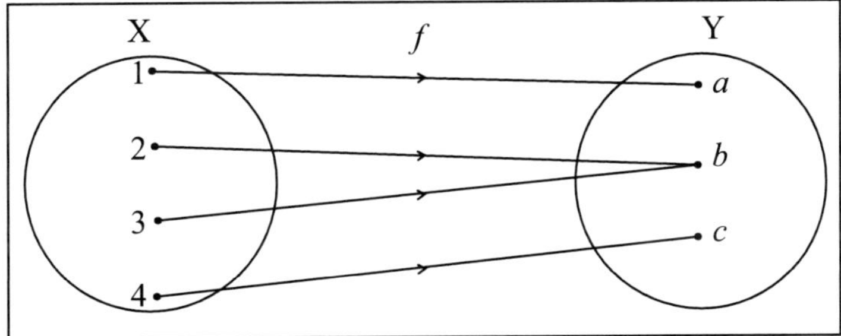

*Solution*

All three elements $a$, $b$, $c$ of Y are $f$-images of the element of X, hence it is onto function.

**One-to-one Correspondence:** A function $f$: X → Y is one-to-one correspondence or bijective if $f$ is both one-one and onto. i.e. each element of X correspond to unique element of Y and vice versa.

(i) $f(x_1) = f(x_2) \Rightarrow x_1 = x_2, x_1, x_2 \in$ X.

(ii) Range of $f$ = co-domain Y, i.e., $\{ f(x)\}$ = Y, $x \in$ X.

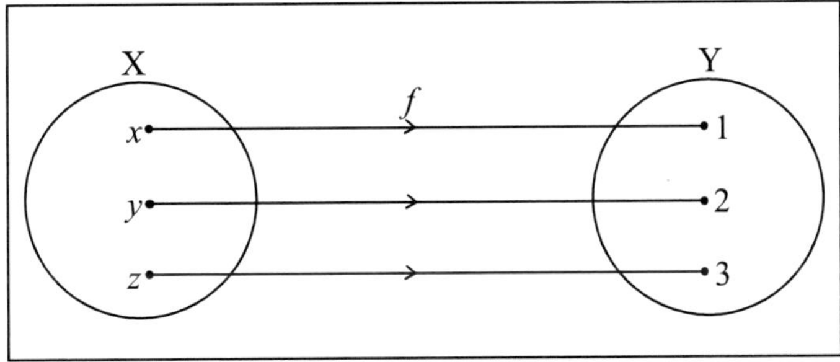

**One-to-one Correspondence (Bijective)**

**Example 27:** Two sets are said to be bijective, written X ~ Y, if there exists a bijection between them. Prove that the relation between sets defined by $\{(A, B) \mid A \sim B\}$ is an equivalence relation.

### *Solution*

$R = \{(A, B) \mid A \sim B\}$

Where, A ~ B denotes that there exists a bijection between A and B, say f: A → B. $f^{-1}$ is also a bijection, $f^{-1}$:B → A. Clearly (A, A) ∈ R, since there always exists a mapping between sets with the same cardinality – for example, the identity function. Thus ~ **is reflexive.**

If A ~ B, then also B ~ A, since if there is a bijection $f$ : A → B, there also exists a bijection $f^{-1}$: B → A. Thus (A, B) ∈ R ⊃ (B, A) ∈ R, i.e. ~ **is symmetric.**

Suppose both $f$ and $g$ are bijections, $f$ : A → B and $g$: B → C. Then $g . f$ : A → C is also a bijection. Hence if A ~ B and B ~ C, then A ~ C, i.e. (A, B) ∈ R ∧ (B, C) ∈ R ⊃ (A, C) ∈ R, so that ~ **is transitive.**

Hence R is an equivalence relation.

**Zero function:** The function $f$: A → B is zero function if the image of each element of A under $f$ is zero i.e. $f(a) = 0$.

**Real valued function:** A function whose domain and range are subsets of R is called a real valued function.

**Absolute value function or Modulus function:** If $f(x) = |x|$, then $f(x)$ is called absolute value or modulus function. D($f$) = R R($f$) = All non-negative real numbers.

**Rational function:** A function of the form $f(x) = g(x)/h(x)$, where $g(x)$ and $h(x)$ are polynomial functions D($f$) = $\{x \mid x \in R$ and $h(x) \neq 0\}$.

**Polynomial function:** If a function $f(x)$ can be written in the form $f(x) = a_0 + a_1 x + a_2 x^2 + a_3 x^3 + \dots + a_n x^n$ where, $x$ is +ve integer and $a_0, a_1, a_2, a_3, \dots , a_n \in R$ then $f$ is called a poly function.

**Identity function:** If X is any set and the mapping $f$ : X → X is defined by $f(x) = x$, i.e., every element of the set X is the image of itself, then the mapping is called the identity mapping. It is always one-one onto or objective mapping.

E.g.:  If X = $\{a, b, c\}$ and $f$ : X→X is defined by

$$f(a) = a, f(b) = b, f(c) = c.$$

Then it is an identity mapping.

**Constant function:** A mapping in which every element of the domain is assigned to the same element of the co-domain is called a constant mapping. The range of the constant mapping is a set with single element. Thus the mapping, $f: X \to Y$, is a constant mapping if $x \in X, f(x) = k, k \in Y$ is a constant.

e.g.:   The mapping, $f: R \to R$ where $f(x) = 3, x \in R$ is a constant mapping.

**Equal functions:** If $f$ and $g$ are two mappings defined on the same domain D and if $f(x) = g(x)$, $x \in D$, then the mappings $f$ and $g$ are said to be equal, i.e., $f = g$.

**Inclusion function:** If $X \subset Y$, then the mapping $f: X \to Y$ defined by $f(x) = x, x \in X$ is called an inclusion mapping of X to Y.

Clearly the inclusion mapping is one-one into, and the inclusion mapping of a set A to itself is the identity mapping on A.

E.g.:   If $X = \{1, 2, 5\}$, and $Y = \{1, 2, 3, 5, 9\}$. Then the mapping defined by $f = \{(1, 1),$ $(2, 2), (5, 5)\}$ is the inclusion mapping of X to Y, because $X \subset Y$, and $f(x) = x, x \in X$.

**Restriction of function:** Let A and B be two sets and let $C \leq A$. Suppose $f: A \to B$ is a mapping. Then the mapping $G : C \to B$ is called restriction of $f$ to C if $f(x) = g(x) \ \forall \ x \in C$, we may also say that $f$ is the extension of $g$ to A.

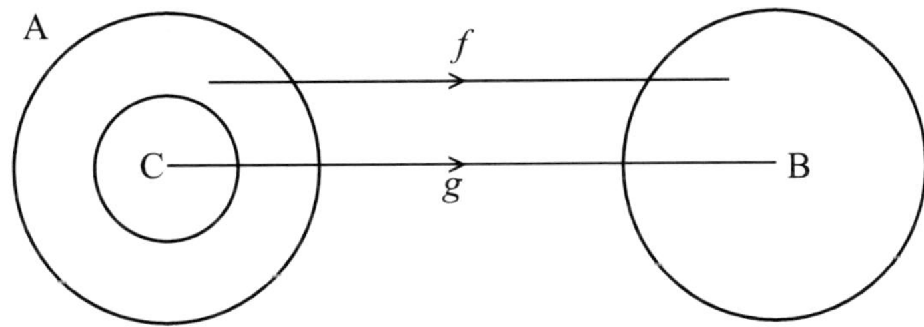

**Example 28:**  Is the function given by the following mapping defined?

*Solution*

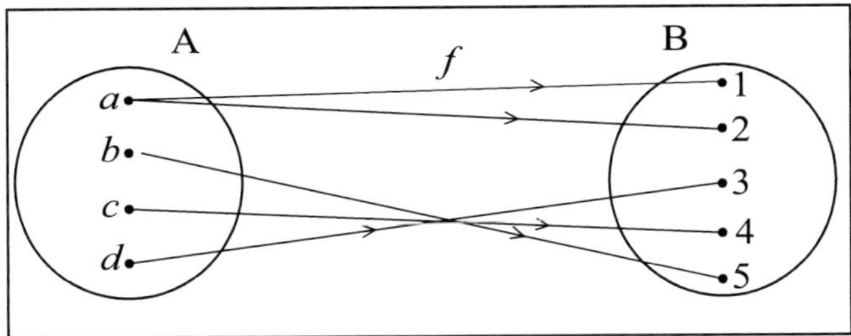

$a \in A \Rightarrow \exists \ 1, 2 \in B$ such that $f(a) = 1 = 2$. Since each element in A is having one or more images in B. This function is not defined.

**Example 29:** Is the function defined by the following figure?

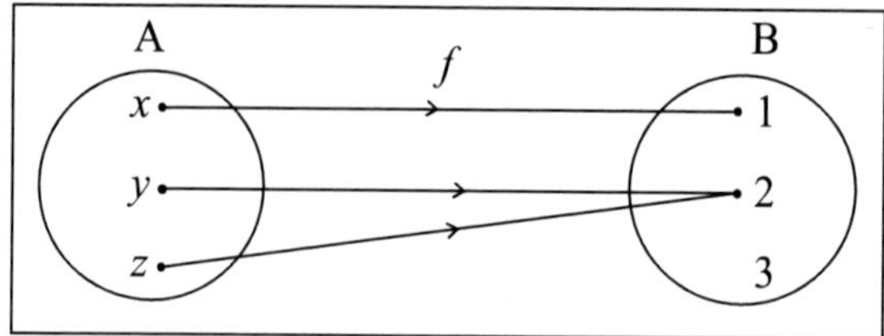

Here

$$\text{domain of } f = \{x, y, z\}$$
$$\text{range of } f = \{1, 2\}$$
$$\text{co-domain of } f = \{1, 2, 3\}$$

Clearly function is well defined.

## 2.11 INVERSE FUNCTIONS

Let $f$ be a one-to-one correspondence from the set A to the set B. The inverse function of $f$ is the function that assigns to each $b \in B$ the unique element $a \in A$ such that $f(a) = b$. The inverse function of $f$ is denoted by $f^{-1}$. Hence $f^{-1}(b) = a$ when $f(a) = b$. A one-to-one correspondence is called invertible.

**Example 30:** Show that the mapping $f : R \rightarrow R$ defined by $f(x) = ax + b$, where $a, b, x \in R$, $a \neq 0$ is invertible. Define its inverse.

### *Solution*

$f(x) = ax + b$

Here $a, b, x \in R$, $a \neq 0$, is one-to-one function, as $ax_1 + b = ax_2 + b$ implies $x_1 = x_2$. $f$ is onto as for each element $y$ of R, there is a $x$ given by $x = (y - b)/a$.

Hence $f$ is one-to-one correspondence and therefore invertible.

$$f^{-1}(y) = (y - b)/a$$

**Theorem** A function $f : A \rightarrow B$ is invertible if and only if both one-to-one and onto, i.e., $f$ is a bijective function.

### *Proof*

$f$ is invertible means $f^{-1}$ is a function from $Y \rightarrow X$, $f^{-1}$ will be a function from $Y \rightarrow X$ if and only if:

1. For every $y \in Y$, there must be an $x \in X$ such that $f^{-1}(y) = x$, i.e., $f(x) = y$, $\Rightarrow f$ must be onto.

2. For every $y \in Y$, $f^{-1}(y)$ must be unique. This is possible if and only if $f$ is one to one.

Hence $f$ is invertible if and only if $f$ is bijective, i.e., one to one and onto.

## 2.12 COMPOSITION OF FUNCTIONS

Let : $X \rightarrow Y$ and $g: Y \rightarrow Z$. Then the composition of the function $g$ with $f$, denoted by $g \circ f$, is the function defined as $g \circ f = g[f(x)]$. The commutative law does not hold good for composition of functions, in general.

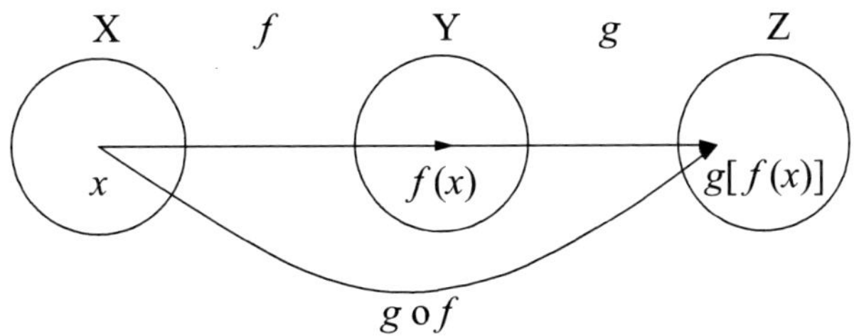


**Example 31:** If $f(x) = x^2$, $g(x) = x + 5$, $x \in$ R. Then find the composition $f \circ g$ and $g \circ f$.

*Solution*

$$(g \circ f)(x) = g(f(x))$$
$$= g(x^2) = x^2 + 5$$

$$(f \circ g)(x) = f(g(x)) = f(x + 5)$$
$$= (x + 5)^2 = x^2 + 10x + 25$$

**Example 32:** Let $f$ and $g$ be real valued functions defined by $f(x) = \sin x$ and $g(x) = x^2$, for $x \in$ R. Find the composition $g \circ f$.

*Solution*

Then the composition $(g \circ f)$ is given

$$g\left[f(x)\right] = g(\sin x)\left[\text{because } f(x) = \sin x\right]$$

$$= (\sin x)^2 \left[\text{because } g(x) = x^2\right]$$

Hence $(g \circ f)(x) = (\sin x)^2$, $x \in$ R.

### Exercise 2.2

1. Differentiate between function and relation.

2. Define a surjective, injective and bijective function.

3. Determine which of the following is surjective, injective, and bijective?

   Let I be set of all integers, and $I_p = \{0, 1, 2 ... p - 1\}$,

   (i) $f: I \rightarrow I, f(j) = \begin{cases} j/2, j \text{ is even} \\ (j-1)/2, j \text{ is odd} \end{cases}$

   (ii) $f: I \rightarrow I f(k) = $ greatest integer $\leq \sqrt{k}$

   (iii) $f: I_7 \rightarrow I_7 f(x) = 3 \text{ MOD } 7$

   (iv) $f: I_4 \rightarrow I_4 f(x) = 3 \text{ MOD } 4$

4. Prove that if $f: X \rightarrow Y$ and $g: Y \rightarrow Z$ are two one-to-one functions, then $g \circ f$ is also one-to-one onto function.

5. Let A = $\{1, 2, 3\}$. Define $f: A \rightarrow A$ by $f(1) = 2, f(2) = 1$ and $f(3) = 3$. Find $f^{-1}$.

**Answers to Selected Problems**

5. $f^{-1}(1) = 2$; $f^{-1}(2) = 1$; $f^{-1}(3) = 3$

## 2.13 SCHRÖDER-BERNSTEIN THEOREM

If there exists injective functions (or one-to-one functions) $f : A \rightarrow B$ and $g : B \rightarrow A$ between the sets A and B, then there exists a bijective function (or one-to-one correspondence) $h : A \rightarrow B$. In terms of cardinality the theorem states that if $|A| \leq |B|$ and $|B| \leq |A|$, then $|A| = |B|$.

This theorem is also known as Cantor-Schröder-Bernstein theorem. Cardinality of a set A is denoted by $|A|$ and is the number of elements in set A. Two sets A and B having the same cardinality are called equipotent.

**Example 33:** Use the Schröder-Bernstein theorem to show that $(0, 1)$ and $[0, 1]$ have the same cardinality.

*Solution:*

A function $f : A \rightarrow B$ is injective if $f(m) = f(n)$ then $m = n$ or $|A| \leq |B|$.

Let $f$ and $g$ be functions,

$f : (0,1) \rightarrow [0,1], f(n) = n$

and

$$g : [0,1] \rightarrow (0.1), \quad g(n) = \frac{n+1}{3}$$

We first prove that $f$ is injective or $|(0,1)| \leq |[0,1]|$

If $f(m) = f(n)$, then by definition of function $f$, $m = n$

Which shows that $f$ is an injective function.

Now we prove that $g$ is injective

If $g(m) = g(n)$, then $\dfrac{m+1}{3} = \dfrac{n+1}{3}$

Multiplying each side by 3.

$m + 1 = n + 1$

subtracting 1 from each side.

$m = n$

which shows that $g$ is injective or $|[0,1]| \leq |(0,1)|$

Now $|(0,1)| \leq |[0,1]|$ and $|[0,1]| \leq |(0,1)|$ and by Schröder-Bernstein theorem,

$|(0,1)| = |[0,1]|$.

# 3

# Techniques of Counting

## 3.1 INTRODUCTION

Combinatories, the study of arrangements of objects, is essential part of discrete mathematics. The counting of objects with certain properties is an important part of combinatories. Further, counting techniques are used extensively when probabilities of events are computed.

### 3.1.1 Equivalent Sets

Two sets A and B are called equivalent if there exists a one-to-one correspondence between them, we write, $A \sim B$.

### 3.1.2 Finite and Infinite Sets

A set A is said to be a finite set if either A is empty or there exists a natural number $m$ such that $N_m \sim A$, where, $N_m = \{1, 2, \dots m\}$. A set which is not finite is called infinite set.

It is important to note that no proper subset of a finite set can be equivalent to the set itself, because a one-to-one correspondence between the elements of such sets is impossible. However, for infinite sets this is not necessarily the case.

### 3.1.3 Countably Infinite/Denumerable/Enumerable Set

A set X is said to be countably infinite (or denumerable or enumerable) if and only if it is equivalent to the set of natural numbers, N.

**Theorem 1** An infinite subset of a denumerable set is also denumerable.

### *Proof*

Let A be a denumerable set and S its infinite subset.

A is denumerable. (Given)

$\therefore \quad A \sim N.$

Let $f(n) = n, f: N \rightarrow A$. The elements of A can be arranged as $f(1), f(2) \dots$. Now, delete from this list those elements which are not present in S. The number of the remaining elements is still infinite because S is infinite. Let us denote these elements by $f(1), f(2) \dots$

Define a function $g: N \rightarrow S$ such that $g(n) = f(i_n)$, then $g$ is one-to-one correspondence between N and S. Hence S is denumerable.

**Example 1:** Show that the set of integers, positive, negative and zero is denumerable.

*Solution*

$I = \{\ldots -2, -1, 0, 1, 2, \ldots\}$.
We define a function $f : N \to I$ where,

$$f(n) = \begin{cases} \dfrac{n}{2} & \text{if } n \text{ is even} \\[2ex] -\dfrac{n+1}{2} & \text{if } n \text{ is odd} \end{cases}$$

so that

$$f(1) = 0, \quad f(2) = 1, \quad f(3) = -2, \quad f(4) = 2, \quad \ldots$$

$\therefore$  $f$ is one-to-one correspondence between N and I.
$\therefore$  I $\sim$ N and hence I is denumerable.

**Corollary A subset of a denumerable set is either finite or denumerable.**

*Proof*

Let A be a denumerable set
$\therefore$  A $\sim$ N
Let B is subset of A. B is either finite or infinite.
If B is infinite, it is denumerable. Because we know that every infinite subset of a denumerable set is denumerable.
Hence a subset of a denumerable set is either finite or denumerable.

**Theorem 2** Union of finite number of denumerable sets is denumerable.

*Proof*

Let A and B be two enumerable sets.
A $\sim$ N and B $\sim$ N.

We can write $A = \{a_1, a_2, a_3, \ldots\}$ and $B = \{b_1, b_2, b_3, \ldots\}$. As every infinite subset of a denumerable set is denumerable, we can suppose that A and B are disjoint sets.

$$A \cup B = \{a_1, a_2, a_3, b_1, b_2, b_3, \ldots\}$$

Consider $f : N \to A \cup B$, where

$$f(n) = \begin{cases} a_{(n+1)/2}, & \text{if } n \text{ is odd} \\ b_n, & \text{if } n \text{ is even} \end{cases}$$

As $f$ is one-to-one and onto, hence $A \cup B \sim N$ and hence $A \cup B$ is denumerable. Repeating this argument with the number of denumerable sets, we can say that union of finite number of denumerable sets is denumerable.

### 3.1.4 Countable Set

A set that is either finite or is denumerable is called countable set. A set that is not countable is called uncountable.

**Example 2:** Prove that the set $X = \{^1/_2, \,^2/_3, \,^3/_4, \, \ldots\}$ is countable.

*Solution*

Let $f: N \to X$, where $f(n) = n/(n + 1)$.
Then

$$f(1) = \,^1/_2, \qquad f(2) = \,^2/_3, \qquad f(3) = \,^3/_4, \quad \cdots$$

$\therefore$  There exists a one-to-one correspondence between N and X.

$$\therefore \quad N \sim X$$

Hence X is countable.

### 3.1.5 Cardinality of Sets

The number of distinct elements in the set A is called cardinality or cardinal number or order of the set A and is denoted by $|A|$ or $O(A)$ or $n(A)$. We can extend the concept of cardinality to all sets finite and infinite. Cardinality of a set is also called order of a set.

Two sets A and B have the same cardinality if and only if there is a one-to-one correspondence from A to B, we write $A \sim B$. Sets having the same cardinality are called **equipotent** or **equivalent** or **similar**.

$\Rightarrow \quad n(A \times B) = n(A) \times n(B)$

$\Rightarrow \quad n(A \cup B) = n(A) + n(B) - n(A \cap B)$

$\Rightarrow \quad n(A) = n(A - B) + n(A \cap B)$

$\Rightarrow \quad n(A \cup B \cup C) = n(A) + n(B) + n(C) - n(A \cap B) - n(A \cap C) - n(B \cap C) + n(A \cap B \cap C)$

**Example 3:** Find the cardinal number of each set.

(*a*) A = {Monday, Tuesday, … , Sunday}.

(*b*) B = $\left\{x : x^2 = 25, \, 3x = 6\right\}$.

(*c*) The power set P(A) of A = {1, 4, 5, 9}.

*Solution*

(*a*) $|A| = 7$, since there are seven days in a week.
(*b*) Here B is empty since no number satisfies both $x^2 = 25$ and $3x = 6$. Thus $|B| = 0$.
(*c*) Here A has 4 elements, so P(A) has $2^4 = 16$ elements or $|P(A)| = 16$.

**Example 4:** The set of all real valued functions defined on [0, 1] has the cardinal number $2^c$, where cardinal number of all real numbers $= c$.

### *Solution*

Let I = [0, 1], recall that
$$R^I = \{f : f \cdot I \to \mathbb{R}\}$$
$$= \text{Set of real valued functions defined on I.}$$

Also, $|R^I| = c^c$, but $2^9 = c$ and $9c = c$

Hence, $|R| = (2^9) = 2^c$ i.e. $|R^I| = 2^c$

## 3.2 PRINCIPLE OF INCLUSION–EXCLUSION

Case (i) for two sets
$$|A \cup B| = |A| + |B| - |A \cap B|$$

Case (ii) for three sets
$$|A \cup B \cup C| = |A| + |B| + |C| - |A \cap B| - |A \cap C| - |B \cap C| + |A \cap B \cap C|$$

Case (iii) In general

For $r$ sets $A_1, A_2, A_3, \ldots A_r$ we have
$$|A_1 \cup A_2 \cup \cdots \cup A_r| = \Sigma |A_i| - \Sigma |A_i \cap A_j| + \Sigma |A_i \cap A_j \cap A_k| + \cdots + (-1)^{r+1} |A_1 \cap A_2 \cap \cdots \cap A_r|$$
$$1 \le i \le r \qquad 1 \le i < j \le r \qquad 1 \le i < j < k \le r$$

**Example 5:** How many positive integers not exceeding 1000 are divisible by 7 or 11.

### *Solution*

Let

A = {$x$: $x$ is a positive integer not exceeding 1000 and divisible by 7}   and

B = {$x$: $x$ is a positive integer not exceeding 1000 and divisible by 11}

Then

$A \cup B$ is the set of integers divisible by 7 or 11.

$n(A \cup B)$ should be there to keep same notation

$$|A \cup B| = |A| + |B| - |A \cap B|$$
$$= [1000/7] + [1000/11] - [1000/77], \text{ where } [x] \text{ is the largest integer smaller than or equal to } x.$$
$$= 142 + 90 - 12 = 232 - 12 = 220$$

**Example 6:** In a class of 60 boys, there are 45 boys who play hockey and 30 boys who play football, find

(a) How many boys play hockey only?
(b) How many boys play football only?

### *Solution*

Let A = Set of boys who play hockey and B = Set of boys who play football

$$|A \cup B| = |A| + |B| - |A \cap B|$$
$$60 = 45 + 30 - |A \cap B|$$
$$|A \cap B| = 75 - 60 = 15$$

No. of boys who play hockey only = $|A| - |A \cap B| = 45 - 15 = 30$

No. of boys who play football only = $|B| - |A \cap B| = 30 - 15 = 15$

**Example 7:** 1232 students have taken a course in Spanish, 879 have taken a course in French and 114 have taken a course in Russian. Further 103 students have taken courses in both Spanish and French, 23 both Spanish and Russian, 14 both French and Russian. If 2092 students have taken courses in at least one of Spanish, French and Russian. How many students have taken a course in all three languages?

### *Solution*

A = Students who have taken course in Spanish

B = Students who have taken course in French

C = Students who have taken course in Russian

$$|A \cup B \cup C| = |A| + |B| + |C| - |A \cap B| - |A \cap C| - |B \cap C| + |A \cap B \cap C|$$
$$2092 = 1232 + 879 + 114 - 103 - 23 - 14 + |A \cap B \cap C|$$

No. of students who have taken a course in all three languages = $|A \cap B \cap C| = 7$

**Example 8:** Evaluate $(A - B) \cup (B - A) \cup (A \cap B)$, where A, B are two sets and $U = A \cup B$.

### *Solution*

$(A - B) \cup (B - A) \cup (A \cap B)$

$\Rightarrow B' \cup A' \cup (A \cap B)$     here   $A - B = B'$   and   $B - A = A'$

$\Rightarrow (A \cap B)' \cup (A \cap B) \Rightarrow U$

i.e. $A \cup B$   $(\because \quad U = A \cup B)$

**Example 9:** Determine set of +ve integers ($\leq 720$) which are not divisible by any of 2, 3, 5.

### *Solution*

Let
A = set of +ve integers $\leq 720$, divisible by 2

B = set of +ve integers $\leq 720$, divisible by 3

C = set of +ve integers $\leq 720$, divisible by 5

$n(A) = 720/2 = 360 \qquad n(B) = 720/3 = 240 \qquad n(C) = 720/5 = 144$

$n(A \cap B) = 720/6 = 120 \qquad n(A \cap C) = 720/10 = 72 \qquad n(B \cap C) = 720/15 = 48$

$n(A \cap B \cap C) = 720/30 = 24$

$n(A \cup B \cup C) = |A| + |B| + |C| - |A \cap B| - |A \cap C| - |B \cap C| + |A \cap B \cap C|$

$$= 360 + 240 + 144 - 120 - 72 - 48 + 24$$

$$= 528$$

$\therefore \quad$ No of +ve integers not divisible by any of 2, 3, 5 $= 720 - 528 = 192$

**Example 10:** In a room containing 28 people, there are 18 people who speak English, 15 speak Hindi and 22 French, 9 persons speak both English and Hindi, 11 speak both Hindi and French whereas 13 speak both French and English. How many people speak all the 3 languages?

### *Solution*

Let
T = set of people in the room
E = set of people who speak English
H = set of people who speak Hindi
F = set of people who speak French

$n(T) = 28 \qquad n(E) = 18 \qquad n(H) = 15 \qquad n(F) = 22 \qquad n(E \cap H) = 9$
$n(H \cap F) = 11 \qquad n(F \cap E) = 13$

$n(E \cup H \cup F) = n(E) + n(H) + n(F) - n(E \cap H) - n(E \cap F) - n(H \cap F) + n(E \cap H \cap F)$

$\therefore \quad 28 = 18 + 15 + 22 - 9 - 13 - 11 + n(E \cap H \cap F)$

$n(E \cap H \cap F) = 28 - 55 + 33$

$$= 6 = \text{people who speak all languages}$$

## 3.3 PIGEONHOLE PRINCIPLE

Pigeonhole principle is also known as Shoe Box Principle or Dirichlet Drawer Principle.

**First Version:** If there are many pigeons and a few pigeonholes then there must be some pigeonholes occupied by two or more pigeons.

### *Proof*

Let there be 1, 2, 3, …, $m$ pigeonholes and 1, 2, 3, …, $m$, $m + 1$, …, $n$ pigeons $(n > m)$. $m$ pigeons will occupy $m$ pigeon holes and $n - m$ pigeons will be left. Thus at least one pigeonhole will be occupied by two or more pigeons.

**Second Version:** (Generalized Pigeonhole Principle): If there are $n$ pigeons and $m$ pigeonholes, $n > m$, then one of the pigeonholes will occupy at least $[(n - 1)/m] + 1$ pigeons.

***Proof***

Suppose none of the pigeonholes contains more than $[(n-1)/m]$ pigeons.

Then the total number of pigeons is at most $m[(n-1)/m] = (n-1)$, which is a contradiction since total number of pigeons is $n$. Hence if there are $n$ pigeons and $m$ pigeonholes, $n > m$, then one of the pigeonholes will occupy at least $[(n-1)/m] + 1$ pigeons.

**Example 11:** How many people among 100 are born in the same month?

***Solution***

Number of pigeons $= n = 100$

Number of pigeonholes $= m = 12$

Hence one of the pigeonholes must contain at least $[(n-1)/m] + 1$ i.e., $\left[\dfrac{99}{12}\right] + 1 (=9)$ pigeons.

Therefore at least 9 persons were born in one of the months.

**Example 12:** How many people among 10000, born on one day, are born in the same hour?

***Solution***

Number of pigeons $= n = 10000$

Number of pigeonholes $= m = 24 =$ number of hours in a day

Hence one of the pigeonholes must contain at least $[(n-1)/m] + 1$ i.e., $\left[\dfrac{9999}{24}\right] + 1 (=417)$ pigeons.

Therefore at least 417 persons were born in the same hour.

**Example 13:** How many people among 10000, born on a day, are born in the same minute?

***Solution***

Number of pigeons $= n = 10000 =$ number of people.

Number of pigeonholes $= m = 1440 =$ number of minutes in a day.

Hence one of the pigeonholes must contain at least $[(n-1)/m] + 1$ i.e., $\left[\dfrac{9999}{1440}\right] + 1 (=7)$ pigeons.

Therefore at least 7 persons were born in the same minute.

**Example 14:** How many people at least in a group of 85 people have the same last initials?

***Solution***

Number of pigeons $= n = 85$

Number of pigeonholes $= m = 26 =$ number of letters.

Hence one of the pigeonholes must contain at least $[(n-1)/m] + 1$ i.e., $\left[\dfrac{85}{26}\right] + 1 (=4)$ pigeons.

Therefore at least 4 persons have the same last initial.

**Example 15:** ABC is an equilateral triangle whose sides are $x$ unit in length. Five points lie on or inside the triangle. Show that at least two of these points must be not more than $x/2$ unit apart.

### *Solution*

Let D, E, F be the midpoints of the sides AB, BC and CA respectively.
Lines DE, EF and FD, divide the triangle into four equilateral triangles with each side of $x/2$ units in length.

Number of pigeons = $n = 5$ = number of points
Number of pigeonholes = $m = 4$ = number of small triangles.

Hence one of the pigeonholes must contain at least $[(n-1)/m] + 1$ i.e. 2 pigeons.
Therefore at least one small triangle out of the four must contain at least two points.
As per our division these points should not be more than $x/2$ unit apart.

### Exercise 3.1

1. A set A is called countable if:
    (i) A is finite or denumerable.
    (ii) A is infinite or non-denumerable.

2. Let $E = \{2, 4, 6, \ldots\}$, the set of even positive integers. Show that $|E| = (X_0)$ {aleph-naught}.

    [**Hint:** Here $|E| = X_0$ if and only if E has the same cardinality as N. A set with cardinality $X_0$ is said to be denumerable or countably infinite.]

3. Find the cardinal number of each set:
    (a) $A = \{a, b, c, \ldots, x, y, z\}$,
    (b) $B = \{1, -3, 5, 10, -10\}$,
    (c) $C = \{x: x \to N, x^2 = 7\}$.

4. Find the cardinal number of each set:
    (a) $A = \{10, 20, 30, 40, \ldots\}$,
    (b) $B = \{7, 8, 9, 10, \ldots\}$.

5. Let $A_1$ $A_2$ be countable number of finite sets. Prove that union $S = U_i A_i$ is countable.

6. Consider the closed unit interval $I = [0, 1]$ and the open unit interval $I^1 = (0, 1)$. Prove that
    (a) $|I| = |I^1|$, and
    (b) $|R| = |I^1| = |I|$ C.
    $I = \{0, 1, \frac{1}{2}, \frac{1}{3}, \ldots\} U A$
    $I^1 = \{\frac{1}{2}, \frac{1}{3}, \frac{1}{4}, \frac{1}{5}, \ldots\} U A$

**Answers to Selected Problems**

3. (a) $|A| = 26$
    (b) $|B| = 5$
    (c) $|C| = 0$

4. (a) $|A| = X_0$
    (b) $|B| = X_0$

## 3.4 PERMUTATIONS AND COMBINATIONS

### 3.4.1 Permutations

**Definition:** The different arrangements which can be made by taking some or all of a number of things are called permutations.

Suppose we have three objects $a, b, c$. If we take two of these at a time, then the arrangements are

$$ab, bc, ca, ba, cb, ac.$$

Thus the number of arrangements of 3 objects taken two at a time is 6. Each of these is called a permutation.

### 3.4.2 Notations of Permutations

If $n$ and $r$ are positive integers such that $1 \leq r \leq n$, then the number of all permutations of $n$ distinct objects, taken $r$ at a time is denoted by the symbol $^{n}P_{r}$ or $P(n, r)$. Thus

$^{n}P_{r}$ or $P(n, r)$ = Total number of permutations of $n$ distinct objects, taken $r$ at a time.

In the examples given above, we have obtained that there are:

(i)  6 permutations on set of 3 letters taken 2 at a time, i.e., $^{3}P_{2} = 6$;

(ii)  12 permutations on a set of 4 objects taken 2 at a time, i.e., $^{4}P_{2} = 12$.

### Factorial Notation

The continued product of first $n$ natural numbers is called the 'factorial $n$' and is denoted by $n!$ or $\lfloor n$. Thus $n! = 1.2.3 \ldots (n-1) \cdot n$

### 3.4.3 Fundamental Principle of Counting

If an operation can be performed in $m$ ways, and when it has been performed in any one of these ways, a second operation can be performed in $n$ ways;

The number of ways of performing the two operations is $m \times n$.

If the first operation is performed in any one way, we can associate with this any of the $n$ ways of performing the second operation; and thus we shall have $n$ ways of performing the two operations. So corresponding to each of the $m$ ways of performing the first operation, we shall have $n$ ways of performing the two. Hence, altogether the number of ways in which the two operations can be performed is $m \times n$.

More generally, if an operation can be performed in $m$ ways, the second operation can be performed in $n$ ways, the third in $p$ ways and so on, then the number of ways in which all the operations can be performed is $m \times n \times p \times \cdots$

### 3.4.4 Permutations of $n$ Dissimilar Things taken $r$ at a Time

The number of permutations of $n$ dissimilar things taken $r$ at a time is the same as the number of ways in which we can fill up $r$ places when we have $n$ different things at our disposal.

In the first place, we may put any one of the $n$ things, and therefore there are $n$ ways of filling the first place. When the first place has been filled up in any one of these ways, the second place can be filled up in $(n-1)$ ways. Since each way of filling the first place can be associated with each way of filling the second place, therefore there are $n(n-1)$ ways of filling the first two places.

Having filled the first two places in any way, there remain $(n-2)$ things to be chosen to fill up the third place. Thus there are $(n-2)$ ways of filling the third place for every way of filling the first two places. Therefore, there are $n(n-1)(n-2)$ ways of filling the first three places.

Proceeding similarly, and noticing that the number, of factors at any stage is the same as the number of places filled up, we find that the number of ways of filling up $r$ places is

$$= n(n - 1)(n - 2) \ldots \text{ to } r \text{ factors}$$
$$= n(n - 1)(n - 2) \ldots (n - (r - 1))$$
$$\text{i.e., } \quad {}^n\mathrm{P}_r = n(n - 1)(n - 2) \ldots (n - r + 1)$$

**Corollary 1 If all $n$ things are taken at a time, i.e., if $r = n$, then**

$$\quad {}^n\mathrm{P}_n = n(n - 1)(n - 2) \ldots (n - n + 1)$$
$$= n(n - 1)(n - 2) \ldots 1$$
$$\Rightarrow \quad {}^n\mathrm{P}_n = n!.$$

**Corollary 2 $^n\mathrm{P}_n = n(n - 1)(n - 2) \cdots (n - r + 1)$**

$$= \frac{\left[n(n-1)(n-2)\cdots(n-r+1)\right]\left[(n-r)(n-r-1)\cdots 3.2.1\right]}{\left[(n-r)(n-r-1)\cdots 3.2.1\right]}$$

$$= \frac{n!}{(n-r)!}$$

$$\text{i.e., } \quad {}^n\mathrm{P}_r = \frac{n!}{(n-r)!}$$

**Corollary 3 We have seen that**

$$\quad {}^n\mathrm{P}_r = \frac{n!}{(n-r)!}$$

$$\therefore \quad {}^n\mathrm{P}_n = \frac{n!}{(n-n)!}$$

$$\Rightarrow \quad n! = \frac{n!}{0!}$$

$$\Rightarrow \quad n!\,0! = n!$$

$$\therefore \quad 0! = 1$$

**Example 16:** In how many ways can 6 students sit on 2 empty chairs?

*Solution*

The first chair can be occupied by 6 students in 6 ways. When it is occupied by any one of the students, then the other chair has to be occupied by remaining 5 students. This can be done in 5 ways.

$\therefore$ The required number of ways $= 6 \times 5 = 30$.

*Aliter:* The total number of ways = permutation of 6 different objects taken 2 at a time

$$= {}^6P_2 = 6 \times 5 = 30.$$

**Example 17:**  How many number of three distinct digits can be formed from 1, 2, 3, 4, 5?

### *Solution*

All the five digit 1, 2, 3, 4, 5 are dissimilar. We have to form number of 3 digits out of these five, digits. The unit place can be filled up in 5 different ways. When the unit place has been filled up, the tenth place can be filled up in 4 different ways. The unit and tenth place can be filled up in 5 × 4 ways. Now, the hundredth place can be filled up in 3 ways.

Hence the required number of ways = 5 × 4 × 3
$$= 60.$$

Alternatively, we have to find the number of permutations of three digits from five digits 1, 2, 3, 4, 5.

Hence, the required number of numbers formed $= {}^5P_3 = 5 \times 4 \times 3$
$$= 60$$

**Example 18:**  Three men have 4 coats, 5 waistcoats and 6 caps. In how many ways can they wear them?

### *Solution*

The total number of ways in which 3 men can wear 4 coats $= {}^4P_3$
The total number of ways of wearing 5 waistcoats by 3 men $= {}^5P_3$
The total number of ways of wearing 6 caps by 3 men $= {}^6P_3$
Hence, by the fundamental principle of counting, the required number of ways

$$= {}^4P_3 \times {}^5P_3 \times {}^6P_3$$
$$= (4 \times 3 \times 2) \times (5 \times 4 \times 3) \times (6 \times 5 \times 4)$$
$$= 24 \times 60 \times 120$$
$$= 172800$$

**Example 19:**  Find the value of $r$ in the following ${}^{12}P_r = 1320$.

### *Solution*

We have

$$^{12}P_r = 1320$$

$$\Rightarrow \quad \frac{12!}{(12-r)!} = 12 \times 110$$

$$\Rightarrow \quad \frac{12!}{(12-r)!} = 12 \times 11 \times 10$$

$$\Rightarrow \quad \frac{12!}{(12-r)!} = \frac{12 \times 11 \times 10 \times 9!}{9!}$$

$$\Rightarrow \quad (12-r)! = 9!$$

$$\Rightarrow \quad (12-r) = 9$$

$$\Rightarrow \quad -r = 9 - 12$$

$$\Rightarrow \quad r = 3$$

### 3.4.5 Restricted Permutations

In the present section, we shall discuss permutations satisfying specific conditions. For example, the permutation where a particular object occurs in every arrangement or some is ignored. Such types of permutations are known as restricted permutations.

**Theorem 3** The number of all permutations of $n$ different objects taken $r$ at a time, when a particular object is always to be included in each arrangement is $r \cdot {}^{n-1}P_{r-1}$.

### *Proof*

Let $a_1, a_2, \ldots a_n$ be $n$ dissimilar objects. Let $a_1$ be the object which has to be taken up every time.

Let $a_1$ be placed at the first place. Then the remaining $(r-1)$ objects out of the $(n-1)$ objects can be arranged in ${}^{n-1}P_{r-1}$ ways. Therefore, the number of placing $a_1$ at the first place is ${}^{n-1}P_{r-1}$. Now $n_1$ can be placed in the second place and the remaining $(r-1)$ objects over of the $(n-1)$ objects can be placed in ${}^{n-1}P_{r-1}$ ways. So the number of ways in which $a_1$ will be at the second, place is again ${}^{n-1}P_{r-1}$.

$\therefore$ Total number of ways in which $a_1$ is placed in the first or the second, ..., or the $r$th place is

$${}^{n-1}P_{r-1} + {}^{n-1}P_{r-1} + \ldots + r \text{ terms} = r \cdot {}^{n-1}P_{r-1}$$

**Corollary: The number of ways in which $r$ things can be taken out of $n$ things and arranged when one particular thing is never taken up is ${}^{n-1}P_{r-1}$.**

**Example 20:** Show $^nP_r = {}^{n-1}P_r + r \cdot {}^{n-1}P_{r-1}$

*Proof*

**First Method**

We have

$$
{}^{n-1}P_r + r \cdot {}^{n-1}P_r = \frac{(n-1)!}{(n-1-r)!} + r \cdot \frac{(n-1)!}{(n-1-r+1)!}
$$

$$
= \frac{(n-1)!}{(n-1-r)!} + r \cdot \frac{(n-1)!}{(n-1)!}
$$

$$
= (n-1)! \left[ \frac{1}{(n-r-1)!} + r \cdot \frac{1}{(n-r)!} \right]
$$

$$
= (n-1)! \left[ \frac{1}{(n-r-1)!} + \frac{1}{(n-r)(n-r-1)!} \right]
$$

$$
= \frac{(n-1)!}{(n-1-r)!} \left[ 1 + \frac{r}{n-r} \right]
$$

$$
= \frac{(n-1)!}{(n-1-r)!} \left[ \frac{n-r+r}{n-r} \right]
$$

$$
= \frac{n \cdot (n-1)!}{(n-1-r)!(n-r)}
$$

$$
= \frac{n!}{(n-r)!}
$$

$$
= {}^nP_r
$$

Hence $\quad {}^nP_r = {}^{n-1}P_r + r \cdot {}^{n-1}P_{r-1}$

Proved.

**Second Method**

The number of permutations of $n$ dissimilar things taken $r$ at a time is $^nP_r$. This number can be divided into two parts, namely,

(i) When a particular thing is never included. In this case, the number of permutations of the remaining $(n-1)$ things taken $r$ at a time is $^{n-1}P_r$.
(ii) When a particular thing is always included. In this case, the number of permutations of the remaining $(n-1)$ things taken $(r-1)$ at a time is $^nP_{r-1}$ each of these permutations, the particular thing can occupy any of the $r$ positions, and therefore the required number of permutations, in these cases is $r \cdot {}^{n-1}P_{r-1}$.

Thus the sum of the permutations in the above two cases is equal to $^nP_r$.
Hence, we have $^nP_r = {}^{n-1}P_r + r \cdot {}^{n-1}P_{r-1}$ **Proved.**

**Example 21:** How many different words can he made out of the letters of the word 'TRIANGLE'? How many of these will begin with T and end with E?

### *Solution*

The word 'TRIANGLE' has 8 different words.

$\therefore$ The number of different words $= {}^8P_8$

$$= 8 \times 7 \times 6 \times 5 \times 4 \times 3 \times 2 \times 1$$
$$= 40320$$

If every word starts with T and ends with E, then the remaining 6 different words can be arranged in ${}^6P_6$, ways.

Therefore, in this case, the number of words formed $= {}^6P_6$

$$= 6 \times 5 \times 4 \times 3 \times 2 \times 1$$
$$= 720$$

**Example 22:** How many numbers between 6000 and 8000 can be formed from the digits 1, 2, 3, 4, 6, 8 when no digit is repeated?

### *Solution*

The number between 6000 and 8000 must contain only 4 digits and must have 6 in the first place. The first place is fixed by 6. So we have to arrange 3 digits out of the remaining 5 digits.

Hence, the required number of numbers $= {}^5P_3$

$$= 5 \times 4 \times 3$$
$$= 60$$

**Example 23:** In how many ways 5 boys and 3 girls can he seated in a row so that no two girls are together?

### *Solution*

We may perform the seating arrangement in two operations:

(i) Seating the 5 boys in a row: $\times$ B $\times$ B $\times$ B $\times$ B $\times$ B $\times$
Clearly, this can be done in ${}^5P_5 = 5!$ ways.

(ii) In between the 5 boys, there are 4 places. Also there are two places at the two ends. Thus there are 6 places in all. If the girls sit anywhere out of these 6 places, no two of them are ever together. Therefore, the number of ways in which the girls can sit is ${}^6P_3$.

Hence, the required number of ways of seating 5 boys and 3 girls are:

$$= 5! \times {}^6P_3$$
$$= (5 \times 4 \times 3 \times 2 \times 1) \times (6 \times 5 \times 4)$$
$$= 120 \times 120$$
$$= 14400$$

**Example 24:** How many different words can be formed from the letters of the word LUCKNOW when

   (i) all the letters are taken,
   (ii) words begin with L,
   (iii) the letters L and W occupy the first and the last places respectively,
   (iv) the vowels are always together.

### Solution

Total number of letters in word 'LUCKNOW is 7 (all distinct) in which there are two vowels (U, O) and 5 consonants.

   (i) When all the letters are taken, the number of words

$$= \text{the number of arrangements of 7 letters}$$
$$= 7!$$
$$= 7 \times 6 \times 5 \times 4 \times 3 \times 2 \times 1$$
$$= 5040.$$

   (ii) When the words begin with L, i.e., L is fixed at the first place, then the number of words formed

$$= \text{the number of arrangements of 6 letters}$$
$$= 6!$$
$$= 6 \times 5 \times 4 \times 3 \times 2 \times 1$$
$$= 720.$$

   (iii) When L is fixed at the first place and W at the last place, then the number of words formed

$$= \text{the number of arrangements of 5 letters}$$
$$= 5!$$
$$= 5 \times 4 \times 3 \times 2 \times 1 = 120.$$

   (iv) Considering the two vowels U and O as one, we are left with 6 distinct letters and the permutation of these 6 letters = 6!.

     But U and O can be interchanged in 2! ways, therefore the number of words formed

$$= 6! \times 2! = 6 \times 5 \times 4 \times 3 \times 2 \times 1 \times 2 = 1440.$$

### 3.4.6 Permutation of Objects Some of Which are Exactly Alike

So far we have discussed permutations of distinct objects. In the present section, we shall discuss the permutation of given number of objects in which the objects are not all different, i.e., some objects are of one kind, some objects are of another kind and so on, and the remaining are all different. For example, the number of permutations in the word 'INDIA', the number of five digit numbers formed by using the digits 1, 2, 1, 4, 3, 2, 4 etc. The following theorem is very helpful to determine the number of such permutations.

**Theorem 4** The number of permutations of $n$ things, taken all at a time, when $p$ of them are exactly alike of one kind, $q$ of them are exactly alike of another kind, $r$ of them exactly alike of a third kind and the rest all different, is

$$\frac{n!}{p!\,q!\,r!}$$

### *Proof*

Let the required number of permutations be $x$. If $p$ like things were replaced by $p$ unlike things different from any one of the rest, from any one of the $x$ permutations, without altering the position of any of the remaining things, we could form $p!$ new permutations. Therefore, if this change were made in each of the $x$ permutations we should obtain $x \times p!$ permutations.

Similarly, if the $q$ like things were replaced by $q$ unlike things, the number of permutations would be

$$= x \times p! \times q!$$

In like manner, by replacing the $r$ like things by $r$ unlike things we should finally obtain

$$x \times p! \times q! \times r!$$

permutations,

Since the things are now all different, and therefore there are $n!$ permutations among themselves.

Hence, it follows that

$$x \times p! \times q! \times r! = n!$$

$$\Rightarrow \quad x = \frac{n!}{p!\,q!\,r!}$$

which is the required number of permutations.

**Example 25:** Find the number of permutations formed by taking all the letters of the following words:

(i) CALCULUS
(ii) MATHEMATICS

### *Solution*

(i) In the word 'CALCULUS', we have
Total number of letters = 8
Number of letter C = 2
Number of letter L = 2
Number of letter U = 2
and remaining letters are different.

Hence, the required number of permutations $= \dfrac{8!}{2!2!2!}$

$$= \dfrac{8 \times 7 \times 6 \times 5 \times 4 \times 3 \times 2 \times 1}{2 \times 1 \times 2 \times 1 \times 2 \times 1}$$

$$= 5040$$

(ii) In the word 'MATHEMATICS', we have

Total number of letters = 11

Number of letter M = 2

Number of letter A = 2

Number of letter T = 2,

and the rest are different.

Hence the required number of permutations $= \dfrac{11!}{2!2!2!} = \dfrac{11 \times 10 \times 9 \times 8 \times 7 \times 6!}{2 \times 2 \times 2}$

$$= 990 \times 7 \times 6!$$

$$= 990 \times 7 \times 6 \times 5 \times 4 \times 3 \times 2 \times 1$$

$$= 4989600.$$

**Example 26:** How many permutations can be made by the letters of the word, 'SERIES'? How many of these will start from S and end with S? In how many of these words, the vowels and the consonants will be situated in alternative order?

***Solution***

(i) In the word 'SERIES', we have

Total number of letters = 6,

Number of letter S = 2

Number of letter E = 2,

and all the remaining letters are different.

$\therefore$ Required number of permutations $= \dfrac{6!}{2!2!}$

$$= \dfrac{6 \times 5 \times 4 \times 3 \times 2!}{2 \times 2!} = 180$$

(ii) If every permutation starts from S and ends with S, then 4 letters remain in which there are 2 E.

Hence, the number of required permutations

$$= \dfrac{4!}{2!} = \dfrac{4 \times 3 \times 2!}{2!} = 12$$

(iii) If the consonants S, R, S come at first, third and fifth places and the vowels E, I, E come at second, fourth and sixth places the number of required permutations

$$= \dfrac{3!}{2!} = \dfrac{3!}{2!} = 9$$

### 3.4.7 Permutations of Repeated Things

**Theorem 5** The number of permutations of $n$ different things, taken $r$ at a time, when each thing may occur any number of times, is $n^r$.

#### *Proof*

The required number is the same as the number of ways in which $r$ places can be filled up out of $n$ things, each thing being used as often as we please in any arrangement.

In the first place we may put any one of the $n$ things. Therefore the first place can be filled up in $n$ ways.

When the first place has been filled up in any one way, the second place may also be filled up in $n$ ways because the thing filling up the first place may be repeated. Thus the first two places can be filled up in $n \times n$ or $n^2$ ways.

Similarly, the third place can be filled up in $n$ ways, and therefore the first three places can be filled up in $n \times n \times n$ or $n^3$ ways.

Proceeding in this manner and noticing that the power of $n$ is the same as the number of places filled up, we find that the number of ways of filling up $r$ places is $n^r$. Hence, the total number of permutations $= n^r$.

**Example 27:** Ekta wrote 7 letters. If there are 4 post offices in the city, in how many ways can she post the letters?

#### *Solution*

Number of post offices = 4.

Since each letter can be posted in any one of the four post offices, therefore a letter can be posted in 4 ways. Similarly, each of the remaining letters can be posted in 4 ways.

Hence, the number of ways of posting 7 letters $= 4 \times 4 \times 4 \times \cdots$ to 7 factors
$$= 16384$$

**Example 28:** How many number of five digits can be formed with the digits 0, 1, 2, 3, 4, 5, 6 when each digit can be repeated any number of times?

#### *Solution*

A number of 5 digits has five, places: ten thousandth, thousandth, hundredth, tenth and units.

In a five digit number 0 cannot be put in the ten thousandth place, therefore this place can be filled up by six digits 1, 2, 3, 4, 5, 6 only in 6 ways.

Since the repetition of digits is allowed, therefore each of the other 4 place's can be filled up by given 7 digits in 7 ways.

Hence the required number of numbers formed $= 6 \times 7 \times 7 \times 7 \times 7$
$$= 14406$$

### 3.4.8 Circular Permutations

We have seen that $n$ dissimilar things, when all arranged in a row in different ways, make $n$ permutations. Such permutations are known as linear permutations. But if we arrange the things around a circle or some other closed curve, the permutations are known as circular permutations. The difference between a linear permutation and a circular permutation is that there are two ends of each arrangement in a linear permutation whereas a circular permutation is endless. Thus in a circular permutation, we consider one thing as fixed and remaining things are arranged as in case of linear permutations.

Hence the number of circular permutations of $n$ dissimilar things taken all at a time is $(n-1)!$. Circular permutations are of two types:

(i) Those circular permutations in which there is no distinction between clockwise and anti-clockwise direction. For example, the arrangement of beads in a necklace, the arrangement of flowers in a garland etc. Here the number of circular permutations of $n$ distinct things is $(n-1)!/2$.

(ii) Those circular permutations in which the clockwise and anti-clockwise are different. For example, permutations of persons sitting on a round table.

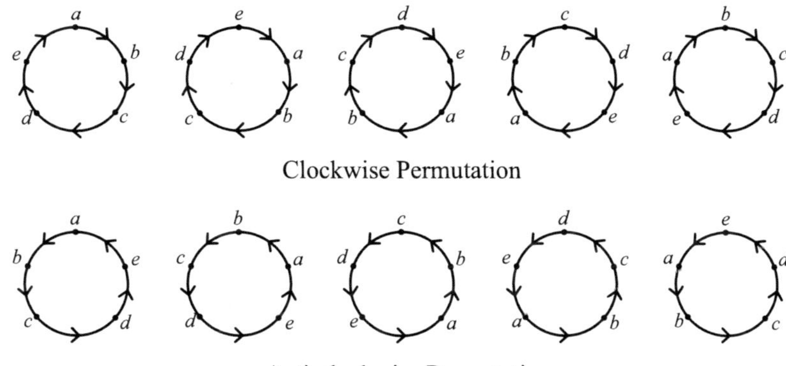

Clockwise Permutation

Anti-clockwise Permutation

From the above two sets of figures, it is clear that the two types of circular permutations are different.

### 3.4.9 Circular Permutations of $n$ Dissimilar Things taken $r$ at a Time

Let $x$ be the number of circular permutations. Then for $x$ circular permutations, there will be $xr$ linear permutations. Since the number of permutations of $n$ different things taken $r$ at a time is $^{n}\mathrm{P}_{r}$, therefore

$$xr = {}^{n}\mathrm{P}_{r}$$

$$x = \frac{{}^{n}\mathrm{P}_{r}}{r}$$

Hence the number of circular permutations of $n$ things taken $r$ at a time is $\dfrac{^n\mathrm{P}_r}{r}$.

When there is no distinction between the clockwise and anti-clockwise permutations. $2r$ permutations will form one circular permutation, and therefore we have

$$x \cdot 2r = {}^n\mathrm{P}_r$$

$$\Rightarrow \quad x = \frac{^n\mathrm{P}_r}{2r}$$

which is the required number of permutations in this case.

**Example 29:** How many garlands can be made out of 10 unlike flowers?

### Solution

Since there is no distinction in clockwise and anti-clockwise circular permutation in the case of a garland, the required number of garlands

$$= \frac{1}{2}(10-1)! = \frac{1}{2} \times 9!$$
$$= 181440.$$

**Example 30:** 6 Indians and 4 Europeans come to a hotel. In how many ways on a round table can they be seated if two Europeans never sit together?

### Solution

Let the Indians sit first on alternate seats.

The number of ways in which the Indians can set on the round table

$$= (6-1)! = 5!$$
$$= 5 \times 4 \times 3 \times 2 \times 1 = 120$$

When Indians have taken their seats, there remain 6 seats for the Europeans, each between two Indians.

The number of ways in which the European can sit

$$= {}^6\mathrm{P}_4 = 6 \times 5 \times 4 \times 3 = 360$$

Hence the required number of ways

$$= 120 \times 360$$
$$= 43200$$

## 3.5 COMBINATION

**Definition and Notation:** The different group or selections which can be made by taking some or all of the number of things are called combination. For example, the combination which can be made by taking the letters $a$, $b$, $c$, two at a time are:

$$ab, bc, ac$$

Thus in combination we are only concerned with selection or group of things, irrespective of the order of the things.

The number of all combinations of $n$ things taken $r$ at a time is denoted by $^nC_r$ or $C(n, r)$. Clearly $^nC_r$ is defined only when $n$ and $r$ are non-negative integers such than $0 \leq r \geq n$.

### 3.5.1 The Number of Combinations of $n$ Dissimilar Things Taken $r$ at a Time

Let the number of combinations of $n$ dissimilar things taken $r$ at a time be denoted by $^nC_r$. Then each of these combinations consists of a group of $r$ dissimilar things, which can be arranged among themselves in $r!$ ways. Therefore $^nC_r \times r!$ is equal to the number of arrangements of $n$ dissimilar things taken $r$ at a time. Consequently,

$$^nC_r \times r! = {}^nP_r$$

$$\Rightarrow \qquad ^nC_r = \frac{^nP_r}{r!}$$

$$\Rightarrow \qquad ^nC_r = \frac{n!}{(n-r)!\,r!}$$

$$^nC_r = \frac{n(n-1)(n-2)\cdots(n-r+1)\cdot(n-r)!}{(n-r)!\,r!}$$

$$^nC_r = \frac{n(n-1)(n-2)\cdots(n-r+1)}{r!}$$

**Remarks**

(i) $\quad ^nC_r = \dfrac{n!}{r!(n-r)!}$

$$^nC_n = \frac{n!}{n!(n-n)!} \qquad (\text{putting } r = n)$$

$$^nC_n = \frac{1}{0!}$$

$$^nC_n = 1, (\because 0! = 1)$$

(ii) $\quad ^nC_r = \dfrac{n!}{r!(n-r)!}$

$$^nC_0 = \frac{n!}{0!(n-0)!}$$

$$^nC_0 = \frac{n!}{n!} \qquad (\because 0! = 1)$$

$$^nC_0 = 1.$$

**Theorem 6** The number of combinations of $n$ things taken $r$ at a time is equal to the number of combinations of $n$ things taken $(n-r)$ at a time.

### *Proof*

$$^nC_{n-r} = \frac{n!}{(n-r)![n-(n-r)]!}$$

$$= \frac{n!}{(n-r)!r!} = {}^nC_r$$

$$^nC_r = {}^nC_{n-r}$$

### Remarks

(i) The formula ${}^nC_r = {}^nC_{n-r}$ simplifies the calculation of ${}^nC_r$, when $r$ is very large. For example,

$$^{50}C_{48} = {}^{50}C_{50-48} = {}^{50}C_2 = \frac{50 \times 49}{1 \times 2} = 1225.$$

(ii)
$$^nC_x = {}^nC_y$$

$$\Rightarrow \quad {}^nC_x = {}^nC_y \quad \text{or} \quad {}^nC_x = {}^nC_{n-y}$$

$$\Rightarrow \quad x = y \quad \text{or} \quad x = n - y$$

$$\Rightarrow \quad x = y \quad \text{or} \quad x + y = n.$$

**Theorem 7** $^nC_r + {}^nC_{r-1} = {}^{n+1}C_r$

### *Proof*

We have

$$^nC_r + {}^nC_{r-1} = \frac{n!}{r!(n-r)!} + \frac{n!}{(r-1)!(n-r+1)!}$$

$$= \frac{n!}{r(r-1)!(n-r)!} + \frac{n!}{(r-1)!(n-r+1)(n-r)!}$$

$$= \frac{n!}{(r-1)!(n-r)!}\left[\frac{1}{r} + \frac{1}{n-r+1}\right]$$

$$= \frac{n!}{(r-1)!(n-r)}\left[\frac{n-r+1+r}{r(n-r+1)}\right]$$

$$= \frac{n!}{(r-1)!(n-r)!} \cdot \frac{(n+1)}{r(n-r+1)}$$

$$= \frac{(n+1)n!}{r(r-1)!(n-r)!(n-r+1)}$$

$$= \frac{(n+1)!}{r!(n-r+1)!} = \frac{(n+1)!}{r!(n+1-r)!}$$

$$= {}^{n+1}C_r$$

Hence, ${}^nC_r + {}^nC_{r-1} = {}^{n+1}C_r$

### 3.5.2 Theorem on Restricted Combination

The combination of $r$ objects out of $n$ objects in which $p$ specific objects are:

    (i) always included is ${}^{n-p}C_{r-p}$
    (ii) never includes is ${}^{n-p}C_r$

### *Proof*

    (i) When $p$ specific objects are always included, then we have to make selection of $(r-p)$ objects out of remaining $(n-p)$. Now, with each combination of $p$ specific objects, we get such combination of $r$ objects which includes $p$ specific objects.

       Hence the number of required combinations $= {}^{n-p}C_{r-p}$

    (ii) When $p$ specific objects are never included in any combination, then we have to form combinations taking $r$ objects out of the remaining $(n-p)$ objects.

       Hence, the required number of combinations $= {}^{n-p}C_r$

**Example 31:** Prove that ${}^nC_r = \dfrac{n-r+1}{r} \times {}^nC_{r-1}$.

### *Solution*

We have

$$\text{R.H.S} = \frac{n-r+1}{r} \times \frac{n!}{(r-1)![n-(r-1)]!}$$

$$= \frac{n-r+1}{r} \times \frac{n!}{(r-1)!(n-r+1)!}$$

$$= \frac{n-r+1}{r} \times \frac{n!}{(r-1)!(n-r+1)\cdot(n-r)!}$$

$$= \frac{n!}{r(r-1)!(n-r)!}$$

$$= \frac{n!}{r!(n-r)!} = {}^nC_r$$

$$= \text{L.H.S.} \qquad\qquad\qquad \textbf{Proved.}$$

**Example 32:** Out of 9 subjects two subjects are compulsory. In how many ways a student can make a selection of 5 subjects out of these 9 subjects?

### *Solution*

Total number of subjects = 9
The number of compulsory subjects = 2

∴  The number of remaining subjects = 9 – 2 = 7
∴  The student has to select (5 – 2), i.e., 3 subjects out of 7.

Hence the required number of selections = $^7C_3 = \dfrac{7\times6\times5}{1\times2\times3}$

$$= 35$$

**Example 33:** In how many ways 11 players can be selected out of 15 players? If
  (i) One particular player is always selected,
  (ii) One particular player is never selected,
 (iii) There is no restriction on selection.

### *Solution*

  (i) When a particular player is always selected, we have to select 10 players out of remaining 14 players.

∴  The number of ways of selection = $^{14}C_{10} = {}^{14}C_4$

$$= \frac{14\times13\times12\times11}{1\times2\times3\times4}$$

$$= 1001$$

  (ii) When a particular player is never selected, then we have to select 11 players out of remaining 14 players.

∴  The required number of ways = $^{14}C_{11} = {}^{14}C_3$

$$= \frac{14\times13\times12}{1\times2\times3} = 364$$

 (iii) When there is no restriction on selection, then we have to select 11 players out of 15.

∴  The required number of ways = $^{15}C_{11} = {}^{15}C_4$

$$= \frac{15\times14\times13\times12}{1\times2\times3\times4} = 1365$$

**Example 34:**  A committee of 4 members is to be formed out of 5 males and 6 females. Find how many committees can be formed consisting of at least one female?

*Solution*

The committee may consist of 1, 2, 3 or 4 females. Therefore, there are following four possibilities of forming the committee
   (i) 3 males and 1 female
  (ii) 2 males and 2 females
 (iii) 1 male and 3 females
 (iv) 4 females

   (i) When the committee consists of 3 males and 1 female
       The number of ways in which 3 males out of 5 males can be taken is $^5C_3$.
       The number of ways in which 1 female can be chosen out of 6 is $^6C_1$.

       $\therefore$   Number of committees consisting of 3 males and 1 female

$$= {}^5C_3 \times {}^6C_1$$
$$= \frac{5 \times 4 \times 3}{1 \times 2 \times 3} \times 6$$
$$= 10 \times 6 = 60.$$

  (ii) When the committee consists of 2 males and 2 females. The number of ways of forming the committee

$$= {}^5C_2 \times {}^6C_2$$
$$= \frac{5 \times 4}{1 \times 2} \times \frac{6 \times 5}{1 \times 2}$$
$$= 10 \times 15 = 150.$$

 (iii) When the committee consists of 1 male and 3 females. The number of ways of forming the committee

$$= {}^5C_1 \times {}^6C_3$$
$$= 5 \times \frac{6 \times 5 \times 4}{1 \times 2 \times 3}$$
$$= 5 \times 20 = 100.$$

 (iv) When the committee consists of 4 females. The number of ways of forming the committee

$$= {}^6C_4 = {}^6C_2$$
$$= \frac{6 \times 5}{1 \times 2} = 15$$

Hence the required number of committees

$$= 60 + 150 + 100 + 15$$
$$= 325$$

**Example 35:** How many diagonals can be drawn in a polygon of $n$ sides?

*Solution*

A straight line is obtained by joining any two points. A polygon of $n$ sides has $n$ angular points. Therefore, the number of straight lines obtained by joining any two of $n$ points is $^nC_2$. But these lines include the $n$ sides of the polygon which are not diagonals.

Hence the number of diagonals $= {}^nC_2 - n$

$$= \frac{n(n-1)}{2} - n$$

$$= \frac{n^2 - n - 2n}{2} = \frac{n(n-3)}{2}$$

**Theorem 8** The total number of ways in which it is possible to make a selection by taking some or all of $n$ things is $2^n - 1$.

*Proof*

Each thing may be dealt with in two ways: it may be taken or left. Since either way of dealing with anything may be associated with either way of dealing with each one of the others, therefore the number of ways of dealing with $n$ things

$$= 2 \times 2 \times 2 \times \cdots \quad \text{to} \quad n \text{ factors} = 2^n.$$

But this includes the case in which all the things are left out, therefore, rejecting this case, the total number of ways is $2^n - 1$.

**Theorem 9** The total number of ways in which it is possible to make a selection by taking some or all out of $p + q + r + \cdots$ things, where $p$ are alike of one kind, $q$ are alike of second kind, $r$ alike of a third kind, and so on, is $(p + 1)(q + 1)(r + 1) \cdots -1$.

*Proof*

Out of $p$ things we may take 0, 1, 2, 3, ... $p$ things. Thus the $p$ things may be disposed of in $(p + 1)$ ways. Similarly, $q$ things may be disposed of in $(q + 1)$ ways, the $r$ things in $(r + 1)$ ways, and so on.

Therefore, the number in which all the things may be disposed of is $(p + 1)(q + 1)(r + 1) \ldots$

But this includes the case of ways in which none of the things are taken. Therefore, rejecting this case, the total number of ways is

$$(p + 1)(q + 1)(r + 1) \cdots -1$$

**Theorem 10 (Division into Groups)** The number of ways in which $m + n$ things can be divided into two groups containing $m$ and $n$ things respectively, is $\dfrac{(m+n)!}{m!\,n!}$

*Proof*

This is clearly equivalent to find the number of combinations of $(m + n)$ things taken $m$ at a time, because every time we select one group of $m$ things and we leave a group of $n$ things behind.

Thus the required number $= \dfrac{(m+n)!}{m!\,n!}$

**Corollary 1: If the groups are equal, i.e., $m = n$, the number of different ways of subdivision is**

$$\frac{(2m)!}{m!\,m!\,2!}$$

because, in anyone way, it is possible to interchange the two groups without obtaining a new distribution.

**Corollary 2: The number of ways in which $m + n + p$ things can be divided into three groups containing $m, n, p$ things is**

$$\frac{(m+n+p)!}{m!\,n!\,p!}$$

*Proof*

First we divide $m + n + p$ things in two groups containing $m$ and $n + p$ things respectively. The number of ways this can be done is

$$\frac{(m+n+p)!}{m!\,(n+p)!}$$

Now the number of ways in which the group of $(n + p)$ things can be divided into two groups containing $n$ and $p$ things respectively, is

$$\frac{(n+p)!}{n!\,p!}$$

Hence, the number of ways in which the sub-division into three groups containing $m, n, p$ things can be made is

$$\frac{(m+n+p)!}{m!\,(n+p)!} \cdot \frac{(n+p)!}{n!\,p!}$$

$$\frac{(m+n+p)!}{m!\,n!\,p!}$$

*Note:* If we have $m = n = p$, the number of ways is

$$\frac{3m!}{m!\,m!\,m!\,3!}$$

But if we have to divide $3m$ things in similar groups, the required number is

$$\frac{3m!}{m!\,m!\,m!}$$

**Example 36:** To pass an examination one has to secure the minimum marks in every subject. If there are 7 subjects, in how many ways can a student fail?

### *Solution*

The student may fail in some subjects or in all the 7 subjects. Therefore, the required number of his failure

$$= 2^7 - 1 = 127$$

### *Alternative method*

The student may fail in 1, 2, 3, 4, 5, 6, 7 subjects. Therefore, the total number of ways in which he may fail

$$= {}^7C_1 + {}^7C_2 + {}^7C_3 + {}^7C_4 + {}^7C_5 + {}^7C_6 + {}^7C_7$$
$$= 7 + {}^7C_2 + {}^7C_3 + {}^7C_3 + {}^7C_2 + 7 + 1$$
$$= 7 + \frac{7\times6}{1\times2} + \frac{7\times6\times5}{1\times2\times3} + \frac{7\times6\times5}{1\times2\times3} + \frac{7\times6}{1\times2} + 7 + 1$$
$$= 127$$

**Example 37:** How many all possible factors may be made of 22680?

### *Solution*

The given number $= 22680$

$$= 2^3 \times 3^4 \times 5 \times 7.$$

The factor 2 can be dealt within four ways because its index may be taken as 0, 1, 2 or 3. Similarly, the factor 3 can be dealt within 5 ways, for its index may be taken as 0, 1, 2, 3 or 4. The factor 5 may be taken in two ways and 7 may be dealt within 2 ways. So the total number of factors is

$$(3 + 1)\,(4 + 1)\,(1 + 1)\,(1 + 1) = 80$$

But this includes the factor 1 and 22680. If these are not taken as factors, then all possible factors are $80 - 2 = 78$.

**Exercise 3.2**

1. Two persons enter into a railway compartment in which 6 seats are vacant. In how many different ways can they sit?

2. A room has 6 doors. In how many ways can a man enter the room through one door and come out through a different door?

3. In a class there are 10 boys and 8 girls. The teacher wants to select a boy and a girl to represent the class in a function. In how may ways can the teacher make this selection?

4. Eight athletes are participating in a race. In how many ways can the first three prizes be won?

5. How many three-digit numbers are there, with distinct digits, with each digit odd?

   [*Hint:* Required number = number of arrangements of digits 1, 3, 5, 7, 9 by taking 3 at a time = $^5P_3$].

6. Find the number of different 4-letter words, with or without meanings, which can be formed from the letters of the word 'NUMBER'.

7. In how many ways can 6 boys and 5 girls be arranged for group of photograph, if the girls are to sit on chairs in a row and the boys are to stand in a row behind them?

   [*Hint:* $^5P_5 \times {}^6P_6$].

8. (i) If $^nP_{n-2} = 60$, find the value of $n$.
   (ii) If $^{10}P_r = 5040$, find the value of $r$.
   (iii) If $^{15}P_r = 2730$, find the value of $r$.

9. (i) If $^{n+1}P_{18} : {}^{n-1}P_{16} = 420 : 1$, find the value of $n$.
   (ii) If $^{2n}P_{n+1} : {}^{2n-2}P_n = 56 : 3$, find the value of $n$.

10. (i) If $^9P_5 + 5 \cdot {}^9P_4 = {}^{10}P_r$, find the value of $r$.
    (ii) Prove that $^{10}P_3 = {}^9P_3 + 3 \cdot {}^9P_2$.
    (iii) Prove that 2, 4, 6, 8, ... in $= 2^n.n!$ ...

11. Prove that
    (i) $^nP_r = n \cdot {}^{n-1}P_{r-1}$
    (ii) $\dfrac{^nP_r}{^nP_{r-2}} = \dfrac{(n-r+2)(n-r+1)}{r(r-1)}$.

12. How many words can be made by the letters of the word 'DELHI' if:
    (i) they start with D?
    (ii) they do not start with D?
    (iii) they start with D and end with I?

13. How many words can be formed from the letters of the word 'ARTICLE' if the vowels A, I, E:
    (i) occur together?
    (ii) never occur together?
    (iii) occur only at odd places?

14. How many numbers between 99 and 1000 can be formed with the digits 0, 1, 2, 3, 4 and 5?

15. How many numbers can be formed by the digits 1, 2, 4, 5, 6, 7 when:
    (i) there is no restriction?
    (ii) 5 always remains at the $n^{th}$ place?
    (iii) all the numbers are divisible by 2?
    (iv) every number starts with 1 and end with 5?
    (v) every number is greater than 4,00,000?

16. Find the number of positive integers which can be formed by using any number of digits from 0, 1, 2, 3, 4, 5 but using each digit not more than once in each number.

    [*Hint:* $^5P_1 + (^6P_2 - {}^5P_1) + (^6P_3 - {}^5P_2) + (^6P_4 - {}^5P_3) + (^6P_5 - {}^5P_4) + (^6P_6 - {}_5P_5)$]

17. Find the number of ways in which 5 boys and 5 girls be seated in a row so that:
    (i) No two girls may sit together.
    (ii) All the girls sit together and all the boys sit together.
    (iii) All the girls are never together.

18. How many four digit numbers divisible by 4 can be made with the digits 1, 2, 3, 4, 5 if the repetition of digits is not allowed?

    [*Hint:* The number divisible by 4 have last two digits 12, 24, 32 and 52.]

    Hence required number of numbers = $^3P_2 \times 4$

19. Find the number of words which can be formed by using all the letters of the following words:
    (i) ALLAHABAD
    (ii) ARRANGE

20. How many arrangements can be made out of the letters of the word 'ALLAHABAD', all arrangements start with A?

21. (i) How many arrangements can be made out of the letters of the word 'HARYANA'?
    (ii) How many of these begin with H and end with N?
    (iii) In how many of these H and N are together?

22. How many arrangements can be made with the letters of the word 'MATHEMATICS' in which vowels are together?

    [*Hint:* Required number of arrangements $= \left[ \dfrac{8!}{2! \times 2!} \times \dfrac{4!}{2!} \right]$

23. In how many ways can the letters of the word 'PLANTAIN' be arranged so that two A's never come together?

24. How many different arrangements can be made out of the letters in the expression $a^3b^2c^4$?

25. Find the number of arrangements which can be made out of the letters of the word 'ALGEBRA' without altering the relative positions of vowels and consonants.

26. If the different permutations of the word 'EXAMINATION' are listed as in a dictionary, how many items are there in the list before the word starting with E?

    [*Hint:* The number of required arrangements

    = the number of arrangements starting from A
    = the number of arrangements from (11 − 1) letters of the word EXAMINATION

    $= \dfrac{10!}{2! \times 2!}$ ].

27. How many numbers greater than a million can be formed with the digits 2, 3, 0, 3, 4, 2, 3?

[***Hint***: The number of required numbers

$$= \frac{7!}{2! \times 3!} - \frac{6!}{2! \times 3!}\Bigg]$$

28. There are 5 letters and 4 letter boxes. Find the number of ways to post the letters.

29. In order to invite 6 friends, in how many ways 3 servants can be sent?

30. In an electric circuit, fourteen switches are arranged such that there are 3 possible positions for each switch. How many switches of this type are there in the electric circuit?

31. How many numbers of 3 digits can be formed by using the digits 0, 1, 3, 5, 7 while each digit may be repeated any number of times?

32. How many numbers of 5 digits can be formed from the digits 0, 1, 2, 3, 4, 5, 6, 7, 8 and 9, when any digit can be repeated as many times as possible?

33. There are 6 multiple choice questions in an examination. How many sequences of answers are possible, if the first three questions have 4 choices each and the next three have 5 each?

[***Hint***: Each one of the first three questions can be answered in 4 ways and each one of the next three in 5 ways. Hence total of different sequences $= (4 \times 4 \times 4) \times (5 \times 5 \times 5)\Big]$

**Answers to Selected Problems**

1. 30

2. 30

6. 260

10. (i) $r = 5$

12. (i) 24
    (ii) 96
    (iii) 06

17. (i) $5! \times 6!$
    (ii) $2 \times (5!)^2$
    (iii) $10! - (5! \times 6!)$

20. 6720

24. $\dfrac{\lfloor 9}{\lfloor 3 \times \lfloor 2 \times \lfloor 4} = 1260$

29. $3^6$

32. $9 \times 10^4$

## 3.6 SEQUENCES AND SERIES

### 3.6.1 Sequences

A sequence is a function used to represent an ordered list, whose domain is the set N of natural numbers. We may also define a sequence as an arrangement of numbers according to a certain fixed rule. Different numbers in a sequence are called its terms and are generally denoted by $a_1$, $a_2$, ... or $T_1$, $T_2$, ... etc.

   Some examples of sequences are:

$$1, 3, 5, 7, 9, \dots \qquad\qquad\text{(i)}$$

$$1, 3, 9, 27, \ldots \qquad\qquad \text{(ii)}$$

$$\frac{1}{2}, \frac{1}{2^2}, \frac{1}{2^3}, \ldots \qquad\qquad \text{(iii)}$$

In (i) the terms are arranged such that the difference between two consecutive terms is the same i.e. $\frac{1}{2}$.

In (ii) and (iii), the terms are arranged in such a way that ratio of two consecutive terms is the same i.e. 3 in (ii) and $\frac{1}{2}$ in (iii).

**General Term:** The $n$th term of a sequence is called the general term and is denoted by $T_n$.

### 3.6.2 Finite and Infinite Sequence

**Finite Sequence:** A sequence having a finite number of terms (i.e. given number of terms) is called a finite sequence.

**Infinite Sequence:** A sequence having infinite number of terms is called an infinite sequence.

### 3.6.3 Series

The terms of a sequence connected by +ve or −ve sign form a series

$$\text{For example,} \quad 1 + 3 + 5 + 7 + \cdots$$
$$1 + \frac{1}{2} + \frac{1}{2}^2 + \frac{1}{2}^3 + \cdots$$

The different terms of the series are denoted by $T_1, T_2, \ldots T_n$

### 3.6.4 Progression

A sequence is said to be in progression if its term increases or decreases numerically.

For example

(1) $2, 4, 6, 8, \ldots$

(2) $1, -\frac{1}{2}, \frac{1}{4}, -\frac{1}{8}$

### 3.6.5 Arithmetic Progression (A.P.)

A sequence is said to be an A.P. if the difference of each term except the first one, from its preceding term is always same. Example $2, 5, 8, 11, \ldots$

General A.P. series

$$a, a + d, a + 2d, \ldots$$

General term $T_n = a + (n - 1)d$

Sum of 1st $n$ terms of A.P. is $\quad S_n = \dfrac{n}{2}\{2a + (n-1)d\}$

### 3.6.6 Arithmetic Mean

If three numbers are in A.P., then the middle one is called arithmetic mean (A.M.) of the other two. If $(n + 2)$ numbers $a, A_1, A_2, ..., A_n, b$ are in A.P., then the numbers $A_1, A_2 ..., A_n$ are said to be $n$ arithmetic mean between $a$ and $b$. e.g., is 2, 4, 6, 8, 10, 12 are in A.P. Then 4, 6, 8, 10 are four arithmetic means between 2 and 12.

**A.M. of two given Numbers**

Let A be the A.M. of $a$ and $b$. Then $a$, A, $b$ are in A.P.

$$\therefore A - a = b - A$$

$$\Rightarrow 2A = a + b$$

or

$$A = \frac{a+b}{2}$$

Hence, the arithmetic mean of two numbers $a$ and $b$ is $\dfrac{a+b}{2}$

### 3.6.7 Geometric Progression

Quantities are said to be in Geometrical Progression (G.P.) when they increase or decrease by a constant factor. The constant factor is also called the common ratio, and it is found by dividing any term by that which immediately proceeds it e.g.,

(i) $1, 2, 4, 8, 16 \ldots$

(ii) $1, -\dfrac{1}{3}, \dfrac{1}{9}, -\dfrac{1}{27}, \ldots$

(iii) $a, ar, ar^2, ar^3, \ldots$

In the first example, the common ratio is 2, in the second it is $-\dfrac{1}{3}$ and in the third it is $r$

The example (iii) is the standard form of geometrical progression, whose first term is '$a$' and common ratio is '$r$'.

**The $n$th term of G.P.**

Let $a$ be the first term and $r$ be the common ratio of a G.P. then its successive terms are

$$a, ar, ar^2, \ldots$$

and we have

$$\text{First term} = u_1 = a = ar^0 = ar^{1-1}$$

$$\text{Second term} = u_2 = ar = ar^{2-1}$$

$$\cdots\cdots\cdots\cdots\cdots\cdots\cdots\cdots\cdots\cdots\cdots\cdots$$

We observe that in any term, the index of $r$ is always less by one by the number of the terms in the series

Hence $\boxed{T_n = ar^{n-1}}$

**The Sum of $n$ Terms of a G.P.**

Let $S_n$ denote the sum of $n$ terms of the G.P. whose first term is $a$ and common ratio is $r$. Then

$$S_n = a + ar + ar^2 + \cdots + ar^{n-1} \tag{1}$$

Multiplying both the sides of (1) by $r$, we have

$$rS_n = ar + ar^2 + ar^3 + \cdots + ar^n \tag{2}$$

Subtracting (2) from (1), we get

$$(1 - r)S_n = a - ar^n$$

$$\Rightarrow \quad \boxed{S_n = \frac{a\left(1 - r^n\right)}{1 - r} \quad \text{for} \quad r < 1}$$

and

$$\boxed{S_n = \frac{a\left(r^n - 1\right)}{r - 1} \quad \text{for} \quad r > 1}$$

*Note:* If $l$ is the last term of the G.P., Then $l = ar^{n-1}$

$$S_n = a\left(\frac{1 - r^n}{1 - r}\right)$$

$$= \frac{a - ar^n}{1 - r} = \frac{a - ar^{n-1} \cdot r}{1 - r}$$

$$= \frac{a - lr}{1 - r}$$

Thus $\quad S_n = \dfrac{a - lr}{1 - r} \quad$ or $\quad S_n = \dfrac{lr - a}{r - 1}(r \neq 1).$

**The Sum of an Infinite G.P.**

The sum of $n$ terms of a G.P. whose first term is $a$ and common ratio is $r$, is given by

$$S_n = \frac{a\left(1 - r^n\right)}{1 - r}$$

or $\quad S_n = \dfrac{a}{1 - r} - \dfrac{ar^n}{1 - r}$, provided $r \neq 1$.

If $-1 < r < 1$, then the value of $r^n$ will decrease with the increase in the value of $n$. Consequently $r^n \to 0$ when $n \to \infty$.

Hence the sum of an infinite G.P. is given by

$$S_\infty = \frac{a}{1 - r}, |r| < 1.$$

*Note:* if $|r| \geq 1$, then the sum of an infinite G.P. tends to infinity.

**Recurring Decimal and Infinite G.P.**

If a number repeats itself, then such repetition is represented by placing a dot on that number e.g.,

$$1.32\dot{5} = 1.32555\ldots$$

$$3.5\dot{3}\dot{6} = 3.5363636.$$

$$6.8\dot{7}\dot{9} = 6.8797979$$

*Note:* A recurring decimal can be expressed as an infinite geometric series whose sum is finite.

**To Find the Value of a Recurring Decimal**

Let P denote the figures which do not recur, and suppose them $p$ in number; let Q denote the recurring period consisting of $q$ figures. Let D denote the value of the recurring decimal. Then

$$\text{D} = \text{PQ} \qquad \text{or}$$
$$\text{D} = .\text{PQQQ} \ldots$$
$$\therefore \quad 10^{p} \times \text{D} = \text{P.QQQ} \ldots$$
$$\text{and} \quad 10^{p+q} \times \text{D} = \text{PQ.QQQ} \ldots$$

$$\therefore \quad \text{By Subtraction,}$$
$$(10^{p+q} - 10^{p})\text{D} = \text{PQ} - \text{P}$$
$$\Rightarrow \quad 10^{p}(10^{q} - 1)\text{D} = \text{PQ} - \text{P}$$

$$\therefore \quad \text{D} = \frac{\text{PQ} - \text{P}}{\left(10^{q} - 1\right)10^{p}}$$

Now $\left(10^{q} - 1\right)$ is a number consisting of $q$ nines. Therefore the denominator consists of $q$ nines followed by $p$ zeros. Hence, we have the following rule for reducing a recurring decimal to a vulgar fraction:

For the numerator subtract the integral number consisting of the non-recurring figures from the integral number consisting of the non-recurring and recurring figures; for the denominator take a number consisting of as many nines as there are recurring figures followed by as many zeros as there are non-recurring figures.

**Example 38:** Find the tenth term of the sequence $2^2, 2^3, 2^4, \ldots$

*Solution*

The given sequence is in G.P. with first term $a = 2^2$ and common ratio $r = \dfrac{2^3}{2^2} = 2$

$$\therefore 10\text{th term} = ar^{10-1}$$

$$= 2^2 \cdot (2)^{10-1}$$

$$= 2^{11}$$

**Example 39:** Find the fourth term from the end of the sequence $8, 4, 2, \supset, \dfrac{1}{128}$.

***Solution***

The given sequence is in G.P. with first term 8 and common ratio $r = \dfrac{4}{8} = \dfrac{1}{2}$.

Hence, the fourth term from the end $= a$

$$\Rightarrow \quad \frac{1}{128} = a(r)^{4-1}$$

$$\Rightarrow \quad a = \frac{1}{128 r^3}$$

$$\Rightarrow \quad a = \frac{1}{128} \times 2^3$$

$$= \frac{8}{128} = \frac{1}{16}$$

**Example 40:** Which term is $\dfrac{1}{128}$ of the sequence $4, 2, 1, \ldots$ ?

***Solution***

The given sequence is in G.P. with first term 4 and common ratio $r = \dfrac{2}{4} = \dfrac{1}{2}$.

Let the $n$th term be $\dfrac{1}{128}$.

$\therefore \quad n$th term $= ar^{n-1}$

$$\Rightarrow \quad \frac{1}{128} = 4 \cdot \left(\frac{1}{2}\right)^{n-1}$$

$$\Rightarrow \quad \frac{1}{128} = 2^2 \cdot \frac{1}{2^{n-1}} = 2^{3-n}$$

$$\Rightarrow \quad \frac{1}{128} = \frac{1}{2^{n-3}}$$

$$\Rightarrow \quad \frac{1}{2^7} = \frac{1}{2^{n-3}}$$

$$\Rightarrow \quad 7 = n - 3 \qquad \text{or} \qquad n = 10$$

Thus $\dfrac{1}{128}$ is the 10th term.

### 3.6.8 Geometric Mean

If three quantities are in G.P. then the middle one is called the geometric mean (G.M.) of the other two. For example the numbers 3, 9, 27 are in G.P. Therefore, 9 is the geometric mean of 3 and 27.

If $(n+2)$ numbers $a, G_1, G_2 ... G_n, b$ are in G.P., then the numbers $G_1, G_2, ... G_n$ are known as $n$ geometric means (G.M.'s) between $a$ and $b$.

**Geometric Mean of Two Given Numbers $a$ and $b$**

Let G be the geometric mean between $a$ and $b$. Then $a$, G, $b$, are in G.P.

$$\therefore \quad \frac{G}{a} = \frac{b}{G} \text{ (common ratio)}$$

$$\Rightarrow \quad G^2 = ab$$

$$\Rightarrow \quad G = \pm\sqrt{ab}$$

Thus the geometric mean between two quantities is the square root of their product.

**Insertion of $n$ Geometric Mean between Two Quantities**

Let $G_1, G_2, ... G_n$ be $n$ geometric mean between $a$ and $b$.
Then $a, G_1, G_2, G_3 ..., G_n, b$ are in G.P. in which $b$ is the $(n+2)$th term. If $r$ is the common ratio of this G.P., then

$$b = ar^{n+1}$$

$$\Rightarrow r^{n+1} = \frac{b}{a}$$

$$\text{or} \quad r = \left(\frac{b}{a}\right)^{1/n+1}$$

$$\therefore \quad G_1 = ar = a\left(\frac{b}{a}\right)^{1/n+1} = a^{n/n+1} \cdot b^{1/n+1},$$

$$G_2 = ar^2 = a\left(\frac{b}{a}\right)^{2/n+1} = a^{n-1/n+1} \cdot b^{2/n+1},$$

$$G_3 = ar^3 = a\left(\frac{b}{a}\right)^{3/n+1} = a^{n-2/n+1} \cdot b^{3/n+1},$$

$$G_n = ar^n = a\left(\frac{b}{a}\right)^{n/n+1} = a^{1/n+1} \cdot b^{n/n+1}.$$

***Important Remark***

(i) Any three consecutive terms of a G.P. whose product is given, should be taken as $\dfrac{a}{r}, a, ar$.

(ii) If the product of four consecutive terms of a G.P. is given, then the numbers should be taken as

$$\frac{a}{r^3}, \frac{a}{r}, ar, ar^3.$$

(iii) If the product of five consecutive terms of a G.P. is given, then the numbers should be taken as

$$\frac{a}{r^2}, \frac{a}{r}, a, ar, ar^2.$$

(iv) If the product of numbers is not given, then we should take numbers in the usual form as $a, ar, ar^2, \ldots$

### 3.6.9 Properties of G.P.

**Property I**

If each term of G.P. is raised to the same power, the resulting sequence is also a G.P.

*Proof*

Let the given G.P. be $a, ar, ar^2, ar^3, \ldots$ and each term be raised to a non-zero real numbers $c$. Then the resulting sequence is

$$a^c, (ar)^c, (ar^2)^c, (ar^3)^c$$

or
$$a^c, a^c r^c, a^c r^{2c}, a^c r^{3c},$$

which is clearly in G.P. with common ratio $r^c$.

**Property II**

If all the terms of a G.P. be multiplied or divided by the same non-zero constant, then the resulting sequence is a G.P. with same constant ratio.

*Proof*

Let the given G.P. be $a, ar, ar^2, ar^3, \ldots$
And its each term is multiplied by a non-zero constant $c$. Then the resulting sequence is

$$ca, c(ar), c(ar^2), \ldots$$

which is clearly in G.P. with common ratio $r$.

Similarly if each term is divided by $c$, then the resulting sequence is

$$\frac{a}{c}, \frac{ar}{c}, \frac{ar^2}{c}, \ldots$$

which is also in G.P. with common ratio $r$.

**Property III**

The reciprocals of the terms of a given G.P. from a G.P. whose common ratio is the reciprocal of the common ratio of the given G.P.

*Proof*

Let the given G.P. be $a, ar, ar^2, \ldots$

Then the sequence formed by the reciprocals of the terms is $\dfrac{1}{a}, \dfrac{1}{ar}, \dfrac{1}{ar^2}, \ldots$ which is clearly in G.P. with common ratio $\dfrac{1}{r}$, the reciprocal of the common ratio of the given G.P.

**To Find the Sum of $n$ Terms of the Series**

$a, (a+d)r, (a+2d)r^2, \ldots$ in which each term is the product of corresponding terms in an arithmetic and geometric series.

Let S be the sum, then

$$S = a + (a+d)r + (a+2d)r' + (a+3d)r^3 + \cdots \left[a + (n-1)d\right]r^n$$

$$\therefore \quad rS = ar + (a+d)r^2 + (a+2d)r^3 + \cdots + \left[a + (n-2)d\right]r^{n-1} + \left[a + (n-1)d\right]r^n$$

Subtracting, we have

$$S(1-r) = a + \left(dr + dr^2 + dr^3 + \cdots + dr^{n-1}\right) - \left[a + (n-1)d\right]r^n$$

$$= a + \frac{dr\left(1 - r^{n-1}\right)}{1-r} - \left[a + (n-1)d\right]r^n$$

$$\Rightarrow S = \frac{a}{(1-r)} + \frac{dr\left(1 - r^{n-1}\right)}{(1-r)^2} - \frac{\left[a + (n-1)d\right]r^n}{1-r}$$

*Note:* Writing S in the form

$$\frac{a}{1-r} + \frac{dr}{(1-r)^2} - \frac{dr^n}{(1-r)^2} - \frac{\left[a + (n-1)d\right]r^n}{(1-r)}$$

Then if $|r| < 1$, we can make $r^n$ as small as we please by taking $n$ sufficiently large. In this case, assuming that all the terms which involve $r^n$ can be made so small that they may be neglected, we have $\dfrac{a}{1-r} + \dfrac{dr}{(1-r)^2}$ for the sum to infinity.

**Example 41:** If the $p$th, $q$th and $r$th terms of a G.P. are $a$, $b$, and $c$ respectively, prove that

(i) $a^{q-r} \cdot b^{r-p} \cdot c^{p-q} = 1$

(ii) $(q-r)\log a + (r-p)\log b + (p-q)\log c = 0$.

### *Solution*

(i) Let $x$ and $y$ be the first term and common ratio of the given G.P. respectively. Then

$$a = xy^{p-1}, \quad b = xy^{q-1} \quad \text{and} \quad c = xy^{r-1}$$

$$\therefore a^{q-r} = x^{q-r} \cdot y^{(p-1)(q-r)}$$

$$b^{r-p} = x^{r-p} \cdot y^{(q-1)(r-p)}$$

and

$$c^{p-q} = x^{p-q} \cdot y^{(r-1)(p-q)}.$$

Thus $a^{q-r} \cdot b^{r-p} \cdot c^{p-q} = x^{q-r} \cdot x^{r-p} \cdot x^{p-q} \cdot y^{(p-1)(q-r)} \cdot y^{(q-1)(r-p)} \cdot y^{(r-1)(p-q)}$

$$= x^{q-r+r-p+p-q} \cdot y^{(p-1)(q-r)+(q-1)(r-p)+(r-1)(p-q)}$$

$$= x^0 \cdot y^0 \quad = 1. \qquad \qquad \text{Proved.}$$

(ii) From result (i) $a^{q-r} \cdot b^{r-p} \cdot c^{p-q} = 1$

Taking log of both sides, we get

$$(q-r)\log a + (r-p)\log b + (p-q)\log c = \log 1 = 0$$
$$\Rightarrow (q-r)\log a + (r-p)\log b + (p-q)\log c = 0$$

**Example 42:** Find the sum of the following series:

(a) $\dfrac{1}{\sqrt{2}} - 1 + \sqrt{2} - \dots$ upto 10 terms,

(b) $\dfrac{x+y}{x-y} + 1 + \dfrac{x-y}{x+y} + \dots$ upto $n$ terms.

### *Solution*

(a) Here $a = \dfrac{1}{\sqrt{2}}, \quad r = -\sqrt{2}, \quad n = 10$

$$\therefore \qquad S = \frac{a(1-r^n)}{1-r}$$

$$\text{or} \qquad S = \frac{\dfrac{1}{\sqrt{2}}\left[1-\left(-\sqrt{2}\right)^{10}\right]}{1+\dfrac{1}{\sqrt{2}}}$$

$$= \dfrac{\dfrac{1}{\sqrt{2}}\left[1-2^5\right]}{1+\sqrt{2}} = \dfrac{\dfrac{1}{\sqrt{2}}\left[1-32\right]}{1+\sqrt{2}}$$

$$= \dfrac{-31}{2+\sqrt{2}} \times \dfrac{2-\sqrt{2}}{2-\sqrt{2}}$$

$$= \dfrac{-31\left(2-\sqrt{2}\right)}{4-2}$$

$$S = \dfrac{-31}{2}\left(2-\sqrt{2}\right)$$

(b)  Here $a = \dfrac{x+y}{x-y}$ and $r = \dfrac{x-y}{x+y}(<1)$

$$\therefore \qquad S_n = \dfrac{a\left(1-r^n\right)}{1-r}$$

$$= \dfrac{\dfrac{x+y}{x-y}\left[1-\left(\dfrac{x-y}{x+y}\right)^n\right]}{1-\dfrac{x-y}{x+y}}$$

$$\text{or} \qquad S_n = \dfrac{x+y}{x-y} \cdot \dfrac{\left[1-\left(\dfrac{x-y}{x+y}\right)^n\right]}{x+y-x+y} \cdot (x+y)$$

$$= \dfrac{(x+y)^2}{2y(x-y)}\left[1-\left(\dfrac{x-y}{x+y}\right)^n\right].$$

**Example 43:**  Find the sum of the series $2 + 6 + 18 + \cdots + 4374$.

***Solution***

Here $a = 2$, $r = 3$, $l = 4374$.

$$\therefore \qquad \text{Sum} = \dfrac{lr-a}{r-1} \qquad \left[\because \ r>1\right]$$

$$= \dfrac{4374 \times 3 - 2}{3-1} = 6560$$

**Example 44:** Find the sum of the series $\dfrac{2}{3} - 1 + \dfrac{3}{2} - \cdots + \dfrac{243}{32}$.

**Solution**

Here $a = \dfrac{2}{3}, r = -\dfrac{3}{2}, l = \dfrac{243}{32}$.

$$\text{Sum} = \frac{a - lr}{1 - r} \qquad (\because \; r < 1)$$

$$= \frac{\dfrac{2}{3} - \left(-\dfrac{3}{2}\right)\left(\dfrac{243}{32}\right)}{1 - \left(-\dfrac{3}{2}\right)}$$

$$= \frac{463}{96}$$

**Example 45:** Find the sum of the following series to infinity

(a) $\left(1 + \dfrac{1}{2^2}\right) + \left(\dfrac{1}{2} + \dfrac{1}{2^4}\right) + \left(\dfrac{1}{2^2} + \dfrac{1}{2^6}\right) + \cdots \infty$

(b) $\dfrac{2}{5} + \dfrac{3}{5^2} + \dfrac{2}{5^3} + \dfrac{3}{5^4} + \cdots \infty$

**Solution**

(a) Required Sum $= \left(1 + \dfrac{1}{2^2}\right) + \left(\dfrac{1}{2} + \dfrac{1}{2^4}\right) + \left(\dfrac{1}{2^2} + \dfrac{1}{2^6}\right) + \cdots$

$$= \left(1 + \dfrac{1}{2} + \dfrac{1}{2^2} + \cdots\right) + \left(\dfrac{1}{2^2} + \dfrac{1}{2^4} + \dfrac{1}{2^6} + \cdots\right)$$

$$= \frac{1}{1 - \dfrac{1}{2}} + \frac{1}{4}\frac{1}{1 - \dfrac{1}{4}} \qquad \left[S = \frac{a}{1 - r}\right]$$

$$= 2 + \frac{1}{4} \cdot \frac{4}{3}$$

$$= 2 + \frac{1}{3} = 2\frac{1}{3}$$

(b) Required Sum $= \dfrac{2}{5} + \dfrac{3}{5^2} + \dfrac{2}{5^3} + \dfrac{3}{5^4} + \dfrac{2}{5^5} + \dfrac{3}{5^6} + \cdots$

$$= \left( \frac{2}{5} + \frac{2}{5^3} + \frac{2}{5^5} + \cdots \right) + \left( \frac{3}{5^2} + \frac{3}{5^4} + \frac{3}{5^6} + \cdots \right)$$

$$= \frac{2}{5}\left( 1 + \frac{1}{5^2} + \frac{1}{5^4} + \cdots \right) + \frac{3}{5^2}\left( 1 + \frac{1}{5^2} + \frac{1}{5^4} + \cdots \right)$$

$$= \left( \frac{2}{5} + \frac{3}{5^2} \right)\left( 1 + \frac{1}{5^2} + \frac{1}{5^4} + \cdots \right)$$

$$= \left( \frac{10+3}{25} \right) \left[ \frac{1}{1 - \left( \dfrac{1}{5} \right)^2} \right]$$

$$= \frac{13}{25} \times \left( \frac{25}{25-1} \right)$$

$$= \frac{13}{25} \times \frac{25}{24} = \frac{13}{24}$$

**Example 46:** Find the sum of the following series to infinity

(a) $\quad 1 - \dfrac{1}{3} + \dfrac{1}{9} - \dfrac{1}{27} + \cdots \infty$

(b) $\quad 0.9 + 0.03 + 0.001 + \cdots \infty$

**Solution**

(a) Here $a = 1, |r| = \left| -\dfrac{1}{3} \right| < 1$

$$\therefore \quad s = \frac{a}{1-r} = \frac{1}{1 + \dfrac{1}{3}} = \frac{1}{4/3} = \frac{3}{4}$$

(b) Here $a = 0.9, |r| = \left| \dfrac{0.03}{0.9} \right| = \left| \dfrac{1}{30} \right| \cdot (<1)$

$$\therefore \quad S = \frac{a}{1-r} = \frac{0.9}{1 - 1/30} = \frac{30 \times 0.9}{29} = \frac{27}{29}$$

**Example 47:** The sum of how many terms of the G.P. $\dfrac{2}{9}-\dfrac{1}{3}+\dfrac{1}{2}-\cdots$ is $\dfrac{55}{72}$

**Solution**

Let the sum of $n$ terms of the given G.P. be $\dfrac{55}{72}$.

Here $a=\dfrac{2}{9}, r=-\dfrac{3}{2}\cdot(<1)$

$$\therefore S_n=\frac{a\left(1-r^n\right)}{1-r}=\frac{\dfrac{2}{9}\left[1-\left(-\dfrac{3}{2}\right)^n\right]}{1-\left(-\dfrac{3}{2}\right)}$$

$$\Rightarrow \quad \frac{55}{72}=\frac{2}{9}\times\frac{2}{5}\left[1-\left(-\frac{3}{2}\right)^n\right]$$

$$\Rightarrow \quad 1-\left(-\frac{3}{2}\right)^n=\frac{55}{72}\times\frac{9\times5}{2\times2}=\frac{275}{32}$$

$$\text{or} \quad -1+\left(-\frac{3}{2}\right)^n=-\frac{275}{32}$$

$$\text{or} \quad \left(-\frac{3}{2}\right)^n=1-\frac{275}{32}=-\frac{243}{32}$$

$$\text{or} \quad \left(-\frac{3}{2}\right)^n=\left(-\frac{3}{2}\right)^5$$

$$\Rightarrow \quad n=5$$

Thus the required number of terms is 5.

**Example 48:** Find the sum of the following series:

(a) $5+55+555+\cdots$ upto $n$ terms,
(b) $0.3+0.33+0.333+\cdots$ upto 100 terms.

**Solution**

(a) $5+55+555+\cdots$ upto $n$ terms
$\quad = 5\left[1+11+111+\cdots \text{ upto } n \text{ terms}\right.$

$\quad = \dfrac{5}{9}\left[9+99+999+\cdots \text{ upto } n \text{ terms}\right]$

$$= \frac{5}{9}\left[(10-1)+(100-1)+(1000-1)+\,\cdots\,\text{upto }n\text{ terms}\right]$$

$$= \frac{5}{9}\left[(10+10^2+10^3+\cdots\text{upto }n\text{ terms})-(1+1+\cdots\text{upto }n\text{ terms})\right]$$

$$= \frac{5}{9}\left[\frac{10(10^n-n)}{10-1}-n\right]$$

$$= \frac{5}{9}\left[\frac{10}{9}(10^n-1)-n\right]$$

(b) $0.3+0.33+0.333+\cdots$ upto 100 terms

$$= 3\,[0.1+0.11+0.111\cdots\text{upto 100 terms}]$$

$$= \frac{3}{9}[0.9+0.99+0.999+\cdots\text{upto 100 terms}]$$

$$= \frac{1}{3}\left[\frac{9}{10}+\frac{99}{100}+\frac{999}{1000}+\cdots\text{upto 100 terms}\right]$$

$$= \frac{1}{3}\left[\left(1-\frac{1}{10}\right)+\left(1-\frac{1}{10^2}\right)+\left(1-\frac{1}{10^3}\right)+\,\cdots\,\text{upto 100 terms}\right]$$

$$= \frac{1}{3}\left[(1+1+\cdots\text{upto 100 terms})-\left(\frac{1}{10}+\frac{1}{10^2}+\frac{1}{10^3}+\cdots\text{upto 100 terms}\right)\right]$$

$$= \frac{1}{3}\left[100-\frac{\frac{1}{10}\left(1-\frac{1}{10^{100}}\right)}{1-\frac{1}{10}}\right]$$

$$= \frac{1}{3}\left[100-\frac{1}{9}\left(1-\frac{1}{10^{100}}\right)\right]$$

$$= \frac{1}{27}\left[900-1+\frac{1}{10^{100}}\right]$$

$$= \frac{1}{27}\left[899+(0.1)^{-100}\right]$$

**Example 49:** If $y = x + x^2 + x^3 + \cdots$ upto $\infty$ and x is a positive number less than 1. Show $x = \dfrac{y}{1+y}$.

**Solution**

$$y = x + x^2 + x^3 + \cdots \text{ upto } \infty$$

$$= \frac{x}{1-x} \qquad \left[ s = \frac{a}{1-r} \right]$$

or $\qquad y(1-x) = x$

or $\qquad y - yx = x$

or $\qquad y = x + yx$

or $\qquad y = x(1+y)$

$\Rightarrow \qquad x = \dfrac{y}{1+y}$

**Example 50:** Find the Geometric Mean of the following numbers:

(a)  162 and 882  (b)  0.2 and 0.002

**Solution**

(a) G.M. of 162 and 882 $= \pm\sqrt{162 \times 882} = \pm\sqrt{(378)^2}$

$$= \pm 378$$

(b) G.M. of 0.2 and 0.002 $= \pm\sqrt{0.2 \times 0.002} = \pm\sqrt{0.0004}$

$$= \pm 0.02$$

**Example 51:** Insert 6 G.M.'s between 27 and $\dfrac{1}{81}$.

**Solution**

Let $G_1, G_2, G_3, G_4, G_5, G_6$, be the 6 geometric means between 27 and $\dfrac{1}{81}$. Let $r$ be the common ratio of the G.P. then 27 is the first term and $\dfrac{1}{81}$ is the last term (8th term).

$$\therefore \quad 27 \cdot r^{8-1} = \frac{1}{81}$$

$$\Rightarrow \quad r^7 = \frac{1}{3^3 \cdot 81} = \left(\frac{1}{3}\right)^7$$

$$\Rightarrow \quad r = \frac{1}{3}$$

$$\therefore \quad G_1 = ar = 27\left(\frac{1}{3}\right) = 9$$

$$G_2 = ar^2 = 27 \cdot \frac{1}{9} = 3$$

$$G_3 = ar^3 = 27 \cdot \frac{1}{27} = 1$$

$$G_4 = ar^4 = 27 \cdot \frac{1}{81} = \frac{1}{3}$$

$$G_5 = ar^5 = 27 \cdot \frac{1}{3^5} = \frac{1}{9}$$

$$G_6 = ar^6 = 27 \cdot \frac{1}{3^6} = \frac{1}{27}$$

Hence, the required G.M. are $9, 3, 1, \dfrac{1}{3}, \dfrac{1}{9}, \dfrac{1}{27}$.

**Example 52:** Three quantities whose sum is 15 are in A.P. If 1, 4 and 19 are added, the resulting quantities are in G.P. Find the three quantities.

***Solution***

Let the three numbers, in A.P. be $a - d, a, a + d$.
Then $(a - d) + a + (a + d) = 15$

$$\Rightarrow \quad a = \frac{15}{3} = 5$$

Thus the numbers are $5 - d, 5, 5 + d$.
When 1, 4, 19 are added respectively to these numbers, then they become
$$5 - d + 1, 5 + 4, 5 + d + 19$$
$$\text{or}$$
$$6 - d, 9, 24 + d.$$

Since these are in G.P. Therefore

$$9^2 = (6 - d) \cdot (24 + d)$$

or $\quad 81 = 144 - 8d - d^2$

or $\quad d^2 + 18d - 63 = 0$

or $\qquad d^2 - 3d + 21d - 63 = 0$

or $\qquad (d-3)(d+21) = 0$

$\Rightarrow \qquad d = 3, -21$

Hence, the required numbers are

$$5 - 3,\ 5,\ 5 + 3 \text{ or } 5 + 21,\ 5,\ 5 - 21$$
$$\therefore \quad 2,\ 5,\ 8 \text{ or } 26,\ 5,\ -16$$

**Example 53:** The product of three consecutive terms of a G.P. is 216 and the sum of their products in pairs is 156. Find the terms.

***Solution***

Let the three consecutive terms of the G.P. be $\dfrac{a}{r}, a, ar$.

Then $\qquad\qquad\qquad\qquad\qquad \dfrac{a}{r} \cdot a \cdot ar = 216 \quad$ (given) $\qquad\qquad\qquad$ (1)

and $\qquad\qquad\qquad \dfrac{a}{r} \cdot a + \dfrac{a}{r} \cdot ar + a \cdot ar = 156 \quad$ (given)

or $\qquad\qquad\qquad\qquad\qquad a^2\left(\dfrac{1}{r} + 1 + r\right) = 156 \qquad\qquad\qquad$ (2)

From (1),
$a^3 = 216 \Rightarrow a = 6$

Putting the value of $a$ in (2), we have

$$36\left(\frac{1}{r} + 1 + r\right) = 156$$

$$3\left(\frac{1}{r} + 1 + r\right) = 13$$

or $\qquad 3r^2 - 10r + 3 = 0$

or $\qquad (r-3)(3r-1) = 0$

$\Rightarrow \qquad r = 3, \dfrac{1}{3}$

$\therefore \quad$ The required terms are $\dfrac{6}{3}, 6, 18;\ \dfrac{6}{1/3}, 6, 6 \times \dfrac{1}{3}$

or $\quad 2, 6, 18;\ 18, 6, 2$

$\Rightarrow \quad 2, 6, 18$

**Example 54:** If $a$, $b$, $c$ are in G.P., prove that $\log a^n, \log b^n, \log c^n$ are in A.P.

*Solution*

$a$, $b$, $c$ are in G.P.

$\therefore \quad b^2 = ac$

$\Rightarrow \qquad 2\log b = \log a + \log c \qquad\qquad$ multiplying both side by $n$

or $\qquad 2n\log b = n\log a + n\log c$

or $\qquad 2\log b^n = \log a^n + \log c^n$

$\Rightarrow \qquad \log a^n, \log b^n, \log c^n \qquad\qquad$ are in A.P.

## Exercise 3.3

1. Find:

    (a) 5th term of the sequence
    $3, 9, 27, 81, \ldots \infty$

    (b) 9th term of the sequence
    $4, 8, 16, 32, \ldots \infty$

2. Find the $n$th terms of the following sequence:

    (a) $\sqrt{3}, \dfrac{1}{\sqrt{3}}, \dfrac{1}{\sqrt[3]{3}}, \ldots$

    (b) $-2, -6, -18, \ldots$

3. (a) Which term of the sequence
    $1, \sqrt{3}, 3, \ldots$ is 81?

    (b) Which term of the sequence
    $\dfrac{1}{4}, -\dfrac{1}{2}, 1, \ldots$ is $-128$?

4. If the first two terms of a series in G.P. are 125 and 25 respectively. Find its 6th term.

5. The first term of a G.P. is 16 and its 5th term is $\dfrac{1}{16}$. Find its 4th term.

6. The product of the 3rd and 8th term of a G.P. is 243. If the 4th term is 3. Find the 7th term.

7. If $(p + q)$ th term of a G.P. is $m$ and the $(p - q)$ th term is $n$ then

    (a) find $p$th term,

    (b) Show that $n$th term is $m\left(\dfrac{n}{m}\right)^{p/2q}$

8. Find the sum of the following series:

    (a) $\dfrac{2}{9} - \dfrac{1}{3} + \dfrac{1}{2} - \cdots + \dfrac{9}{8}$

    (b) $\sqrt{2} - 2 + 2\sqrt{2} - \cdots + 64\sqrt{2}$

9. Find the sum of the following series:

    (a) $\sqrt{3} + \dfrac{1}{\sqrt{3}} + \dfrac{1}{3\sqrt{3}} + \cdots \infty$

    (b) $\dfrac{2}{3} + \dfrac{4}{9} + \dfrac{8}{27} + \cdots \infty$

    (c) $7 + 7.7 + 7.77 + \cdots$ upto $n$ terms

10. The sum of the first 6 terms of a G.P. is 9 times the sum of the first three terms. Find the common ratio.

11. The sum of a G.P. whose common ratio is 3, is 728. The last term is 486. Find the first term.

12. If $x = 1 + a + a^2 + \dots \infty, y = 1 + b + b^2 + \dots \infty$; prove that

$$1 + ab + a^2b^2 + \dots \infty = \frac{xy}{x+y-1}$$

Here $|a| < 1$ and $|b| < 1$.

13. If $S_1, S_2, S_3$, are respectively the sums of $n$ terms, $2n$ terms and to infinity of a G.P. Prove that $S_1(S_1 - S_3) = S_3(S_1 - S_2)$.

14. Find the value of $n$ so that $\dfrac{a^{n+1} + b^{n+1}}{a^n + b^n}$ may be the geometric mean of $a$ and $b$.

15. If $a$, $b$, $c$ are in G.P. and $x$ and $y$ are the A.M. of $a$, $b$ and $c$ respectively. Prove that

   (i) $\dfrac{a}{x} + \dfrac{c}{y} = 2,$     (ii) $\dfrac{1}{x} + \dfrac{1}{y} = \dfrac{2}{b}$

16. If the A.M. between $a$ and $b$ is to their G.P. as $m : n$. Show that

$$a : b = \left(m + \sqrt{m^2 - n^2}\right) : \left(m - \sqrt{m^2 - n^2}\right).$$

17. (a) Insert 3 G.M. between $\dfrac{1}{3}$ and 432.

   (b) Insert 5 G.M. between 3 and 192.

18. (a) The A.M. and G.M. of two numbers are 10 and 8 respectively. Find the numbers.

   (b) The sum of three numbers which are consecutive terms of an A.P. is 15. If 8, 6, 4 be added to these terms respectively, the resulting numbers are in G.P. Find the numbers.

19. Find two numbers whose difference is 2 and whose A.M. exceeds the G.M. by $\dfrac{1}{2}$.

20. If the A.M. between $a$ and $b$ be two times of their G.M. Show that

$$a : b = \left(2 + \sqrt{3}\right) : \left(2 - \sqrt{3}\right).$$

21. The product of three consecutive terms of a G.P. is 216 and their sum is 19. Find the terms.

22. Find three numbers is G.P. whose sum is $\dfrac{13}{12}$ and their product is $-1$.

23. The sum of four numbers in G.P. is 60 and the A.M. between the first and the last is 18. Find the numbers.

24. If $a$, $b$, $c$ are in G.P, prove that

   (i) $\dfrac{1}{a^2 - b^2} + \dfrac{1}{b^2} = \dfrac{1}{b^2 - c^2}$

   (ii) $(a + 2b + 2c)(a - 2b - 2c) = a^2 + 4c^2.$

**Answers to Selected Problems**

1. (a) 243
   (b) 1024

3. (a) 9th term
   (b) 10th

5. $\dfrac{1}{4}$

8. (a) $\dfrac{1}{72}$   (b) $\dfrac{128 + \sqrt{2}}{1 + \sqrt{2}}$

10. 2

17. (a) 2, 12, 72
    (b) 6, 12, 24, 48, 96

21. 9, 6, 4

23. 4, 8, 16, 32

## 3.7 THE WELL-ORDERING PRINCIPLE

A nonempty set S of numbers is well-ordered when each of its nonempty subset has the smallest element. A set is well-ordered if it has a least element.

The following sets are well-ordered:

1. $N \cup \{0\}$, with smallest element 0.
2. $N \cup \{-1, 0\}$, with smallest element $-1$.
3. $N \cup \{-2, -1\}$, with smallest element $-2$.
4. $\{n \in N : n > 1\}$, with smallest element 2.
5. The integers $\geq -2$, with smallest element $-2$.

**Example 55:** Prove that set of rational numbers of the form $\dfrac{1}{n}$, where $n$ is a positive integer $\leq 10^{100}$, is well-ordered.

### *Solution:*

Let $A = \{\dfrac{1}{n}$, where $n$ is a +ve integer $\leq 10^{100}\}$

Let $g = 10^{100}$.

$$A = \left\{1, \frac{1}{2}, \frac{1}{3}, ..., \frac{1}{g}\right\}$$

The least element of the set is $\dfrac{1}{g}$, $g = 10^{100}$

Hence, it is well-ordered.

**Example 56:** Prove that the set of rational numbers of the form $\dfrac{1}{n}$, where $n$ is a positive integer, is not well-ordered.

### *Solution:*

Let $A = \{\dfrac{1}{n}$, $n$ is a +ve integer$\}$

$$= \left\{1, \frac{1}{2}, \frac{1}{3}, ..., \frac{1}{100}, \frac{1}{1000} ...\right\}$$

The set has no least element.

Hence, the set is not well-ordered.

**Example 57:** Prove that the set of integers $Z$ is not well-ordered.

### *Solution:*

We prove it by contradiction, suppose $Z$ has the smallest element.

Let $x \in Z$ is the smallest element of $Z$.

Then $(x - 1) \in Z$ and $(x - 1) < x$, which contradicts that $x$ is the smallest element.

Hence, there can be no smallest element of $Z$ and hence $Z$ is not well-ordered.

## 3.8 MATHEMATICAL INDUCTION

**Mathematical induction** is a technique of proving a statement, theorem or formula which is thought to be true, for each and every natural number $n$.

## WORKING RULE

The technique involves two steps:

1. **Base Step:** In base step it is proved that the statement is true for 1, the initial value $n$.
2. **Induction Step:** Assuming the statement is true for $n = k$, prove that the statement is true for $n = k + 1$.

**Example 58:** Prove by induction that $3^n - 1$ is a multiple of 2 for every natural number $n$.

*Solution:*

**Base Step**
For $n = 1$,
$= 3^1 - 1 = 3 - 1 = 2$, is a multiple of 2.
So the statement is true for $n = 1$
**Induction Step**
Let the statement be true for $n = k$, so that $3^k - 1$ is a multiple of 2.
We now prove that the statement is true for $n = k + 1$
$3^{k+1} - 1 = 3.3^k - 1 = 3.3^k - 3 + 2 = 3\,(3^k - 1) + 2$ is a multiple of 2.
Hence, by induction the statement is true.

**Example 59:** Prove that the sum of first $n$ natural numbers is $\dfrac{n(n+1)}{2}$

*Solution:*

**Base Step**
For $n = 1$,
$$\frac{n(n+1)}{2} = \frac{1(2)}{2} = 1 = \text{sum of first 1 natural number.}$$
So the statement is true for $n = 1$.
**Induction Step**
Let the statement be true for $n = k$, so that $\dfrac{k(k+1)}{2}$ is the sum of first $k$ natural numbers.

$$1 + 2 + 3 + \ldots.. + k = \frac{k(k+1)}{2}.$$

We now prove that the statement is true for $n = k + 1$.
Sum of first $k + 1$ numbers $= 1 + 2 + \ldots + (k + 1)$
$= (1 + 2 + \ldots. + k) + (k + 1)$

$$= \frac{k(k+1)}{2} + (k+1) = \frac{k(k+1) + 2(k+1)}{2} = \frac{(k+1)(k+2)}{2}$$

Which proves that the statement is true for $n = k + 1$

Hence by induction the statement is true for all natural numbers.

**Example 60:**  Prove that the sum of first $n$ odd natural numbers is $n^2$.

*Solution:*

For $n = 1$

$n^2 = 1 =$ sum of first 1 odd integer.

So the statement is true for $n = 1$.

Let the statement be true for $n = k$, so that $k^2$ is the sum of first $k$ odd numbers.

$1 + 3 + 5 + \ldots\ldots + (2k - 1) = k^2$ .

We now prove that the statement is true for $n = k + 1$.

Sum of first $k + 1$ odd numbers

$= 1 + 3 + 5 + \ldots + (2k - 1) + (2k + 1)$.

$= k^2 + 2k + 1 = (k + 1)^2$ .

Which proves that the statement is true for $n = k + 1$

Hence by induction the statement is true for all natural numbers.

**Example 61:**  Prove that the sum of squares of first $n$ natural numbers is $\dfrac{n(n+1))(2n+1)}{6}$ .

*Solution:*

For $n = 1$

$$\frac{n(n+1)(2n+1)}{6} = \frac{1.2.3}{6} = 1 = 1^2 = \text{sum of square of first natural number.}$$

So the statement is true for $n = 1$.

Let the statement be true for $n = k$, so that

$$1^2 + 2^2 + 3^2 + \ldots.. + k^2 = \frac{k(k+1)(2k+1)}{6}$$

We now prove that the statement is true for $n = k + 1$

Sum of squares of $k + 1$ natural numbers

$$1^2 + 2^2 + 3^2 + \ldots.. + k^2 + (k+1)^2 = \frac{k(k+1)(2k+1)}{6} + (k+1)^2 .$$

$$= \frac{k(k+1)(2k+1) + 6(k+1)^2}{6}$$

$$= \frac{(k+1)(2k^2 + k + 6k + 6)}{6}$$

$$= \frac{(k+1)(k+2)(2k+3)}{6}$$

$$= \frac{[(k+1)(k+1)+1][2(k+1)+1]}{6}$$

Which proves that the statement is true for $n = k + 1$.

Hence by induction the statement is true for all natural numbers.

**Example 62:** Prove that $1 + a + a^2 + a^3 + \ldots + a^n = \dfrac{a^{n+1} - 1}{a - 1}$.

*Solution:*

For $n = 1$

$$1 + a + a^2 + a^3 + \ldots + a^n = 1 + a = \frac{a^{1+1} - 1}{a - 1}$$

So the statement is true for $n = 1$.

Let the statement be true for $n = k$, so that

$$1 + a + a^2 + a^3 + \ldots + a^k = \frac{a^{k+1} - 1}{a - 1}$$

We now prove that the statement is true for $n = k + 1$

$$1 + a + a^2 + a^3 + \ldots a^k + a^{k+1}$$

$$= \frac{a^{k+1} - 1}{a - 1} + a^{k+1} = \frac{a^{k+1} - 1 + (a-1)a^{k+1}}{a - 1} = \frac{a^{(k+1)+1} - 1}{a - 1}$$

Which shows that the statement is true for $n = k + 1$.

Hence by induction the statement is true for all natural numbers.

## 3.9 THE DIVISION ALGORITHM

Let $a$ and $b$ be integers with $b \neq 0$ then there exists unique integers $q$ and $r$ such that $a = bq + r$ and $0 \leq r < |b|$, that is, $r$ is non-negative. $q$ is called quotient and $r$ is called remainder. They are generally of two types, i.e., slow algorithm and fast algorithm. Slow division algorithms produce one digit of the final quotient per iteration, e.g., restoring, non-restoring and SRT division algorithms.

Fast division algorithms start with a close approximation to the final quotient and produce twice as many digits of the final quotient in each iteration, e.g., Newton-Raphson and Goldschmidt algorithms.

**Example 63:** Express $-16$ as $bq + r$

*Solution:*

Let $a = -16$ and $b = 3$

$-16 = 3(-5) - 1 = 3(-5) - 1 + 3 - 3 = 3(-6) + 2$

So that $q = -6$ and $r = 2$

**Example 64:** Prove that any integer can be written in the form $2k, 2k + 1$.

*Solution:*

By division theorem
$a = 2k + r$ with $0 \leq r < 2$
from the last inequality , $r = 0$ or $1$.
Hence, $a = 2k, a = 2k + 1$

**Example 65:** Prove that any integer can be written in the form $3k, 3k + 1,$ or $3k + 2$

*Solution:*

By division theorem
$a = 3k + r$ with $0 \leq r < 3$
from the last inequality, $r = 0, 1$ or $2$
Hence, $a = 3k, a = 3k + 1$ or $a = 3k + 2$

**Example 66:** Prove that one of the three consecutive integers is a multiple of 3.

*Solution:*

By division theorem
$a = 3k + r$ with $0 \leq r < 3$
from the last inequality, $r = 0, 1$ or $2$
Hence, $a = 3k, a = 3k + 1$ or $a = 3k + 2$
Let three consecutive integers be $a, a + 1, a + 2$
If $a = 3k,$
then $a$ is a multiple of 3.
If $a = 3k +1,$
then $a + 2 = 3k + 1 + 2 = 3k + 3 = 3(k + 1)$, is a multiple of 3.
Hence, $a + 2$ is a multiple of 3
If $a = 3k + 2,$
then $a + 1 = 3k + 2 + 1 = 3(k + 1)$ , is a multiple of 3.
Hence, $a + 1$ is a multiple of 3.

## 3.10 PRIME NUMBERS

A prime number (or a prime) is a natural number greater than 1, that is, not a product of two smaller natural numbers. Prime numbers have only two factors 1 and themselves. A natural number greater than 1 that is not prime is called a composite number.
First seven prime numbers are 2, 3, 5, 7, 11, 13, and 17.

## 3.11 THE GREATEST COMMON DIVISOR

The greatest common divisor (*gcd*) of two or more integers, which are not all zero, is the greatest positive integer that divides each of the integers.

Let $a$ and $b$ be integers, not both 0. A common divisor of $a$ and $b$ is any nonzero integer that divides both $a$ and $b$. The largest natural number that divides both $a$ and $b$ is called the greatest common divisor of $a$ and $b$. The greatest common divisor of $a$ and $b$ is denoted by $gcd(a, b)$. Greatest common divisor is also known as greatest common factor ($gcf$).

## 3.12 EUCLIDEAN ALGORITHM

Euclidean algorithm is an efficient method to compute greatest common divisor of two integers. The Euclidian algorithm to find greatest common divisor of two integers $a$ and b is as follows:

If $a = 0$ then $gcd(a, b) = a$, and we can stop.

If $b = 0$ then $gcd(a, b) = b$, and we can stop.

If $a \neq 0$ and $b \neq 0$, $a > b$, let $a = bq + r$

Find $gcd(b, r)$, since $gcd(a, b) = gcd(b, r)$

Repeat until $r = 0$

**Example 67:** Find $gcd(24,57)$

*Solution:*

$57 = 24 \times 2 + 9$

$gcd(24, 57) = gcd(24,9)$

$24 = 9 \times 2 + 6$

$gcd(24, 9) = gcd(9,6)$

$9 = 6 \times 1 + 3$

$gcd(9, 6) = gcd(6,3)$

$6 = 3 \times 2 + 0$

$gcd(3,0) = 3$

$gcd(24,57) = gcd(24,9) = gcd(9,6) = gcd(6,3) = gcd(3.0) = 3$

To find $gcd(a, b)$, divide $b$ by $a$ by long division and then repeatedly divide the divisor by remainder until we get zero remainder.

## 3.13 THE FUNDAMENTAL THEOREM OF ARITHMETIC

The fundamental theorem of arithmetic state that every integer $n > 1$ is either itself a prime number or can be represented as a product of prime numbers and this representation is unique except for the order of factors. The primes in the factorization need not be distinct. The fundamental theorem of arithmetic is also known as unique factorization theorem or unique-prime-factorization theorem of arithmetic.

$n$ can be expressed uniquely in the form

$$n = p_1^{m_1} p_2^{m_2} \ldots\ldots\ldots p_r^{m_r}$$

where $m_i$ are positive integer and $p_1, p_2, \ldots p_r$ are primes $p_1 < p_2 < \ldots < p_r$. This form is called canonical factorization of $n$.

**Example 68:** Find *gcd* (24, 57) using fundamental theorem of arithmetic.

***Solution:***

$24 = 2^3 . 3.$
$57 = 3. 19$
The prime 3 appears in both 22 and 57.
Hence, *gcd* (24, 57) = 3

**Example 69: Find *gcd* (72, 342) using fundamental theorem of arithmetic.**

***Solution:***

$72 = 2^3 . 3^2$
$342 = 2.3^2.19$
Hence, *gcd* (72, 342) = 2. $3^2$ . = 18

## 3.14 RECURSIVE DEFINITIONS

Recursion occurs when a thing is defined in terms of itself or of its type.
In mathematics and computer science, a class of objects or methods exhibits recursive behaviour when it can be defined by two properties:

- A simple *base case* (or cases) — a terminating scenario that does not use recursion to produce an answer
- A *recursive step* — a set of rules that reduces all successive cases towards the base case.

**Example 70:** The following is a recursive definition of a person's *ancestor*. One's ancestor is either:

***Solution:***
(i)  One's parent (*base case),* or
(ii)  One's parent's ancestor (*recursive step*).

**Example 71:** The Fibonacci sequence is an example of recursion. The Fibonacci sequence is a series of numbers where a number is the addition of the last two numbers, starting with 0 and 1.

***Solution:***

The Fibonacci sequence: 0, 1, 1, 2, 3, 5, 8, 13, 21, …
In recursive form
$a_0$ = first term = 0
$a_1$ = second term = 1
$a_n = a_{n-2} + a_{n-1}.$

Writing recursive formulas

Suppose we want to find recursive formula for the sequence 1, 5, 9, 13 ….

Base case

$a_1$ = first term = 1

Recursive step

$a_n = a_{n-1} + 4$

we can generate the series

$a_2 = a_1 + 4 = 1 + 4 = 5$

$a_3 = a_2 + 4 = 5 + 4 = 9$

and so on.

## Example 72

The factorial function is an example of recursion.

$n! = 1. 2. 3. ..(n-2)(n-1)n$ in recursive form it is

If $n = 0$, then $n! = 1$

If $n > 0$, then $n! = n.(n-1)!$

Let us find 5!

$5! = 5. 4!$

$\quad = 5. 4. 3!$

$\quad = 5. 4\ \ 3. 2!$

$\quad = 5. 4. 3. 2. 1!$

$\quad = 5. 4. 3 . 2. 1. 0!$

$\quad = 5.4 .3.2.1 .1 = 5. 4. 3. 2. 1$

---

### Exercise 3.4

1. Prove that the sum of first $n$ even positive numbers is $n(n + 1)$

2. Prove that

$$\frac{1}{1.2} + \frac{1}{2.3} + \frac{1}{3.4} + ... + \frac{1}{n(n+1)} = \frac{n}{n+1}$$

3. Prove that the sum of cubes of first $n$ natural numbers is $\left[\dfrac{n(n+1)}{2}\right]^2$.

4. Prove that $x^{n+1} + y^{n+1}$

$= (x + y)(x^n - x^{n-1}y + x^{n-2}y^2 - ...$
$\qquad\qquad - ...\, x\, y_{n-1} + y^n)$

5. For the following values of integers, $a$ and $b$, find the values of quotient and integer.

(i) $a = 370$ and $b = 13$

(ii) $a = -210$ and $b = 12$

6. Find all prime numbers between 50 and 100.

7. Express 372 as a product of prime numbers.

8. Prove that any integer $a$ can be of the form $5k, 5k + 1, 5k + 2, 5k + 3$. or $5k + 4$.

9. Prove that one of the five consecutive integers is a multiple of 5.

10. Prove that the product of three consecutive integers is divisible by 6.

11. Prove that if $a = bq + r$, then

$$gcd(a, b) = gcd(b, r).$$

# 4

# Partial Ordering Relations

## 4.1 INTRODUCTION

The role of partial ordering is significant in the study of algebraic systems. This chapter covers the basic concepts involving partially ordered sets, partial ordering relations and their important properties.

## 4.2 PARTIAL ORDERING RELATIONS

A relation R on a set S is **partially ordering** if it is reflexive, anti-symmetric and transitive.

(i) **Reflexive**: $a\,R\,a$ for every $a \in S$
(ii) **Anti-symmetric**: If $a\,R\,b$ and $b\,R\,a$, then $a = b$
(iii) **Transitive**: If $a\,R\,b$ and $b\,R\,c$, then $a\,R\,c$.

The usual notation for partially ordered relation is $\leq$. The statement $a \leq b$ is read "$a$ precedes $b$". For example, consider the relation $(\geq)$ on the set of integers. Since $a \geq a$ for every integer $a$, $\geq$ is reflexive. If $a \geq b$ and $b \geq a$, then $a = b$. Hence, $\geq$ is anti-symmetric. Finally, $\geq$ is transitive since $a \geq b$ and $b \geq c$ imply that $a \geq c$. It follows that $\geq$ is a partial ordering on the set of integers.

## 4.3 PARTIAL ORDERED SET (POSET)

A set S together with partially ordering relation R $(\leq)$ on it is called partially ordered set or simply POSET and is denoted by (S, R) or (S, $\leq$)

**Example 1:** Show that the divisibility relation $/$, is a partial ordering on the set of positive integers.

### Solution

$I_+ = \{1, 2, 3, 4 \ldots\}$

R is divisibility relation $/$.

$a/a$ for every $a \in I_+$

Hence R is reflexive.

If $a/b$ and $b/a$ then $a = b$, for every $a, b \in I_+$.

Hence R is anti-symmetric.

$a/b$ and $b/c$ implies $a/c$ for every $a, b, c \in I_+$

∴　The divisibility relation $/$ is a partial ordering on the set of positive integers.

**Example 2:** Show that the inclusion relation $\subseteq$ is a partial ordering on the power set of a set S.

### *Solution*

Let A is a subset of S

Since $A \subseteq A$ whenever A is a subset of S, $\subseteq$ is reflexive.

It is antisymmetric since $A \subseteq$ Band $B \subseteq$ A imply that $A = B$.

Finally, $\subseteq$ is transitive, since $A \subseteq$ B and $B \subseteq$ C imply that $A \subseteq$ C.

Hence, $\subseteq$ is a partial ordering on the power set of a set S.

**Example 3:** Consider a set $A = \{4, 9, 16, 36\}$. Is the relation 'divides' a partial order?

### *Solution*

1. Since for every $a \in A$, we have $a/a$. Hence, 'divides' is reflexive.
2. If $a/b$ and $b/a$, we have $a = b$ for any $a, b \in A$. Hence, 'divides' is anti-symmetric.
3. If $a/b$ and $b/c$, we have $a/c$ for any $a, b, c \in A$.

Hence, the relation 'divides' is a partial order and $(A, /)$ is a poset.

## 4.4 DUAL PARTIAL ORDERED SET

If R is a partial order on a set A and let $R^{-1}$ be the inverse relation of R then $R^{-1}$ is also partial order relation.

### *Proof*

**Reflexivity:** $a R a \ \forall \ a \in A$
$$\Rightarrow \quad a R^{-1} a$$

∴　$R^{-1}$ is reflexive.

**Antisymmetry:** Let $a R^{-1} b$ and $b R^{-1} a$
$$\Rightarrow \quad b R a \quad \text{and} \quad a R b$$

as R is partial order relation this implies $a = b$

∴　$R^{-1}$ is antisymmetric

**Transitivity:** Let $a\mathrm{R}^{-1}b$ and $b\mathrm{R}^{-1}c$

$\Rightarrow\quad b\mathrm{R}a$ and $c\mathrm{R}b$

$\Rightarrow\quad c\mathrm{R}b$ and $b\mathrm{R}a$

As R is transitive this implies that $c\mathrm{R}a$

$\Rightarrow\quad a\mathrm{R}^{-1}c$

$\therefore\quad \mathrm{R}^{-1}$ is transitive.

Therefore $\mathrm{R}^{-1}$ is a partial order relation if R is a partial order relation.

The poset $(\mathrm{A}, \mathrm{R}^{-1})$ is called the **dual of the poset** $(\mathrm{A}, \mathrm{R})$ and the partial order relation $\mathrm{R}^{-1}$ is called the **dual partial order relation of R**. For instance '$\geq$' and '$\leq$' both are partial order relations and are dual of each other.

## 4.5 COMPARABILITY OF ELEMENTS OF A POSET

Two distinct elements $a$ and $b$ of a poset $(\mathrm{A}, \mathrm{R})$ are called comparable if either $a\mathrm{R}b$ or $b\mathrm{R}a$. When $a$ and $b$ are elements of A such that neither $a\mathrm{R}b$ nor $b\mathrm{R}a$, $a$ and $b$ are called incomparable.

**Example 4:** Consider $\mathrm{A} = \{1, 2, 3, 5, 6)$ is ordered by divisibility. Determine all the comparable and non-comparable pairs of elements of A.

### *Solution*

The comparable pairs of elements of A are: $(1, 2), (1, 3), (1, 5), (1, 6), (2, 6), (3, 6)$.

The non-comparable pair of elements of A are: $(2, 3), (2, 5), (3, 5), (5, 6)$.

## 4.6 TOTALLY ORDERED SET

If $(\mathrm{S}, \leq)$ is a poset and every two elements of S are comparable, S is called a **totally ordered** or **linearly ordered set**, and $\leq$ is called a *total order* or a *linear order*.

**For example**, the set N of natural numbers is a totally ordered set with the usual relation '$\leq$ in numbers'.

## 4.7 WELL ORDERED SET

$(\mathrm{S}, \leq)$ is a *well-ordered set* if it is a poset such that $\leq$ is a total ordering and such that every nonempty subset of S has a least element.

## 4.8 CHAIN

Let A be a poset, and B be a subset of A, such that B is totally ordered, with the induced partially ordered relation on B, then B is called **chain** in A. The number of elements in chain is called length of chain. **For example**, the set N of natural numbers is a partially ordered set with the relation of divisibility.

N is not a totally ordered set because 3, 5 $\in$ N and neither 3/5 or 5/3.

Let A = {5, 20, 40, 80}.

A is chain in N because 5/20, 5/40, 5/80, 20/40, 20/80, 40/80

## 4.9 ANTI-CHAIN

A subset of a poset is called **Anti-Chain** if no two distinct elements in the subset are comparable.

**Example 5:** Let A be a set containing at least two elements. The power set P(A) of A is a partially ordered set with the relation $\subseteq$ of set inclusion. Let $a, b \in$ A.

$\therefore$    $\{a\}, \{b\} \in$ P(A) and neither $\{a\} \subseteq \{b\}$ nor $\{b\} \subseteq \{a\}$, and are not comparable.

$\therefore$    The subset $\{\{a\}, \{b\})$ of P (A) is an anti-chain in P(A)

## 4.10 REPRESENTATION OF A POSET (HASSE DIAGRAM)

A graphical representation of a Poset (A, R), in which all arrowheads are understood to be pointing upward is known as **Hasse diagram** named after the German mathematician Helmut Hasse. The elements of set S are represented by **vertices** in the Hasse diagram. For $a, b \in$ A, if $a$ is an immediate predecessor of $b$ if $a < b$ and no element in A lies between $a$ and $b$, we place $b$ higher than $a$ and draw a line between them. Such lines are called the **edges** of the Hasse Diagram.

Directed graph of a poset can be simplified and converted to Hasse diagram. Since the order is reflexive, we can omit loops. Since the order is transitive, we can omit arrows between points that are connected by sequence of arrows. In cases in which all arrowheads point in one direction (upward, downward, left, right), we can omit the arrowheads.

**Immediate predecessor:** Let A be a poset (A, R) and Let $a, b \in$ A, $a$ is said to be immediate predecessor of $b$ if $a$ R $b$, and there exists no element in A such that $a$ R $c$ and $c$ R $b$, where $a \neq b$, $b \neq c$ and $a \neq c$. If $a$ is immediate predecessor of $b$ we write $a << b$.

**Immediate Successor:** If $a$ is an immediate predecessor of $b$ then $b$ is said be an immediate successor of $a$ or cover of $a$.

**Example 6:** Let X = {2, 3, 6, 12, 24}, let R, $\leq$ be the partial order defined by $x \leq y$. Draw the directed graph and the Hasse diagram of R.

### *Solution*

We can omit all loops, implied by the reflexive property and omit edges, implied by the transitive property. Arrange all edges to point upward. Omit all arrowheads pointing in the upward direction to get the Hasse diagram.

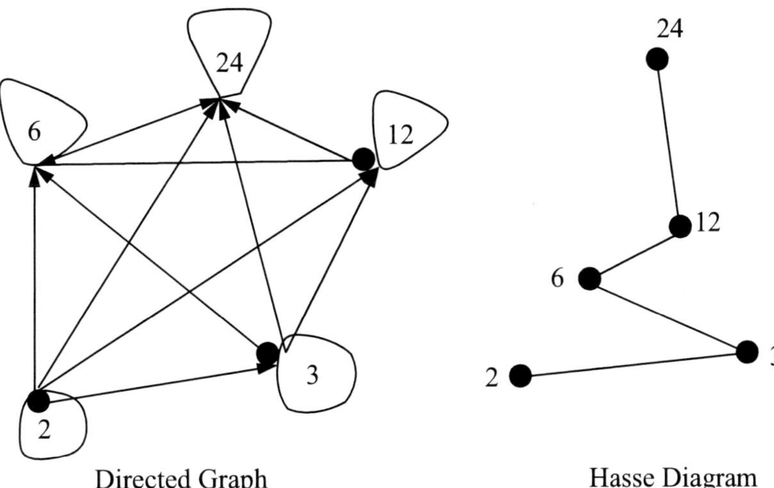

Directed Graph
Hasse Diagram

**Example 7:**  Consider the set A = {4, 5, 6, 7}. Let R be the relation ≤ on A. Draw the directed graph and the Hasse diagram of R.

***Solution***

R = {(4, 5}, {4, 6}, {4, 7}, {5, 6}, {5, 7}, {6, 7}, {4, 4}, {5, 5}, {6, 6}, (7, 7}} the directed graph of the relation R is

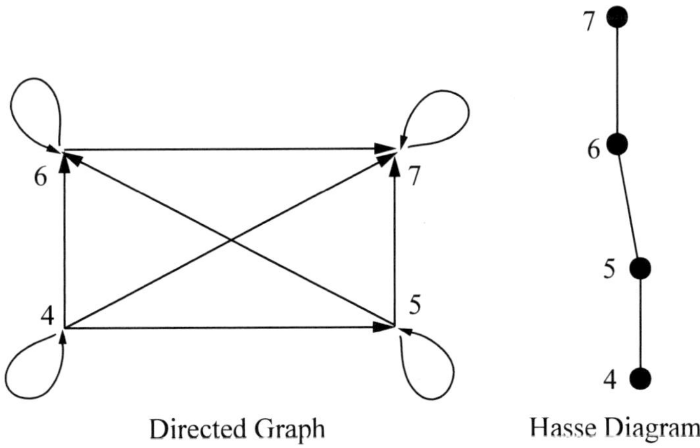

Directed Graph
Hasse Diagram

**Example 8:** Let A = {2, 3, 4, 6, 8, 24, 48} be the partially ordered set with the relation R '*x* divides *y*'. Draw the Hasse diagram.

***Solution***

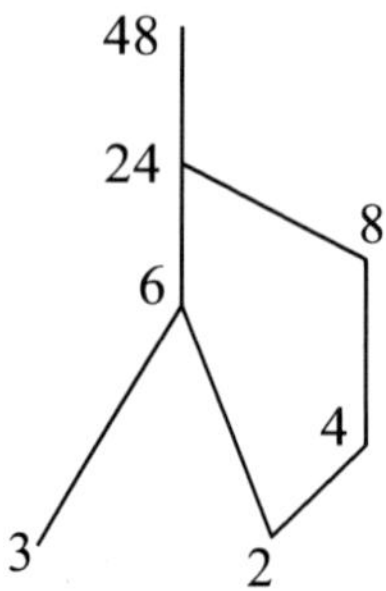

Hasse Diagram

**Example 9:** Determine the matrix of the partial order whose Hasse diagrams are given as:

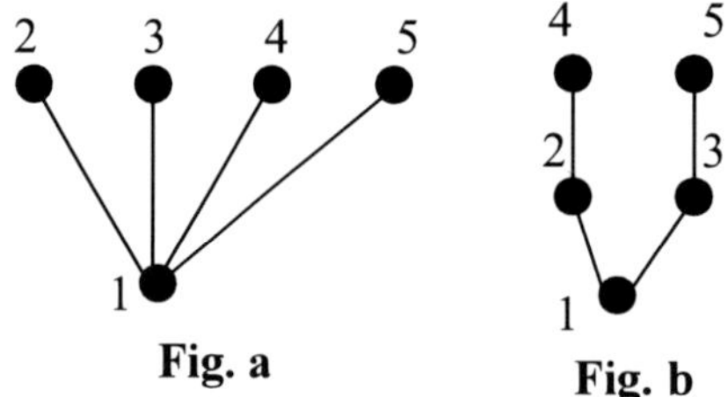


***Solution***

From Fig., Let A = {1, 2, 3, 4, 5}. Matrix of the relation may be determined using the given Hasse diagram and considering the properties of partial ordering relation i.e. reflexive, anti-symmetric and transitive.

| R | 1 | 2 | 3 | 4 | 5 |
|---|---|---|---|---|---|
| 1 | 1 | 1 | 1 | 1 | 1 |
| 2 |   | 1 |   |   |   |
| 3 |   |   | 1 |   |   |
| 4 |   |   |   | 1 |   |
| 5 |   |   |   |   | 1 |

**Relation Matrix for fig. a**

| R | 1 | 2 | 3 | 4 | 5 |
|---|---|---|---|---|---|
| 1 | 1 | 1 | 1 | 1 | 1 |
| 2 |   | 1 |   | 1 |   |
| 3 |   |   | 1 |   | 1 |
| 4 |   |   |   | 1 |   |
| 5 |   |   |   |   | 1 |

**Relation Matrix for fig. b**

**Exercise 4.1**

1. Show that the relation ($\geq$) is a partial ordering on the set of integers.

2. Show that the divisibility relation/is a partial ordering on the set of positive integers.

3. Show that the inclusion relation $\subseteq$ is a partial ordering on the power set of a set S.

4. Let R be a binary relation on the set of all strings of $0_s$ and $1_s$ such that R = {$(a, b)|$ $a$ and $b$ are strings having same number of $0_s$}. Is R a partial order relation?

5. In the poset $(Z^+, /)$, are the integers 3 and 9 comparable? Are 5 and 7 comparable?

6. Draw the directed graph and Hasse diagram of R, where R, $\leq$, is the partial order relation on X = {2, 3, 6, 12, 24}

7. Consider the set A = {4, 5, 6, 7}. Let R be the relation $\leq$ on A. Draw the directed graph and the Hasse diagram of R.

8. Draw the Hasse diagram for the "greater than or equal to' relation on {0, 1, 2, 3, 4, 5}.

9. Let X = {1, 2, 3, 4} then draw the Hasse diagram for poset (P($x$), $\subseteq$).

**Answers to Selected Problems**

4. No, R is not a partial ordering relation.

5. 3 and 9 are comparable, 5 and 7 are non-comparable.

6. 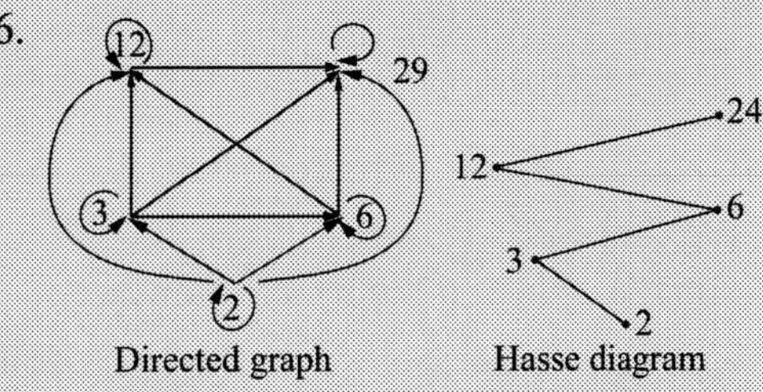
Directed graph     Hasse diagram

8. 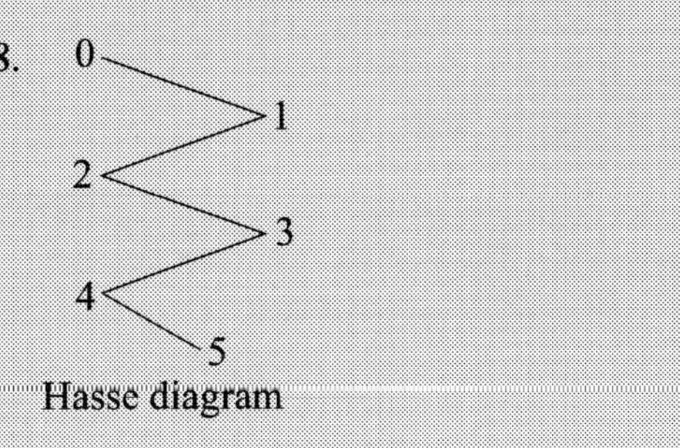
Hasse diagram

## 4.11 MAXIMAL AND MINIMAL ELEMENTS

**Maximal element:** An element $b \in$ A, of a poset (A, R) or (A, $\leq$), is called maximal if for no $a \in$ A, $b \neq a$, $b$ R $a$ or $b \leq a$.

**Minimal element:** An element $b \in$ A, of a poset (A, R), is called minimal if for no $a \in$ A, $b \neq a$, $a$ R $b$ or $a \leq b$.

(i) Maximal and minimal elements are the top and bottom in the Hasse diagram.

(ii) There can be more than one maximal or more than one minimal element.

(iii) A finite non-empty poset always has at least one maximal element and one minimal element

## 4.12 GREATEST AND LEAST ELEMENTS

**Greatest element (last element):** An element $a$ in A is called greatest element of A if and only if for all $b$ in A, $b \leq a$. A greatest element is maximal but $a$ maximal element need not be the greatest element.

**Least element (first element):** An element $a$ in A is called least element of A if and only if for all $b$ in A, $a \leq b$. A least element is minimal but $a$ minimal element need not be $a$ least element.

**Theorem 1** A poset $(A, \leq)$ has at most one greatest element and at most one least element.

### Proof

Let $a$ and $b$ be greatest elements of poset $(A, \leq)$

Since $b$ is the greatest element, $a \leq b$            (i)

Also, since $a$ is the greatest element $b \leq a$        (ii)

From (i) and (ii) $a = b$ by antisymmetry

Hence, the greatest element if exists is unique

By the same argument, it can be proved that least element if exists is unique. Hence $a$ poset has at most one greatest element and at most one least element.

**Example 10:** Let A = {2, 3, 4, 6, 8, 24, 48} be the partially ordered set with the relation R '$x$ divides $y$'. Determine the greatest element, least element and all the maximal and minimal elements of A.

### Solution

R = {(2, 2), (2, 4), (2, 6), (2, 8), (2, 24), (2, 48), (3, 3), (3, 6), (3, 24), (3, 48), (4, 4), (4, 8), (4, 24), (4, 48), (6, 6), (6, 24), (6, 48), (8, 8), (8, 24), (8, 48), (24, 24), (24, 48), (48, 48)}

1. Maximal element is 48 as there exist no element $a \in$ A, such that 48 divides $a$ for $a \neq 48$.
2. Minimal elements are 2 and 3 as there exist no element $a \in$ A, such that $a$ divides 2 for $a \neq 2$, also there exists no such element $a \in$ A, such that $a$ divides 3 for $a \neq 3$, $a \in$ A.
3. Greatest element is 48 as for every element $a \in$ A, $a$ divides 48.
4. There is no least element as there is no such element $a$ such that $a \,/\, b$ for all $b \in$ A.

## 4.13 LOWER BOUND AND UPPER BOUND

**Lower bound of a set:** Let $a, b \in$ S where S is a poset under the relation R then an element $c \in$ S is called lower bound of $a$ and $b$ if $c\,\mathrm{R}\,a$ and $c\,\mathrm{R}\,b$.

**Upper bound of:** Let $a, b \in$ S where S is a poset under the relation R then an element $d \in$ S is called upper bound of $a$ and $b$ if $a\,\mathrm{R}\,d$ and $b\,\mathrm{R}\,d$.

(i) A subset B of poset A, may or may not contain upper bounds or lower bounds.
(ii) An upper bound or lower bound of B may or may not belong to subset B itself

## 4.14 GREATEST LOWER BOUND AND LEAST UPPER BOUND

**Greatest lower bound (glb) or infimum of a set**: Lets $a, b \in$ S where S is a poset under the relation R then an element $g \in$ S is called greatest lower bound of $a$ and $b$ if and only if

(i) $g$ R $a$ and $g$ R $b$ i.e., $g$ is the lower bound of $a$ and $b$.
(ii) If there exits one element $g' \in$ S such that $g'$ is also a lower bound of $a$, $b$ and $g'$ R $g$.

A lower bound of the set that is greater than all other lower bounds. $g$ is called the greatest lower bound of $a$ and $b$.

**Least upper bound (lub) or supremum a set**: Let $a, b \in$ S where S is a poset under the relation R then an element $l \in$ S is called least upper bound of $a$ and $b$ if and only if

(i) $a$ R $l$ and $b$ R $l$ i.e., $l$ is the upper bound of $a$ and $b$.
(ii) If there exists one element $l' \in$ S such that $l'$ is also a upper bound of $a$, $b$ and $l$ R $l'$.

$l$ is called the least upper bound of $a$ and $b$.

**Example 11:** For the poset (X, R), where $x = \{1, 3, 5, 7, 15, 21, 35, 105\}$ and R is the relation '/' divides, find lub of 3 and 7.

We see all $x$ such that $3/x$ and $7/x$. We find the values of $x \in$ X, are 21 and 105. Hence 21 and 105 are upper bounds of 3 and 7. Now 21/105, therefore 21 is lub of 3 and 7.

## 4.15 JOIN AND MEET IN A POSET

Let S be a poset under ordering $\geq$ or R. Let $a, b \in$ S, we define

Join $(a, b) =$ lub of $a$, $b = a \vee b$

Meet $(a, b) = g'b$ of $a$, $b = a \wedge b$.

**Details about a ∨ b and a ∧ b**

(i) If $a \leq (a \vee b)$ and $b \leq (a \vee b)$, then $(a \vee b)$ will be an upper bound of $a$ and $b$.
(ii) If $a \leq c$ and $b \leq c$, then $(a \vee b) \leq c$ and $(a \vee b)$ will be the lub of $a$ and $b$.
(iii) If $(a \vee b) \leq a$ and $(a \vee b) \leq b$, then $(a \vee b)$ will be a lower bound of $a$ and $b$.
(iv) If $c \leq a$ and $c \leq b$, then $c \leq (a \vee b)$ and $(a \vee b)$ will be the glb of $a$ and $b$.
(v) If $a, b, c$ are natural numbers $a \vee b$ is the least common multiple of $a$ and $a \vee b$ is the greatest common multiple of $a$.

**Example 12:** Let X = {1, 3, 5, 7, 15, 21, 35, 105} and let R be the relation '/' (divides) on the set X, and X is the poset. Determine

(a) lub of 3 and 5

(b) Greatest element and least element of X

### *Solution*

To determine the lub of 3 and 5. We look for all $x \in X$ such that $3/x$ and $5/x$ both $x = 15$ and $x = 105$ work, this implies 15 and 105 are the upper bounds [see fig.] of 3 and 5 both.

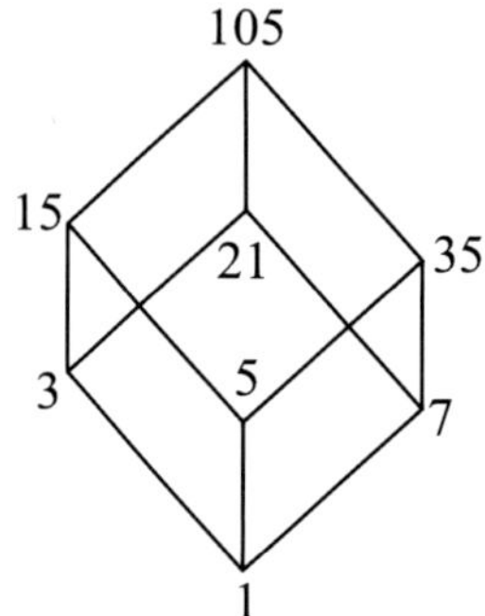


Now 15/105 implies 15 is the lub of (3, 5)

The greatest element of X is 105 as $a/105 \ \forall \ a \in X$

The least element of X is 1 as $1/a \ \forall \ a \in X$

**Example 13:** $D_n = \{x: x / n \ \forall \ x \in N\}$, Consider $D_{30} = \{1, 2, 3, 5, 6, 10, 15, 30\}$

(a) Find all the lower bounds of 10 and 15.

(b) Determine GLB of 10 and 15.

(c) Find all the upper bounds of 10 and 15.

(d) Determine LUB of 10 and 15.

(e) Find greatest element of $D_{30}$.

(f) Find least element of $D_{30}$.

### *Solution*

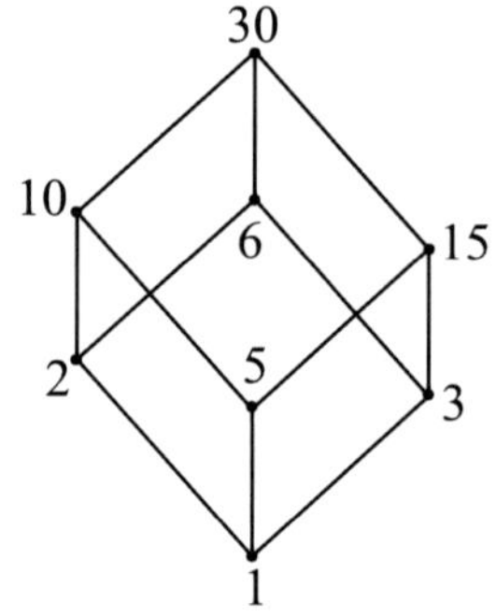


(a) The lower bounds of 10 are 10, 2, 5, 1.

   The lower bounds of 15 are 15, 5, 3, 1.

   Therefore, the lower bounds of 10 & 15 are 5, 1.

(b) The greatest lower bound of 10 & 15 are 5;

(c) Upper bounds of 10 are 10, 30.

   Upper bounds of 15 are 15, 30.

   Therefore, the upper bounds of 10 & 15 are 30.

(d) As there is only one upper bound of 10 and 15 which is the least upper bound.

(e) The greatest element of $D_{30}$ is 30.

(f) The least element of $D_{30}$ is 1

## 4.16 ISOMORPHISM OF ORDERED SETS

A one-to-one function $f: A \rightarrow A'$ is called an isomorphism, where $(A, R)$ and $(A', R')$ are two posets, if for any $a$ and $b$ in A

(i) $a$ R $b$ iff $f(a)$ R$'$ $f(b)$
(ii) If $a$ and $b$ are incomparable then $f(a)$ and $f(b)$ are incomparable.

Obviously if A and A$'$ are linearly ordered then only condition (i) is needed for isomorphism. If $f: A \rightarrow A'$ is isomorphism then A and A$'$ are called **Isomorphic** written as $A \cong A'$.

**Example 14:** If $(A, R)$ be poset of positive integers ordered with $\leq$ and $(A', R')$ be a poset of even positive integers with order $\leq$, then the function $f: A \rightarrow A'$ given by $f(a) = 2a \ \forall \ a \in A$. Then, prove that $f$ is an isomorphism from A to A$'$.

### Solution

$$f(a) = f(b) \Rightarrow 2a = 2b \Rightarrow a = b$$

Hence $f$ is one-to-one

(a) Domain of $f$ is the set of +ve integers and range of $f$ is the set of even +ve integers, Hence $f$ is onto.
(b) $A$ R $b \Leftrightarrow 2a$ R$'$ $2b$

Also if $a$ and $b$ are incomparable then $2a$ and $2b$ are incomparable. Hence $f$ is an isomorphism.

**Theorem 2** Let A and B be order isomorphic partially ordered sets. If A is totally ordered then prove that B is also totally ordered.

### Proof

Let $f: A \rightarrow B$ be the order isomorphism between A and B

$\therefore$ $f$ is bijective and preserves partial order relation.

Let $b_1$ and $b_2$ be any elements of B.

$\therefore$ There exists $a_1$ and $a_2$ in A such that $f(a_1) = b_1$ and $f(a_2) = b_2$. Since A is totally ordered and $a_1, a_2 \in A$, we have $a_1 \leq a_2$ or $a_2 \leq a_1$

$\Rightarrow \quad f(a_1) \leq f(a_2) \quad$ or $\quad f(a_2) \leq f(a_1) \qquad (\therefore \quad f$ is an order isomorphism$)$

$\Rightarrow \quad b_1 \leq b_2 \quad$ or $\quad \quad b_2 \leq b_1$

$\therefore$ Every two elements of B are comparable and hence B is also totally ordered.

**Theorem 3** Let A and B be order isomorphic partially ordered sets. Prove the following:

(i) $a \in A$ is the first element of A iff $f(a)$ is the first element of B.
(ii) $a \in A$ is the last clement of A iff $f(a)$ is the last element of B.

### *Proof*

Let $f: A \to B$ be the order isomorphism between partially ordered sets A and B.

$\therefore$  $f$ is bijective and preserves partial order relation.

(i) Let $a \in A$ be the first element of A. If possible let $f(a)$ be not the first element of B.

$\Rightarrow$  $f(a) \leq b$ for every $b \in B$ is false

$\therefore$  There exists $y \in B$ such that $f(a) \leq y$ is not true.

$\Rightarrow$  $f(a) \leq f(x)$ is not true.   (Taking $f(x) = y$)

$\Rightarrow$  $a \leq x$ is not true for otherwise we would have $f(a) \leq f(x)$.

$\Rightarrow$  $a$ is not the first element of A This is not possible.

$\therefore$  $f(a)$ is the first element of B.

Similarly, the converse holds.

The proof of part (b) is left for the reader as exercises.

**Theorem 4** Let A and B be order isomorphic partially ordered sets. Prove the following:

(a) $a \in A$ is a minimal element of A iff $f(a)$ is a minimal element of B.
(b) $a \in A$ is a maximal element of A iff $f(a)$ is a maximal element of B.

### *Proof*

Let $a \in A$ be a minimal element of A. If possible let $f(a)$ be not a minimal element of B.

$\therefore$  There exists at least one $x \in A$ such that $f(x) \leq f(a)$ and $f(x) \neq f(a)$.

$\Rightarrow$  $x \leq a$ and $x \neq a => x$ cannot be a minimal element of A. This is not possible.

$\therefore$  $f(a)$ is a minimal element of B.

Similarly, the converse holds.

The proof of part (b) is left for the reader as exercises.

**Example 15:**  Let $A = \{a, b, c, d, e\}$ be a partially ordered set with the given Hasse diagram. Let $B = \{1, 2, 3, 4, 5\}$ be order isomorphic to A.

Let $f: A \to B$, defined by $f(a) = 2, f(b) = 1, f(c) = 3, f(d) = 4, f(e) = 5$ be an order isomorphic. Draw the Hasse diagram of the Poset B.

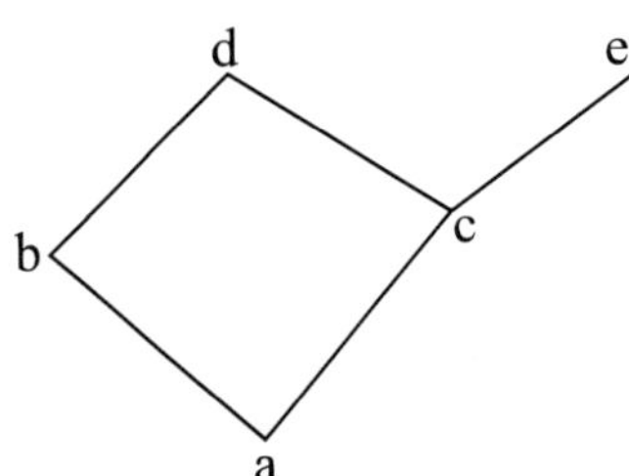

### *Solution*

Using the Hasse diagram of A, we have

$a << b$, $a<<c$, $b << d$, $c << d$ and $c << e$

Since, $f$ is order isomorphism, the only immediate precedence in B are

$f(a) << f(b), f(a) << f(c), f(b) << f(d), f(c) << f(d)$ and $f(c) << f(e)$

i.e. $2 << 1$, $2 << 3$, $1 << 4$, $3 << 4$, $3 << 5$

Using these Hasse diagrams of B can be drawn as below

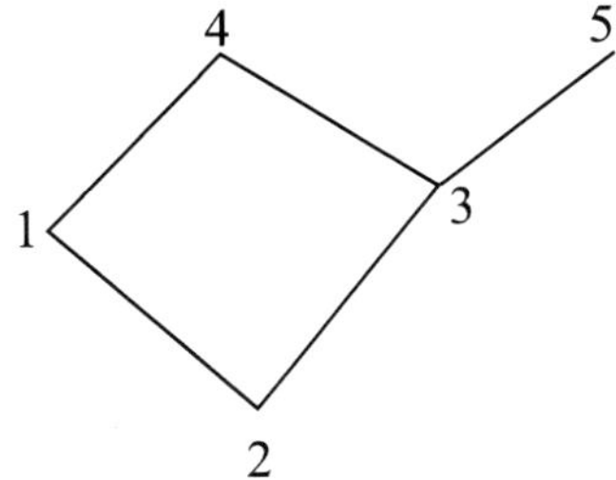

---

### Exercise 4.2

1. For the poset (X, R), where $x = \{1, 3, 5, 7, 15, 21, 35, 105\}$ and R is the relation '/' divides, find glb of 3 and 7.

2. Consider $D_{30} = \{1, 2, 3, 5, 6, 10, 15, 30\}$

   (a) Find all the lower bounds of 10 and 15.
   (b) Determine GLB of 10 and 15.
   (c) Find all the upper bounds of 10 and 15.
   (d) Determine *LUB* of 10 and 15.
   (e) Find greatest element of $D_{30}$.
   (f) Find least element of $D_{30}$.

3. Let $A = \{2, 3, 4, 6, 8, 24, 48\}$ with partial ordering of divisibility. Determine all the maximal and minimal points of A.

4. If $A = \{1, 2, 4, 8\}$ and let $\leq$ be the partial order of divisibility on A. Let $A' = \{0, 1, 2, 3\}$ and let '$\leq$' be the usual relation "less than or equal to" on integers. Show that $(A, \leq)$ and $(A', \leq')$ are isomorphic posets.

5. Let $A = \{a, b, c, d, e\}$ be a partially ordered set with the given Hasse diagram. Let $B = \{1, 2, 3, 4, 5\}$ be order isomorphic to A. Let $f: A \to B$, defined by $f(a) = 1, f(b) = 3, f(c) = 5, f(d) = 2, f(e) = 4$ be an order isomorphic. Draw the Hasse diagram of the Poset B

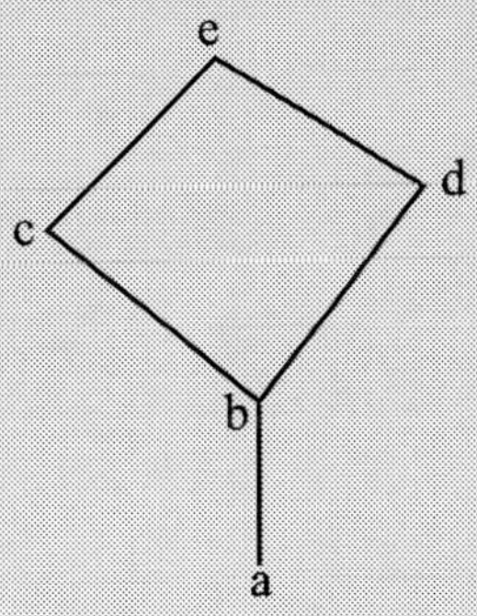

**Answers to Selected Problems**

1. glb of 3 and 7 is 1

2.   (i)  lower bounds of 10 & 15 are 1 and 5
  (ii)  glb of 10 & 15 is 5
  (iii)  upper bounds of 10 & 15 is 30
  (iv)  lub of 10 & 15 is 30
  (v)  greatest element of $D_{30}$ is 30
  (vi)  least element of $D_{30}$ is 1

3. Maximal element is 48
Minimal elements are 2 and 3

4.

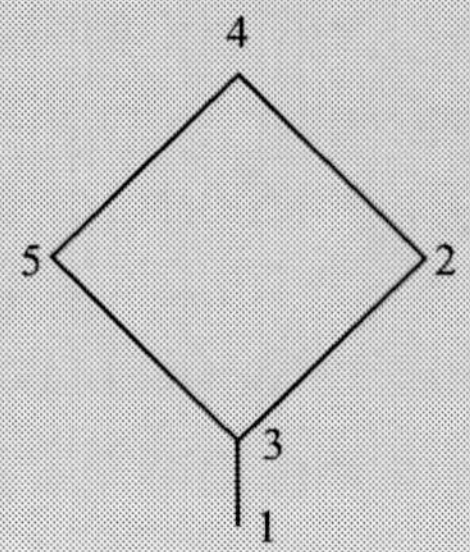

# 5

# Discrete Probability

---

## 5.1 PROBABILITY THEORY

In the random phenomenon, past information no matter how voluminous, will not allow formulating a rule to determine precisely what happens in the future. The theory of probability is the study of such random phenomenon which is not deterministic. Probability theory is a mathematical modelling of the phenomenon of chance. If a coin is tossed in a random manner, it can land heads or tails, but we do not know which of these will occur in a single toss. However, suppose we let $s$ be the number of times head appears when the coin is tossed $n$ times. As $n$ increases, the ratio $f = s / n$, called the relative frequency of the outcome, becomes more stable. If the coin is perfectly balanced, then we expect that the coin will land heads approximately 50% of the time or, in other words, the relative frequency will approach ½. That is, any side of the coin is as likely to occur as the other; hence the chance of getting a head is ½. Although the specific outcome on any one toss is unknown, the behaviour over the long run is determined. This stable long-run behaviour of random phenomena forms the basis of probability theory. A probabilistic mathematical model of random phenomena is defined by assigning "probabilities" to all the possible outcomes of an experiment.

## 5.2 SAMPLE SPACE

Probability theory is rooted in the real life situations where the engineers and scientists perform experiments the outcomes of which may not be certain, these experiments are described as random experiments.

A set S that consists of all possible outcomes of a random experiment is called the **sample space**, each outcome of the experiment is called sample point.

It is useful to think of the outcomes of an experiment. For example, if the experiment consists of examining the state of a single component, it may function properly or not. It has only two possible outcomes. This kind of sample space is one dimensional. In general if the system has $n$ components then there are $2^n$ possible outcomes, each of which can be regarded as a sample point in $n$-dimensional sample space. If the set of all possible outcomes of the experiment is finite, then the associated sample space is known as finite sample space.

**Example:**

1. In tossing a die set of all possible outcomes is given by the sample space
   $S = \{1, 2, 3, 4, 5, 6\}$.
2. In tossing a pair of dice the set of all possible outcomes is given by the sample space

$$S = \{(1, 1), (1, 2), (1, 3), (1, 4), (1, 5), (1, 6),$$
$$(2, 1), (2, 2), (2, 3), (2, 4), (2, 5), (2, 6),$$
$$(3, 1), (3, 2), (3, 3), (3, 4), (3, 5), (3, 6),$$
$$(4, 1), (4, 2), (4, 3), (4, 4), (4, 5), (4, 6),$$
$$(5, 1), (5, 2), (5, 3), (5, 4), (5, 5), (5, 6),$$
$$(6, 1), (6, 2), (6, 3), (6, 4), (6, 5), (6, 6)\}$$

Total number of outcomes are $2^6 = 36$.

3. Tossing of a coin until a head appears and count the number of times coin toss is having a infinite sample space $S = \{1, 2, 3, \ldots\}$ since all positive integers can be element of this set.

## 5.3 EVENTS

An event is simply a collection of certain sample points, that is, a subset A of the sample space S. An event consisting of a single point of S is called a simple or elementary event. The whole sample space S itself is known as a sure or certain event and a empty set $\phi$ is called the impossible event.

By using set operation we can define other events as

1. $A \cup B$ is the event "either A or B or both".
2. $A \cap B$ is the event "A and B".
3. $\overline{A}$ is the event "not A".
4. $A - B = A \cap B'$ is the event "A but not B".

**Example:**

1. Drawing two cards from a pack of well-shuffled cards is an experiment and getting a king and queen is an event.
2. In the tossing of a coin the sample space $S = \{H, T\}$ and the coming out head is an event $A = \{H\}$.
3. In tossing of pair of die the event coming out a doublet then A will be given as
   $A = \{(1, 1), (2, 2), (3, 3), (4, 4), (5, 5), (6, 6)\}$.
4. Let us consider the experiment of tossing a coin twice. Let the event A be the first toss results into head and the event B the second toss results in a head.

$$A = \{HH, HT\} \qquad B = \{HH, TH\}$$
$$A \cup B = \{HH, HT, TH\} \qquad A \cap B = \{HH\}$$
$$\overline{A} = \{TT, TH\} \qquad A - B = \{HT\}$$

5. If someone takes three shots at a target and we care only whether each shot is hit or miss, describe a suitable sample space the element of the sample space that constitute

event M that the person will miss the target three times, and the elements of the sample space that constitute event N that the person will hit the target once and miss it twice.

If we let 0 and 1 represent a miss and a hit respectively, then the sample space is given by

$S = \{ (0,0,0), (0,0,1), (0,1,0), (0,1,1), (1,0,0), (1,0,1), (1,1,0), (1,1,1)\}$ and the event

$M = \{(0,0,0)\}$   and   $N = \{(0,0,1), (1,0,0), (0,1,0)\}$.

### 5.3.1 Exhaustive Events

A system of events is exhaustive if it includes all the possible events.

### Example:

1. In tossing of a coin there are two exhaustive cases, viz., head and tail (the possibility of the coin standing on the edge being ignored).
2. In throwing of a die, there are six exhaustive cases since any one of the 6 faces 1, 2, 3, 4, 5, 6 may turn up.
3. In drawing two cards from a pack of cards the exhaustive number of cases is $^{52}C_2$ ways.
4. In throwing of two dice, the exhaustive number of cases is $6^2 = 36$, since any of the six numbers 1 to 6 on the first die can be associated with any of the six numbers on the other die.

### 5.3.2 Mutually Exclusive Events

A system of events in which the occurrence of one of the events precludes the occurrence of all others is called mutually exclusive or if set corresponding to the events are disjoint, i.e., $A \cap B = \phi$. e.g., in tossing a coin head and tail events are mutually exclusive. Either head occurs or tail occurs and both cannot happen.

### Example:

1. In throwing a die the events of coming six faces numbered 1 to 6 are mutually exclusive since if any one of these faces comes, the possibility of others, in the same trial/experiment, is ruled out.
2. Similarly in tossing a coin the head and tail are mutually exclusive.

### 5.3.3 Equally Likely Events

If one of the events cannot be expected to happen in preference to another then such events are called equally likely.

### Example:

1. In tossing an unbiased or uniform coin, head or tail are equally likely events.
2. In throwing an unbiased die, all the six faces are equally likely to come.

### 5.3.4 Independent Events

Two events are called independent if happening or failure of one does not affect the happening or failure of the other. Events which are not independent are called dependent.

**Example:**

1. In tossing an unbiased coin the event of getting a head in the first toss is independent of getting a head in the second, third and subsequent throws.
2. If we draw a card from a pack of well-shuffled cards and replace it before drawing the second card, the result of the second draw is independent of the first draw. But, however, if the first card is not replaced then the second draw is dependent on the first draw.

## 5.4 PROBABILITY

There are two important ways to calculate the probability of an event.

**Mathematical or classical or a priori probability:**

If there are $n$ mutually exclusive, collectively exhaustive and equally likely cases, out of which $m$ are favourable to an event A. Then

$$P(A) = p = \text{probability of happening of A} = \frac{m}{n} = \frac{\text{Favourable cases for A}}{\text{Total cases}}$$

The probability of non-occurrence of event A is given by

$$P(\overline{A}) = q = \text{probability of not happening of A} = \frac{n-m}{n} = 1 - \frac{m}{n} = 1 - p$$

$$p + q = 1 \quad \text{and} \quad 0 \leq p \leq 1 \quad \text{and} \quad 0 \leq q \leq 1$$

If an event is certain to happen then its probability is 1. If an event is certainly not to happen then its probability is zero.

**Limitation of classical definition:**

The above stated definition of classical probability fails if the outcomes of the experiment are not equally likely or if the exhaustive number of cases in a trail is infinite.

**Statistical or Empirical probability:**

$$p = P(A) = \lim_{n \to \infty} \frac{m}{n}$$

Where $m$ is the number of times event A happens in $n$-trials assuming that the trials are performed under essentially homogeneous and identical conditions.

## 5.5 AXIOM OF PROBABILITY

The theory of probability starts with the assumption that the probability can be assigned so as to satisfy the following three basic axioms of the probability.

Let S be a finite sample space of the random experiment. Let P(A) be the probability associated with the event A. Then the probability must satisfy the following axioms or postulates:

A.1   For any event A, $P(A) \geq 0$.

A.2   $P(S) = 1$.

A.3   $P(A \cup B) = P(A) + P(B)$. Whenever A and B are mutually exclusive events.

$$P\left(\sum_{i=1}^{n} A_i\right) = \sum_{i=1}^{n} P(A_i)$$

### Example 1:

An experiment has 4 possible outcomes, A, B, C and D, which are mutually exclusive. Explain why the following assignments of probabilities are not permissible.

(a)  $P(A) = 0.12$,   $P(B) = 0.63$,   $P(C) = 0.45$,   $P(D) = -0.20$.

(b)  $P(A) = \dfrac{9}{120}$,   $P(B) = \dfrac{45}{120}$,   $P(C) = \dfrac{27}{120}$,   $P(D) = \dfrac{46}{120}$

### Solution

a.  P(D) is –ve, which is not permissible

b.  $P(A) + P(B) + P(C) + P(D) \neq 1$, which is not permissible

### Basic procedure to calculate the probability

1.  Identify the sample space: The sample space must be chosen in such a way so that their entire elements are mutually exclusive and collectively exhaustive.
    e.g. in tossing of two coins the sample space S = {HH, HT, TH, TT}.
2.  Assign the probabilities: In the second step we assign the probability to each sample point strictly inconsistent with our probability axiom. i.e. P(HH) = 1/4, P(HT) = 1/4, P(TH) = 1/4, P(TT) = 1/4.
3.  Identify the event of interest: Let here the desired event be that "at least one head".
4.  Compute the desired probability: Then the desired event is A = {HT, TH, HH} and the associated probability is P(A) = 3/4.

## 5.6 NOTATIONS

P(A) denotes probability of the happening of event A.

$P(\bar{A})$ denotes probability of not happening of the event A.

$P(A + B)$ or $P(A \cup B)$ denotes probability of happening one of the events A and B.

P(AB) or $P(A \cap B)$ denotes probability of happening of both the events A and B.

P(A/B) denotes probability of happening of B, when A has already happened.

**Elementary results for calculating the probability:**

1. For any event A, $P(\bar{A}) = 1 - P(A)$

   ***Proof***

   We know that A and $\bar{A}$ are mutually exclusive events therefore using the A.3

   $P(A) + P(\bar{A}) = 1$ and hence the result.

2. Addition law of probability or theorem of total probability

   $$P(A + B) = P(A) + P(B) - P(A \cap B)$$

   ***Proof***

   Let $n$ be the total number of equally likely cases out of which $m_1$ is favourable to an event A and $m_2$ is favourable to event B. Then

   Number of cases favourable to A or B = $m_1 + m_2$

   Let $m_3$ cases be favourable to both the events A and B. These cases are included in $m_1$ and $m_2$.

   Number of cases favourable to A or B or both = $m_1 + m_2 - m_3$

   $$P(A \cup B) = (m_1 + m_2 - m_3)/n$$
   $$= m_1/n + m_2/n - m_3/n$$
   $$= P(A) + P(B) - P(A \cap B)$$

   In case A and B are mutually exclusive $P(A \cap B)$ is zero and

   $$P(A \cup B) = P(A) + P(B)$$

   In general if $A_1, A_2,..., A_n$ are mutually exclusive events. Then

   $$P(A_1 \cup A_2 \cup ... \cup A_n) = P(A_1) + P(A_2) + \cdots + P(A_n)$$

2. If $B \subset A$, then $P(B) \leq P(A)$

   ***Proof***

   $A = B \cup (A \cap \bar{B})$, B and $A \cap \bar{B}$ are mutually exclusive.

   So $P(A) = P(B) + P(A \cap \bar{B})$. Since $P(A \cap \bar{B}) \geq 1$

   $\Rightarrow P(B) \leq P(A)$

**Example 1:** Find the chance of throwing 2 with an ordinary six faces dice.

***Solution***

$n$ = total ways = 6
$m$ = favourable cases of throwing 2 = 1
$p = 1/6$

**Example 2:** A single die is tossed once. Find the chance of getting 2 or 5.

*Solution*

$S = \{1, 2, 3, 4, 5, 6\}$

If we assign equal possibilities to the sample points, i.e., if we assume that the die is fair, then

$$P(1) = P(2) = P(3) = P(4) = P(5) = P(6) = \frac{1}{6}$$

$$P(2 \cup 5) = P(2) + P(5) = \frac{1}{6} + \frac{1}{6} = \frac{1}{3}.$$

**Example 3:** What is the chance that a leap year selected at random will contain 53 Sundays?

*Solution*

The number of days in a leap year = 366

The number of complete weeks are 52 and there are 2 remaining days.

The possible combinations for these two days are given as:

(1) Sunday and Monday      (2) Monday and Tuesday      (3) Tuesday and Wednesday
(4) Wednesday and Thursday      (5) Thursday and Friday      (6) Friday and Saturday
(7) Saturday and Sunday.

Therefore, out of these seven possibilities defining above only 2 are favourable to get 53 Sundays i.e. (1) and (7)

$\therefore$     Probability (53 Sundays) = 2/7.

**Example 4:** A committee consists of 9 students. Two of them are from the first semester, three from the second semester, and four from the third semester. Three students are to be removed at random. Find the probability that the three students belong to different classes.

*Solution*

$n$ = total ways = number of ways of choosing 3 students out of 9

$$= {}^{9}C_3 = 84$$

$m$ = favourable cases that one student is from the first semester, one student from the second semester and one student from the third semester

$$= 2 \times 3 \times 4 = 24$$

$$p = m/n = 24/84 = 2/7$$

**Example 5:** Find the probability of drawing an ace or a spade or both from a pack of cards.

### *Solution*

Card being an ace and a card being a spade are not mutually exclusive.

Let the event of getting an ace be A and the event of getting a spade be B.

$P(A) = 4/52 = 1/13$
$P(B) = 13/52 = 1/4$

$P(A \cap B) =$ Probability of getting an ace of spade $= 1/52$

$$P(A \cup B) = P(A) + P(B) - P(A \cap B)$$
$$= 1/13 + 1/4 - 1/52 = 4/13$$

**Example 6:** A problem is given to three students A, B and C whose chances of solving it are $\frac{1}{2}, \frac{1}{3}, \frac{1}{4}$. Find the probability that the problem is solved.

### *Solution*

Probability that A can solve is $\frac{1}{2}$

Probability that A cannot solve $= 1 - \frac{1}{2} = \frac{1}{2}$

Similarly,

Probability that B cannot solve $= 1 - \frac{1}{3} = \frac{2}{3}$

Probability that C cannot solve $= 1 - \frac{1}{4} = \frac{3}{4}$

Probability that A, B and C cannot solve the problem $= \frac{1}{2} \times \frac{2}{3} \times \frac{3}{4}$
$$= 6/24 = 1/4$$

i.e. $P(A \cap B \cap C) = P(A) \cdot P(B) \cdot P(C)$ since A, B and C are independent events.

Probability that the problem is solved $= 1 - \frac{1}{4} = \frac{3}{4}$

**Example 7:** If a person visits his dentist, suppose that the probability that he will have the teeth cleaned is 0.44, the probability that he will have a cavity filled is 0.24, the probability that he will have a tooth extracted is 0.21, the probability that he will have tooth cleaned and cavity filled is 0.08, the probability that he will have his teeth cleaned and a tooth extracted is 0.11, the probability that he will have cavity filled and a tooth extracted is 0.07 and the probability that he will have his tooth cleaned, a cavity filled, and a tooth extracted is 0.03. What is the probability that a person visiting his dentist will have at least on of these things done to him.

### *Solution*

Let C be the event that the person will have his tooth cleaned.

F is the event that he will have a cavity filled.

E is the event that he will have a tooth extracted

so we are given

$$P(C) = 0.44, \quad P(F) = 0.24, \quad P(E) = 0.21, \quad P(C \cap F) = 0.08$$

$$P(C \cap E) = 0.11, \quad P(F \cap E) = 0.07, \quad \text{and} \quad P(C \cap F \cap E) = 0.03.$$

$$P(C \cup F \cup E) = 0.44 + 0.24 + 0.21 - 0.08 - 0.11 - 0.07 + 0.03$$
$$= 0.66.$$

## 5.7 CONDITIONAL PROBABILITY

If A and B are any two events in a sample space S and probability $P(A) \neq 0$, the conditional probability of B given A is

$$P\left(\frac{B}{A}\right) = \frac{P(A \cap B)}{P(A)}$$ and if $P(B) \neq 0$, the conditional probability of A given B is

$$P\left(\frac{A}{B}\right) = \frac{P(A \cap B)}{P(B)}.$$

## 5.8 MULTIPLICATION LAW OF PROBABILITY OR THEOREM OF COMPOUND PROBABILITY

For two events A and B

$$P(A \cap B) = \begin{cases} P(A) \cdot P\left(\dfrac{B}{A}\right), & \text{if } P(A) > 0 \\ \\ P(B) \cdot P\left(\dfrac{A}{B}\right), & \text{if } P(B) > 0 \end{cases} \tag{1}$$

**For independent events:**

If A and B are independent then

$P(A/B) = P(A)$ and $P(B/A) = P(B)$

Hence from (1) we have $P(A \cap B) = P(A)P(B)$.

**Extension of multiplication law of probability:**

For $n$ events $A_1, A_2, A_3, ..., A_n$, we have

$P(A_1 \cap A_2 \cap A_3 \cap ... A_n)$
$$= P(A_1)P(A_2 \mid A_1)P(A_3 \mid A_1 \cap A_2) ... P(A_n \mid A_1 \cap A_2 \cap A_3 \cap ... A_{n-1})$$

where $P(A_i \mid A_j \cap A_k \cap ... A_l)$ represents the conditional probability of the event $A_i$ given that $A_j, A_k, ... A_l$ have already happened.

**Example 8:** A manufacturer of airplane parts knows from the past experience that the probability is 0.80 that an order will be ready for shipment in time, and it is 0.72 that an order will be ready for shipment in time and will also be delivered in time. What is the probability that such an order will be delivered in time given that it was ready for shipment in time?

### Solution

If we let R stand for the event that an order is ready for shipment in time and D be the event that is delivered in time, we have $P(R) = 0.80$ and $P(R \cap D) = 0.72$, and it follows that

$$P\left(\frac{D}{R}\right) = \frac{P(R \cap D)}{P(R)} = \frac{0.72}{0.80} = 0.90$$

**Example 9:** A box of fuses contains 20 fuses, of which five are defective. If three of the fuses are selected at random and removed from the box in succession without replacement, what is the probability that all three fuses are defective?

### Solution

If A is the event that the first fuse is defective, B is the event that the second fuse is defective, and C is the event that the third fuse is defective.

$$P(A) = 5/20 \quad P(B/A) = 4/19 \quad P(C/(A \cap B)) = 3/18$$

Probability that all the three fuses are defective is given as under

$$P(A \cap B \cap C) = P(A) \cdot P(B/A) \cdot P(C/(A \cap B)) = 5/20 \times 4/19 \times 3/18 = 1/114.$$

(Using the theorem of compound probability).

**Example 10:** An unbiased coin is tossed three times. If A is the event that a head occurs on each of the first two tosses, B is the event that a tail occurs on third toss, and C is the event that exactly two tails occur in the three tosses, show that

(a)  Event A and B are independent.
(b)  Event B and C are dependent.

### Solution

$S = \{HHH, HHT, HTH, HTT, THH, THT, TTH, TTT\}$

$A = \{HHH, HHT\}$

$B = \{HHT, HTT, THT, TTT\}$

$C = \{HTT, THT, TTH\}$

$A \cap B = \{HHT\}$

$B \cap C = \{HTT, THT\}$

$\therefore \quad P(A) = 2/8 = 1/4$

P(B) = 4/8 = 1/2

P(C) = 3/8,

$$P(A \cap B) = 1/8 = P(A) \cdot P(B),$$

And $P(B \cap C) = 1/4 \neq P(B) \cdot P(C)$

So it is evident from the above results that A and B are independent and B and C are dependent.

**Example 11:** A die is rolled in such a way that each odd number is twice as likely to occur as each even number. What is the probability that the number of points rolled is a perfect square? Also what is the probability that it is a perfect square given that it is greater than 3?

*Solution*

The sample space is S = {1, 2, 3, 4, 5, 6}.

Let $x$ be the probability assigned to each even number then $2x$ be the probability for odd number. Then according to the postulate A.2

$$P(S) = P(1) + P(2) + P(3) + P(4) + P(5) + P(6) = 2x + x + 2x + x + 2x + x = 9x = 1$$

$\therefore$ $x = 1/9$

$\therefore$ $P(1) = P(3) = P(5) = 2/9$ and $P(2) = P(4) = P(6) = 1/9$

If A is the event that the number of points rolled is greater than 3 and B is the event that it is a perfect square, we have A = {4, 5, 6} B = {1, 4} and $A \cap B = \{4\}$.

P(B) = 2/9 + 1/9 = 1/3

P(A) = 1/9 + 2/9 + 1/9 = 4/9

$P(A \cap B) = 1/9$

$$\therefore \quad P(B/A) = \frac{P(A \cap B)}{P(A)} = \frac{1/9}{4/9} = 1/4.$$

**Example 12:** A bag contains 10 gold and 8 silver coins. Two successive drawings of 4 coins are made such that: (i) coins are replaced before the second trial, (ii) the coins are not replaced before the second trial. Find the probability that the first drawing will give 4 gold and the second 4 silver coins.

*Solution*

Let A denote the event of drawing 4 gold coins in the first draw and B denote the drawing 4 silver coins in the second draw. Then we have to find the probability of $P(A \cap B)$.

(i) (*with replacement*) If the coins drawn in the first draw are replaced back in the bag before the second draw then the events A and B are independent and the required probability is given

Total number of coins = 18

Let A be event of drawing 4 gold coin then $P(A) = \dfrac{^{10}C_4}{^{18}C_4}$ and

Let B be the event of drawing 4 silver coins $P(B) = \dfrac{^{8}C_4}{^{18}C_4}$

Since we are replacing the coin before making the second draw therefore the drawing 4 silver coins in the second draw is independent to the first draw.

$$\therefore \qquad P(A \cap B) = P(A) \cdot P(B) = \left( \dfrac{^{10}C_4}{^{18}C_4} \right) \cdot \left( \dfrac{^{8}C_4}{^{18}C_4} \right)$$

(ii) (*without replacement*) If the coins drawn are not replaced back before the second draw, then the events A and B are not independent and the required probability is given by

$$P(A \cap B) = P(A) \cdot P\left( \dfrac{B}{A} \right)$$

$$P\left( \dfrac{B}{A} \right) = \dfrac{^{8}C_4}{^{14}C_4}$$

$$P(A \cap B) = \left( \dfrac{^{10}C_4}{^{18}C_4} \right) \cdot \left( \dfrac{^{8}C_4}{^{14}C_4} \right)$$

## 5.9 BAYE'S THEOREM

If $B_1, B_2, \ldots B_n$ are mutually disjoint events with probability $P(B_i) \neq 0$, $(i = 1, 2, \ldots n)$ then for any arbitrary event A which is a subset of $\bigcup\limits_{i=1}^{n} B_i$ such that $P(A) > 0$, we have

$$P\left( \dfrac{B_i}{A} \right) = \dfrac{P(B_i) P\left( \dfrac{A}{B_i} \right)}{\sum\limits_{i=1}^{n} P(B_i) P\left( \dfrac{A}{B_i} \right)}, \; i = 1, 2, \ldots n.$$

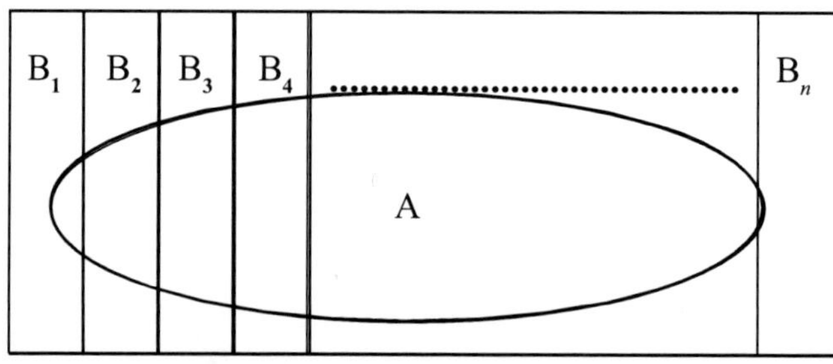

## *Proof*

Since $A \subset \bigcup\limits_{i=1}^{n} B_i$, we have

$$A = A \cap \left( \bigcup_{i=1}^{n} B_i \right) = \bigcup_{i=1}^{n} (A \cap B_i) \quad \text{by distributive law}$$

Since $(A \cap B_i) \subset B_i$ are mutually disjoint events therefore using the postulate A.3

$$P(A) = P\left[ \bigcup_{i=1}^{n} (A \cap B_i) \right] = \sum_{i=1}^{n} P(A \cap B_i) = \sum_{i=1}^{n} P(B_i) \cdot P\left( \frac{A}{B_i} \right)$$

by compound theorem of probability.

*Note:* The probabilities $P(B_i)$, $i = 1.2, \ldots, n$ are called a priori probabilities, while $P(A/B_i)$ are called posterior probabilities.

**Example 13:** In a bolt factory machines A, B and C manufacture respectively 25%, 35% and 40% of the total. Of their output 5%, 4%, 2% are defective bolts. A bolt is drawn at random from the product and is found to be defective. What are the probabilities that it was manufactured by machines A, B and C?

## *Solution*

Let $B_1$, $B_2$ and $B_3$ denote the events that a bolt selected at random is manufactured by the machines A, B and C respectively and let A denote the event of its being defective. Then we have

$$P(B_1) = 0.25, \quad P(B_2) = 0.35, \quad P(B_3) = 0.40$$

The probability of drawing A defective bolt manufactured by machine A is $P(A/B_1) = 0.05$. Similarly, we have $P(A/B_2) = 0.04, \quad P(A/B_3) = 0.02$

$$\therefore \quad P\left( \frac{B_1}{A} \right) = \frac{P(B_1) P\left( \dfrac{A}{B_1} \right)}{\sum\limits_{i=1}^{3} P(B_i) P\left( \dfrac{A}{B_i} \right)}$$

$$= \frac{0.25 \times 0.05}{0.25 \times 0.05 + 0.35 \times 0.04 + 0.40 \times 0.02} = \frac{25}{69}$$

similarly

$$P\left(\frac{B_2}{A}\right) = \frac{0.35 \times 0.04}{0.25 \times 0.05 + 0.35 \times 0.04 + 0.40 \times 0.02} = \frac{28}{69}$$

$$P\left(\frac{B_3}{A}\right) = \frac{0.40 \times 0.02}{0.25 \times 0.05 + 0.35 \times 0.04 + 0.40 \times 0.02} = \frac{16}{69}$$

**Example 14:** The members of a consulting firm rent cars from three agencies: 60% from 1st agency, 30% from 2nd agency and 10% from 3rd agency. If 9% of the cars from 1st agency need a tune up, 20% of the cars from 2nd agency need a tune up and 6% of the cars from 3rd agency need a tune up, what is the probability that a car delivered to the firm will need a tune up? If a car delivered to the firm needs a tune up, what is the probability that it came from 2nd agency?

### *Solution*

Let A be the event that the car needs a tune up and $B_1$, $B_2$ and $B_3$ are the events that the cars come from 1st, 2nd or 3rd agencies

$$P(B_1) = 0.60, \quad P(B_2) = 0.30, \quad P(B_3) = 0.10$$
$$P(A/B_1) = 0.09, \quad P(A/B_2) = 0.20, \quad P(A/B_3) = 0.06.$$

$$\therefore \quad P(A) = 0.60 \times 0.09 + 0.30 \times 0.20 + 0.10 \times 0.06$$
$$= 0.12$$

12% of all the car delivered to this firm will need a tune up.

and   $P(B_2/A) = (0.30 \times 0.20)/0.12 = 0.5.$

Observe that out of 30% of the cars come from 2nd agency, 50% of those requires a tune up.

**Example 15:** Three machines A, B, and C produce identical items. Of their respective outputs 5%, 4% and 3% are faulty. On a certain day A produced 25% of the total output, B produced 30% and C the remainder. An item selected at random is found to be faulty. What is the chance that it was produced by the machine with the highest output?

### *Solution*

Let E be the event that the selected item is found to be faulty.

|  | **A** | **B** | **C (highest production)** |  |
| --- | --- | --- | --- | --- |
| $P(B_i)$ | 0.25 | 0.30 | 0.45 |  |
| $P(E/B_i)$ | 0.05 | 0.04 | 0.03 |  |
| $P(B_i)\,P(E/B_i)$ | 0.0125 | 0.012 | 0.0135 | Sum = 0.038 |
| $P(B_i/E)$ |  |  | 0.0135/0.038 = 0.355 | By Baye's theorem |

## 5.10 RANDOM VARIABLE

If the values of a real variable, associated with a random experiment, depend on chance it is called a random variable or a stochastic variable or simply a variate. If the variate takes a finite set of values it is called a discrete variate. If the variate takes an infinite number of values it is called a continuous variate.

Random variable is a real number X connected with the outcome of a random experiment S. For example, in tossing of two coins the values of the random variable associated with the number of heads is given by

$$S = \{HH, HT, TH, TT\} \quad \text{and} \quad X = (0, 1, 2).$$

## 5.11 DISCRETE PROBABILITY DISTRIBUTION

Let a discrete variate X take values $x_1, x_2, x_3, \ldots, x_n$ with probabilities $p_1, p_2, \ldots, p_n$, where

(i) $P(X = x_i) = p(x_i) \geq 0$ for all values of $i$.
(ii) $\Sigma p(x_i) = 1$

The set of values $x_i$ with their probabilities $p_i$ constitute a probability distribution of the discrete variate X.

This function $p$ is called the probability mass function (pmf) of the random variable X.

## 5.12 DISTRIBUTION FUNCTION

The function $F(x)$ defined by

$F(x) = P(X \leq x)$ is called distribution function. The distribution function is also known as **cumulative distribution function**.

$$F(x_i) = P(X \leq x_i) = \Sigma p(x_i), i = 1, 2, \ldots, n$$

**Example 16:** Find the distribution function of the total number of heads obtained in four tosses of a balanced coin.

### *Solution*

Let $x$ be the corresponding values of the random variable X, the total number of heads.

| Outcome of the sample space | Probability $p(x)$ | $x$ |
|---|---|---|
| HHHH | 1/16 | 4 |
| HHHT | 1/16 | 3 |
| HHTH | 1/16 | 3 |
| HHTT | 1/16 | 2 |
| HTHH | 1/16 | 3 |

| | | |
|---|---|---|
| HTHT | 1/16 | 2 |
| HTTH | 1/16 | 2 |
| HTTT | 1/16 | 1 |
| THHH | 1/16 | 3 |
| THHT | 1/16 | 2 |
| THTH | 1/16 | 2 |
| THTT | 1/16 | 1 |
| TTHH | 1/16 | 2 |
| TTHT | 1/16 | 1 |
| TTTH | 1/16 | 1 |
| TTTT | 1/16 | 0 |

p(0) = 1/16
p(1) = 1/16 + 1/16 + 1/16 + 1/16 = 4/16
p(2) = 1/16 + 1/16 + 1/16 + 1/16 + 1/16 + 1/16 = 6/16
p(3) = 1/16 + 1/16 + 1/16 + 1/16 = 4/16
p(4) = 1/16

So the distribution function is given by:

F(0) = P (X ≤ 0) = p(0) = 1/16
F(1) = P (X ≤ 1) = p(0) + p(1) = 5/16
F(2) = P (X ≤ 2) = p(0) + p(1) + p(2) = 11/16
F(3) = P (X ≤ 3) = p(0) + p(1) + p(2) + p(3) = 15/16
F(4) = P (X ≤ 3) = p(0) + p(1) + p(2) + p(3) + p(4) = 1.

**Example 17:** The probability mass function of a variate X is

| X | 0 | 1 | 2 | 3 | 4 | 5 | 6 |
|---|---|---|---|---|---|---|---|
| P(X) | $k$ | $3k$ | $5k$ | $7k$ | $9k$ | $11k$ | $13k$ |

Find P(X < 4).

### *Solution:*

$\sum p(X) = k + 3k + 5k + 7k + 9k + 11k + 13k = 49k = 1$

$k = 1/49$

$P(X < 4) = k + 3k + 5k + 7k = 16k = 16 \cdot (1/49) = 16/49$

## 5.13 BERNOULLI DISTRIBUTION

A random variable X which takes two values 0 and 1, with probabilities $q$ and $p$ respectively, i.e., P(X = 1) = $p$, P(X = 0) = $q$, $q = 1 - p$ is called a Bernoulli variate and is said to have Bernoulli distribution.

## 5.14 REPEATED TRIALS

If $p$ is the probability of success and $q$ is the probability of failure in a trial and the trial is repeated $n$ times, then the probability of $r$ successes and $n - r$ failures is $p^r q^{(n-r)}$. But the $r$ successes and $n - r$ failures can occur in $^nC_r$ ways in each of which the probability is the same.

Hence the probability of r success is $^nC_r p^r q^{(n-r)}$.

## 5.15 BINOMIAL DISTRIBUTION

It is concerned with trials of repetitive nature in which the occurrence or non-occurrence of a particular event, success or failure, is of interest. If in a series of $n$ independent trials such that for each trial $p$ is the probability of success and $q$ that of a failure, then the probability of $r$ successes in $n$ trials is given by

$$P(X = r) = {}^nC_r\, p^r q^{(n-r)} \text{ where } r \text{ takes any integral value from 0 to } n.$$

The two independent constants $n$ and $p$ in this distribution are known as the parameters of the distribution.

The probability distribution is

| r | 0 | 1 | 2 | .. | r | .. | n |
|---|---|---|---|---|---|---|---|
| $p(r)$ | $p^n$ | $^nC_1 pq^{(n-1)}$ | $^nC_2 p^2 q^{(n-2)}$ | .. | $^nC_r\, p^r q^{(n-r)}$ | .. | $p^n$ |

The probability distribution is called binomial distribution because the probabilities are the successive terms in the expansion of the binomial $(q + p)^n$.

$$\text{Sum of probabilities} = \sum {}^nC_r\, p^r\, q^{(n-r)}$$
$$= (q + p)^n = 1$$

### Conditions for Binomial Distribution

1. Each trial results in two mutually disjoint outcomes known as success and failure.
2. The number of trial $n$ is finite.
3. The trails are independent of each other.
4. The probability of success $p$ is constant for each trial.

### 5.15.1 Applications of Binomial Distribution

It is applied to the problems concerning

   (i) Number of defectives in a sample
   (ii) Estimation of reliability of systems
   (iii) Number of rounds fired from a gun hitting a target
   (iv) Radar detection

## 5.15.2 Mean of a Binomial Distribution

$$\text{Mean} = \mu = \sum r \cdot p(r) = \sum r \cdot {}^nC_r\, p^r q^{(n-r)}$$

$$= \sum \frac{r \cdot n!\, p^r q^{(n-r)}}{r!(n-r)!}$$

$$= np \sum \frac{(n-1)!\, p^{r-1} q^{(n-r)}}{(r-1)!(n-r)}$$

$$= np \sum {}^{(n-1)}C_{(r-1)}\, p^{(r-1)} q^{(n-r)} = np\,(q+p)^{(n-1)}$$

$$= np(1)^{(n-1)} = np$$

The mean value is commonly known as expectation and is denoted by $E(x)$

## 5.15.3 Variance and Standard Deviation of a Binomial Distribution

$$\textbf{Variance} = \sigma^2 = \sum (r-\mu)^2\, p(r) = \sum (r-np)^2\, {}^nC_r p^r q^{(n-r)}$$

$$= \sum (r^2 + n^2p^2 - 2rnp)\, {}^nC_r p^r q^{(n-r)}$$

$$= \sum (r(r-1) + r + n^2p^2 - 2rnp)\, {}^nC_r p^r q^{(n-r)}$$

$$= \sum (r(r-1)\, {}^nC_r p^r q^{(n-r)} + \sum r\, {}^nC_r p^r q^{(n-r)} + n^2p^2 \sum {}^nC_r p^r q^{(n-r)} - 2np \sum r\, {}^nC_r\, p^r q^{(n-r)}$$

$$= n(n-1)p^2\, (q+p)^{(n-2)} + np + n^2p^2 - 2np + np$$

$$= n(n-1)p^2 + np + n^2p^2 - 2n^2p^2$$

$$= -np^2 + np = np(1-p)$$

$$= npq$$

$$\textbf{Standard Deviation} = \sigma = \sqrt{npq}$$

## 5.16 BINOMIAL FREQUENCY DISTRIBUTION

If $n$ independent trials constitute one experiment and the experiment is repeated $n$ times, then the frequency of $r$ successes is $r.\ {}^nC_r p r q^{(n-r)}$.

The possible number of successes together with their respective frequencies constitutes the Binomial frequency distribution.

**Example 18:** The probability that a fan manufactured by a company will be defective is 1/100. If 10 fans are manufactured find that none will be defective.

### *Solution*

$n = 10$

$p$ = probability of a defective fan = $1/100 = 0.01$

$q = 1 - p = 0.99$

Probability of $r$ defective $= P(X = r) = {}^{10}C_r \, p^r \, {}^{(10-r)}$

Probability of zero defective $= P(X = 0) = {}^{10}C_0 \, p^0 q^{10}$
$$= q^{10}$$
$$= (0.99)^{10}$$

**Example 19:** In a company manufacturing fans, the mean number of defectives in a sample of 20 is 2. Out of 1000 such samples how many would be expected to contain at least 3 defective fans.

***Solution***

It is binomial distribution

$n = 20$

Mean $= np = 2 = 20p$

$P = 2/20 = 1/10 = $ probability that a fan is defective

$q = 1 - p = 9/10 = $ probability that a fan is non-defective

Probability of at least 3 defective fans in 20

$$= 1 - \text{probability of none, or one, or two non-defective fans}$$
$$= 1 - [(0.9)^{20} + 20(0.1)(0.9)^{19} + {}^{20}C_2 \cdot (0.1)^2 (0.9)^{18}]$$
$$= 0.323$$

Number of samples having at least 3 defective fans out of 1000 samples
$$= 1000 \times 0.323 = 323 = 100 \times 0.323 = 32.3$$

**Example 20:** Ten coins are tossed simultaneously. Find the probability of getting at least seven heads.

***Solution***

$p = $ probability of getting a head $= 1/2$

$q = $ probability of not getting a head $= 1/2$

The probability of getting $x$ heads in a random throws of 10 coins is

$$p(x) = \binom{10}{x}\left(\frac{1}{2}\right)^x \left(\frac{1}{2}\right)^{10-x} = \binom{10}{x}\left(\frac{1}{2}\right)^{10}; \quad x = 0, 1, 2, \ldots, 10$$

$\therefore$ Probability of getting at least seven heads is given by

$$P(X \geq 7) = p(7) + p(8) + p(9) + p(10)$$

$$= \left(\frac{1}{2}\right)^{10} \left\{ \binom{10}{7} + \binom{10}{8} + \binom{10}{9} + \binom{10}{10} \right\}$$

$$= \frac{120 + 45 + 10 + 1}{1024} = \frac{176}{1024}$$

**Example 21:** An irregular six faced die is thrown and the expectation that in 10 throws it will give five even numbers is twice the expectation that it will give four even numbers. How many times in 10,000 sets of 10 throws each, would you expect it to give no even numbers.

*Solution*

Let $p$ be the probability of getting an even number in a throw of a die. Then the probability of getting $x$ even numbers in 10 throws of a die is

$$P(X = x) = \binom{10}{x} p^x q^{10-x}; \; x = 0, 1, 2, \ldots 10$$

We are given that

$$P(X = 5) = 2P(X = 4)$$

i.e., $$\binom{10}{x} p^5 q^5 = 2 \binom{10}{x} p^4 q^6$$

$$\Rightarrow \quad \frac{10! \, p}{5!5!} = 2 \frac{10! \, q}{4!6!}$$

$$\Rightarrow \quad \frac{p}{5} = \frac{2q}{6} = \frac{q}{3}$$

$\therefore \quad 3p = 5q = 5(1-p) \quad \Rightarrow \quad 8p = 5 \quad \Rightarrow \quad p = 5/8 \quad$ and $\quad q = 3/8$

$\therefore \quad P(X = x) = \binom{10}{x} \left(\frac{5}{8}\right)^x \left(\frac{3}{8}\right)^{10-x}$

Hence the required number of times that in 10,000 sets of 10 throws each, we get no even number

$$= 10,000 \times P(X = 0) = 10,000 \times \left(\frac{3}{8}\right)^{10}$$

**Example 22:** In a bow and arrow game, the probability of hitting a target by an arrow is 0.3. Find the number of arrows that should be targeted so that at least 80% probability of hitting the target.

### *Solution*

The probability of hitting a target is $p = 0.3$

The probability of missing a target is $q = 1 - p = 1 - 0.3 = 0.7$

let $n$ be the number of arrows required to hit the target

Probability of $n$ arrows missing the target is $= (0.7)^n$

Thus in order to find out the minimum of $n$ in hitting the target with probability of 80% we have

$$1 - (0.7)^n > 0.8 \text{ i.e. } (0.7)^n < 0.2 \text{ so by hit and trial } n = 5.$$

---

### Exercise 5.1

1. A bag contains 3 red, 6 white and 7 blue balls. What is the probability that two balls drawn are white and blue?

   **Answer:** 7/20

2. (a) Two cards are drawn at random from a well shuffled pack of 52 cards. Show that the chance of drawing two aces is 1/221.

   (b) From a pack of 52 cards, three are drawn at random. Find the chance that they are a King, a Queen and a Knave.

   (c) Four cards are drawn from a pack of cards. Find the probability that (i) All are Diamonds. (ii) There is one card of each suit. (iii) There are two spades and two Hearts.

   **Answer:** (b) 16/5525  (c) (i) 11/4165 (ii) 2197/20825 (iii) 468/20825.

3. Out of $(2n + 1)$ tickets consecutively numbered three are drawn at random. Find the chance that numbers on them are in A.P.

   **Answer:** $\dfrac{3n}{4n^2 - 1}$.

4. A committee consists of 9 students. Two of them are from the first year, three from the second year, and four from the third year. Three students are to be removed at random. Find the probability that

   (i) Two students belong to the same class and third to different class.

   (ii) The three students belong to the same class.

5. Find the probability of drawing an ace or a diamond or both from a pack of cards.

6. Three machines X, Y, and Z produce identical items. Of their respective outputs 5%, 4% and 3% are faulty. On a certain day A produced 25% of the total output, B produced 30% and C the remainder. An item selected at random is found to be faulty. What is the chance that it was produced by the machine with the highest output?

7. A five figure number is formed by the digits 0, 1, 2, 3, 4 without repetition. Find the probability that the number formed is divisible by 4.

8. X has one share in a lottery in which there is 1 prize and 2 blanks, Y has 3 shares in a lottery in which there are 3 prizes and 6 blanks. Compare the probability of X's success to that of Y's success.

9. A bag contains 8 white and 6 red balls. Find the probability of drawing two balls of the same colour.

10. Two cards are drawn in succession from a pack of 52 cards. Find the chance that the first is a king and the second a queen if the first card is (i) replaced (ii) not replaced.

11. A pair of dice is tossed twice. Find the chance of scoring 7 (i) once (ii) at least once (iii) twice.

12. A box contains 2 white and 4 black balls. Another box B contains 5 white and 7 black balls. A ball is transferred from the box A to the box B. Then a ball is drawn from the box B. Find the probability that it is white.

13. The probability that a pen manufactured by a company will be defective is 1/10. If 12 such pens are manufactured, find the chance that

    (i) exactly two are defective

    (ii) at least two are defective

    (iii) none defective.

14. In sampling a large number of parts manufactured by a company, the mean number of defectives in a sample of 20 is 2. Out of 1000 such samples, how many would be expected to contain at least three defective parts.

15. What is the probability that a non-leap year will have 53 Sundays?

16. What is the chance of 4 turning up at least once in two tosses of a fair dice?

17. A coin is tossed 5 times. What is the chance of getting at least three heads?

18. The mean and variance of a binomial distribution 2 and 1 respectively. What is the probability that the variate takes a value greater than 1?

19. In an examination the probabilities of guessing, copying and knowing the answer of a multiple choice question with four choices are 1/3, 1/6 and 1/8 respectively. Find the probability that he knew the answer to the question given that he correctly solved it.

20. A box contains 2 red and 3 blue marbles. Find the probability that if two marbles are drawn at random (without replacement) (a) both are blue (b) both are red (c) one is red and one is blue.

    **Answer:**  (a) 3/10 (b) 1/10 (c) 3/5.

21. A binary communication channel carries data as one of two types of signal denoted by 0 and 1. Owing to noise, a transmitted 0 is sometimes received as 1 and a transmitted 1 is sometimes received as 0. For a given channel, assume the probability of 0.94 that a transmitted 0 is correctly received as a 0 and a probability of 0.91 that a transmitted 1 is received as 1. Further, assume a probability of 0.45 of transmitting a 0. If a signal is sent determine:

    (i) Probability that 1 is received.

    (ii) Probability that 0 is received.

    (iii) Probability that 1 was transmitted given that a 1 was received.

(iv) Probability that 0 was transmitted given that a 0 was received.

(v) Probability of an error.

**Answer:**  (i) 0.5275 (ii) 0.4725 (iii) 0.9488 (iv) 0.8952 (v) 0.0765

22. A random variable X has the following probability function:

| $x$: | 0 | 1 | 2 | 3 | 4 | 5 | 6 | 7 |
|------|---|---|---|---|---|---|---|---|
| $p(x)$: | 0 | $k$ | $2k$ | $2k$ | $3k$ | $k^2$ | $2k^2$ | $7k^2 + k$ |

(i) Find $k$, (ii) Evaluate $P(X < 6)$, $P(X \geq 6)$, and $P(0 < X < 5)$, (iii) if $P(X \leq k) > 1/2$, find the minimum value of $k$ and (iv) Determine the distribution function of X.

**Answer:** (i) 1/10 (ii) 81/100, 19/100, 4/5 (iii) $k = 4$

23. In a precision bombing attack there is a 50% chance that any one bomb will strike the target. two direct hits are required to destroy the target completely. how many bombs must be dropped to give 99% chance or better of completely destroying the target?

**Answer:** $n = 11$.

# Section 2

# 6

# Recurrence Relations

## 6.1 INTRODUCTION

The rule for finding terms from terms that precedes is called recurrence relation. Recurrence relations and generating functions are excellent techniques to solve many different types of counting problems. One such problem is how many bit strings of length $r$ do not contain two consecutive zeros? This can be solved by solving the recurrence relation $a_{r+1} = a_r + a_{r-1}$, given $a_1 = 2$ and $a_2 = 3$, where $a_r$ is the number of such strings of length $r$. Another example is Fibonacci series i.e. 1, 1, 2, 3, 5, 8, 13, 21, ..., $n^{\text{th}}$ term, $a_n$ can be expressed as

$$a_n = a_{n-1} + a_{n-2} \quad n \geq 3; \; a_1 = 1, a_2 = 1.$$

Generating functions are formal power series in which the coefficients of various powers of $z$ represent terms of a sequence. Generating functions are used to solve recurrence relations and to prove combinatorial identities.

## 6.2 DISCRETE NUMERIC FUNCTION

A function $a_r: \text{N} \rightarrow \text{R}$ is called discrete numeric function or simply numeric function. The values of the function at $r = 0, 1, 2, 3, ..., n$ are denoted by $a_0, a_1, a_2, ..., a_n$ respectively. The sequence $a_0, a_1, a_2, ..., a_n$ is a numeric function $a_r$.

## 6.3 GENERATING FUNCTION

The infinite series $A(z) = a_0 + a_1 z + a_2 z^2 + a_3 z^3 + \cdots + a_n z^n +, \ldots$ is called the generating function of the numeric function $a_0, a_1, a_2, a_3 ..., a_n.$ The coefficient of $z^n$ in the generating function is the value of the numeric function at $n$.

## 6.4 PROPERTIES OF GENERATING FUNCTION

   (i)  Let $a$ and $b$ be numeric functions and $\alpha$ a scalar such that $b = \alpha\, a.$ Then $\text{B}(z) = \alpha\, \text{A}(z)$

  (ii)  Let $a, b, c$ be numeric functions such that $c = a + b.$ Then $\text{C}(z) = \text{A}(z) + \text{B}(z)$

 (iii)  Let $a_n$ and $b_n$ be numeric functions and $\alpha$ a scalar such that $b_n = \alpha^n\, a_n.$ Then
     $\text{B}(z) = \text{A}(\alpha\, z)$

**Example 1:** Find the generating function of

    (i) $1, 1, 1, 1, \ldots$
    (ii) $1, 2^1, 2^2, 2^3, \ldots$
    (iii) $1, 3^1, 3^2, 3^3, \ldots$
    (iv) $1^n + 2^n + 3^n + \cdots, n \geq 0$
    (v) $3.2^n, n \geq 0$

*Solution*

    (i) $a_n = 1, 1, 1, 1, \ldots$
        $A(z) = 1 + z + z^2 + z^3 + \cdots = 1/1 - z$      (sum of infinite G.P.)

    (ii) $a_n = 1, 2^1, 2^2, 2^3, \ldots$
        $A(z) = 1 + 2^1 z + 2^2 z^2 + 2^3 z^3 + \cdots = 1/1 - 2z$

    (iii) $a_n = 1, 3^1, 3^2, 3^3, \ldots$
        $A(z) = 1 + 3^1 z + 3^2 z^2 + 3^3 z^3 + \cdots = 1/1 - 3z$

    (iv) $a_n = 1^n + 2^n + 3^n + \cdots, n \geq 0$
        $A(z) = 1/1 - z + 1/1 - 2z + 1/1 - 3z + \cdots$

    (v) $a_n = 3.2^n, \quad n \geq 0$
        Let $b_n = 2^n, \quad n \geq 0$
        Therefore $a_n = 3.b^n$

$$\text{Hence } A(z) = 3B(z) = 3\left[\frac{1}{1 - 2z}\right]$$

## 6.5 CONVOLUTION OF NUMERIC FUNCTIONS

Let $a$ and $b$ be two numeric functions, then their convolution is defined as $a*b$ denoted as $c$, where

$$c_r = a_0 b_r + a_1 b_{r-1} + a_2 b_{r-2} + \cdots + a_{r-1} b_1 + a_r b_0 = \sum_{j=0}^{r} a_j b_{r-j}$$

is the coefficient of $z^r$ in the product

$$(a_0 + a_1 z + a_2 z^2 + a_3 z^3 + \cdots + a_r z^r +, \ldots)(b_0 + b_1 z + b_2 z^2 + b_3 z^3 + \cdots + b_r z^r +, \ldots)$$

## 6.6 GENERATING FUNCTION OF THE CONVOLUTION OF TWO NUMERIC FUNCTIONS

Let $a$ and $b$ be two numeric functions having generating functions $A(z)$ and $B(z)$ respectively. $c = a*b$ is their convolution having generating function $C(z)$. Then

$$C(z) = A(z)\, B(z)$$

***Proof***

$$A(z) = \sum_{i=0}^{\infty} a_i z^i$$

$$B(z) = \sum_{j=0}^{\infty} b_j z^j$$

$$A(z)B(z) = \left[\sum_{i=0}^{\infty} a_i z^i\right]\left[\sum_{j=0}^{\infty} b_j z^j\right]$$

$$= \left[a_0 \sum_{i=0}^{\infty} b_j z^j\right] + \left[a_1 \sum_{j=0}^{\infty} b_j z^{j+1}\right] + \cdots$$

$$= \sum_{i=0}^{\infty} a_i \sum_{j=0}^{\infty} b_j z^{i+j}$$

Recurrence Relations:

$$= \sum_{n=0}^{n} \left[\sum_{j=0}^{n} a_j b_{n-j}\right] z^n = C(z)$$

**Example 2:** Find the generating function of the numeric function $a_n = 3^{n+5}, n \geq 0$

***Solution***

$$a^n = 3^{n+5}, \quad n \geq 0$$
$$= 3^5 . 3^n$$

$$A(z) = 3^5 \left[\frac{1}{1-3z}\right]$$

**Example 3:** Find the numeric function corresponding to the generating function

(a) $A(z) = \dfrac{1 + 2z - 6z^2}{1 - 3z}$

(b) $A(z) = \dfrac{2}{1 - 4z^2}$

***Solution***

(a) $A(z) = 1 + 2z - 6z^2/1 - 3z$

$\quad\quad = 2z + [1/1 - 3z]$

$\quad\quad = 2z + [1 - 3z]^{-1}$

$\quad\quad = 2z + 1 + 3z + 9z^2 + 27z^3 + \cdots$

$\quad\quad = 1 + 5z + 9z^2 + 27z^3 + \cdots$

Hence

$\quad a_0 = 1$

$\quad a_1 = 5$

$\quad a_2 = 9 = 3^2$

$\quad a_3 = 27 = 3^3$

$\quad \ldots\ldots\ldots$

$\quad a_r = 3^r$

Therefore

$$a_r = \begin{cases} 1 & r = 0 \\ 5 & r = 1 \\ 3^r & r \geq 2 \end{cases}$$

(b) $A(z) = \dfrac{2}{1 - 4z^2}$

$\quad\quad\quad = 2/(1 - 2z)(1 + 2z)$

$\quad\quad\quad = 1/(1 - 2z) + 1/(1 + 2z)$

$\quad\quad\quad = [1 + 2z + 4z^2 + 8z^3 \ldots] + [1 - 2z + 4z^2 - 8z^3 \ldots]$

$\quad\quad\quad = 2[1 + 4z^2 + 16z^4 \ldots]$

Hence,

$\quad a_0 = 2$

$\quad a_1 = 0$

$\quad a_2 = 8$

$\quad a_3 = 0$

$\quad a_4 = 2.2^4 = 32$

$\quad \ldots\ldots\ldots$

$\quad a_r = 2.2^r = 2^{r+1}$

Therefore

$$a_r = \begin{cases} 0 & r \text{ is odd} \\ 2^{r+1} & r \text{ is even} \end{cases}$$

**Example 4:** Determine the numeric function $a * b$, where the numeric functions $a$ and $b$ are defined as:

$$a_r = \begin{cases} 4 & 0 \le r \le 2 \\ 0 & r \ge 3 \end{cases}$$

$$b_r = \begin{cases} 17 & 0 \le r \le 1 \\ 0 & r \ge 2 \end{cases}$$

***Solution***

Since we have

$$a_0 = a_1 = a_2 = 4, \quad a_3 = a_4 = \cdots = 0$$

and

$$b_0 = b_1 = 17, \qquad b_2 = b_3 = b_4 = \cdots = 0$$

Hence,

$$c_0 = a_0 b_0 = 4 \times 17 = 68$$

$$c_1 = a_0 b_1 + a_1 b_0 = 4 \cdot 17 + 4 \cdot 17 = 136$$

$$c_2 = a_0 b_2 + a_1 b_1 + a_2 b_0$$

$$= 4 \cdot 0 + 4 \cdot 17 + 4 \cdot 17 = 136$$

$$c_3 = a_0 b_3 + a_1 b_2 + a_2 b_1 + a_3 b_0$$

$$= 4 \cdot 0 + 4 \cdot 0 + 4 \cdot 17 + 0 \cdot 17$$

$$= 68$$

$$c_4 = a_0 b_4 + a_1 b_3 + a_2 b_2 + a_3 b_1 + a_4 b_0$$

$$= 4 \cdot 0 + 4 \cdot 0 + 4 \cdot 0 + 0 \cdot 17 + 0 \cdot 17$$

$$= 0$$

Similarly we find that values of $C$ for $n > 4$ as all zero

$$C_r = \begin{cases} 68 & r = 0 \\ 136 & r = 1 \\ 136 & r = 2 \\ 68 & r = 3 \\ 0 & r \ge 4 \end{cases}$$

## 6.7 TABLE OF SOME IMPORTANT GENERATING FUNCTIONS

| Numeric function | Generating function |
|---|---|
| $1^r$ | $\dfrac{1}{(1-z)}$ |
| $r+1$ | $\dfrac{1}{(1-z)^2}$ |
| $r$ | $\dfrac{z}{(1-z)^2}$ |
| $r(r+1)$ | $\dfrac{2z}{(1-z)^3}$ |
| $k^r,\ k$ is constant | $\dfrac{1}{(1-zk)}$ |
| $C(n, k)$ | $(1+z)^n$ |
| $C(n, k)\,a^k$ | $(1+az)^n$ |
| $\dfrac{1}{k!}$ | $e^z$ |
| $\dfrac{(-1)^{k+1}}{k}$ | $\ln(1+z)$ |
| $a_r$ | $A(z)$ |
| $a_{r+1}$ | $\dfrac{A(z)-a_0}{z}$ |
| $a_{r+2}$ | $\dfrac{A(z)-a_0-a_1 z}{z^2}$ |
| $a_{r+n}$ | $\dfrac{A(z)-a_0-a_1 z \ldots -a_{n-1}z^{n-1}}{z^n}$ |

## 6.8 RECURRENCE RELATION

A recurrence relation for a sequence $a_0, a_1, \ldots$ is a formula that relates $a_n$ to certain of its predecessors $a_0, a_1, \ldots, a_{n-1}$. e.g., $a_{r+2} - 2a_{r+1} + 7 = 0$. The **initial conditions**, also called **boundary conditions**, for a sequence specify the term that precedes the first term where the recurrence relation takes effect. The recurrence relation together with initial conditions determines the sequence uniquely. A sequence is called a solution of the recurrence relation if its terms satisfy the recurrence relation. A recurrence is also called a **difference equation**.

Alternately, a recurrence relation is a functional relation between independent variable $x$, dependent variable $f(x)$ and the differences of various orders of $f(x)$ e.g., $f(x + 2h) - 2f(x + h) + 7f(x) = 0$ is a recurrence relation which can also be written as $a_{r+2} - 2a_{r+1} + 7 = 0$.

**Example 5:** The Tower of Hanoi consists of three pegs mounted on a board together with $n$ disks of different sizes. Initially these disks are placed on the first peg in order of size, with the largest on the bottom. The disks can be moved one at a time from one peg to another as long as a disk is never placed on top of a smaller disk. Find the number of moves needed to have all the disks on the second peg in order of size, with the largest on the bottom.

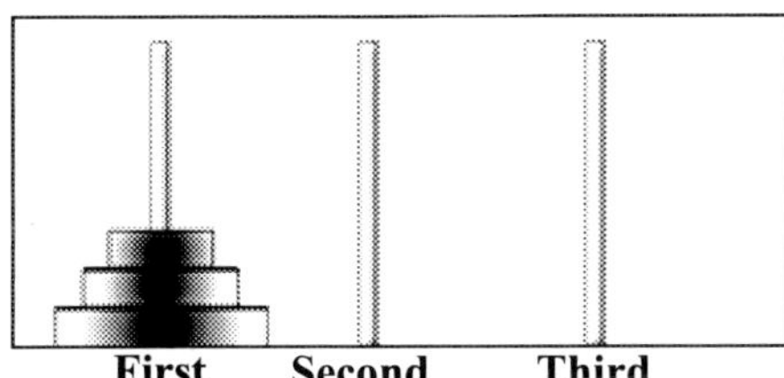


### Solution

Transfer the top $n - 1$ disks to peg 3 using $a_{n-1}$ moves, as per rules of the puzzle. Then transfer the largest disk on the first peg to the second peg. Now transfer $n-1$ disks from the third peg to the second peg by $a_{n-1}$ moves, placing them on the top of the largest disk.

Hence $a_n = 2a_{n-1} + 1$, $a_1 = 1$, because one disk can be transferred from peg 1 to peg 2, as per rules, in one move.

$$a_n = 2a_{n-1} + 1$$
$$= 2(2a_{n-2} + 1) + 1 = 2^2 a_{n-2} + 2 + 1 = 2^2(2a_{n-3} + 1) + 2 + 1 = 2^3 a_{n-3} + 2^2 + 2 + 1$$

$$\ldots\ldots\ldots$$

$$= 2^{n-1} a_1 + 2^{n-2} + \cdots + 2 + 1$$
$$= 2^{n-1} + 2^{n-2} + \cdots + 2 + 1 \qquad \text{(G.P. with first term 1 and common ratio 2)}$$
$$= 1(2^n - 1) / (2 - 1)$$
$$= 2^n - 1$$

**Example 6:** Find a sequence from the recurrence relation $a_n = 3a_{n-1}$, $n \leq 5$, with the initial condition $a_0 = 2$.

***Solution***

$a_1 = 3\,a_0 = 6$
$a_2 = 3\,a_1 = 18$

. . . . . . . . . . . . . . .

$a_5 = 3\,a_4 = 486$

The solution sequence is 2, 6, 18, 54, 162, 486

**Example 7:** Find a sequence from the recurrence relation $a_n = a_{n-1} + a_{n-2}, n \leq 8$, with the initial condition $a_0 = 1 = a_1$.

***Solution***

$a_n = a_{n-1} + a_{n-2}$
$a_2 = a_1 + a_0 = 2$
$a_3 = a_2 + a_1 = 3$

. . . . . . . . . . . . . . .

. . . . . . . . . . . . . . .

$a_7 = a_6 + a_5 = 21$
$a_8 = a_7 + a_6 = 34$

The solution sequence is 1, 1, 2, 3, 5, 8, 13, 21, 34, which is a portion of **Fibonacci sequence of numbers**.

## 6.9 EXPLICIT FORMULA FOR A SEQUENCE

When in a sequence $\{a_n\}$, $a_n$ can be found only with the help of positive integer, $n$, the formula for $a_n$ is called explicit formula for the sequence, e.g., $a_n = n^3$ is an explicit formula for the sequence 0, 1, 8, 27...

**Example 8:** Find the explicit formula for the sequence 67, 65, 63, 61, 59, ... Can this sequence be described by a recurrence relation?

***Solution***

$67 = 69 - 2$
$65 = 69 - 4$
$63 = 69 - 6$

. . . . . . . . . . . . .

$a_n = 69 - 2n$

The explicit formula is $a_n = 69 - 2n, \quad n = 1, 2, 3, \ldots$

The sequence can also be found by the recurrence relation $a_n = a_{n-1} - 2, a_1 = 67$

**Example 9:** Fibonacci posed the following problem:

"A single pair of rabbits (male and female) is born at the beginning of a year."

Find the number of rabbits at the end of the year subject to the following conditions:

  (i) Rabbit pairs are not fertile during the first month of life but thereafter give birth to one new male/female pair at the end of every month.
  (ii) no rabbit die.

### Solution

$a_0$ = number of rabbit pairs in the beginning = 1

$a_1$ = number of rabbit pairs at the end of the first month = 1

$a_2$ = number of rabbit pairs at the end of the second month = $a_1 + 1 = 1 + 1 = 2$

$a_3$ = number of rabbit pairs at the end of the third month = $a_2 + a_0 = 2 + 1 = 3$

......................................................

......................................................

$a_{10}$ = number of rabbit pairs at the end of the tenth month = $a_9 + a_8 = 55 + 34 = 89$

$a_{11}$ = number of rabbit pairs at the end of the eleventh month = $a_{10} + a_{11} = 89 + 55 = 144$

$a_{12}$ = number of rabbit pairs at the end of the twelfth month = $a_{11} + a_{10} = 144 + 89 = 233$

The recurrence relation for the sequence is $a_n = a_{n-1} + a_{n-2}$, with the initial condition $a_0 = 1 = a_1$, is called **Fibonacci sequence**.

Hence number of rabbit pairs at the end of 12th month = $2 \times 233 = 466$

## 6.10 SOLUTION OF RECURRENCE RELATIONS USING BACK TRACKING

A technique for finding explicit formula for the sequence whose recurrence relation is given, is called back tracking. The value of $a_n$ is back tracked, substituting the values of $a_{n-1}, a_{n-2} \ldots$ till a pattern is clear.

**Example 10:** Backtrack the sequence defined by the recurrence relation $a_n = a_{n-1} - 2$, $a_1 = 67$

### Solution

$a_n = a_{n-1} - 1.2$

$\quad = a_{n-2} - 2.2 = a_{n-3} - 3.2 = a_{n-4} - 4.2 = \ldots = a_{n-(n-1)} - (n-1)2$

$\quad = 67 - 2n + 2$

$\quad = 69 - 2n$

Hence explicit formula for the sequence is $a_n = 67 - 2n + 2 = 69 - 2n$

**Example 11:** Backtrack the sequence defined by the recurrence relation $a_n = 2a_{n-1} + 1$, $a_1 = 1$.

*Solution*

$$a_n = 2a_{n-1} + 1 = 2(2a_{n-2} + 1) + 1 = 2^2 a_{n-2} + 2^1 + 1 = 2^2(2a_{n-3} + 1)$$
$$+ 2.1 + 1 = 2^3 a_{n-3} + 2^2 + 2^1 + 1$$

$$\dots\dots\dots\dots$$

$$= 2^k a_{k-(k-1)} + 1 + 2^{(k-1)} + 2^{(k-2)} + \cdots + 1$$
$$= 2^k a_1 + 2^{(k-1)} + 2^{(k-2)} + \cdots + 1$$
$$a_n = 2^{n-1} + 2^{(n-2)} + 2^{(n-3)} + \cdots + 1 \quad \text{(G.P. with first term 1, common ratio 2 and number of terms } n\text{)}$$

$$a_n = \frac{2^n - 1}{2 - 1} = 2^n - 1$$

Hence explicit formula for the sequence is $a_n = 2^n - 1$

## 6.11 HOMOGENEOUS LINEAR RECURRENCE RELATION WITH CONSTANT COEFFICIENTS

The relation $a_r = c_1 a_{r-1} + c_2 a_{r-2} + \cdots + c_k a_{r-k}$, where $c_i$'s are constants, is called a linear recurrence relation with constant coefficients. If $c_k$ is non-zero the above relation is called $k$th order recurrence relation e.g.,

First order recurrence relation: $a_r = c_1 a_{r-1}$
Second order recurrence relation: $a_r = c_1 a_{r-1} + c_2 a_{r-2}$
Third order recurrence relation: $a_r = c_1 a_{r-1} + c_2 a_{r-2} + c_3 a_{r-3}$.

## 6.12 SOLUTION OF HOMOGENEOUS LINEAR DIFFERENCE EQUATION OF SECOND ORDER

i.e. $a_r = c_1 a_{r-1} + c_2 a_{r-2}$

$a_r = c_1 a_{r-1} + c_2 a_{r-2}$ is linear second order homogeneous linear difference equation. $\quad$ (1)

The equation, $x^2 - c_1 x - c_2 = 0$ is called characteristic equation of (1) $\quad$ (2)

**Case (i) when the roots of (1) are distinct**

Let $s_1$ and $s_2$ are the roots of (1), called characteristic roots,

Therefore

$$s_1^2 - c_1 s_1 - c_2 = 0 \qquad (3)$$
$$s_2^2 - c_1 s_2 - c_2 = 0 \qquad (4)$$

Let $a_r = u_1 s_1{}^r + u_2 s_2{}^r$      (5)

From (5)

$$a_1 = u_1 s_1 + u_2 s_2$$
$$a_2 = u_1 s_1{}^2 + u_2 s_2{}^2$$

Equation (5) can be written as

$$a_r = u_1 s_1{}^{r-2} s_1{}^2 + u_2 s_2{}^{r-2} s_2{}^2$$

using (3) and (4)

$$
\begin{aligned}
a_r &= u_1 s_1{}^{r-2} (c_1 s_1 + c_2) + u_2 s_2{}^{r-2}(c_1 s_2 + c_2) \\
&= c_1 u_1 s_1{}^{r-1} + c_2 u_1 s_1{}^{r-2} + c_1 u_2 s_2{}^{r-1} + c_2 u_2 s_2{}^{r-2} \\
&= c_1(u_1 s_1{}^{r-1} + u_2 s_2{}^{r-1}) + c_2(u_1 s_1{}^{r-2} + u_2 s_2{}^{r-2}) \\
&= c_1 a_{r-1} + c_2 a_{r-2}
\end{aligned}
$$

which is (1).

Hence we can say that if $s_1$ and $s_2$ are characteristic roots then the solution is $a_r = u_1 s_1{}^r + u_2 s_2{}^r$, where $u_1$ and $u_2$ depend on the initial conditions.

**Case (ii)   when $s$ is the characteristic root of multiplicity 2.**

Therefore

$$s^2 - c_1 s - c_2 = 0 \tag{6}$$

Let

$$a_r = u_1 s^r + u_2 r s^r \tag{7}$$

From (7)

$$a_1 = u_1 s + u_2 s r$$
$$a_2 = u_1 s^2 + u_2 r s^2$$

Equation (7) can be written as

$$ar = u_1 s^{r-2} s^2 + u_2 s^{r-2} s^2$$

Using (6)

$$
\begin{aligned}
ar &= u_1 s^{r-2} (c_1 s + c_2) + u_2 s^{r-2}(c_1 s + c_2) \\
&= c_1 u_1 s^{r-1} + c_2 u_1 s^{r-2} + c_1 u_2 s^{r-2} + c_2 u_2 s^{r-2} \\
&= c_1(u_1 s^{r-1} + u_2 s^{r-1}) + c_2(u_1 s^{r-2} + u_2 s^{r-2}) \\
&= c_1 a_{r-1} + c_2 a_{r-2}
\end{aligned}
$$

which is (1).

Hence we can say that if $s$ is a characteristic root of multiplicity 2, then the solution is

$$a_r = u_1 s^r + u_2 r s^r, \text{ where } u_1 \text{ and } u_2 \text{ depend on the initial conditions.}$$

**Example 12 (distinct roots):** Solve the following recurrence relation

$$a_n = a_{n-1} + 2a_{n-2}, a_0 = 2, a_1 = 7$$

***Solution***

Characteristic equation is

$$t^2 - t - 2 = 0$$

Which gives roots $s_1 = 2, s_2 = -1$

The roots are different, so we have the equation:

$$a_n = u_1 2^n + u_2(-1)^n$$

Using our initial conditions, we have

$$a_0 = u_1 + u_2 = 2,$$
$$a_1 = 2u_1 - u_2 = 7$$

Solving for $u_1$ and $u_2$, we get

$$u_1 = 3, u_2 = -1$$

Solution is:

$$a_n = 3*2^n - (-1)^n$$

**Example 13 (one root):** Solve the following recurrence relation

$$a_n = 6a_{n-1} - 9a_{n-2}, a_0 = 1, a_1 = 6$$

***Solution***

Characteristic equation is

$$t^2 - 6t + 9 = 0$$

Which gives roots $s_1 = s_2 = 3$

The roots are the same, so we have the equation:

$$an = u_1 3^n + u_2 n 3^n$$

Using our initial conditions, we have

$$a_0 = u_1 = 1,$$
$$a_1 = 3u_1 + 3 u_2 = 6$$

Solving for $u_1$ and $u_2$, we get

$$u_1 = 1, u_2 = 1$$

Solution is:

$$a_n = 3^n + n3^n$$

## 6.13 SOLUTION OF HOMOGENEOUS LINEAR DIFFERENCE EQUATION OF $k^{TH}$ ORDER

### *Solution*

Let

$$c_0 a_r + c_1 a_{r-1} + c_2 a_{r-2} + \cdots + c_k a_{r-k} = 0 \tag{1}$$

is a homogeneous linear difference equation of order $k$.

The equation $c_0 x^k + c_1 x^{k-1} + c_2 x^{k-2} + \cdots + c_k = 0$ is called characteristic equation of equation (1).

It is a polynomial equation of order $k$ and has $k$ roots.

### Case (i)  when $k$ characteristic roots are distinct

Let roots be $s_1, s_2, s_3 \ldots s_k$,

The solution is

$$a_r = u_1 s_1{}^r + u_2 s_2{}^r + \ldots + u_k s_k{}^r$$

where, $u_1, u_2, u_3, \cdots, u_k$ are determined using initial conditions. It can be verified that this solution satisfies equation (1).

### Case (ii)  when roots are not distinct

Let $s_1 = s_2 = s_3$ and all other roots be distinct

The solution is

$$a_r = (u_1 + nu_2 + n^2 u_3)s_1{}^r + u_4 s_4{}^r + \cdots + u_k s_k{}^r$$

where, $u_1, u_2, u_3, \cdots, u_k$ are determined using initial conditions. It can be verified that this solution satisfies equation (1).

**Example 14:** Find explicit formula for sequence defined by $C_n = 3C_{n-1} - 2C_{n-2}$, $C_1 = 5$ and $C_2 = 3$;

### *Solution*

$$x^2 - 3x + 2 = 0$$

Characteristic roots are

$$s_1 = 1 \text{ and } s_2 = 2$$

The solution is:

$$C_n = u_1 1^r + u_2 2^r$$

Now from initial conditions

$$C_1 = u_1 + 2u_2 = 5$$
$$C_2 = u_1 + 4u_2 = 3$$

We get,

$$u_1 = 7 \text{ and } u_2 = -1$$

Hence the explicit formula is

$$C_n = 7.1^r - 2^r = 7 - 2^r$$

**Example 15:** Solve the recurrence relation:
$$f_n = f_{n-1} + f_{n-2}, \, n \geq 2 \text{ with initial conditions } f_0 = f_1 = 1$$

***Solution***

Homogeneous solution:

$$f_n = f_{n-1} + f_{n-2}$$

The difference equation is of $2^{nd}$ order
The characteristic equation is $\alpha^2 = \alpha + 1$ i.e. $\alpha^2 - \alpha - 1 = 0$

$\therefore$   roots of the difference equation are given by

$$\alpha = \frac{1 + \sqrt{5}}{2}, \quad \beta = \frac{1 - \sqrt{5}}{2}$$

Homogeneous solution is

$$f_r = A_1(\alpha)^r + A_2(\beta)^r \qquad n \geq 0$$

where, $A_1$ and $A_2$ are arbitrary constants

$$f_0 = 1 \qquad \rightarrow A_1 + A_2 = 1 \tag{1}$$
$$f_1 = 1 \qquad \rightarrow A_1\alpha + A_2\beta = 1 \tag{2}$$

from (1) and (2)

$$A_2(\alpha - \beta) = \alpha - 1$$

$$\therefore \quad A_2 = \frac{\alpha - 1}{\alpha - \beta} = -\frac{1 - \sqrt{5}}{2\sqrt{5}}$$

$$A_1 = 1 - A_2 = 1 - \left(\frac{\alpha - 1}{\alpha - \beta}\right)$$

$$= \frac{1 - \beta}{\alpha - \beta} = \frac{1 + \sqrt{5}}{2\sqrt{5}}$$

$$f_r = \frac{1 + \sqrt{5}}{2\sqrt{5}}\left(\frac{1 + \sqrt{5}}{2}\right)^r - \frac{1 - \sqrt{5}}{2\sqrt{5}}\left(\frac{1 - \sqrt{5}}{2}\right)^r, \quad n \geq 0$$

## 6.14 LINEAR RECURRENCE RELATION WITH CONSTANT COEFFICIENTS

The relation

$$c_0 a_r + c_1 a_{r-1} + c_2 a_{r-2} + \cdots + c_k a_{r-k} = f(r),$$

where $c_i$'s are constants and $f(r)$ is a function of $r$ only, is called a linear recurrence relation with constant coefficients.

## 6.15 SOLUTION OF LINEAR DIFFERENCE EQUATION WITH CONSTANT COEFFICIENTS

Total solution = Homogeneous solution + Particular Solution = $a_r^{(h)} + a_r^{(p)}$

**The solution of**

$$c_0 a_r + c_1 a_{r-1} + c_2 a_{r-2} + \ldots + c_k a_{r-k} = 0 \text{ is called homogeneous solution}$$

while the solution which satisfies the equation

$$c_0 a_r + c_1 a_{r-1} + c_2 a_{r-2} + \ldots + c_k a_{r-k} = f(r) \text{ is called particular solution of the linear differential equation.}$$

## 6.16 METHOD TO FIND PARTICULAR SOLUTION

There is no general method to find a particular solution. It is found by inspection. When $f(r)$ is a polynomial of degree $t$ in $r$, the particular solution will be of the form

$$P_1 r^t + P_2 r^{t-1} + \ldots, + P_t r + P_{t+1}$$

where $P_i$'s are determined by substituting the particular solution in the given difference equation and then comparing the two sides.

| $f(r)$ | Particular solution $a_r^{(p)}$ |
|---|---|
| $f(r)$ is a polynomial of degree $t$ i.e., $f(r)$ is of the form $c_1 r^t + c_2 r^{t-1} + \ldots + c_t r + c_{t+1}$, where $c_1, c_2, \ldots$ $c_t, c_{t+1}$ are constants | $P_1 r^t + P_2 r^{t-1} + \ldots + P^t r + P_{t+1}$<br><br>$P_1, P_2, \ldots P_t, P_{t+1}$ are constants |
| $(c_1 r^t + c_2 r^{t-1} + \ldots + c_t r + c_{t+1})\beta^r$, where $\beta$ is not a characteristic root | $(P_1 r^t + P_2 r^{t-1} + \ldots + Pt\, r + P_{t+1})\beta^r$ |
| $(F_1 r^t + F_2 r^{t-1} + \ldots + F^t r + F_{t+1})\beta^r$, where $\beta$ is a characteristic root of multiplicity $m$ | $r^m(P_1 r^t + P_2 r^{t-1} + \ldots + P_t r + P_{t+1})\beta^r$ |
| F, where 1 is not a characteristic root. | P |
| F.$k^r$, where 1 is not a characteristic root. e.g. $5^r$, $2.5^r$ | P$k^r$ |

| $f(r)$ | Particular solution $a_r^{(p)}$ |
|---|---|
| F, where 1 is a characteristic root of multiplicity $m$. | $r^m P$ |
| $F_1 r + F_2$ e.g. $r + 5$ | $P_1 r + P_2$ |
| $F_1 r$ e.g. $2r$ | $P_1 r + P_2$ |
| $F_1 r^2 + F_2 r + F_3$ | $P_1 r^2 + P_2 r + P_3$ |
| $F_1 r^2 + F_2 r$ | $P_1 r^2 + P_2 r + P_3$ |
| $F_1 r^2$ | $P_1 r^2 + P_2 r + P_3$ |

**Example 16:** Solve the difference equation

$$a_n = 5a_{n-1} - 6a_{n-2} + 7^n$$

*Solution*

**Homogeneous Solution:**
The difference equation is of $2^{nd}$ order

The characteristic equation is
$$\alpha^2 - 5\alpha + 6 = 0$$
$$\text{i.e. } (\alpha - 2)(\alpha - 3) = 0$$

Roots are 2, 3

Homogeneous solution is
$$a_n^{(h)} = c_1 2^n + c_2 3^n$$

**Particular Solution:**

$f(n) = 7^n$
Particular solution $= d7^n$
Substituting in (1),

$$d7^n = 5d7^{n-1} - 6d7^{n-2} + 7^n$$

Equating coefficients of like powers of $n$ on both sides and solving for $d$,

$$d7^n = 7^{n-2}\left(5d7 - 6d + 7^2\right)$$
$$49d7^{n-2} = 7^{n-2}\left(5d7 - 6d + 7^2\right)$$
$$49d = 5d7 - 6d + 7^2$$
$$49d = 29d + 49$$
$$20d = 49$$
$$d = \frac{49}{20}$$

Therefore the particular solution is

$$a_n^{(p)} = \left(49/20\right)7^n$$

Total or general solution is

$$a_n = c_1 2^n + c_2 3^n + \left(49/20\right)7^n$$

**Example 17:** Solve the difference equation $a_r + 5a_{r-1} + 6a_{r-2} = 3r^2$

***Solution***

**Homogeneous Solution**:

$$a_r + 5\,a_{r-1} + 6\,a_{r-2} = 3r^2 \tag{1}$$

The difference equation is of $2^{nd}$ order

The characteristic equation is $\alpha^2 + 5\alpha + 6 = 0$

    i.e. $(\alpha + 2)(\alpha + 3) = 0$

Roots are $-2, -3$

Homogeneous solution is

$$a^{(h)}{}_r = A_1(-2)^r + A_2(-3)^r$$

**Particular Solution**:

$f(r) = 3r^2$, polynomial of second degree in $r$.

Hence particular solution $= P_1 r^2 + P_2 r + P_3$
Substituting in (1),

$$(P_1 r^2 + P_2 r + P_3) + 5[P_1(r-1)^2 + P_2(r-1) + P_3] + 6[P_1(r-2)^2 + P_2(r-2) + P_3] = 3r^2$$
$$12P_1 r^2 - (34P_1 - 12P_2)\,r + (29P_1 - 17P_2 + 12\,P_3) = 3\,r^2$$

Equating coefficients of like powers of $r$ on both sides,

    $12P_1 = 3$
    $34P_1 - 12P_2 = 0$
    $29P_1 - 17P_2 + 12P_3 = 0$

Hence $P_1 = 1/4$, $P_2 = 17/24$, $P_3 = 115/288$

Therefore the particular solution is

$a_n^{(p)}{}_r = 1/4 r^2 + 17/24 r + 115/288$

Total solution is

$a_r = a^{(h)}{}_r + a^{(p)}{}_r = A_1(-2)^r + A_2(-3)^r + 1/4r^2 + 17/24r + 115/288$

**Example 18:** Solve the difference equation $a_r - 4a_{r-1} + 4a_{r-2} = 3r^2$

*Solution*

**Homogeneous Solution:**

$$a_r - 4a_{r-1} + 4a_{r-2} = 3r^2$$

The difference equation is of $2^{rd}$ order

The characteristic equation is: $\alpha^2 - 4\alpha + 4 = 0$

roots are 2, 2

Homogeneous solution is

$a^{(h)}{}_r = (A_1 r + A_2)(2)^r$

**Particular Solution:**

$f(r) = 3r^2$, polynomial of second degree in $r$.

Hence Particular Solution $= P_1 r^2 + P_2 r + P_3$

Substituting in (1),

$$(P_1 r^2 + P_2 r + P_3) - 4[P_1(r-1)^2 + P_2(r-1) + P_3] + 4[P_1(r-2)^2 + P_2(r-2) + P_3] = 3r^2$$
$$P_1 r^2 - (8P_1 - P_2) r + P_3 = 3r^2$$

Equating coefficients of like powers of $r$ on both sides, $P_1 = 3$

$$8P_1 - P_2 = 0$$
$$P_3 = 0$$

Hence $P_1 = 3$, $P_2 = 24$, $P_3 = 0$. Therefore the particular solution is

$$a^{(p)}{}_r = 3r^2 + 24r + 0$$

Total solution is $ar = a^{(h)}{}_r + a^{(p)}{}_r = (A_1 r + A_2)(5)^r + 3r^2 + 24r$

**Example 19:** Solve the difference equation: $a_n - 4a_{n-1} + 4a_n - 2 = 3n + 2n$

*Solution*

$$a_n - 4a_{n-1} + 4a_{n-2} = 3n + 2^n \tag{1}$$

It is a linear difference equation of order 2.

Characteristics equation is

$$x^2 - 4x + 4 = 0$$
$$(x - 2)^2 = 0$$

2 is a characteristic root of multiplicity 2

Homogeneous solution $= ar^{(h)} = (u_1 + u_2 n)2^n$

Particular solution $= a_r^{(p)} = (p_1 n + p_2) + n^2 p_3 2^n$

It satisfies equation (1), therefore

$$(p_1 n + p_2) + n^2 p_3 2^n - 4[p_1(n-1) + p_2 + (n-1)^2 p_3 2^{n-1}] + 4[(p_1(n-2) + p_2) + (n-2)^2 p_3 2^{n-2}]$$
$$= 3n + 2^n$$

**Comparing both sides**

$$p_1 n - 4p_1(n-1) + 4p_1(n-2) = 3$$
$$p_1 = 3$$
$$p_2 + 4p_1 - 8p_1 = 0$$
$$p_2 - 4p_1 = 0$$
$$p_2 = 4p_1 = 12$$
$$n^2 p_3 2^n - 2(n-1)^2 p_3 2^n + (n-2)^2 p_3 2^n = 2^n$$

Dividing throughout by $2^n$

$$p_3[n^2 - 2(n-1)^2 + (n-2)^2] = 1$$
$$2p_3 = 1$$
$$p_3 = 1/2$$
$$ar^{(p)} = 3n + 12 + n^2 *(1/2) * 2^n = 3n + 12 + n^2 * 2^{n-1}$$

Total Solution $= a_r = ar^{(h)} + ar^{(p)} = (u_1 + u_2 n)2^n + 3n + 12 + n^2 * 2^{n-1}$

**Example 20:** Solve the recurrence relation $a_{n+2} + a_{n+1} + a_n = n.2^n$

***Solution***

$$a_{n+2} + a_{n+1} + a_n = n.2^n \tag{1}$$
Putting $n = r - 2$, the above equation becomes

$$a_r + a_{r-1} + a_{r-2} = (r-2)/4 * 2^r = (1/4 * r - 1/2)\, 2^r \tag{2}$$

It is a linear difference equation of order 2

Characteristic equation is $x^2 + x + 1 = 0$,

Characteristic roots $(\alpha, \beta)$ are $\alpha = (-1 + \sqrt{3}i)/2$, $\beta = (-1 - \sqrt{3}i)/2$

$$a_r^{(h)} = u_1\,\alpha^r + u_2\,\beta^r$$

Particular solution $= ar^{(p)} = (p_1r + p_2)2^r$

It satisfies equation (2)

$$\therefore\quad (p_1r + p_2)\,2^r + [p_1(r-1) + p_2]\,2^{r-1} + [p_1(r-2) + p_2]\,2^{r-2} = (\tfrac{1}{4}*\,r - \tfrac{1}{2})\,2^r$$

Equating both sides and simplifying we get

$$p_1 = 1/7;\quad p_2 = -10/49$$

**Hence**

$$a_r^{(p)} = (1/7\,r - 10/49)2^r$$

Total solution $= a_r^{(h)} + a_r^{(p)} = u_1\,\alpha^r + u_2\,\beta^r + (1/7\,r - 10/49)2^r$

$$\therefore\quad a_r = u_1\,\alpha^r + u_2\,\beta^r + (1/7\,r - 10/49)2^r$$

Putting the values of $\alpha$ and $\beta$ and changing $r$ to $n + 2$, the solution of the given equation is

$$a_{n+2} = u_1\left((-1 + \sqrt{3i})/2\right)^r + u_2\left((-1 - \sqrt{3i})/2\right)^r + (1/7\,r - 10/49)2^r$$

**Example 21:** Find particular solution of the difference equation $a_r + a_{r-1} = 5r.2^r$

***Solution***

$$a_r + a_{r-1} = 5r.2^r \tag{1}$$

Particular solution $= a_r^{(p)} = (p_1r + p_2)2^r$

Using 1,

$$(p_1r + p_2)2^r + (\,p_1(r-1) + p_2)\,2^{r-1} = 5r.2^r$$

Equating both sides and simplifying, we get

$$p_1 = 10/3;\quad p_2 = 10/9$$

Particular solution is

$$a_r^{(p)} = 10/3\,(r + 1/3)2^r;\quad r \geq 0$$

## 6.17 SOLUTION BY THE METHOD OF GENERATING FUNCTIONS

In this method, we shall first find the generating function of the required solution and using this generating function, we shall find the desired solution of the given recurrence relation.

**To Solve** $C_0a_r + C_1a_{r-1} + C_2a_{r-2} + \cdots + C_ka_{r-k} = f(r)$, valid for $r \geq s$, where $s \geq k$, by the method of generating functions. Multiplying both sides of the above equation by $zr$ and summing from $r = s$ to $r = \infty$

$$\sum_{r=s}^{\infty}\left(C_0 a_r + C_1 a_{r-1} + C_2 a_{r-2} + \cdots + C_k a_{r-k}\right)z^r = \sum_{r=s}^{\infty} f(r)z^r$$

**Because**

$$\sum_{r=s}^{\infty} C_0 a_r z^r = C_0\left[A(z) - a_0 - a_1 z - a_2 z^2 - \cdots - a_{s-1}z^{s-1}\right]$$

$$\sum_{r=s}^{\infty} C_1 a_{r-1} z^r = C_1 z\left[A(z) - a_0 - a_1 z - a_2 z^2 - \cdots - a_{s-2}z^{s-2}\right]$$

$$\cdots\cdots\cdots\cdots\cdots\cdots\cdots\cdots\cdots\cdots\cdots\cdots\cdots$$

$$\sum_{r=s}^{\infty} C_1 a_{r-1} z^r = C_k z^k\left[A(z) - a_0 - a_1 z - a_2 z^2 - \cdots - a_{s-k-1}z^{s-k-1}\right]$$

we get

$$A(z) = \frac{1}{C_0 + C_1 z + \cdots + C_k z^k}\sum_{r=s}^{\infty} f(r)z^r$$

$$+ C_0(a_0 + a_1 z + a_2 z^2 + \cdots + a_{s-1}z^{s-1}) + C_1 z(a_0 + a_1 z + a_2 z^2 + \cdots + a_{s-1}z^{s-1})$$

$$+ \cdots\cdots\cdots\cdots\cdots\cdots\cdots\cdots\cdots\cdots\cdots + C_k z^k(a_0 + a_1 z + a_2 z^2 + \cdots + a_{s-k-1}z^{s-k-1})]$$

**Example 22:** Solve the difference equation $a_r = 3a_{r-1} + 1$, $a_0 = 0$, $a_1 = 1$

*Solution*

The difference equation is

$$a_r - 3a_{r-1} - 1 = 0, r \geq 2$$

Multiplying both sides by $z^r$ and summing from $r = 2$ to $r = \infty$

$$\sum_{r=2}^{\infty} a_r z^r - 3\sum_{r=2}^{\infty} a_{r-1} z^r - \sum_{r=2}^{\infty} z^r = 0$$

$$A(z) - a_1 z - a_0 - 3(A(z) - a_0)z - z^2/1 - z = 0$$

Using initial conditions

$$A(z) - z - 3z \cdot A(z) - z^2/1 - z = 0$$

$$A(z) = \frac{\dfrac{z^2}{1-z} + z}{1 - 3z} = \frac{z}{(1-3z)(1-z)}$$

Using partial fractions

$$A(z) = \frac{1}{2.(1-3z)} - \frac{1}{2.(1-z)}$$

$$A(z) = \frac{1}{2}\left(\frac{1}{(1-3z)} - \frac{1}{(1-z)}\right)$$

Since

Generating function for $k^r$ is given by $\dfrac{1}{(1-kz)}$ and

Generating function for $1^r$ is given by $\dfrac{1}{(1-z)}$

We get

$a_r = \frac{1}{2}(3^r - 1), \quad r \geq 0$

**Example 23:** Solve the difference equation $a_r = 2a_{r-1} + 3a_{r-2}$, $a_0 = 1$, $a_1 = 1$ for $r \geq 2$, using method of generating functions.

***Solution***

The difference equation is

$a_r - 2a_{r-1} - 3a_{r-2} = 0, \quad r \geq 2$

Multiplying both sides by $z^r$ and summing from $r = 2$ to $r = \infty$,

$$\sum_{r=2}^{\infty} a_r z^r - \sum_{r=2}^{\infty} 2a_{r-1} z^r - \sum_{r=2}^{\infty} 3a_{r-2} z^r = 0$$

We get

$$A(z) - a_0 - a_1 z) - 2z(A(z) - a_0) - 3z^2 A(z) = 0$$
$$A(z) - 2z\,A(z) - 3z^2 A(z) = a_0 + a_1 z - 2a_0 z$$

Using initial conditions

$$A(z) = \frac{1-z}{1-2z-3z^2}$$

$$A(z) = \frac{1-z}{(1-3z)(1+z)}$$

by the method of partial fractions we get:

$$A(z) = \frac{1}{2} \times \frac{1}{1-3z} + \frac{1}{2} \times \frac{1}{1+z}$$

We get

$$a_r = \tfrac{1}{2}\,3r + \tfrac{1}{2}(-1)^r$$

**Example 24:** Solve the difference equation $a_r = a_{r-1} + a_{r-2}, \quad a_1 = 2, a_2 = 3$

***Solution***

$a_r = a_{r-1} + a_{r-2}, \quad a_1 = 2, a_2 = 3$

The difference equation is

$a_r - a_{r-1} - a_{r-2} = 0, r \geq 3$

Multiplying both sides by $z^r$ and summing from $r = 3$ to $r = \infty$,

$$\sum_{r=3}^{\infty} a_r z^r - \sum_{r=3}^{\infty} a_{r-1} z^r - \sum_{r=3}^{\infty} a_{r-2} z^r = 0$$

$$A(z) - a_2 z^2 - a_1 z - a_0 - z[A(z) - a_1 z - a_0] - z^2[A(z) - a_0] = 0$$

Using initial conditions

$$A(z) = \frac{z^2 + 2z}{1 - z - z^2} = -1 + \frac{1 + z}{1 - z - z^2}$$

Let

$$1 - z - z^2 = 0$$
$$z^2 + z - 1 = 0$$
$$z = (-1 + \sqrt{5})/2, \ (-1 - \sqrt{5})/2$$

$$1 - z - z^2 = \left(1 - \frac{1 - \sqrt{5}}{2}z\right)\left(1 - \frac{1 + \sqrt{5}}{2}z\right)$$

Hence

$$A(z) = -1 + \frac{\dfrac{5 + 3\sqrt{5}}{10}}{1 - \dfrac{\left(1 + \sqrt{5}\right)}{2}z} - \frac{\dfrac{-5 + 3\sqrt{5}}{10}}{1 - \dfrac{\left(1 - \sqrt{5}\right)}{2}z}$$

We get

$a_r = 0, \quad r = 0$

$$= \left(\frac{5 + 3\sqrt{5}}{10}\right)\left(\frac{1 + \sqrt{5}}{2}\right)^r - \left(\frac{-5 + 3\sqrt{5}}{10}\right)\left(\frac{1 - \sqrt{5}}{2}\right)^r, \quad r \geq 1$$

**Exercise 6.1**

1. Find the generating function of the numeric function where

$$a_n = \begin{cases} 2^n & \text{if } n \text{ is even} \\ -2^n & \text{if } n \text{ is odd} \end{cases}$$

2. Find the numeric function corresponding to the generating function

$$A(z) = \frac{1}{5 - 6z + z^2}$$

3. Backtrack the sequence defined by the recurrence relation $a_n = a_{n-1} + 5$, $n \geq 2$, $a_1 = 7$.

4. Backtrack the sequence defined by the recurrence relation $a_n = 3\,a_{n-1} + 1$, $n \geq 2$, $a_1 = 4$.

5. Find the solution of homogeneous linear difference equation
   (a) $a_r = c_1 a_{r-1} + c_2 a_{r-2}$
   (b) $a_n = 5a_{n-1} - 6a_{n-2}$,
   where $a_0 = 1$, $a_1 = 4$

6. Solve the following difference equations
   (a) $a_r - 4\,a_{r-1} + 4a_{r-2} = 3r^2$
   (b) $a_{r+2} - a_{r+1} - 2a_r = r^2$
   (c) $a_r = 6a_{r-1} - 12a_{r-2} + 8a_{r-3} + 3^r$

7. The Tower of Hanoi consists of three pegs mounted on a board together with 64 disks of different sizes. Initially these disks are placed on the first peg in order of size, with the largest on the bottom. The disks can be moved one at a time from one peg to another as long as a disk is never placed on top of a smaller disk. Find the number of moves needed to have all the disks on the second peg in order of size, with the largest on the bottom. If one move takes 1 second, then find the time in years to solve the problem.

8. Find particular solution of the difference equation

$$a_r + a_{r-1} = 3r \cdot 2^r, \quad r \geq 1$$

9. Find the solution of the recurrence relation

   $a_n = 3a_{n-1} + 2n$, with initial condition $a_1 = 3$.

10. Let $a_r$ denote the number of edges in a complete graph on $r$ vertices:
    (a) Derive a recurrence relation for $a_r$ in terms of $a_{r-1}$
    (b) Solve the recurrence relation.

11. Solve the difference equation $a_{r+2} - 5a_{r+1} + 6a_r = 2$ by the method of generating functions satisfying the initial conditions $a_0 = 1$, $a_1 = 2$.

**Answers to Selected Problems**

1. $\dfrac{1}{1 + 2z}$

2. $\dfrac{1}{4}\left[1^r - \left(\dfrac{1}{5}\right)^{r+1}\right]$

3. $a_n = 5n + 2, \quad n \geq 1$

4. $a_n = \dfrac{3^{n+1} - 1}{2}, \quad n \geq 1$

5(b). $a_n = 2 \cdot 3^n - 2^n$

7. $2^{64} - 1$ moves

8. $a_r = \left[2r + \dfrac{2}{3}\right] \cdot 2^r, \quad r \geq 0$

9. $a_n = A \cdot 3^n + \displaystyle\sum_{K=1}^{n} 2 \cdot K$

   $= 3^n + n(n+1)$

# 7

# Logic

## 7.1 INTRODUCTION

*Mathematical Logic* or *logic* is the study of reasoning. It is the basis on which sentences are built. In this chapter, we introduce the study of logic from a mathematical point of view. Logic has practical applications to the Computer Programming, Artificial Intelligence and other areas of computer science. The main objective of logic is to formulate rules which may be helpful in taking decisions whether any argument or reasoning is "valid".

## 7.2 PROPOSITIONS

The fundamental objects we work with in arithmetic are numbers. In a similar way, the fundamental objects in logic are propositions. A **proposition** is a declarative sentence that is either true or false, but not both.

**Example 1:** All the following declarative sentences are propositions

1. Delhi, is the capital of India
2. $1 + 1 = 2$
3. $2 + 1 = 4$

Propositions 1 and 2 are true, whereas 3 is false.

**Example 2:** Consider the following sentences

1. What time is it?
2. Read this carefully.
3. $x + 1 = 2$.
4. $x + y = z.-$

Sentences 1 and 2 are not propositions because they are not declarative sentences. Sentences 3 and 4 are not propositions because they are neither true nor false, since the variables in these sentences have not been assigned values.

Letters are used to denote propositions, just as letters are used to denote variables. The conventional letters used for this purpose are $p, q, r, s, \ldots$ . The **truth value** of a proposition is true, denoted by T, if it is a true proposition and false, denoted by F, if it is a false proposition.

The area of logic that deals with propositions is called the **propositional calculus** or **propositional logic**. It was first developed systematically by the Greek philosopher Aristotle more than 2300 years ago.

We now turn our attention to methods for producing new propositions from those that we already have. These methods were discussed by the English mathematician George Boole in 1854 in his book *The Laws of Thought*. Many mathematical statements are constructed by combining one or more propositions. New propositions, called **compound propositions**, are formed from existing propositions using logical operators.

---

**GEORGE BOOLE (1815–1864)** George Boole, was born in Lincoln, England, in 1815. He became one of the most important mathematicians of the 1800s. Although he considered a career as a clergyman, he decided instead to go into teaching and soon afterward opened a school of his own. In his preparation for teaching mathematics, Boole-unsatisfied with textbooks of his day decided to read the works of the great mathematicians. In 1854 he published *The Laws of Thought*, his most famous work. In this book Boole introduced what is now called *Boolean algebra* in his honor.

---

## 7.3 BASIC LOGICAL OPERATIONS

**Negation:** Let $p$ be a proposition. The statement "It is not the case that $p$" is another proposition, called the *negation* of $p$. The negation of $p$ is denoted by $\neg p$ or $\bar{p}$ or $\sim p$, read as "not $p$".

**Example 3:** Find the negation of the proposition "Today is Friday" and express this in simple English.

### *Solution*

The negation is "It is not the case that today is Friday." This negation can be more simply expressed by "Today is not Friday," Or "It is not Friday today."

A **truth table** displays the relationships between the truth values of propositions. Truth tables are especially valuable in the determination of the truth values of propositions constructed from simpler propositions. **Table 1** displays the two possible truth values of a proposition $p$ and the corresponding truth values of its $\sim p$.

The negation of a proposition can also be considered the result of the operation of the **negation operator** on a proposition. The negation operator constructs a new proposition from a single existing proposition. We will now introduce the logical operators that are used

**Table 1:** Truth Table for the
Negation of a Proposition

| $p$ | $\sim p$ |
|:---:|:---:|
| T | F |
| F | T |

to form new propositions from two or more existing propositions. These logical operators are also called **connectives**.

**Conjunction:** Let $p$ and $q$ be propositions. The proposition "$p$ and $q$" denoted $p \wedge q$, is the proposition that is true when both $p$ and $q$ are true and is false otherwise. The proposition $p \wedge q$ is called the *conjunction* of $p$ and $q$.

The truth table for $p \wedge q$ is shown in **Table 2**. Note that there are four rows in this truth table, one row for each possible combination of truth values for the propositions $p$ and $q$.

**Table 2:** Truth Table for the Conjunction of
Two Propositions

| $p$ | $q$ | $p \wedge q$ |
|:---:|:---:|:---:|
| T | T | T |
| T | F | F |
| F | T | F |
| F | F | F |

**Example 4:** Find the conjunction of the propositions $p$ and $q$ where $p$ is the proposition "Today is Friday" and $q$ is the proposition "It is raining today."

*Solution*

The conjunction of these propositions, $p \wedge q$, is the proposition "Today is Friday and it is raining today." This proposition is true on rainy Fridays and is false on any day that is not a Friday and on Fridays when it does not rain.

**Disjunction:** Let $p$ and $q$ be propositions. The proposition "$p$ or $q$," denoted $p \vee q$, is the proposition that is false when $p$ and $q$ are both false and true otherwise. The proposition $p \vee q$ is called the *disjunction* of $p$ and $q$. The truth table for $p \vee q$ is shown in **Table 3**.

**Consider the sentence**

"Students who have taken calculus or computer science can take this class."

Here, we mean that students who have taken both calculus and computer science can take the class, as well as the students who have taken only one of the two subjects. On the other hand,

**Table 3:** The Truth Table for the Disjunction of Two Propositions

| $p$ | $q$ | $p \wedge q$ |
|---|---|---|
| T | T | T |
| T | F | T |
| F | T | T |
| F | F | F |

we are using the exclusive or when we say "Students who have taken calculus or computer science, but not both, can enroll in this class."

Here, we mean that students who have taken both calculus and a computer science course cannot take the class. Only those who have taken exactly one of the two courses can take the class. Hence, this is an exclusive, rather than an inclusive.

**Example 5:** What is the disjunction of the propositions $p$ and $q$ where $p$ is the proposition "Today is Friday" and $q$ is the proposition "It is raining today?"

### *Solution*

The disjunction of $p$ and $q$, $p \vee q$, is the proposition "Today is Friday or it is raining today." This proposition is true on any day that is either a Friday or a rainy day (including rainy Fridays). It is only false on days that are not Fridays when it also does not rain.

A disjunction is true when at least one of the two propositions in it is true. Sometimes, we use *or* in an exclusive sense. When the **exclusive or** is used to connect the propositions $p$ and $q$, the proposition "$p$ or $q$ (but not both)" is obtained. This proposition is true when $p$ is true and $q$ is false, and when $p$ is false and $q$ is true. It is false when both $p$ and $q$ are false and when both are true.

**Table 4:** The Truth Table for the Exclusive OR of Two Propositions

| $p$ | $q$ | $p \oplus q$ |
|---|---|---|
| T | T | F |
| T | F | T |
| F | T | T |
| F | F | F |

**Table 5:** The Truth Table for the Implication $p \rightarrow q$

| $p$ | $q$ | $p \rightarrow q$ |
|---|---|---|
| T | T | T |
| T | F | F |
| F | T | T |
| F | F | T |

## 7.4 EXCLUSIVE OR

Let $p$ and $q$ be propositions. The *exclusive or* of $p$ and $q$, denoted by $p \oplus q$, is the proposition that is true when exactly one of $p$ and $q$ is true and is false otherwise.

## 7.5 IMPLICATION

Let $p$ and $q$ be propositions. The *implication* $p \rightarrow q$ *is* the proposition that is false when $p$ is true and $q$ is false, and true otherwise. $p$ is called the *hypothesis* (or *antecedent* or *premise*) and $q$ *is* called conclusion or *consequence*.

The truth table for the implication $p \rightarrow q$ is shown above. An implication is sometimes called a **conditional statement**. A variety of terminology is used to express $p \rightarrow q$. You will encounter most if not all of the following ways to express this implication:

| | |
|---|---|
| "if $p$, then $q$" | "$p$ implies $q$" |
| "if $p$, $q$" | "$p$ only if $q$" |
| "$p$ is sufficient for $q$" | "a sufficient condition for $q$ is $p$" |
| "$q$ if $p$" | "$q$ whenever $p$" |
| "$q$ *when p*" | "$q$ is necessary for $p$" |
| "a necessary condition for $p$ *is q*" | "$q$ follows from $p$" |

The implication $p \rightarrow q$ is false only in the case that $p$ is true, but $q$ is false. It is true when both $p$ and $q$ are true, and when $p$ is false (no matter what truth value $q$ has).

A useful way to understand the truth value of an implication is to think of an obligation or a contract. For example, the pledge many politicians make when running for office is: "If I am elected, then I will lower taxes."

If the politician is elected, voters would expect this politician to lower taxes. Furthermore, if the politician is not elected, then voters will not have any expectation that this person will lower taxes, although the person may have sufficient influence to cause those in power to lower taxes. It is only when the politician is elected but does not lower tax; than voters can say that the politician has broken the campaign pledge. This last scenario corresponds to the case when $p$ is true, but $q$ is false in $p \rightarrow q$.

Similarly, for instance, the implication "If it is sunny today, then we will go to the beach" is an implication used in normal language, since there is a relationship between the hypothesis and the conclusion. Further, this implication is considered valid unless it is indeed sunny today, but we do not go to the beach. On the other hand, the implication "If today is Friday, then $2 + 3 = 5$" is true from the definition of implication, since its conclusion is true. (The truth value of the hypothesis does not matter then.) The implication "If today is Friday, then $2 + 3 = 6$" is true every day except Friday, even though $2 + 3 = 6$ is false.

We would not use these last two implications in natural language, since there is no relationship between the hypothesis and the conclusion in either implication. The if-then construction used in many programming languages is different from that used in logic. Most programming languages contain statements such as **if $p$ then** S, where $p$ is a proposition and S is a program segment (one or more statements to be executed). When execution of a program encounters such a statement, S is executed if $p$ is true, but S is not executed if $p$ is false. For example, consider following code segment

$$x = 0;$$
**if** $(2 + 2 = = 4)$ **then** $x = x + 1.$

Since $2 + 2 = 4$ is true, the assignment statement $x = x + 1$ is executed. Hence, $x$ has the value $0 + 1 = 1$ after this statement is encountered.

## 7.6 CONVERSE, CONTRAPOSITIVE AND INVERSE

There are some related implications that can be formed from $p \rightarrow q$. The proposition $q \rightarrow p$ is called the **converse** of $p \rightarrow q$. The **contrapositive** of $p \rightarrow q$ is the proposition $\sim q \rightarrow \sim p$. The proposition $\sim p \rightarrow \sim q$ is called the **inverse** of $p \rightarrow q$. The contrapositive, $\sim q \rightarrow \sim p$ of an implication $p \rightarrow q$ has the same truth value as $p \rightarrow q$. To see this, note that the contrapositive is false only when $\sim p$ is false and $\sim q$ is true, that is, only when $p$ is true and $q$ is false. On the other hand, neither the converse, $q \rightarrow p$, nor the inverse $\sim p \rightarrow \sim q$, has the same truth value as $p \rightarrow q$ for all possible truth values of $p$ and $q$. To see this, note that when $p$ is true and $q$ is false, the original implication is false, but the converse and the inverse are both true. When two compound propositions always have the same truth value we call them **equivalent**, so that an implication and its contrapositive are equivalent. The converse and the inverse of an implication are also equivalent, as the reader can verify. One of the most common logical errors is to assume that the converse or the inverse of an implication is equivalent to this implication.

**Example 7:** What are the contrapositive, the converse, and the inverse of the implication "The home team wins whenever it is raining."?

### *Solution*

Because "$q$ whenever $p$" is one of the ways to express the implication $p \rightarrow q$, the original statement can be rewritten as "If it is raining, then the home team wins." Consequently, the contrapositive of this implication is "If the home team does not win, then it is not raining." The converse is "If the home team wins, then it is raining."

The inverse is "If it is not raining, then the home team does not win." Only the contrapositive is equivalent to the original statement. We now introduce another way to combine propositions.

**Example 8:** Consider the following conditional statement:

"If the flood destroys my house or the fire destroys my house, then my insurance company will pay me." Find the converse and contrapositive of the statement $p$.

### *Solution*

Let $p$: "If the flood destroys my house or the fire destroys my house and $q$: my insurance company will pay me".

$\therefore$    Given statement is $p \rightarrow q$

The converse of $p \rightarrow q$ is $q \rightarrow p$ i.e.

Converse of the statement is "If my insurance company pays me, then the flood destroys my house or the fire destroys my house"

The contrapositive of $p \rightarrow q$ is $\sim q \rightarrow \sim p$ i.e.

Contrapositive of the statement is "If my insurance company does not pay me, then the flood does not destroy my house and the fire does not destroy my house".

## 7.7 BICONDITIONAL

Let $p$ and $q$ be propositions. The *biconditional* $p \leftrightarrow q$ is the proposition that is true when $p$ and $q$ have the same truth values, and is false otherwise.

The truth table for $p \leftrightarrow q$ is shown in **Table 6**. Note that the biconditional $p \leftrightarrow q$ is true precisely when both the implications $p \rightarrow q$ and $q \leftarrow p$ are true. Because of this, the terminology

"$p$ if and only if $q$"

is used for this biconditional and it is symbolically written by combining the symbols $\rightarrow$ and $\leftarrow$. There are some other common ways to express $p \leftrightarrow q$:

**Table 6:** The Truth Table for the Biconditional $p \leftrightarrow q$

| $p$ | $q$ | $p \leftrightarrow q$ |
| --- | --- | --- |
| T | T | T |
| T | F | F |
| F | T | F |
| F | F | T |

"$p$ is necessary and sufficient for $q$"
"if $P$ then $q$, and conversely"
"$p$ If $q$".

The last way of expressing the biconditional uses the abbreviation "if" for "if and only if." Note that $p \leftrightarrow q$ has exactly the same truth value as $(p \rightarrow q) \wedge (q \rightarrow p)$.

**Example 9:** Let $p$ be the statement "You can take the flight" and let $q$ be the statement "You buy a ticket." Then $p \leftrightarrow q$ is the statement "You can take the flight if and only if you buy a ticket."

### *Solution*

This statement is true if $p$ and $q$ are either both true or both false, that is, if you buy a ticket and can take the flight or if you do not buy a ticket and you cannot take the flight. It is false when $p$ and $q$ have opposite truth values, that is, when you do not buy a ticket, but you can take the flight (such as when you get a free trip) and when you buy a ticket and cannot take the flight (such as when the airline bumps you).

## 7.8 PRECEDENCE OF LOGICAL OPERATORS

We can construct compound propositions using the negation operator and the logical operators defined so far. We will generally use parentheses to specify the order in which logical operators in a compound proposition are to be applied. For instance, $(p \vee q) \wedge (\sim r)$ is the conjunction of $p \vee q$ and $\sim r$. However, to reduce the number of parentheses, we specify that the negation operator is applied before all other logical operators. This means that $\sim p \wedge q$ is the conjunction of $\sim p$ and $q$, namely, $(\sim p) \wedge q$, not the negation of the conjunction of $p$ and $q$, namely $\sim(p \wedge q)$.

Another general rule of precedence is that the conjunction operator takes precedence over the disjunction operator, so that $p \wedge q \vee r$ means $(p \wedge q) \vee r$ rather than $p \wedge (q \vee r)$. Because this rule may be difficult to remember, we will continue to use parentheses so that the order of the disjunction and conjunction operators is clear.

Finally, it is an accepted rule that the conditional and biconditional operators $\rightarrow$ and $\leftrightarrow$ have lower precedence than the conjunction and disjunction operators, $\wedge$ and $\vee$. Consequently, $p \vee q \rightarrow r$ is the same as $(p \vee q) \rightarrow r$. We will use parentheses when the order of the conditional operator and biconditional operator is at issue, although the conditional operator has precedence over the biconditional operator. **Table 7**, displays the precedence levels of the logical operators, $\sim$, $\wedge$, $\vee$, $\rightarrow$, and $\leftrightarrow$.

**Table 7:** Precedence of Logical Operators.

| Operator | Precedence |
| --- | --- |
| $\sim$ | 1 |
| $\wedge$ | 2 |
| $\vee$ | 3 |
| $\rightarrow$ | 4 |
| $\leftrightarrow$ | 5 |

## 7.9 TRANSLATING ENGLISH SENTENCES TO SYMBOLS

There are many reasons to translate English sentences into expressions involving propositional variables and logical connectives. In particular, English (and every other human language) is often ambiguous. Translating sentences into logical expressions removes the ambiguity. Note that this may involve making a set of reasonable assumptions based on the intended meaning of the sentence. Moreover, once we have translated sentences from English into logical expressions we can analyze these logical expressions to determine their truth values, we can manipulate them, and we can use rules of inference (discussed later) to reason about them. To illustrate the process, consider following examples.

**Example 10:** How can this English sentence be translated into a logical expression?

"You can access the Internet from campus only if you are a computer science major or you are not a freshman."

### Solution

Although it is possible to represent the sentence by a single propositional variable, such as $p$, this would not be useful when analyzing its meaning or reasoning with it. Instead, we will use propositional variables to represent each sentence part and determine the appropriate logical connectives between them. In particular, we let $a$, $c$, and $f$ represent "You can access the Internet from campus," "You are a computer science major," and "You are a freshman," respectively. Noting that "only if" is one way an implication can be expressed; this sentence can be represented as

$$a \rightarrow (c \vee {\sim}f).$$

**Example 11:** How can this English sentence be translated into a logical expression?

"You cannot ride the roller coaster if you are less than 4 feet tall unless you are older than 16 years old."

### Solution

There are many ways to translate this sentence into a logical expression. The simplest but least useful way is simply to represent the sentence by a single propositional variable, say, $p$. Although this is not wrong, doing this would not assist us when we try to analyze the sentence or reason using it. More appropriately, what we can do is to use propositional variables to represent each of the sentence parts and to decide on the appropriate logical connectives between them. In particular, we let $q$, $r$, and $s$ represent.

"You can ride the roller coaster," "you are under 4 feet tall" and "you are older than 16 years old" respectively. Then the sentence can be translated to

$$(r \wedge {\sim}s) \rightarrow {\sim}q.$$

## 7.10 PROPOSITIONAL EQUIVALENCE

An important type of step used in a mathematical argument is the replacement of a statement with the same truth values. Because of this, methods that produce propositions with the same truth value as a given compound proposition are used extensively in the construction of mathematical-arguments.

We begin our discussion with a classification of compound propositions according to their possible truth values.

### *Tautology*

A compound proposition that is always true, no matter what the truth values of the propositions that occur in it, is called a *tautology*. i.e. if each entry in the final column of the truth table of a statement formula is T, then it is called *tautology*.

### *Contradiction*

A compound proposition that is always false is called a *contradiction*. i.e. if each entry in the final column of the truth table of a statement formula is F, then it is called *contradiction*.

### *Contingency*

A proposition that is neither a tautology nor a contradiction is called a *contingency*.

Tautologies and contradictions are often important in mathematical reasoning. The following example illustrates these types of propositions.

We can construct examples of tautologies and contradictions using just one proposition. Consider the truth tables of $p \vee \sim p$ and $p \wedge \sim p$, shown in **Table 8**. Since $p \vee \sim p$ is always true, it is a tautology. Since $p \wedge \sim p$ is always false, it is a contradiction.

**Table 8:** Examples of a Tautology and a Contradiction

| $p$ | $\sim p$ | $p \vee \sim p$ | $p \wedge \sim p$ |
|---|---|---|---|
| T | F | T | F |
| F | T | T | F |

**Example 13:** Prove that each of the following is a tautology:

    a. $[(p \rightarrow q) \wedge (q \rightarrow r)] \rightarrow (p \rightarrow r)$
    b. $\sim (p \rightarrow q) \rightarrow p$

### *Solution*

**a.** Truth Table for $[(p \rightarrow q) \wedge (q \rightarrow r)] \rightarrow (p \rightarrow r)$ *is*

**Table 9**

| $p$ | $q$ | $r$ | $p \to q$ | $q \to r$ | $(p \to q) \wedge (q \to r)$ | $p \to r$ | $[(p \to q) \wedge (q \to r)] \to (p \to r)$ |
|---|---|---|---|---|---|---|---|
| F | F | F | T | T | T | T | T |
| F | F | T | T | T | T | T | T |
| F | T | F | T | F | F | T | T |
| F | T | T | T | T | T | T | T |
| T | F | F | F | T | F | F | T |
| T | F | T | F | T | F | T | T |
| T | T | F | T | F | F | F | T |
| T | T | T | T | T | T | T | T |

We see that for all possible truth values of $p$, $q$ and $r$, the compound statement
$[(p \to q) \wedge (q \to r)] \to (p \to r)$ is true, hence the statement is a tautology.

**b.** Truth Table for $\sim (p \to q) \to p$

**Table 10**

| $p$ | $q$ | $p \to q$ | $\sim(p \to q)$ | $\sim(p \to q) \to p$ |
|---|---|---|---|---|
| F | F | T | F | T |
| F | T | T | F | T |
| T | F | F | T | T |
| T | T | T | F | T |

We see that for all possible truth values of $p$ and $q$, the compound statement $\sim (p \to q) \to p$ is true, hence the statement is a tautology.

## 7.11 LOGICAL EQUIVALENCE

Compound propositions that have the same truth values in all possible cases are called **logically equivalent**.

Consider a rather complicated proposition:

'It is not the case that both the input file and the output file are not on the disk.'

The proposition below expresses the same idea more simply:

'Either the input file or the output file is on the disk.'

If we were to express these propositions symbolically, we would expect the resulting logical expressions to have the same truth table. Let $p$ and $q$ denote respectively the propositions 'The input file is on the disk' and 'The output file is on the disk'. Then we have the following result, in which for convenience we have combined the truth tables for the two expressions into a single table (**Table 11**).

**Table 11**

| $p$ | $q$ | $\neg p$ | $\neg q$ | $\neg p \wedge \neg q$ | $\neg(\neg p \wedge \neg q)$ | $p \vee q$ |
|---|---|---|---|---|---|---|
| T | T | F | F | F | T | T |
| T | F | F | T | F | T | T |
| F | T | T | F | F | T | T |
| F | F | T | T | T | F | F |

The sixth and seventh columns are the truth tables for the first and second expressions respectively, and we can see that their truth values are the same. We can also define this notion as follows.

**Definition:**  The propositions $p$ and $q$ are called *logically equivalent if $p \leftrightarrow q$* is a tautology. The notation $p \equiv q$ denotes that $p$ and $q$ are logically equivalent.

**Remark:** The symbol $\equiv$ is not a logical connective since $p \equiv q$ is not a compound proposition, but rather is the statement that $p \leftrightarrow q$ is a tautology. The symbol $\Leftrightarrow$ is sometimes used instead of $\equiv$ to denote logical equivalence.

One way to determine whether two propositions are equivalent is to use a truth table. In particular, the propositions $p$ and $q$ are equivalent if and only if the columns giving their truth values agree. The following example illustrates this method.

**Remark:** In general, $2^n$ rows are required in the truth table if a compound proposition involves $n$ propositions.

## 7.12 LAWS OF LOGIC

A statement of the form $P \equiv Q$, where P and Q are logical expressions is called a law of logic. A list of the most important laws of logic is given in **Table 12**.

In these equivalences, **T** denotes any proposition that is always true and **F** denotes any proposition that is always false. We also display some useful equivalences for compound propositions involving implications and biconditionals in Tables A and B below, respectively.

The associative law for disjunction shows that the expression $p \vee q \vee r$ is well defined, in the sense that it does not matter whether we first take the disjunction of $p$ and $q$ and then the disjunction of $p \vee q$ with $r$, or if we first take the disjunction of $q$ and $r$ and then take the disjunction of p and $q \vee r$. Similarly, the expression $p \wedge q \wedge r$ is well defined. By extending this reasoning, it follows that $p_1 \vee p_2 \vee \ldots \vee p_n$ and $p_1 \wedge p_2 \wedge \ldots \wedge pn$ are well defined whenever $p_1$, $p_2$, ..., $p_n$ are propositions.

**Table 12:** Logical Equivalences

| Equivalence | Name |
|---|---|
| $p \wedge T \equiv p$<br>$p \vee F \equiv p$ | Identity laws |
| $p \vee T \equiv T$<br>$p \wedge F \equiv F$ | Domination laws |
| $p \vee p \equiv p$<br>$p \wedge p \equiv p$ | Idempotent laws |
| $\sim(\sim p) \equiv p$ | Double negation law |
| $p \vee q \equiv q \vee p$<br>$p \wedge q \equiv q \wedge p$ | Commutative laws |
| $(p \vee q) \vee r \equiv p \vee (q \vee r)$<br>$(p \wedge q) \wedge r \equiv p \wedge (q \wedge r)$ | Associative laws |
| $p \vee (q \wedge r) \equiv (p \wedge q) \wedge (p \vee r)$<br>$p \wedge (q \vee r) \equiv (p \vee q) \vee (p \vee r)$ | Distributive laws |
| $\sim(p \wedge q) \equiv \sim p \vee \sim q$<br>$\sim(p \vee q) \equiv \sim p \wedge \sim q$ | De Morgan's laws |
| $p \vee (p \wedge q) \equiv p$<br>$p \wedge (p \vee q) \equiv p$ | Absorption laws |
| $p \vee \sim p \equiv T$<br>$p \wedge \sim p \equiv F$ | Negation laws |

**AUGUSTUS DE MORGAN (1806–1871)** Augustus De Morgan was born in India, De Morgan was a noted teacher who stressed principles over techniques. His students included many famous mathematicians, including Ada Augusta, Countess of Lovelace, who was Charles Babbage's collaborator in his work on computing machines. De Morgan was an extremely prolific writer. In 1838 he presented what was perhaps the first clear explanation of an important proof technique known as *mathematical induction*, a term he coined. He invented notations that helped him prove propositional equivalences, such as the laws that are named after him. In 1842 De Morgan presented what was perhaps the first precise definition of a limit and developed some tests for convergence of infinite series. De Morgan was also interested in the history of mathematics and wrote biographies of Newton and Halley.

**Table 13:** Logical Equivalences
Involving Implications

$$p \to q \equiv {\sim} p \vee q$$
$$p \to q \equiv {\sim} q \to {\sim} p$$
$$p \vee q \equiv {\sim} p \to q$$
$$p \wedge q \equiv {\sim}(p \to {\sim} q)$$
$$(p \to q) \wedge (p \to r) \equiv p \to (q \wedge r)$$
$$(p \to r) \wedge (q \to r) \equiv (p \vee q) \to r$$
$$(p \to q) \vee (p \to r) \equiv p \to (q \vee r)$$
$$(p \to r) \vee (q \to r) \equiv (p \wedge q) \to r$$

**Table 14:** Logical Equivalences
Involving Implications

$$p \leftrightarrow q \equiv (p \to q) \wedge (q \to p)$$
$$p \leftrightarrow q \equiv {\sim} p \leftrightarrow {\sim} q$$
$$p \leftrightarrow q \equiv (p \wedge q) \vee ({\sim} p \wedge {\sim} q)$$
$${\sim}(p \leftrightarrow q) \equiv p \leftrightarrow {\sim} q$$

The logical equivalences in above tables can be used to construct additional logical equivalences. The reason for this is that a proposition in a compound proposition can be replaced by one that is logically equivalent to it without changing the truth value of the compound proposition. This technique is illustrated in Examples below, where we also use the fact that if $p$ and $q$ are logically equivalent and $q$ and $r$ are logically equivalent then $p$ and $r$ are logically equivalent.

**Example 14:** Show that ${\sim}(p \vee q)$ and ${\sim}p \wedge {\sim}q$ are logically equivalent. This equivalence is one of *De Morgan's laws* for propositions, named after the English mathematician Augustus De Morgan, of the mid-nineteenth century.

**Table 15:** Truth Tables for ${\sim}(p \vee q)$ and ${\sim}p \wedge {\sim}q$.

| **P** | $q$ | $p \vee q$ | ${\sim}(p \vee q)$ | ${\sim}p$ | ${\sim}q$ | ${\sim}p \wedge {\sim}q$ |
|---|---|---|---|---|---|---|
| T | T | T | F | F | F | F |
| T | F | T | F | F | T | F |
| F | T | T | F | T | F | F |
| F | F | F | T | T | T | T |

***Solution***

The truth tables for these propositions are displayed in **Table 15**. Since the truth values of the propositions ${\sim}(p \vee q)$ and ${\sim}p \wedge {\sim}q$ agree for all possible combinations of the truth values of $p$ and $q$, it follows that ${\sim}(p \vee q) \leftrightarrow ({\sim}p \wedge {\sim}q)$ is a tautology and that these propositions are logically equivalent.

**Example 15:** Show that the propositions $p \to q$ and ${\sim}p \vee q$ are logically equivalent.

***Solution***

We construct the truth table for these propositions in **Table 16**. Since the truth values of ${\sim}p \vee q$ and $p \to q$ agree, these propositions are logically equivalent.

**Table 16:** Truth Table for
$\sim p \vee q$ and $p \rightarrow q$.

| $p$ | $Q$ | $\sim p$ | $\sim p \vee q$ | $p \rightarrow \sim q$ |
|---|---|---|---|---|
| T | T | F | T | T |
| T | F | F | F | F |
| F | T | T | T | T |
| F | F | T | T | T |

**Example 16:** Show that the propositions $p \vee (q \wedge r)$ and $(p \vee q) \wedge (p \vee r)$ are logically equivalent. i.e. Prove the *distributive law* of disjunction over conjunction.

***Solution***

We construct the truth table for these propositions in **Table 17**. Since the truth values of $p \vee (q \vee r)$ and $(p \vee q) \wedge (p \vee r)$ agree, these propositions are logically equivalent.

**TABLE 17:** A Demonstration equivalence of $p \vee (q \wedge r)$ and $(p \vee q) \wedge (p \vee r)$

| $p$ | $q$ | $r$ | $q \wedge r$ | $p \vee (q \wedge r)$ | $p \vee q$ | $p \vee r$ | $(p \vee q) \wedge (p \wedge r)$ |
|---|---|---|---|---|---|---|---|
| T | T | T | T | T | T | T | T |
| T | T | F | F | T | T | T | T |
| T | F | T | F | T | T | T | T |
| T | F | F | F | T | T | T | T |
| F | T | T | T | T | T | T | T |
| F | T | F | F | F | T | F | F |
| F | F | T | F | F | F | T | F |
| F | F | F | F | F | F | F | F |

**Example 17:** Show that $\sim (p \vee (\sim p \wedge q))$ and $\sim p \wedge \sim q$ are logically equivalent.

***Solution***

We could use a truth table to show that these compound propositions are equivalent. Instead, we will establish this equivalence by developing a series of logical equivalences, using one of the equivalences in above tables at a time, starting with $\sim (p \vee (\sim p \wedge q))$ and ending with $\sim p \wedge \sim q$. We have the following equivalences.

$$\sim (p \vee (\sim p \wedge q)) \equiv \sim p \wedge \sim (\sim p \wedge q) \qquad \text{from the second De Morgan's law}$$
$$\equiv \sim p \wedge (\sim (\sim p) \vee \sim q) \qquad \text{from the first De Morgan's law}$$

$$\equiv \sim p \wedge (p \vee -q) \qquad \text{from the double negation law}$$
$$\equiv (\sim p \wedge p) \vee (\sim p \wedge \sim q) \qquad \text{from the second distributive law}$$
$$\equiv F \vee (\sim p \wedge \sim q) \qquad \text{since } \sim p \wedge p \equiv F$$
$$\equiv (\sim p \wedge \sim q) \vee F \qquad \text{from the commutative law for disjunction}$$
$$\equiv \sim p \wedge \sim q \qquad \text{from the identity law for F}$$

Consequently $\sim(p \vee (\sim p \wedge q))$ and $\sim p \wedge \sim q$ are logically equivalent.

**Example 18:** Show that $(p \wedge q) \to (p \vee q)$ is a tautology.

*Solution*

To show that this statement is a tautology, we will use logical equivalences to demonstrate that it is logically equivalent to T. (*Note*: This could also be done using a truth table.)

$$(p \wedge q) \to (p \vee q) \equiv \sim(p \wedge q) \vee (p \vee q)$$
$$\equiv (\sim p \vee \sim q) \vee (p \vee q) \qquad \text{by the first De Morgan's law}$$
$$\equiv (\sim p \vee p) \vee (\sim q \vee q) \qquad \text{by the associative and commutative}$$
$$\equiv T \vee T \qquad \text{laws for disjunction}$$
$$\equiv T \qquad \text{by the domination law}$$

A truth table can be used to determine whether a compound proposition is a tautology, but when the number of variables grows, this becomes impractical. For instance, there are $2^{20} = 1,048,576$ rows in the truth value table for a proposition with 20 variables it becomes difficult to construct a truth table. But when there are 1000 variables, checking everyone of the $2^{1000}$ (a number with more than 300 decimal digits) possible combinations of truth values simply cannot be done by a computer in even trillions of years.

## 7.13 PREDICATES

Propositional logic provides a useful setting in which we can analyze many types of logical argument. There are situations, however, where propositional logic is inadequate, because it cannot deal with the logical structure that is sometimes present within atomic propositions. Predicate logic allows us to do this. The logic based on analysis of predicates in any statement is called *predicate logic*. A **predicate** is a statement containing one or more variables. If values are assigned to all the variables in a predicate, the resulting statement is a proposition.

For example, '$x < 5$' is a predicate, where $x$ is a variable denoting any real number. If we substitute a real number for $x$, we obtain a proposition; for example, '$3 < 5$' and '$6 < 5$' are propositions with truth values T and F respectively.

A variable need not be a number. For example, '$x$ is an employee of the Ezisoft Software Company' becomes a proposition with a well defined truth value when $x$ is replaced by a person's name: 'Sumit is an employee of the Ezisoft Software Company.'

## 7.14 QUANTIFIERS

The predicate '$x < 5$' is not always true; it *is* true for some values of $x$, so we can form a true proposition by writing:

'There exists an $x$ such that $x < 5$.' (Here, 'an' means 'at least one'.)

The expressions 'for all' and 'there exists' are called **quantifiers**. The process of applying a quantifier to a variable is called **quantifying** the variable. A variable which has been quantified is said to be **bound**. For example, the variable $x$ in 'There exists an $x$ such that $x < 5$' is bound by the quantifier 'there exists'. A variable that appears in a predicate but is not bound is said to be **free**.

We will use capital letters to denote predicates. A predicate P that contains a variable $x$ can be written symbolically as $P(x)$. A predicate can contain more than one variable; a predicate P with two variables, $x$ and $y$ for example, can be written $P(x, y)$. In general, a predicate with $n$ variables, $x_1, x_2, x_n$, can be written $P(x_1, x_2, ..., x_n)$. The quantifiers 'for all' and 'there exists' are denoted by the symbols $\forall$ and $\exists$ respectively. With this notation, expressions containing predicates and quantifiers can be written symbolically.

**Example 19:**  Write the following sentences in symbolic form:

'For every number $x$ there is a number $y$ such that $y = x + 1$.'
'There is a number $y$ such that, for every number $x$, $y = x + 1$.'

***Solution***

Let $P(x, y)$ denote the predicate '$y = x + 1$'. The first proposition is:

$$\forall x \, \exists y \, P(x, y)$$

The second proposition is:

$$\exists y \, \forall x \, P(x, y)$$

---

### Exercise 7.1

1. Use truth tables to verify these equivalences.

   (a) $p \wedge T \equiv p$
   (b) $p \vee F \equiv p$
   (c) $p \wedge F \equiv F$
   (d) $p \vee T \equiv T$
   (e) $p \vee p \equiv p$
   (f) $p \wedge p \equiv p$

2. Show that $\sim (\sim p)$ and $p$ are logically equivalent.

3. Use truth tables to verify the commutative laws

   (a) $p \vee q \equiv q \vee P$
   (b) $p \wedge q \equiv q \wedge p$

4. Use truth tables to verify the associative laws

   (a) $(p \vee q) \vee r \equiv p \vee (q \vee r)$
   (b) $(p \wedge q) \wedge r \equiv p \wedge (q \wedge r)$

5. Use a truth table to verify the distributive law

$$p \wedge (q \vee r) = (p \wedge q) \vee (p \wedge r).$$

6. Use a truth table to verify the equivalence $\sim (p \wedge q) = \sim p \vee \sim q$.

7. Show that each of these implications is a tautology by using truth tables.

   (a) $(p \wedge q) \to p$
   (b) $p \to (p \vee q)$
   (c) $\sim p \to (p \to q)$
   (d) $(p \wedge q) \to (p \to q)$
   (e) $\sim (p \to q) \to \sim q$

8. Show that each of these implications is a tautology by using truth tables.

   (a) $[\sim p \wedge (p \vee q)] \to q$
   (b) $[p \wedge (p \to q)] \to q$
   (c) $[(p \vee q) \wedge (p \to r) \vee (q \to r)] \to r$

9. Show that each implication in Exercise 7 is a tautology without using truth tables.

10. Show that each implication in Exercise 8 is a tautology without using truth tables.

11. Use truth tables to verify the absorption laws.

   (a) $p \vee (p \wedge q) = p$
   (b) $p \wedge (p \vee q) = p$

12. Determine whether the following are tautology

   (a) $(\sim p \wedge (p \to q)) \to \sim q$
   (b) $(p \vee q) \to p$

**Answers to Selected Problems**

12. (a) Not a tautology
    (b) Not a tautology

# Section 3

# 8

# Lattices

## 8.1 LATTICES

We have already studied the concept of partial order relation and partially ordered set (poset) in chapter 4. We also studied that the supremum (l.u.b) and infimum (g.l.b) of a poset may or may not exist in a poset. If in a poset every pair of elements has a unique supremum and infimum, then the poset forms a lattice. Now, in the present chapter we shall study the concept of lattice.

A lattice is a special type of a poset with two binary operations $\vee$ and $\wedge$, namely, join and meet respectively. Lattice structures are used in many different applications such as models of information flow and play an important role in Boolean algebra.

## 8.2 LATTICE AS A POSET

Let $(L, \leq)$ be a poset. If every two elements of L has supremum (l.u.b) and infimum (g.l.b) in L, then we say $(L, \leq)$ is a **lattice.**

In other words, a poset $(L, \leq)$ is a lattice, if for any $a, b \in L$, sup. $\{a, b\} \in L$ and inf. $\{a, b\} \in L$.

For the sake of convenience, will denote sup. $\{a, b\}$ by $a \vee b$ (read as '$a$ join $b$') and inf. $\{a, b\}$ by $a \wedge b$ (read as '$a$ meet $b$').

Other notations like $a + b$ and $a \cdot b$ or $a \cup b$ and $a \cap b$ are also used for sup. $\{a, b\}$ and inf. $\{a, b\}$ in certain cases.

*Note*:
1. P(X), power set of non-empty set X is a poset with the relation '$\subseteq$', then inf. $\{A, B\} = A \cap B$ and sup. $\{A, B\} = A \cup B$.
2. If set A is poset with the relation divisibility, then inf. $\{a, b\} = $ g.c.d $\{a, b\}$ and sup. $\{a, b\} = $ l.c.m $\{a, b\}$, for $a, b \in A$.

**Illustrations:**

1. Let X be any non-empty set and $P(X)$ be power set of X with usual inclusion relation '$\subseteq$'. Then $(P(X), \subseteq)$ is a poset.

   Now, let A, B $\in$ $P(X)$ be any subsets of X. Then inf. $\{A, B\} = A \cap B$ and sup. $\{A, B\} = A \cup B$ both are exist and subsets of X.

   $\therefore$ sup. $\{A, B\} \in P(X)$ and inf. $\{A, B\} \in P(X)$. Hence $(P(X), \subseteq)$ is a lattice.

2. Let $(S, \leq)$ is a chain with the partial order relation '$\leq$'.

   Since any pair of two elements $a$ and $b$ in a chain are comparable.

   $\therefore$ either $a \leq b$ or $b \leq a$

   Then, $\quad$ inf. $\{a, b\} = \begin{cases} a, \text{if } a \leq b \\ b, \text{if } b \leq a \end{cases}$ $\quad$ and $\quad$ sup. $\{a, b\} = \begin{cases} b, \text{if } a \leq b \\ a, \text{if } b \leq a \end{cases}$

   Hence, every chain is a lattice.

**Example 1:** Let $D_{10}$ be the set of all positive factors of 10, then prove that $D_{10}$ forms a lattice with the relation of divisibility. Also, draw the Hasse diagram of the lattice $D_{10}$.

*Solution*

Here $D_{10} = \{1, 2, 5, 10\}$

Clearly, $D_{10}$ is a partially ordered set with the relation of divisibility. [*verify!*]

Now, we know that, when the partial order, relation is divisibility, then for any $a, b \in D_{10}$, sup. $\{a, b\} = $ l.c.m $\{a, b\}$ and inf. $\{a, b\} = $ g.c.d $\{a, b\}$.

Then the following tables give the sup. $\{a, b\}$ and inf. $\{a, b\}$ of any pair of elements of $D_{10}$.

|  | | $b$ | | |
|---|---|---|---|---|
| sup. $\{a, b\}$ | 1 | 2 | 5 | 10 |
| 1 | 1 | 2 | 5 | 10 |
| 2 | 2 | 2 | 10 | 10 |
| 5 | 5 | 10 | 5 | 10 |
| 10 | 10 | 10 | 10 | 10 |

(column $a$ labels the rows: 1, 2, 5, 10)

|  | | $b$ | | |
|---|---|---|---|---|
| inf. $\{a, b\}$ | 1 | 2 | 5 | 10 |
| 1 | 1 | 1 | 1 | 1 |
| 2 | 1 | 2 | 1 | 2 |
| 5 | 1 | 1 | 5 | 5 |
| 10 | 1 | 2 | 5 | 10 |

(column $a$ labels the rows: 1, 2, 5, 10)

From the above tables, we observe that, for any $a, b \in D_{10}$ sup. $\{a, b\}$ and inf. $\{a, b\}$ both are exists in $D_{10}$.

Hence, $D_{10}$ is a lattice with the relation of divisibility.

The Hasse diagram of lattice $D_{10}$ is shown in figure below.

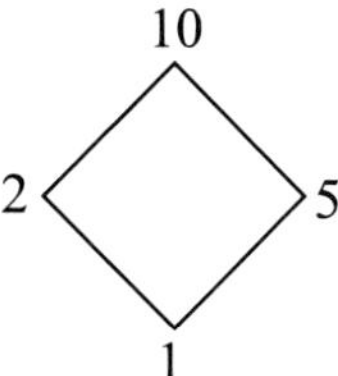

**Example 2:** Show that the set A = {2, 3, 4, 6} is not a lattice with the relation of divisibility. Also, draw the Hasse diagram of the poset A.

*Solution*

Here A = {2, 3, 4, 6}

Clearly, A is a poset with the relation of divisibility on A.

Now, for 2, 3, $\in$ A

sup. {2, 3} = l.c.m {2, 3} = 6 $\in$ A   and   inf. {2, 3} = g.c.d {2, 3} = 1 $\notin$ A.

$\therefore$   inf. {2, 3} = 1 $\notin$ A.

Again, for 4, 6 $\in$ A,

sup. {4, 6} = 12 $\notin$ A and inf. {4, 6} = 2 $\in$ A.

$\therefore$   sup. {4, 6} $\notin$ A. Hence, A is not a lattice.

The Hasse diagram of the poset A is shown in the figure below.

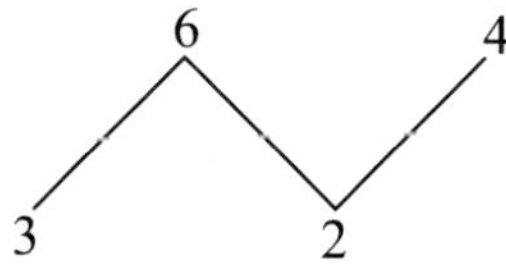

*Note*: From the above two examples, it can be noted that, if the Hasse diagram of a poset is a closed figure. Then the poset is a lattice. But if the Hasse diagram of a poset is an open figure, then the poset may or may not be a lattice.

**Remark:** In the above example, if we insert two numbers 1 and 12 in the poset A, then the poset A = {1, 2, 3, 4, 6, 12} forms a lattice and the Hasse diagram of A become a closed figure shown in below figure.

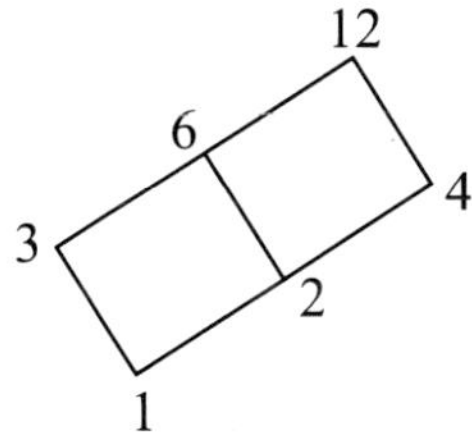

**Example 3:** Prove that $(N, \leq)$ is a lattice where '$\leq$' is the relation of divisibility. [*KU 2006*]

### *Solution*

Here the relation '$\leq$' is the relation of divisibility on N.

$\therefore$   $x \leq y$ means $x$ divides $y$, for $x, y \in N$

### Reflexivity

Since $\dfrac{x}{x}$, for all $x \in N$

$\Rightarrow$   $x \leq x$. Hence, $\leq$ is reflexive.

### Anti-symmetry

Let $x \leq y$ and $y \leq x$, for $x, y \in N$

$\Rightarrow$   $\dfrac{x}{y}$   and   $\dfrac{y}{x}$

$\therefore$   There exists some $u$ and $v$ in N such that $y = xu$ and $x = yv$

$\Rightarrow$   $y = (yv)u$   $\Rightarrow$   $y = y(uv)$   $\Rightarrow$   $1 = uv$   $\Rightarrow$   $u = 1, v = 1$

$\Rightarrow$   $y = x.1$      $\Rightarrow$   $y = x$

$\therefore$   $x \leq y$   and   $y \leq x$   $\Rightarrow$   $x = y$

Hence, '$\leq$' is anti-symmetric.

### Transitivity

Let $x \leq y$ and $y \leq z$, for $x, y, z \in N$

$\Rightarrow$   $\dfrac{x}{y}$   and   $\dfrac{y}{z}$

$\therefore$   There exist some $l$ and $m$ in N such that $y = xl$ and $z = ym$

$\Rightarrow$   $z = (xl)m$   $\Rightarrow$   $z = x(lm)$

$\Rightarrow$   $\dfrac{x}{z}$      $\Rightarrow$   $x \leq z$

$\therefore$   $x \leq y$   and   $y \leq z$   $\Rightarrow$   $x \leq z$

Hence, '$\leq$' is transitive.

$\therefore$   '$\leq$' is a partial order relation on N.

Hence $(N, \leq)$ is a poset, where, '$\leq$' is the relation of divisibility.

Now, we know that, when the partial order relation is divisibility, then

sup. $\{x, y\} = $ l.c.m $\{x, y\}$   and   inf. $\{x, y\} = $ g.c.d $\{x, y\}$   for   $x, y \in N$

Let, for $x, y \in N$,

l.c.m $\{x, y\} = a$ and g.c.d $\{x, y\} = b$

$\therefore$ $a, b \in N$

Since l.c.m $\{x, y\} = a$

$\Rightarrow$ $\dfrac{x}{a}$ and $\dfrac{y}{a}$ $\Rightarrow$ $x \le a$ and $y \le a$

$\therefore$ $a$ is an upper bound of $\{x, y\}$.

Let $a'$ be any upper bound of $\{x, y\}$.

$\therefore$ $x \le a'$ and $y \le a'$ $\Rightarrow$ $\dfrac{x}{a'}$ and $\dfrac{y}{a'}$

$\Rightarrow$ $\dfrac{a}{a'}$ $\Rightarrow$ $a \le a'$ $\qquad\qquad$ [$\because$ l.c.m $\{x, y\} = a$]

$\therefore$ $a$ is a least upper bound of $\{x, y\}$.

$\therefore$ sup. $\{x, y\} = a$.

Similarly, g.c.d $\{x, y\} = b$ $\Rightarrow$ $\dfrac{b}{x}$ and $\dfrac{b}{y}$

$\Rightarrow$ $b \le x$ and $b \le y$

$\therefore$ $b$ is an lower bound of $\{x, y\}$.

Let $b'$ be any lower bound of $\{x, y\}$.

$\therefore$ $b' \le x$ and $b' \le y$ $\Rightarrow$ $\dfrac{b'}{x}$ and $\dfrac{b'}{y}$

$\Rightarrow$ $\dfrac{b'}{b}$ $\Rightarrow$ $b' \le b$ $\qquad\qquad$ [$\because$ g.c.d $\{x, y\} = b$]

$\therefore$ $b$ is a greatest lower bound of $\{x, y\}$.

$\therefore$ inf. $\{x, y\} = b$

Hence, $x, y \in N$, sup. $\{x, y\}$ and inf. $\{x, y\}$ both are exist.

Hence, $(N, \le)$ forms a lattice, where '$\le$' is the relation of divisibility.

**Example 4:** (i) Show that the set $D_{20}$ of all positive divisors of 20 form a lattice with the relation of divisibility. (ii) Represent the lattice $D_{20}$ by a Hasse diagram. $\qquad$ [*KU 2006*]

***Solution***

(i) Here $D_{20} = \{1, 2, 4, 5, 10, 20\}$

Clearly, $D_{20}$ is a poset with the relation divisibility.

Now, we know that, when the partial order relation is divisibility, then for $a, b \in D_{20}$,

$$\text{sup. } \{a, b\} = \text{l.c.m } \{a, b\} \quad \text{and} \quad \text{inf. } \{a, b\} = \text{g.c.d } \{a, b\}$$

Then the following tables given below give the sup. $\{a, b\}$ and inf. $\{a, b\}$ of any pair of elements of $D_{20}$.

$b$

| sup. {a, b} | 1 | 2 | 4 | 5 | 10 | 20 |
|---|---|---|---|---|---|---|
| 1 | 1 | 2 | 4 | 5 | 10 | 20 |
| 2 | 2 | 2 | 4 | 10 | 10 | 20 |
| 4 | 4 | 4 | 4 | 20 | 20 | 20 |
| 5 | 5 | 10 | 20 | 5 | 10 | 20 |
| 10 | 10 | 10 | 20 | 10 | 10 | 20 |
| 20 | 20 | 20 | 20 | 20 | 20 | 20 |

| inf. {a, b} | 1 | 2 | 4 | 5 | 10 | 20 |
|---|---|---|---|---|---|---|
| 1 | 1 | 1 | 1 | 1 | 1 | 1 |
| 2 | 1 | 2 | 2 | 1 | 2 | 2 |
| 4 | 1 | 2 | 4 | 1 | 2 | 4 |
| 5 | 1 | 1 | 1 | 5 | 5 | 5 |
| 10 | 1 | 2 | 2 | 5 | 10 | 10 |
| 20 | 1 | 2 | 4 | 5 | 10 | 20 |

From the above tables, we observe that for any $a, b \in D_{20}$,

sup. {a, b} and inf. {a, b} exist in $D_{20}$.

Hence, $D_{20}$ is a lattice with the relation divisibility.

(ii) The Hasse diagram of the lattice $D_{20}$ is shown in the below figure.

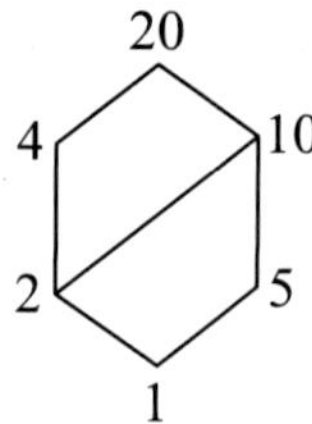


**Example 5:** Check whether the following posets shown by Hasse diagram are lattices or not (give reason).  [*MDU 2008*]

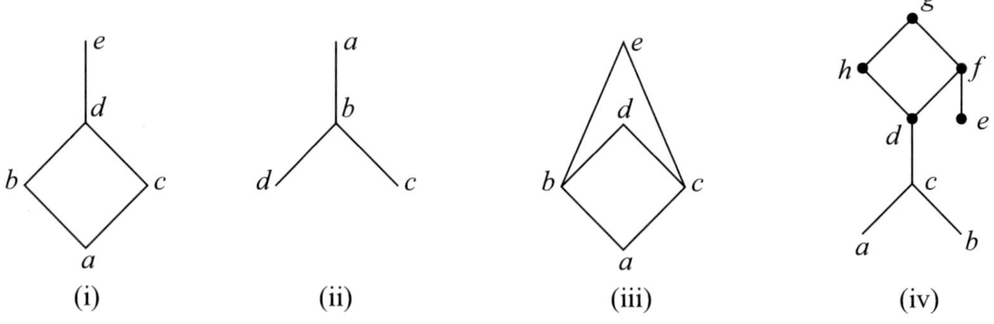


***Solution***

(i) From the given Hasse diagram (i), let A = {a, b, c, d, e}.

For $x, y \in$ A, let sup. {x, y} = $x \vee y$ and inf. {x, y} = $x \wedge y$ where $x \vee y$ is successor of $x$ and $y$ and $x \wedge y$ is predecessor of $x$ and $y$.

For example, $\quad a \vee b = b \quad$ and $\quad a \wedge b = a$

$\therefore \quad$ sup. $\{a, b\} = b \quad$ and $\quad$ inf. $\{a, b\} = a$

Now, the tables for sup. $\{x, y\}$ and inf. $\{x, y\}$ are given below.

<table>
<tr><td></td><td colspan="6" align="center">y</td></tr>
<tr><td>$x \vee y$</td><td>a</td><td>b</td><td>c</td><td>d</td><td>e</td></tr>
<tr><td>a</td><td>a</td><td>b</td><td>c</td><td>d</td><td>e</td></tr>
<tr><td>b</td><td>b</td><td>b</td><td>d</td><td>d</td><td>e</td></tr>
<tr><td>c</td><td>c</td><td>d</td><td>c</td><td>d</td><td>e</td></tr>
<tr><td>d</td><td>d</td><td>d</td><td>d</td><td>d</td><td>e</td></tr>
<tr><td>e</td><td>e</td><td>e</td><td>e</td><td>e</td><td>e</td></tr>
</table>

$x$ labels the rows.

<table>
<tr><td></td><td colspan="6" align="center">y</td></tr>
<tr><td>$x \wedge y$</td><td>a</td><td>b</td><td>c</td><td>d</td><td>e</td></tr>
<tr><td>a</td><td>a</td><td>a</td><td>a</td><td>a</td><td>a</td></tr>
<tr><td>b</td><td>a</td><td>b</td><td>a</td><td>b</td><td>b</td></tr>
<tr><td>c</td><td>a</td><td>a</td><td>c</td><td>c</td><td>c</td></tr>
<tr><td>d</td><td>a</td><td>b</td><td>c</td><td>d</td><td>d</td></tr>
<tr><td>e</td><td>a</td><td>b</td><td>c</td><td>d</td><td>e</td></tr>
</table>

$x$ labels the rows.

Using the above tables, we have sup. $\{x, y\}$ and inf. $\{x, y\}$ both exist for every $x, y \in A$.

$\therefore \quad$ The poset A with the Hasse diagram (i) is a lattice.

(ii) From the given Hasse diagram (ii), let A $= \{a, b, c, d\}$.

For $x, y \in A$, let sup. $\{x, y\} = x \vee y$ and inf. $\{x, y\} = x \wedge y$

Now, the tables for sup. $\{x, y\}$ and inf. $\{x, y\}$ are given below.

<table>
<tr><td></td><td colspan="5" align="center">y</td></tr>
<tr><td>$x \vee y$</td><td>a</td><td>b</td><td>c</td><td>d</td></tr>
<tr><td>a</td><td>a</td><td>a</td><td>a</td><td>a</td></tr>
<tr><td>b</td><td>a</td><td>b</td><td>b</td><td>b</td></tr>
<tr><td>c</td><td>a</td><td>b</td><td>c</td><td>b</td></tr>
<tr><td>d</td><td>a</td><td>b</td><td>b</td><td>d</td></tr>
</table>

$x$ labels the rows.

<table>
<tr><td></td><td colspan="5" align="center">y</td></tr>
<tr><td>$x \wedge y$</td><td>a</td><td>b</td><td>c</td><td>d</td></tr>
<tr><td>a</td><td>a</td><td>b</td><td>c</td><td>d</td></tr>
<tr><td>b</td><td>b</td><td>b</td><td>c</td><td>d</td></tr>
<tr><td>c</td><td>c</td><td>c</td><td>c</td><td></td></tr>
<tr><td>d</td><td>d</td><td>d</td><td></td><td>d</td></tr>
</table>

$x$ labels the rows.

Using the above tables, we observe that inf. $\{c, d\}$ does not exist.

$\therefore \quad$ The poset A with Hasse diagram (ii) is not a lattice.

(iii) From the given Hasse diagram (iii), let A $= \{a, b, c, d, e\}$.

For $x, y \in A$, let sup. $\{x, y\} = x \vee y$ and inf. $\{x, y\} = x \wedge y$

Now, the tables for sup. $\{x, y\}$ and inf. $\{x, y\}$ are given below

<table>
<tr><td></td><td colspan="6" align="center">y</td></tr>
<tr><td>$x \vee y$</td><td>a</td><td>b</td><td>c</td><td>d</td><td>e</td></tr>
<tr><td>a</td><td>a</td><td>b</td><td>c</td><td>d</td><td>e</td></tr>
<tr><td>b</td><td>b</td><td>b</td><td>d</td><td>d</td><td>e</td></tr>
<tr><td>c</td><td>c</td><td>d</td><td>c</td><td>d</td><td>e</td></tr>
<tr><td>d</td><td>d</td><td>d</td><td>d</td><td>d</td><td></td></tr>
<tr><td>e</td><td>e</td><td>e</td><td>e</td><td></td><td>e</td></tr>
</table>

$x$ labels the rows.

<table>
<tr><td></td><td colspan="6" align="center">y</td></tr>
<tr><td>$x \wedge y$</td><td>a</td><td>b</td><td>c</td><td>d</td><td>e</td></tr>
<tr><td>a</td><td>a</td><td>a</td><td>a</td><td>a</td><td>a</td></tr>
<tr><td>b</td><td>a</td><td>b</td><td>a</td><td>b</td><td>b</td></tr>
<tr><td>c</td><td>a</td><td>a</td><td>c</td><td>c</td><td>c</td></tr>
<tr><td>d</td><td>a</td><td>b</td><td>c</td><td>d</td><td>b</td></tr>
<tr><td>e</td><td>a</td><td>b</td><td>c</td><td>b</td><td>e</td></tr>
</table>

$x$ labels the rows.

Using the above tables, we have sup. $\{d, e\}$ does not exist.

∴    The partially ordered set A with the Hasse diagram (iii) is not a lattice.

(iv)  From the given Hasse diagram (iv), let A = $\{a, b, c, d, e, f, g, h\}$.

For $x, y \in$ A, let sup. $\{x, y\} = x \vee y$ and inf. $\{x, y\} = x \wedge y$

We see that inf. $\{a, b\}$, inf. $\{d, e\}$, inf. $\{c, e\}$, inf. $\{a, b\}$ and inf. $\{b, e\}$ does not exist.

∴    The poset A with Hasse diagram (iv) is not a lattice.

**Remark:** If (L, $\leq$) is a lattice, then from the definitions of sup. $\{x, y\} = x \vee y$ and inf. $\{x, y\} = x \wedge y$, we have

$$x \leq x \vee y, \ y \leq x \vee y \text{ as } x \vee y \text{ is an upper bound of } x \text{ and } y$$

and        $$x \wedge y \leq x, \ x \wedge y \leq y \text{ as } x \wedge y \text{ is a lower bound of } x \text{ and } y.$$

**Theorem 1** A poset (L, $\leq$) is a lattice iff every non-empty finite subset of L has supremum and infimum.

### *Proof*

Let (L, $\leq$) be a lattice and S be a non-empty finite subset of L.

Then the following cases will arise:

  (i)  If S has only one element $a$, then inf. S = sup. S = $a$
  (ii)  If S contains two elements $a$ and $b$. Then by definition of lattice L sup. $\{a, b\}$ and inf. $\{a, b\}$ both exist.
  (iii)  If S contains three elements $a$, $b$ and $c$ i.e., S = $\{a, b, c\}$

Since S $\subset$ L and L is a lattice.

∴    Any two elements of L have supremum and infimum.

Let inf. $\{a, b\} = d$ and inf. $\{c, d\} = e$

∴    By definition of inf. $\{a, b\}$ and inf. $\{c, d\}$, we have

$$d \leq a, \ d \leq b \quad \text{and} \quad e \leq c, \ e \leq d$$

$\Rightarrow$    $e \leq a, e \leq b, e \leq c$

$\Rightarrow$    $e$ is a lower bound of $\{a, b, c\}$.

Let $f$ be any lower bound of $\{a, b, c\}$, then

$\quad f \leq a, \ f \leq b, \ f \leq c \ \Rightarrow \ f$ is a lower bound of $\{a, b\}$   and   $f \leq c$

$\Rightarrow$    $f \leq d, \ f \leq c$ $\hspace{4cm}$ $[\because \ d = \text{inf. } \{a, b\} = \text{g.l.b } \{a, b\}]$

$\Rightarrow$    $f$ is a lower bound of $\{c, d\}$

$\Rightarrow \quad f \le e$ $\qquad\qquad\qquad$ $[\because \quad e = \inf. \{c, d\} = \text{g.l.b } \{c, d\}]$

$\Rightarrow \quad e = \text{g.l.b } \{a, b, c\}$

Hence, $e = \inf. \{a, b, c\} = \inf. S$

Similarly, we can show that sup. S exists

By continuing like this, the result can be proven for any finite number of elements in S.

If $S = \{a, a_2, a_3, \ldots, a_n\}$, then

$$\inf. S = \inf. \{\ldots \inf. \{\inf. \{a_1, a_2\}, a_3\}, \ldots, a_n\}$$

and $\qquad\qquad \text{sup. } S = \text{sup. } \{\ldots \text{sup. } \{\text{sup. } \{a_1, a_2\}, a_3\}, \ldots, a_n\}$

*Conversely,* let every finite subset S of L have supremum and infimum

$\therefore \quad$ For every $a, b \in L$, sup. $\{a, b\}$ and inf. $\{a, b\}$ both exist.

Hence, $(L, \le)$ is a lattice.

**Theorem 2** Let L be a lattice and S be any set. Let $\Im$ be the set of all functions from S to L. Define a relation '$\le$' on $\Im$ by $f \le g \Leftrightarrow f(x) \le g(x)$ for all $x \in$ S and $f, g \in \Im$, where £ in $f$ £ $g$ represents a relation on $\Im$ and $\le$ in $f(x) \le g(x)$ represents a relation on L. Then, $(\Im, \le)$ is a lattice.

***Proof***

Here $\Im = \left\{ \dfrac{f}{f} : S \to L \right\}$

**Reflexivity**

Since $f(x) \le f(x)$ for all $x \in$ S and $f \in \Im$

$\Rightarrow \quad f \le f$, for all $f \in \Im$

$\therefore \quad$ '$\le$' is reflexive.

**Anti-symmetry**

Let $f, g \in \Im$ such that $f \le g$ and $g \le f$

$\Rightarrow \quad f(x) \le g(x) \quad$ and $\qquad g(x) \le f(x)$ for all $x \in$ S

$\Rightarrow \quad f(x) = g(x) \quad$ for all $\quad x \in S \Rightarrow f = g$

$\therefore \quad$ '$\le$' is anti-symmetric.

**Transitivity**

Let $f, g, h \in \Im$ such that $f \le g$ and $g \le h$

$\Rightarrow \quad f(x) \le g(x) \quad$ and $\qquad g(x) \le h(x)$ for all $x \in$ S

$\Rightarrow \quad f(x) \le h(x) \quad$ for all $\quad x \in$ S

$\Rightarrow \quad f \leq h$

$\therefore \quad$ '$\leq$' is transitive.

$\therefore \quad$ '$\leq$' is a partial order relation on $\Im$.

Hence, $(\Im, \leq)$ forms a poset.

Now, let $f, g \in \Im$

$\therefore \quad f: S \to L \quad$ and $\quad g: S \to L$

We define a function $\theta: S \to L$ by

$$\theta(x) = f(x) \vee g(x) \text{ for all } x \in S$$

For $x \in S \Rightarrow f(x) \in L \quad$ and $\quad g(x) \in L \Rightarrow f(x) \vee g(x) \in L \qquad\qquad [\because \ L \text{ is a lattice}]$

We claim that $\theta = \sup. \{f, g\} = f \vee g$

Since $\qquad\qquad\qquad\qquad \theta(x) = f(x) \vee g(x) \text{ for all } x \in S$

$\Rightarrow \qquad\qquad\qquad\qquad f(x) \leq f(x) \vee g(x) \quad$ and $\quad g(x) \leq f(x) \vee g(x)$

i.e., $\qquad\qquad\qquad\qquad f(x) \leq \theta(x) \qquad\qquad$ and $\quad g(x) \leq \theta(x)$

$\therefore \quad$ For $x \in S$, $\theta(x)$ is an upper bound of $f(x)$ and $g(x)$.

$\Rightarrow \quad f \leq \theta \quad$ and $\quad g \leq \theta$

$\therefore \quad \theta$ is an upper bound of $f$ and $g$ in $\Im$

Let $\theta'$ be any other upper bound of $f$ and $g$ in $\Im$

$\therefore \quad f \leq \theta' \quad$ and $\quad g \leq \theta'$

$\Rightarrow \quad f(x) \leq \theta'(x) \quad$ and $\quad g(x) \leq \theta'(x)$ for all $x \in S$

$\Rightarrow \quad f(x) \vee g(x) \leq \theta'(x) \quad \Rightarrow \quad \theta(x) \leq \theta'(x)$ for all $x \in S \qquad [\because \ \theta(x) = f(x) \vee g(x)]$

$\Rightarrow \quad \theta \leq \theta'$

$\therefore \quad \theta$ is a least upper bound (l.u.b) of $f$ and $g$ in $\Im$.

$\therefore \quad \theta = \sup. \{f, g\}$.

Similarly, we define a function $\phi: S \to L$ by

$\qquad \phi(x) = f(x) \wedge g(x)$, for all $x \in S$

Then, we can show that l.b. $\{f, g\}$

$\therefore \quad \phi = \inf. \{f, g\}$ is the greatest lower bound (g.l.b) of $f$ and $g$.

$\therefore \quad$ For any $f, g \in \Im$, $\sup. \{f, g\}$ and $\inf. \{f, g\}$ both are exist in $\Im$.

$\therefore \quad (\Im, \leq)$ is a lattice.

**Exercise 8.1**

1. Let $X = \{1, 2, 3\}$ and $P(X)$ be the power set of $X$, show that $(P(X), \subseteq)$ is a lattice, where '$\subseteq$' is the relation of inclusion.

2. Determine whether the set $\{1, 2, 4, 8, 16\}$ is a lattice or not with the relation of divisibility.

3. Show that the set of integers is not a lattice with the relation of divisibility.

4. Determine whether each of the following partially ordered set under the relation divisibility is a lattice or not:

   (i) $L_1 = \{2, 4, 8, 10, 20, 40\}$

   (ii) $L_2 = \{1, 2, 4, 5, 20\}$

5. Which of the following Hasse diagrams represent lattices?　　　[*MDU 2008*]

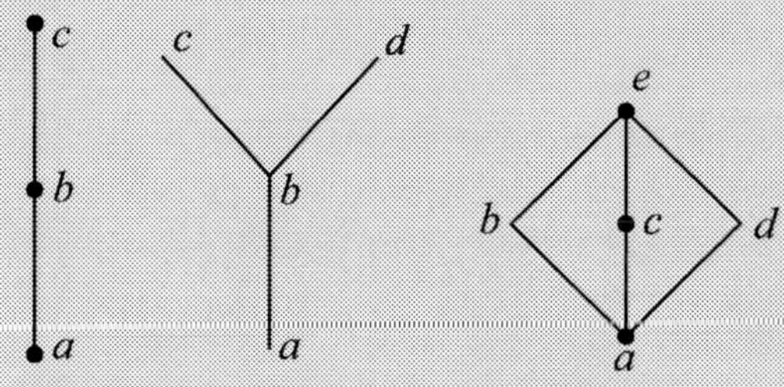

(i)　　　　　(ii)　　　　　(iii)

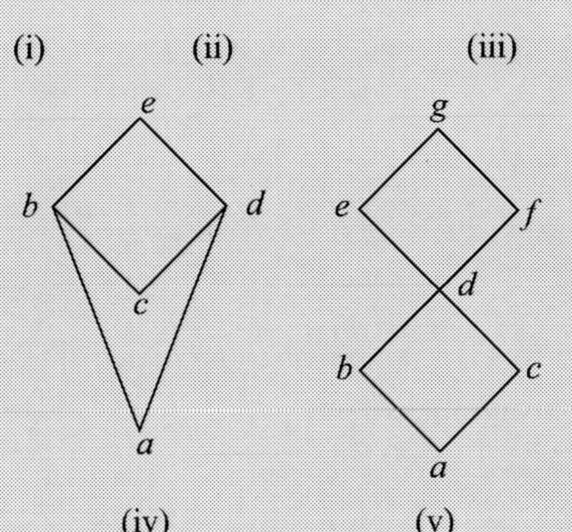

(iv)　　　　　(v)

6. Show that a lattice with three or fewer elements is a chain.

7. (i) Let $A = \{1, 2, 3, 5, 30\}$. Determine whether the poset A with the relation of divisibility is a lattice or not.

   (ii) Draw the Hasse diagram of poset A.

8. (i) Show that the set $D_{12}$ of all positive factors of 12 forms a lattice with the relation of divisibility?

   (ii) Represent the lattice $D_{12}$ by a Hasse diagram.

9. (i) Show that the set $D_{30}$ of all positive divisors of 30 forms a lattice with the relation of divisibility.

   (ii) Represent the lattice $D_{30}$ by a Hasse diagram.

10. Let $A = [0, 1]$ and '$\leq$' be the operation 'less than or equal to' on A. Show that $(A, \leq)$ is a lattice.

**Answers to Selected Problems**

  2. Lattice

  4. (i) Lattice　　(ii) Not a lattice

  5. (i), (iii), (v)

  7. (i) Lattice

    (ii) 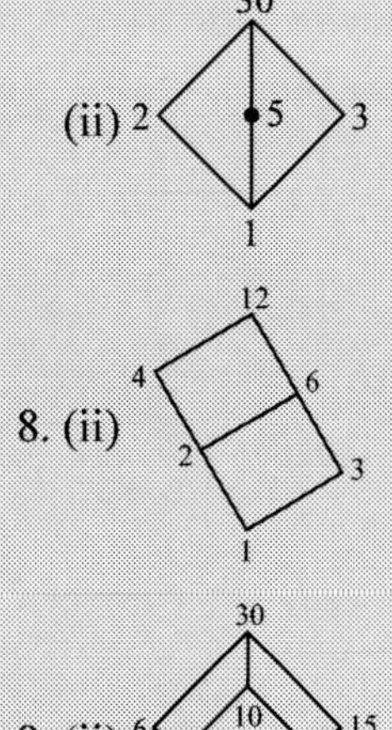

  8. (ii) 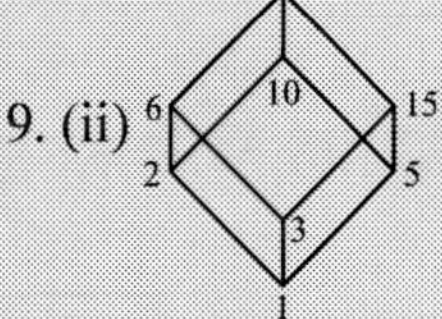

  9. (ii)

## 8.3 PROPERTIES OF A LATTICE

If L is any lattice. Then for any $a, b, c \in L$, the following properties hold true:

**I.** *Idempotency:*  (i) $a \wedge a = a$  (ii) $a \vee a = a$
**II.** *Commutativity:*  (i) $a \wedge b = b \wedge a$  (ii) $a \vee b = b \vee a$
**III.** *Associativity:*  (i) $a \wedge (b \wedge c) = (a \wedge b) \wedge c$  (ii) $a \vee (b \vee c) = (a \vee b) \vee c$
**IV.** *Consistency:*  (i) $a \leq b \Leftrightarrow a \wedge b = a$  (ii) $a \leq b \Leftrightarrow a \vee b = b$
**V.** If $o$ and $u$ respectively are the least and greatest elements of L. Then
(i) $o \wedge a = o, \quad o \vee a = a$  (ii) $u \wedge a = a, \quad u \vee a = u$
**VI.** *Absorption:*  (i) $a \wedge (a \vee b) = a$  (ii) $a \vee (a \wedge b) = a$

### *Proof*

We know that in a lattice L, for $a, b \in L$
$a \wedge b = \inf. \{a, b\}$ and $a \vee b = \sup. \{a, b\}$

**I.**  (i) $a \wedge a = \inf. \{a, a\} = \inf. \{a\} = a$
(ii) $a \vee a = \sup. \{a, a\} = \sup. \{a\} = a.$

**II.**  (i) $a \wedge b = \inf. \{a, b\} = \inf. \{b, a\} = b \wedge a$
(ii) $a \vee b = \sup. \{a, b\} = \sup. \{b, a\} = b \vee a.$

**III.**  (i) Let $b \wedge c = d$, then $d = \inf. \{b, c\}$

$\Rightarrow \quad d \leq b \quad$ and $\quad d \leq c$  (1)

Let $e = \inf. \{a, d\}$, then $e \leq a$ and $e \leq d$

$\Rightarrow \quad e \leq a, \quad e \leq b, \quad e \leq c$  [*Using* (1)]

$\Rightarrow \quad e$ is a lower bound of $\{a, b, c\}$.

If $f$ is any lower bound of $\{a, b, c\}$, then $f \leq a, f \leq b, f \leq c$.

$\Rightarrow \quad f \leq a \quad$ and $\quad f \leq d$  [$\because \quad d = \inf. \{b, c\} = $ g.l.b $\{b, a\}$]

$\Rightarrow \quad f \leq e$  [$\because \quad e \, \inf. \{a, d\} = $ g.l.b $\{a, d\}$]

$\therefore \quad e = \inf. \{a, b, c\}$

Thus, $e = a \wedge d = a \wedge (b \wedge c) = \inf. \{a, b, c\}$

Similarly, we can show that $(a \wedge b) \wedge c = \inf. \{a, b, c\}$

Hence, $a \wedge (b \wedge c) = (a \wedge b) \wedge c$  [$\because \quad$ *infimum is unique*]

Similarly, students can prove this property for $\vee$ (join operations).

**IV.**  (i) Let $a \leq b \quad \Rightarrow \quad a \leq a, \ a \leq b$

$\Rightarrow \quad a$ is a lower bound of $\{a, b\}$.

$\Rightarrow \quad a \leq a \wedge b$  [$\because \quad a \wedge b = \inf. \{a, b\} = $ g.l.b $\{a, b\}$]

Also, $a \wedge b$ being infimum, is a lower bound of $\{a, b\}$

$\therefore \quad a \wedge b \leq a$

$\therefore \quad$ By anti-symmetry, we have $a = a \wedge b$

*Conversely*, let $a \wedge b = a$

Since $a \wedge b = \text{inf.} \{a, b\}$ is also a lower bound of $\{a, b\}$.

$a \wedge b \leq b \quad \Rightarrow \quad a \leq b$ $\hfill [\because \quad a \wedge b = a]$

Hence, $a \leq b \quad \Leftrightarrow \quad a \wedge b = a.$

(ii) Let $a \leq b \quad \Rightarrow \quad a \leq b, \; b \leq b$

$\Rightarrow \quad b$ is an upper bound of $\{a, b\}$

$\Rightarrow \quad a \vee b \leq b$ $\hfill [\because \quad a \vee b = \text{sup.} \{a, b\} = \text{l.u.b} \{a, b\}\}$

Also, $a \vee b$ being supremum, is an upper bound of $\{a, b\}$.

$\therefore \quad b \leq a \vee b$

$\therefore \quad$ By anti-symmetry, we have $b = a \vee b$

*Conversely*, let $a \vee b = b$

Since $a \vee b = \text{sup.} \{a, b\}$ is also an upper bound of $\{a, b\}$

$\therefore \quad a \leq a \vee b \quad \Rightarrow \quad a \leq b$ $\hfill [\because \quad a \vee b = b]$

Hence, $a \leq b \quad \Leftrightarrow \quad a \vee b = b.$

**V.** (i) Since $o$ is a least element of L

$\therefore \quad o \leq a$, for all $a \in$ L

Then by property IV, we have $o \wedge a = 0, \; 0 \vee a = a$

(ii) Since $u$ is a greatest element of L

$\qquad a \leq u$, for all $a \in$ L

Then by property IV, we have $o \wedge u = a, \; a \vee u = u$

**VI.** (i) Let $a \vee b = c,$ then $\quad c = \text{sup.} \{a, b\}$

$\Rightarrow \quad a \leq c$

Then by using property IV (i), we have

$\qquad a \wedge c = a$ i.e., $\quad a \wedge (a \vee b) = a$

(ii) Let $a \wedge b = d,$ then $\quad d = \text{inf.} \{a, b\}$

$\Rightarrow \quad d \leq a$

Then, by using property (IV) (ii), we have $d \vee a = a$ i.e. $\quad (a \wedge b) \vee a = a.$

***Cor.*** In a lattice $(L, \leq)$, for $a, b \in L$

$$a \vee b = b \iff a \wedge b = a$$

***Proof.*** From property IV, we have

$$a \leq b \iff a \vee b = b$$

and
$$a \leq b \iff a \wedge b = a$$

$\Rightarrow$
$$a \vee b = b \iff a \leq b \iff a \wedge b = a$$

$\Rightarrow$
$$a \vee b = b \iff a \wedge b = a.$$

**Theorem 3** Let $(L, \leq)$ be a lattice and $a, b, c, d, \in L$, then prove that $a \leq b$ and $c \leq d \Rightarrow$ $a \wedge c \leq b \wedge d$ and $a \vee c \leq b \wedge d$

[Isotone laws]

***Proof***

We know that in a lattice,

$$a \wedge c = \text{inf. } \{a, c\}$$

$\therefore$
$$a \wedge c \leq a \quad \text{and} \quad a \wedge c \leq c$$

Also, $a \leq b$  and  $c \leq d$
[Given]

$\therefore$
$$a \wedge c \leq a \leq b \quad \text{and} \quad a \wedge c \leq c \leq d$$

$\Rightarrow$
$$a \wedge c \leq b \quad \text{and} \quad a \leq c \leq d$$

$\Rightarrow \quad a \wedge c$ is a lower bound of $\{b, d\}$

But $b \wedge d = \text{inf. } \{b, d\}$ is the greatest lower bound of $\{b, d\}$

$\therefore$
$$a \wedge c \leq b \wedge d$$

Hence, $a \leq b, c \leq d \Rightarrow a \wedge c \leq b \wedge d.$

Again, we know that in a lattice $b \vee d = \text{sup. } \{b, d\}$

$\therefore$
$$b \leq b \vee d \quad \text{and} \quad d \leq b \vee d$$

Also, $a \leq b$  and  $c \leq d$
[Given]

$\Rightarrow$
$$a \leq b \leq b \vee d \quad \text{and} \quad c \leq d \leq b \vee d$$

$\Rightarrow$
$$a \leq b \vee d \quad \text{and} \quad c \leq b \vee d$$

$\Rightarrow \quad b \vee d$ is an upper bound of $\{a, c\}$.

But $a \vee c$ is the least upper bound of $\{a, c\}$.

$\Rightarrow$
$$a \vee c \leq b \vee d$$

Hence, $a \leq b$  and  $c \leq d \Rightarrow a \vee c \leq b \vee d$

***Cor.*** In a lattice $(L, \leq)$. If $a \leq b$, then $a \wedge x \leq b \wedge x$ and $a \vee x \leq b \vee x$ for all $x \in L$

***Proof.*** Here $x \leq x$, for all $x \in L$
[By reflexivity]

$\therefore \quad a \le b \quad$ and $\quad x \le x$

Now, proceeding as in above theorem, we have

$\quad a \wedge x \le b \wedge x \quad$ and $\quad a \vee x \le b \vee x.$

## 8.4 LATTICE AS AN ALGEBRAIC STRUCTURE

**Definition.** A non-empty set L together with two binary operations $\vee$ (join) and $\wedge$ (meet) is said to form a *lattice*, if it satisfies the following axioms:

1. **Commutativity:** (i) $a \vee b = b \vee a$      (ii) $a \wedge b = b \wedge a$, for all $a, b \in L$
2. **Associativity:**     (i) $(a \vee b) \vee c = a \vee (b \vee c)$    (ii) $(a \wedge b) \wedge c = a \wedge (b \wedge c)$, for all $a, b, c \in L$
3. **Absorption:**       (i) $a \wedge (a \vee b) = a$      (ii) $a \wedge (a \wedge b) = a$, for all $a, b \in L$

The lattice L with binary operations $\vee$ and $\wedge$ is denoted by $(L, \vee, \wedge)$.

## 8.5 EQUIVALENCE OF TWO DEFINITIONS OF LATTICE

**Theorem 4** If a poset $(L, \le)$ is a lattice, then it is also a lattice as an algebraic structure.

### Proof

Let a poset $(L, \le)$ is a lattice with the relation $\le$.

$\therefore \quad$ For any pair of two elements $x$ and $y$ in L, sup. $\{x, y\}$ and inf. $\{x, y\}$ exist in L.

Let us define two binary operations $\wedge$ and $\vee$ in L as

$x \wedge y = $ inf. $\{x, y\}$ and $x \vee y = $ sup. $\{x, y\}$

Now, we show that the poset L with binary operation $\wedge$ and $\vee$ forms a lattice i.e., the algebraic structure $(L, \wedge, \vee)$ is a lattice. For this we will show that $(L, \wedge, \vee)$ satisfies commutativity, associativity and absorption laws.

### 1. Commutativity

For $x, y \in L$

     (i) $x \wedge y = $ inf. $(x, y) = $ inf. $(y, x) = y \wedge x$

     (ii) $x \vee y = $ sup. $(x, y) = $ sup. $(y, x) = y \vee x$

Hence, $x \wedge y = y \wedge x$ and $x \vee y = y \vee x$ for all $x, y \in L$

### 2. Associativity

For $x, y, z \in L$

To show $(x \vee y) \vee z = x \vee (y \vee z)$ and $(x \wedge y) \wedge z = x \wedge (y \wedge z)$

Let $x \vee y = u$, then $u = $ sup. $\{x, y\}$

$\Rightarrow \quad x \le u \quad$ and $\quad y \le u$                   (1)

Let $u \vee z = v$, then $v = \sup. \{u, z\}$

$\Rightarrow \qquad\qquad u \leq v \quad$ and $\quad z \leq v$

$\Rightarrow \qquad\qquad x \leq v, y \leq v \quad$ and $\quad z \leq v$      [*By transitivity, using* (1)]

$\Rightarrow \quad v$ is an upper bound of $\{x, y, z\}$

Let $s$ be any other upper bound of $\{x, y, z\}$

$\Rightarrow \qquad\qquad x \leq s, y \leq s \quad$ and $\quad z \leq s$

$\Rightarrow \quad s$ is an upper bound of $\{x, y\}$ and $z \leq s$

$\Rightarrow \quad u \leq s, z \leq s = \text{l.u.b } \{x, y\}$      $[\because \quad u = \sup. \{x, y\} = \text{l.u.b } \{x, y\}]$

$\Rightarrow \quad s$ is an upper bound of $\{u, z\}$

$\Rightarrow \qquad v \leq s$      $[\because \quad v = \sup. \{u, z\} = \text{l.u.b } \{u, z\}]$

$\Rightarrow \qquad v = \text{l.u.b } \{x, y, z\} = \sup. \{x, y, z\}$

$\therefore \quad v = u \vee z = (x \vee y) \vee z = \sup. \{x, y, z\}$      (2)

Similarly, we can also show that $x \vee (y \vee z) = \sup. \{x, y, z\}$      (3)

From (2) and (3), we have

$(x \vee y) \vee z = x \vee (y \vee z)$      $[\because \quad Supremum\ is\ unique]$

Similarly, students can also show that $(x \wedge y) \wedge z = x \wedge (y \wedge z)$.

## 3. Absorption

For $x, y \in L$, $x \wedge y = \inf. \{x, y\}$

$\Rightarrow \quad x \wedge y \leq x$ and by reflexivity $x \leq x$

$\Rightarrow \quad x$ is an upper bound of $\{x, x \wedge y\}$

$\Rightarrow \quad \sup. \{x, x \wedge y) \leq x$      $[\because \quad \sup. \{x, x \wedge y\} = \text{l.u.b. } \{x, x \wedge y\}]$

$\Rightarrow \quad x \vee (x \wedge y) \leq x$      (1)

Also, $x \vee (x \wedge y) = \sup. \{x, x \wedge y\} \Rightarrow x \leq x \vee (x \wedge y)$      (2)

From (1) and (2), $x \vee (x \wedge y) = x$      [*By anti-symmetry*]

Similarly, we can show that

$$x \wedge (x \vee y) = x$$

Hence, $\quad x \vee (x \wedge y) = x \quad$ and $\quad x \wedge (x \vee y) = x$

Hence, the poset $(L, \leq)$ is a lattice as an algebraic structure.

**Theorem 5** If $(L, \wedge, \vee)$ is a lattice, then prove that it is also a lattice as a poset.

*Proof*

Let $(L, \wedge, \vee)$ be a lattice as an algebraic structure.

∴ The commutative laws, associative laws and absorption laws w.r.t. binary operations $\vee$ and $\wedge$ are satisfies in L.

Let us define a relation '$\leq$' on L,

$$x \leq y \text{ if } x \vee y = y \quad \text{and} \quad x \wedge y = x, \text{ for } x, y \in L$$

Now,    if $x \vee y = y$,   then $x \wedge y = x \wedge (x \vee y) = x$          *[By absorption law]*

and      if $x \wedge y = x$,   then $x \vee y = (x \wedge y) \vee y = y \vee (y \wedge x) = y$

                                                   *[By absorption law]*

∴         $x \vee y = y$    iff    $x \vee y = x$

∴       $x \leq y$ if $x \vee y = y$ (or $x \wedge y = x$)

We now prove that '$\leq$' is reflexive, anti-symmetric and transitive.

### (a) Reflexivity

For $x \in L$ and $y \in L$, we have $x \vee x = x \vee (x \wedge (x \vee y)) = x$

                                 *[By Absorption laws $x \wedge (x \vee y) = x$]*

$\Rightarrow$   $x \leq x$ for all $x \in L$

∴    '$\leq$' is reflexive.

### (b) Anti-symmetry

Let     $x \leq y$   and   $y \leq x$   $\Rightarrow$   $x \vee y = y$   and   $y \vee x = x$

$\Rightarrow$   $x = y$

∴    '$\leq$' is anti-symmetric.

### (c) Transitivity

Let $x \leq y$ and $y \leq z$

∴   $x \vee y = y$   and   $y \vee z = z$

Now,             $x \vee z = x \vee (y \vee z)$

                      $= (x \vee y) \vee z$                      *[By associativity]*

                      $= y \vee z = z$

$\Rightarrow$                   $x \leq z$

∴    '$\leq$' is transitive.

Hence, $(L, \leq)$ is a poset.

Now, we show that $(L, \leq)$ is a lattice i.e., sup. $\{x, y\} \in L$ and $g$: inf. $\{x, y\} \in L$

We claim that sup. $\{x, y\} = x \vee y$ and inf. $\{x, y\} = x \wedge y$

By absorption laws, we have

$$x \wedge (x \vee y) = x \quad \text{and} \quad y \wedge (x \vee y) = y$$

$\Rightarrow \qquad\qquad\qquad\qquad\qquad x \le x \vee y \quad \text{and} \quad y \le x \vee y$

$\Rightarrow \quad x \vee y$ is an upper bound of $x$ and $y$

Let $u$ be any other upper bound of $x$ and $y$

$\therefore \qquad\qquad\qquad\qquad\qquad x \le u \quad \text{and} \quad y \le u$

$\Rightarrow \qquad\qquad\qquad\qquad\quad x \vee u = u \quad \text{and} \quad y \vee u = u$

$\therefore \qquad\qquad\qquad\quad (x \vee y) \vee u = x \vee (y \vee u) = x \vee u = u$

$\Rightarrow \qquad\qquad\qquad\qquad\quad (x \vee y) \le u$

$\Rightarrow \quad x \vee y$ is the least upper bound of $x$ and $y$

$\therefore \quad$ sup. $\{x, y\} = x \vee y.$

Similarly, we can show that $x \wedge y$ is the infimum of $x$ and $y$ i.e., inf. $\{x, y\} = x \wedge y.$

Hence, a lattice $(L, \le)$ is a poset, in which every sub-set $\{a, b\}$ consisting of two elements has supremum and infimum.

$\therefore \quad$ The lattice $(L, \wedge, \vee)$ is also a lattice as a poset.

## 8.6 PRINCIPLE OF DUALITY

The dual of any statement in a lattice $(L, \wedge, \vee)$ is obtained by interchanging the operations $\vee$ and $\wedge$ in the given statement. The dual of an axiom of L is again an axiom of L. Also dual of any theorem in L can be proved by using the dual of each step in the proof of the theorem. Thus, the principle of duality says that the dual of any theorem in L is also a theorem.

**Theorem 6** Dual of a lattice is also a lattice. Let $(L, R)$ be a poset and let $(L, R')$ be its dual. or If $(L, R)$ is a lattice, then $(L, R')$ is also a lattice.

### *Proof*

Here $(L, R)$ is a lattice and $(L, R')$ be its dual. We have already proved that dual of a poset is a poset.

Let $x, y \in L$. Since $(L, R)$ is a lattice.

$\therefore \quad$ sup. $\{x, y\}$ exists in L.

Let sup. $\{x, y\} = x \vee y \Rightarrow x \vee y$ is the l.u.b of $x$ and $y$ in $(L, R)$, then

$\qquad x \, R \, x \vee y \quad \text{and} \quad y \, R \, x \vee y$

$\Rightarrow \quad x \vee y \, R' \, x \quad \text{and} \quad x \vee y \, R' \, y$

$\Rightarrow \quad x \vee y$ is the lower bound of $x$ and $y$ in $(L, R')$

Now, we need to show that $x \vee y$ is the greatest lower bound of $x$ and $y$ in $(L, R')$.

Let $z \in L$ is any other lower bound of $x$ and $y$ in $(L, R')$, then

$$z\,R'\,x \quad \text{and} \quad z\,R'\,y$$

$\Rightarrow \quad x\,R\,z \quad \text{and} \quad y\,R\,z$

$\Rightarrow \quad z$ is the upper bound of $x$ and $y$ in (L, R).

$\Rightarrow \quad (x \vee y)\,R\,z$ $\qquad\qquad\qquad\qquad [\because \quad x \vee y = \text{sup. } \{x, y\}]$

$\Rightarrow \quad z\,R'\,(x \vee y)$

$\Rightarrow \quad x \vee y$ is the greatest lower bound of $x$ and $y$ in (L, R′).

Similarly, we can show that $x \wedge y$ is the least upper bound of $x$ and $y$ in (L, R′).

$\therefore \quad$ inf. $\{x, y\}$ and sup. $\{x, y\}$ exist in (L, R′).

Hence, (L, R′) is a lattice.

**Theorem 7** In any lattice L, the distributive inequalities. $a \wedge (b \vee c) \leq (a \wedge b) \vee (a \wedge c)$ and $a \vee (b \wedge c) \leq (a \vee b) \wedge (a \vee c)$ holds true for any $a, b, c \in$ L

***Proof***

We know that $a \wedge b \leq a$ and $a \wedge b \leq b$

Also, $\qquad\qquad\qquad\qquad\qquad b \leq b \vee c$

$\Rightarrow \qquad\qquad\qquad\qquad a \wedge b \leq b \leq b \vee c$

$\Rightarrow \quad a \wedge b$ is a lower bound of $\{a, b \vee c\}$

But, $a \wedge (b \vee c) = $ inf. $(a, b \vee c)$ is a greatest lower bound of $\{a, b \vee c\}$

$\Rightarrow \qquad\qquad\qquad\qquad a \wedge b \leq a \wedge (b \vee c)$ $\qquad\qquad\qquad (1)$

Again, $a \wedge c \leq a$ and $a \wedge c \leq c$

Also, $\qquad\qquad\qquad\qquad\qquad c \leq b \vee c$

$\Rightarrow \qquad\qquad\qquad\qquad a \wedge c \leq c \leq b \vee c$

$\Rightarrow \quad a \wedge c$ is a lower bound of $\{a, b \vee c\}$

But $a \wedge (b \vee c) = $ inf. $\{a, b \vee c\}$ is the greatest lower bound of $\{a, b \vee c\}$.

$\Rightarrow \qquad\qquad\qquad\qquad a \wedge c \leq a \wedge (b \vee c)$ $\qquad\qquad\qquad (2)$

From (1) and (2), we have $a \wedge (b \vee c)$ is an upper bound of $a \wedge b$ and $a \wedge$ c.

But $(a \wedge b) \vee (a \wedge c) = $ sup. $\{a \wedge b, a \wedge c\}$ is a least upper bound of $\{a \wedge b, a \wedge c\}$

$\Rightarrow \qquad\qquad\qquad (a \wedge b) \vee (a \wedge c) \leq a \wedge (b \vee c)$.

Similarly, we can show that

$$a \vee (b \wedge c) \leq (a \vee b) \wedge (a \vee c).$$

**Note:** The above inequalities are also called semi-distributive laws.

**Theorem 8** In any lattice L, the modular inequality $a \wedge (b \vee c) \geq (a \wedge b) \vee c$ holds for all $a, b, c \in$ L and $a \geq c$.

### *Proof*

From the distributive inequality, we have

$$a \wedge (b \vee c) \geq (a \wedge b) \vee (a \wedge c) \tag{1}$$

Now, $a \geq c$ (given) $\Rightarrow$ $a \wedge c = c$

From (1), we have

$$a \wedge (b \vee c) \geq (a \wedge b) \vee c$$

The dual of modular inequality is $a \vee (b \wedge c) \leq (a \vee b) \wedge c$ for all $a, b, c \in$ L, $a \leq c$.

**Theorem 9** Product of two lattices is also a lattice. Or If $(L_1, \leq)$ and $(L_2, \leq)$ are two lattices, then $L_1 \times L_2$ is also a lattice.

### *Proof*

Let $L_1$ and $L_2$ be two lattices, then $L_1 \times L_2 = \{(a, b) : a \in L_1, b \in L_2\}$

We have already proved that $L_1 \times L_2$ is a poset with the relation $\leq$ defined by

$(a_1, b_1) \leq (a_2, b_2)$ iff $a_1 \leq a_2$ in $L_1$ and $b_1 \leq b_2$ in $L_2$.

Now, we need to show that $L_1 \times L_2$ forms a lattice.

Let $(a_1, b_1), (a_2, b_2) \in L_1 \times L_2$ be any elements

$\Rightarrow$ $a_1, a_2 \in L_1$ and $b_1, b_2 \in L_2$.

Since $L_1$ and $L_2$ are lattices.

$\therefore$ $\{a_1, a_2\}$ and $\{b_1, b_2\}$ have supremum and infimum in $L_1$ and $L_2$ respectively.

Let $a_1 \vee a_2 =$ sup. $\{a_1, a_2\}$ and $b_1 \vee b_2 =$ sup. $\{b_1, b_2\}$ in $L_1$ and $L_2$ respectively.

$\Rightarrow \qquad a_1 \leq a_1 \vee a_2, a_2 \leq a_1 \vee a_2 \qquad$ and $\quad b_1 \leq b_1 \vee b_2, b_2 \leq b_1 \vee b_2$

$\Rightarrow \qquad a_1 \leq a_1 \vee a_2, b_1 \leq b_1 \vee b_2 \qquad$ and $\quad a_2 \leq a_1 \vee a_2, b_2 \leq b_1 \vee b_2$

$\Rightarrow \qquad (a_1, b_1) \leq (a_1 \vee a_2, b_1 \vee b_2) \quad$ and $\quad (a_2, b_2) \leq (a_1 \vee a_2, b_1 \vee b_2)$

$\Rightarrow \quad (a_1 \vee a_2, b_1 \vee b_2)$ is an upper bound of $\{(a_1, b_1), (a_2, b_2)\}$

Suppose $(c, d)$ is any upper bound of $\{(a_1, b_1), (a_2, b_2)\}$.

Then $(a_1, b_1) \leq (c, d)$ and $(a_2, b_2) \leq (c, d)$

$\Rightarrow \qquad\qquad a_1 \leq c, b_1 \leq d \quad$ and $\quad a_2 \leq c, b_2 \leq d$

$\Rightarrow \qquad\qquad a_1 \leq c, a_2 \leq c \quad$ and $\quad b_1 \leq d, b_2 \leq d$

$\Rightarrow \quad c$ is an upper bound of $\{a_1, a_2\}$ in L and $d$ is an upper bound of $\{b_1, b_2\}$ in $L_2$.

$\Rightarrow \quad a_1 \vee a_2 \leq c$ and $b_1 \vee b_2 \leq d \quad [\because \quad a_1 \vee a_2 =$ sup. $\{a_1, a_2\}$ and $b_1 \vee b_2 =$ sup. $\{b_1, b_2\}]$

$\Rightarrow \quad (a_1 \vee a_2, b_1 \vee b_2) \leq (c, d)$

$\therefore$    $(a_1 \vee a_2, b_1 \vee b_2)$ is a least upper bound of $\{(a_1, b_1), (a_2, b_2)\}$

$\therefore$    sup. $\{(a_1, b_1), (a_2, b_2)\} = (a_1 \vee a_2, b_1 \vee b_2) \in L_1 \times L_2$

Similarly, by principle of duality a lattice, we can show that

inf. $\{(a_1, b_1), (a_2, b_2)\} = (a_1 \wedge a_2, b_1 \wedge b_2) \in L_1 \times L_2$

Hence, $(L_1 \times L_2, \leq)$ is a lattice.

**Example 6:** Let $D_5$ and $D_9$ are lattices with the relation '$\leq$' where '$\leq$' is the relation of divisibility. Draw the Hasse diagram of the lattice $D_5 \times D_9$ under the product partial order.

***Solution***

Here $D_5 = [1,5]$ and $D_9 = \{1, 3, 9\}$

Also, $1 \leq 5, 1 \leq 3, 1 \leq 9, 3 \leq 9$

The Hasse diagrams of the lattices $D_5$ and $D_9$ with the relation '$\leq$' are shown in the figures

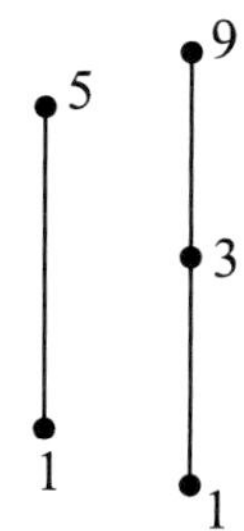

$\therefore$          $D_5 \times D_9 = \{(1, 1), (1, 3), (1, 9), (5, 1), (5, 3), (5, 9)\}$

For $(a, b), (c, d) \in D_4 \times D_9$

$(a, b) \leq (c, d)$ iff $\dfrac{a}{c}$ and $\dfrac{b}{d}$ under the product partial order, $D_5 \times D_9$ is a lattice.

Also, $(a, b) << (c, d)$ if there exist no $(x, y) \in D_4 \times D_9$ such that

$(a, b) \leq (x, y) \leq (c, d)$

$\therefore$      $(1, 1) << (1, 3) << (1, 9) << (5, 9)$

and     $(1, 1) << (5, 1) << (5, 3) << (5, 9)$

The Hasse diagram of $D_5 \times D_9$ is shown in the figure below.

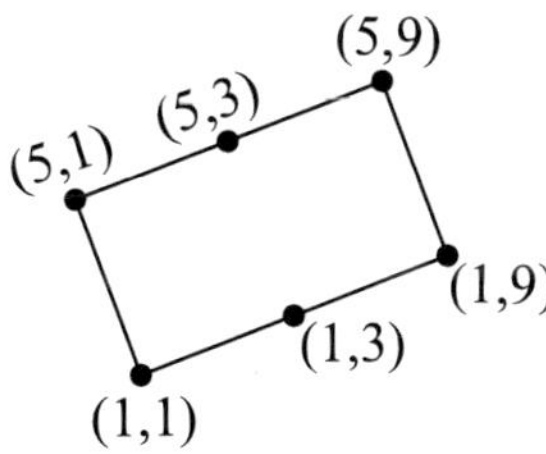

## Exercise 8.2

1. In any lattice L, prove that for $a, b, c \in$ L.

   $(a \wedge b) \vee (b \wedge c) \vee (c \wedge a) \leq (a \vee b) \wedge (b \vee c) \wedge (c \wedge a)$

2. If L is a lattice then show that $a \vee (b \wedge c) = (a \wedge b) \wedge (a \wedge c)$, when $a, b, c \in$ L and any two of $a, b, c$, are equal.

3. Let $(L, \leq)$ be a lattice and $a, b, c \in$ L, Prove that $a \vee (b \vee c) = (a \vee b) \vee c$.

4. Let $(L, \leq)$ be a lattice and $a, b, c \in$ L, Prove that

   (i) $a \leq c$ and $b \leq c \Leftrightarrow a \vee b \leq c$
   (ii) $c \leq a$ and $c \leq b \Leftrightarrow c < a \wedge b$
   (iii) $a \leq b \Rightarrow a \vee c \leq b \vee c$ and $a \wedge c \leq b \wedge c$.

5. If L is a lattice then show that $a \wedge (b \vee c) = (a \wedge b) \vee (a \wedge c)$, when $a, b, c \in$ L and any two of $a, b, c$ are equal.

6. Let $(L, \leq)$ be a lattice and $a, b, c \in$ L. If $a \leq b \leq c$ show that:

   (a) $a \vee b = b \wedge c$    (b) $(a \wedge b) \vee (b \wedge c) = (a \vee b) \wedge (a \vee c)$

7. Let $L = \{a, b, c, d, e, f, g\}$, be a lattice with the Hasse diagram shown in the figure. Which of the following subsets of L are sublattice of L?

   (i) $A = \{a, b, c\}$
   (ii) $B = \{a, b, c, d\}$

(iii) $C = \{c, d, f, g\}$
(iv) $D = \{b, c, d, e, f, g\}$

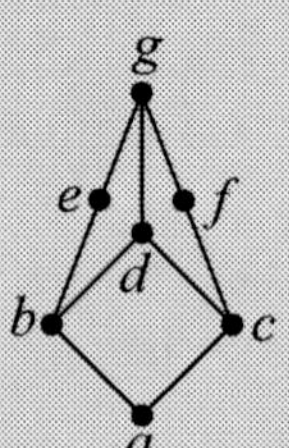

8. Let $L_1, L_2$ be lattices shown in the figure. Represent the lattice $L_1 \times L_2$ under the product partial order by a Hasse diagram.

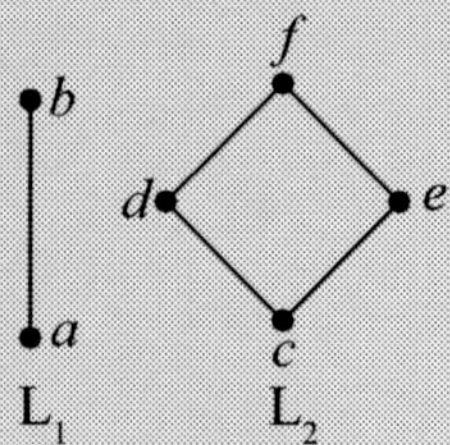


### Answers to Selected Problems

8.

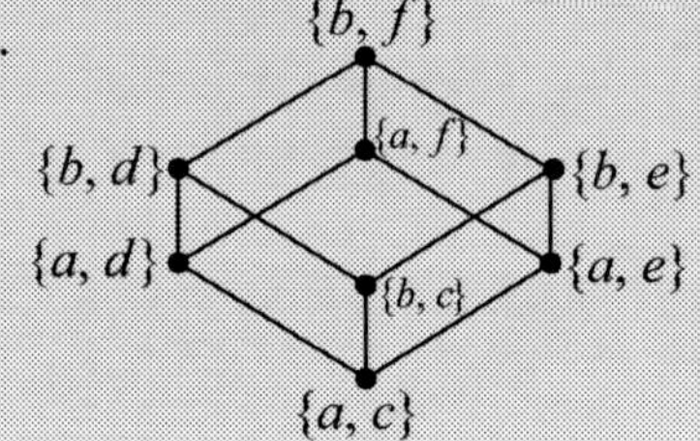


## 8.7 SEMI LATTICES

### Join Semi Lattice

**Definition 1.** A poset $(P, \leq)$ is said to be a join-semi lattice, if for $a, b \in$ P, sup. $\{a, b\} \in$ P.

**Definition 2.** A non-empty set P together with a binary operation $\vee$ (join) is called a join-semi lattice if for all $a, b, c \in$ P

   (i) $a \vee a = a$    (ii) $a \vee b = b \vee a$    (iii) $a \vee (b \vee c) = (a \vee b) \vee c$

The two definitions of join-semi lattice are equivalent and it can be easily proved with the help of theorems on equivalence of lattices.

### Meet-semi Lattice

**Definition 1.** A poset $(P, \leq)$ is said to be a meet-semi lattice, if for all $a, b \in P$, inf.$\{a, b\} \in P$.

**Definition 2.** A non-empty set P together with a binary operation $\vee$ (meet) is called a meet-semi lattice if for all $a, b, c \in P$

$$(i)\ a \wedge a = a \quad (ii)\ a \wedge b = b \wedge a \quad (iii)\ a \wedge (b \wedge c) = (a \wedge b) \wedge c$$

**Remark.** A poset $(P, \leq)$ is a lattice iff it is a join-semi lattice and a meet-semi lattice.

**Example 7:** A finite lattice has least and greatest elements.

### *Solution*

Let $L = \{x_1, x_2, x_3, \ldots, x_n\}$ be the given finite lattice, where the elements $x_1, x_2, x_3, \ldots, x_n$ are written in any random order.

Putting

$$a_0 = x_1$$
$$a_1 = a_0 \wedge x_1$$
$$a_2 = a_1 \wedge x_2$$
$$a_3 = a_2 \wedge x_3$$
$$\ldots\ldots\ldots\ldots$$
$$\ldots\ldots\ldots\ldots$$
$$a_n = a_{n-1} \wedge x_n$$

$\Rightarrow \quad a_n \leq a_{n-1} \leq a_{n-2} \leq \ldots \leq a_1 \leq a_0$

i.e., $a_i \leq x_i$, for all $i = 1, 2, 3, \ldots, n$

$\Rightarrow \quad a_n \leq x_i$, for all $i = 1$ to $n$

$\Rightarrow \quad a_n$ is the least element of L.

Similarly, we can find greatest element of L.

**Cor.** A finite chain has least and greatest element.

## 8.8 COMPLETE LATTICE

A lattice L is a complete lattice if every non-empty subset of L has its supremum and infimum in L.

Obviously, every finite lattice is a complete lattice.      [*Refer theorem-1 on page 196*]

The lattice $(Z, \leq)$ of integers is not complete as the subset $A = \{x \in Z : x > 0\}$ does not have an upper bound and supremum in Z.

The real interval $[0, 1]$ with usual relation '$\leq$' forms a complete lattice.

Also, it can be noted that if L is a complete lattice, then L must have the least and greatest elements.      [$\because$   L *have supremum and infimum*]

Now, since Dual of any lattice is also a lattice, therefore Dual of a complete lattice is complete.

**Theorem 10** If $(P, \leq)$ is a poset with greatest element $u$ such that every non-empty subset S of P has infimum, then P is a complete lattice.

### *Proof*

Let S be any non-empty subset of P.

For $a, b \in$ S, inf.$\{a, b\} \in$ S                                    [*Given*]

We have to show that sup.$\{a, b\} \in$ S

Here $u$ is the greatest element of P

$\therefore \quad x \leq u, \quad$ for all $\quad x \in$ P

$\Rightarrow \quad s \leq u, \quad$ for all $\quad s \in$ S                                    [$\because \quad s \subseteq$ P]

$\Rightarrow \quad u$ is an upper bound of S

Let T be the set of all upper bounds of S.

$\Rightarrow \quad$ T is non-empty subset of P

$\therefore \quad$ By given condition, inf. T exists in T

Let $k =$ inf. T

Now, $s \in$ S $\quad \Rightarrow \quad s \leq y, \quad$ for all $y \in$ T            [$\because \quad$ T *is a set of upper bounds of* S]

$\Rightarrow \quad$ Each element of S is a lower bound T

$\Rightarrow \quad s \leq k, \quad$ for all $s \in$ S

$\Rightarrow \quad k$ is an upper bound of S. But $k$ being infimum the greatest lower bound of T

$\therefore \quad k \leq y, \quad$ for all $y \in$ T

i.e., $\quad k \leq y$, for all upper bounds of S

$\Rightarrow \quad k =$ sup. S

$\therefore \quad$ P is a poset in which every non-empty subset has supremum and infimum.

Hence, P is a complete lattice.


## 8.9 SUBLATTICES

A non-empty subset S of a lattice $(L, \wedge, \vee)$ is called a sublattice of L, if for $a, b \in$ S $\Rightarrow a \wedge b \in$ S and $a \vee b \in$ S where $\wedge$ and $\vee$ are binary operations taken in L.

Clearly, any lattice L is a sublattice of itself.

It is understood that $a \wedge b =$ inf.$\{a, b\}$ and $a \vee b =$ sup.$\{a, b\}$.

$\therefore \quad$ A sublattice is a non-empty subset S of a lattice $(L, \leq)$ in which sup.$\{a, b\}$ and inf.$\{a, b\}$ are in S, for all $a, b \in$ S.

**Illustration.**  If L is any lattice and $a \in$ L, be any element, then $\{a\}$ is a sublattice of L.

**Note.**  A sublattice is itself a lattice but any subset of a lattice need not be a sublattice.

For  example, the set $D_{10}$ of positive factors of 10 is a lattice under the relation divisibility and $A = \{1, 2, 5\}$ is a subset of $D_{10}$.

Then, the subset $A = \{1, 2, 5\}$ is not a sublattice of $D_{10}$ as sup.$\{2, 5\} = 10 \notin$ A.

**Theorem 11** Prove that intersection of two sublattices is a sublattice.

***Proof***

Let L be any lattice under the relation R.

Let $K_1$ and $K_2$ are sublattices of L.

We need to show that $K_1 \cap K_2$ is a sublattice of L.

Let $x, y \in K_1 \cap K_2$

$\Rightarrow$　$x, y \in K_1$　and　$x, y \in K_2$ as $K_1$ and $K_2$ are sublattices.

$\therefore$　$x \wedge y \in K_1$　and　$x \vee y \in K_1$

Similarly, $x \wedge y \in K_2$ and $x \vee y \in K_2$

Therefore, by definition of intersection

$$x \wedge y \in K_1 \cap K_2 \quad \text{and} \quad x \vee y \in K_1 \cap K_2$$

Since $K_1$ and $K_2$ are arbitrary. Therefore, $K_1 \cap K_2$ is a sublattice.

**Example 8:** Show that union of two sublattices need not be a sublattice.

***Solution***

Consider the lattice L = {1, 2, 3, 5, 6, 10, 15, 30} of all positive factors of 30 under divisibility.

Then, S = {1, 2}　and　T = {1, 5} are sublattices of L.

But S $\cup$ T = {1, 2, 5} is not a sublattice of L, as 2, 5 $\in$ S $\cup$ T, but $2 \vee 5 = 10 \notin$ S $\cup$ T.

**Example 9:** Every non-empty subset of a chain is a sublattice.

***Solution***

Let S be any non-empty subset of a chain L, then $a, b \in$ S $\Rightarrow a, b \in$ L

Since L is a chain

$\therefore$　$a$ and $b$ are comparable elements.

$\therefore$　either $a \leq b$　or　$b \leq a$

If $a \leq b$, then　　$a \wedge b = a \in$ S　　　and　$a \vee b = b \in$ S

If $b \leq a$, then　　$a \wedge b = b \in$ S　　　and　$a \vee b = a \in$ S

$\therefore$　For any　　$a, b \in$ S $a \vee b \in$ S　and　$a \wedge b \in$ S

Hence, S is a sublattice.

**Example 10:** Show that a lattice L is a chain iff every non-empty subset of it is a sublattice.

### Solution

If lattice L is a chain, then we have already shown in the above example that every non-empty subset of a chain is a sublattice.

*Conversely*, let L be a lattice in which every non-empty subset of L is a sublattice.

Let, $a, b \in L$, then $\{a, b\}$ being a non-empty subset of L, is a sublattice of L

$\therefore \quad a \wedge b \in \{a, b\} \quad$ and $\quad a \vee b \in \{a, b\}$

$\therefore \quad a \wedge b \in \{a, b\} \quad$ and $\quad a \vee b \in \{a, b\} \quad \Rightarrow \quad$ either $a \wedge b = a \quad$ or $\quad a \wedge b = b$

$\Rightarrow \quad$ either $a \leq b \quad$ or $\quad b \leq a$

$\Rightarrow \quad a$ and $b$ are comparable in L

Hence, L is a chain.

**Example 11:** Let L $= \{a, b, c, d\}$ be a lattice with the Hasse diagram shown in the figure which of the following subsets of L are sublattices of L?

(i) $\{a, b\}$       (ii) $\{a, b, c\}$
(iii) $\{a, b, d\}$     (iv) $\{a, c, d\}$

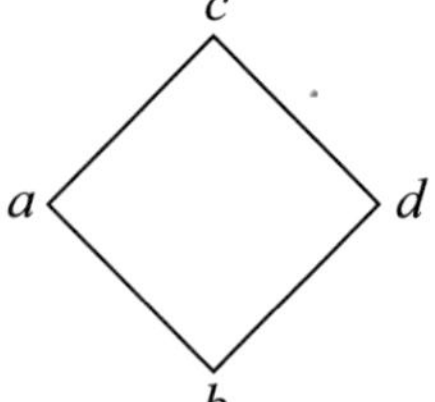

### Solution

Here L $= \{a, b, c, d\}$

(i) Let A $= \{a, b\}$

Clearly, A is a poset with the Hasse diagram shown in the adjoining figure.

Here $\quad a \wedge b = b \in A$

and $\quad a \vee b = a \in A$

$\therefore \quad$ A is a sublattice of L.

(ii) Let B $= \{a, b, c\}$

Clearly, B is a poset with the Hasse diagram shown in the adjoining figure

Now,       for $a, b \in B, \quad a \wedge b = b \in B \quad$ and $\quad a \vee b = a \in B$

         for $a, c \in B, \quad a \wedge c = a \in B \quad$ and $\quad a \vee c = c \in B$

         for $b, c \in B, \quad b \wedge c = b \in B \quad$ and $\quad b \vee c = c \in B$

$[\because \quad$ *There is a directed line from b to c*$]$

$\therefore \quad x \wedge y \in B \quad$ and $\quad x \vee y \in B, \quad$ for all $x, y \in B$

$\therefore \quad$ B is a sublattice of L.

(iii) Let C = {$a, b, d$}

Clearly, C is a poset with the Hasse diagram shown in the adjoining figure.

Now, for $a, d, \in$ C, we have

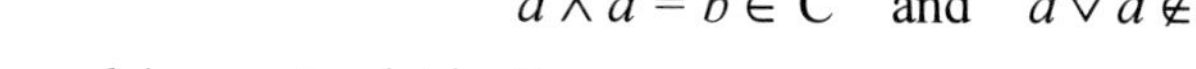

$$a \wedge d = b \in C \quad \text{and} \quad a \vee d \notin C$$

∴   $a \vee d$ does not exist in C.

∴   C is not a sublattice of L.

(iv) Let D = {$a, c, d$}

Clearly, D is a poset with the Hasse diagram shown in the adjoining figure.

Now, for $a, d \in$ D, we have

$$a \vee d = c \in D \quad \text{and} \quad a \wedge d \notin D$$

∴   $a \wedge d$ does not exist in D.

∴   D is not a sublattice of L.

## 8.10 CONVEX SUBLATTICE

A sublattice S of a lattice is said to be a convex sublattice if, for all $[a \wedge b, a \vee b] \subseteq S$, $a, b \in$ S.

**Theorem 12** A sublattice S of a lattice L is a convex sublattice iff for all $a, b \in$ S, ($a \leq b$), $[a, b] \subseteq$ S.

### *Proof*

Let S be a convex sublattice of L

∴   For $a, b \in$ S   and   $a \leq b, [a \wedge b, a \vee b] \subseteq$ S                [*By definition*]

Since $a \leq b$ (given)   $\Rightarrow$   $a \wedge b = a$   and   $a \vee b = b$

$\Rightarrow$   $[a, b] \in$ S

Conversely, Let $[a, b] \subseteq$ S for all $a, b \in$ S, $a \leq b$                (1)

Let $a, b \in$ S be any elements. Since S is a sublattice

$\Rightarrow$   $a \wedge b \in$ S and $a \vee b \in$ S

Also, we know that $a \wedge b \leq a \vee b$

∴     $[a \wedge b, a \vee b] \subseteq$ S                [*Using (1)*]

Hence, S is a convex sublattice.

### **Illustrations**

1.  Any interval $[a, b]$ in a lattice is a convex sublattice.
2.  In the lattice {1, 2, 3, 4, 6, 12} with the relation divisibility. The subset {1, 6} is a sublattice which is not a convex sublattice as

$$2, 3 \in [1, 6], \text{ but } 2, 3 \notin \{1, 6\}$$
$$[1, 6] \not\subseteq \{1, 6\}.$$

## 8.11 LATTICE ISOMORPHISM

**Definition 1.** Let L and L′ be two lattices as an algebraic structure, then L and L′ are said to be isomorphic, if there exists a bijective function $f: L \to L'$ such that

$$f(a \wedge b) = f(a) \wedge f(b)$$

and $\qquad f(a \vee b) = f(a) \vee f(b), \qquad$ for any $a, b \in L$

**Definition 2.** Two lattice $(L, \le)$ and $(L', \le')$ are said to be isomorphic, if there exists a bijective function $f: L \to L'$ such that

$$a \le b \text{ iff } f(a) \le' f(b), \quad \text{for } a, b \in L$$

If L and L′ are isomorphic lattices, then it is denoted as $L \cong L'$.

## 8.12 TWO DEFINITIONS OF ISOMORPHIC LATTICES ARE EQUIVALENT

**Theorem 13** If $(L, \le)$ and $(L', \le')$ are two isomorphic lattices and $f: L \to L'$ is a lattice isomorphism, then

$$f(a \wedge b) = f(a) \wedge f(b)$$
$$f(a \vee b) = f(a) \vee f(b), \text{ for } a, b \in L$$

### *Proof*

Here $(L, \le)$ and $(L', \le')$ are two as posets isomorphic lattices as posets.

$\therefore \quad f: L \to L'$ is a bijective function such that

$$a \le b \text{ in L} \quad \Leftrightarrow \quad f(a) \le' f(b) \text{ in L'}.$$

Let $a, b \in L$

$\therefore \quad$ sup. $\{a, b\}$ and inf. $\{a, b\}$ exist in L.

Let inf. $\{a, b\} = u \in L \quad \Rightarrow \quad u \le a, u \le b \qquad\qquad (1)$

Since $f$ is an isomorphism function.

$\therefore \quad f(a) \in L'$ and $f(b) \in L'$

$\therefore \quad$ inf. $\{f(a), f(b)\}$ exists in L′.

Let inf. $\{f(a), f(b)\} = f(v)$

$\Rightarrow \quad f(v) \le' f(a)$ and $f(v) \le' f(b) \qquad\qquad (2)$

Now from (1), we have

$u \le a, u \le b \quad \Rightarrow \quad f(u) \le' f(a), \quad f(u) \le' f(b)$

$\Rightarrow \quad f(u)$ is a lower bound of $f(a)$ and $f(b)$

$\qquad f(v) \le' f(u) \qquad\qquad\qquad [\because \; f(v) = \text{g.l.b } \{f(a), f(b)\}] \qquad (3)$

Also from (2), we have

$\qquad f(v) \le' f(a)$ and $f(v) \le' f(b)$

$\Rightarrow \quad u \le a$ and $u \le b \qquad\qquad\qquad\qquad [\because \; f \text{ is an isomorphism}]$

$\Rightarrow \quad v$ is a lower bound of $a$ and $b$

$\Rightarrow \quad u \le v,$ $\qquad\qquad\qquad\qquad\qquad\qquad [\because \quad u = \text{g.l.b } \{a, b\}$

$\Rightarrow \quad f(u) \le' f(v)$ $\qquad\qquad\qquad\qquad [\because \quad a \le b \Leftrightarrow f(a) \le' f(b)] \qquad (4)$

$\therefore \quad$ By anti-symmetry, from (3) and (4), we have

$$f(u) = f(v)$$

$\Rightarrow \qquad\qquad\qquad f(a \wedge b) = f(a) \wedge f(b)$

Hence, $\qquad\qquad\quad f(a \wedge b) = f(a) \wedge f(b)$, for $a, b \in L$

Similarly, we can show that

$$f(a \vee b) = f(a) \vee f(b), \text{ for } a, b \in L$$

**Theorem 14** Let L and L′ be two lattices as an algebraic structures and $f$: L → L′ be a lattice isomorphism. If (L, $\le$) and (L′, $\le'$) are posets, then show that $a \le b$ iff $f(a) \le' f(b)$ for any $a, b \in L$.

### *Proof*

Here L and L′ are lattices as algebraic structures and $f$ is a lattice isomorphism

$f(a \vee b) = f(a) \vee f(b) \quad$ and $\quad f(a \wedge b) = f(a) \wedge f(b)$

Since L and L′ are lattices as an algebraic,

$\therefore \quad$ (L, $\le$) and (L′, $\le'$) are lattices as a poset with partial order relation '$\le$' and '$\le'$' repectively.

Let $a \le b$ in L $\Rightarrow a \vee b = b$

$\Rightarrow \quad f(a \vee b) = f(b)$ $\qquad\qquad\qquad\qquad\qquad\qquad [\because \quad f$ is an isomorphism]

$\Rightarrow \quad f(b) \vee f(b) = f(b) \quad \Rightarrow \quad f(a) \le' f(b)$ in L′

*Conversely*, let $f(a) \le' f(b)$

$\Rightarrow \quad f(a) \vee f(b) = f(b)$

$\Rightarrow \quad f(a \vee b) = f(b) \quad \Rightarrow \quad a \vee b = b$ $\qquad\qquad\qquad [\because \quad f$ is an isomorphism]

$\Rightarrow \quad a \le b$ in L

Hence, $a \le b$ iff $f(a) \le' f(b)$ for $a, b \in L$.

**Example 12:** Determine whether the lattices shown in the figure below are isomorphic or not.

*Solution*

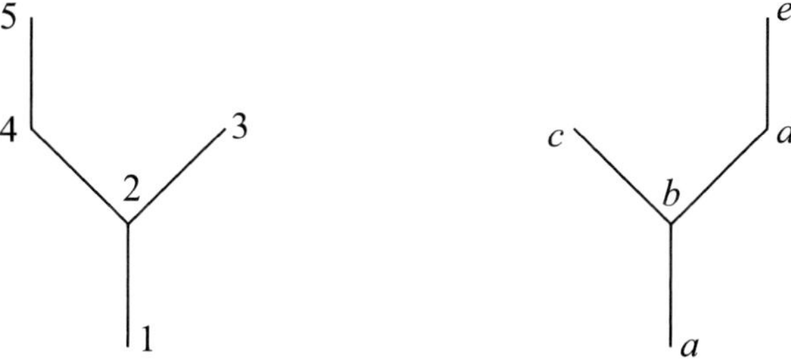

From the Hasse diagrams,

Let $L_1 = \{1, 2, 3, 4, 5\}$   and   $L_2 = \{a, b, c, d, e\}$

Define a function $f\colon L_1 \rightarrow L_2$ by

$$f(1) = a, \quad f(2) = b, \quad f(3) = c, \quad f(4) = d \quad \text{and} \quad f(5) = e.$$

Clearly, each element of $L_1$ has unique and distinct image in $L_2$

∴   The function $f$ is one-one and onto.

⇒  $f$ is a bijective function.

Again, $x \leq y$ in $L_1$  ⇔  $f(x) \leq f(y)$ in $L_2$ for all $x, y \in L_1$,

For example, $1 \leq 2$ in $L_1$  ⇔  $a \leq b$ in $L_2$

∴   The lattices $L_1$ and $L_2$ are isomorphic lattices.

---

**Exercise 8.3**

1. Show that dual of a meet-semi lattice is a join-semi lattice and conversely.

2. Dual of a complete lattice is complete.

3. If $(P, \leq)$ is a poset with least element $o$ such that every non-empty $S$ of $P$ has supremum, then show that $P$ is a complete lattice.

4. We know that $D_4$ and $D_9$ are lattices under the relation of divisibility. Draw the Hasse diagram of the lattice $D_4 \times D_9$ under the product partial order.

5. Consider the lattice $L$ as shown in the figure. Determine whether or not each of the following is the sublattice of $L$?

    $A = \{\phi, \{1, 2\}, \{2, 3\}, \{1, 2, 3\}\}$

    $B = \{\phi, \{1\}, \{1, 2\}, \{1, 2, 3\}\}$

    $C = \{\phi, \{3\}, \{1, 3\}, \{1, 2, 3\}\}$

    $D = \{\{1\}, \{3\}, \{1, 3\}, \{1, 2, 3\}\}$

    $E = \{\phi, \{3\}, \{1, 2\}, \{1, 2, 3\}\}$

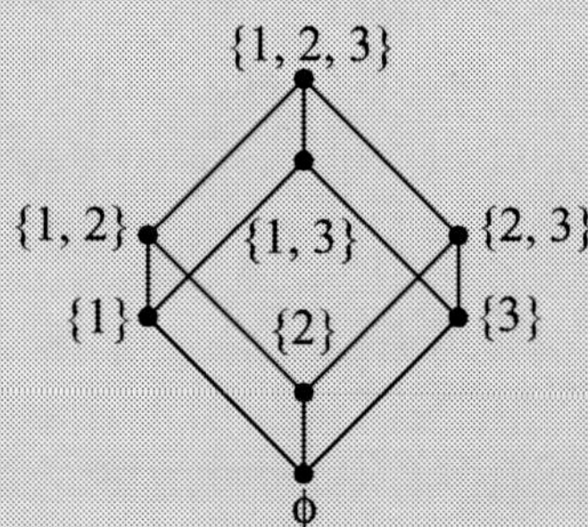


6. Let $L$ be a lattice and let $a$ and $b$ be elements of $L$ such that $a \leq b$. The interval $[a, b]$ is defined as the set of all $x \in L$ such that $a \leq x \leq b$. Prove that $[a, b]$ is a sublattice of $L$.

7. Let $D_{12}$ be a lattice with the relation divisibility and $A = \{1, 2\}$ and $B = \{1, 3\}$ are sublattices of $L$, then determine whether or not $A \cup B$ is a sublattice of $D_{12}$.

8. Let $L = P(S)$ be the lattice of all subsets of a non-empty set $S$ under the relation

of containment. Let T be a non-empty subset of S. Show that P(T) is a sublattice of L.

9. Show that intersection of two convex sublattices is a convex sublattice.

10 If $f: (L, \leq) \to (L', \leq')$ be a lattice isomorphism, then show that $f(a \vee b) = f(a) \vee f(b)$ for $a, b \in L$.

11. Determine whether the lattices shown in the figures are isomorphic or not.

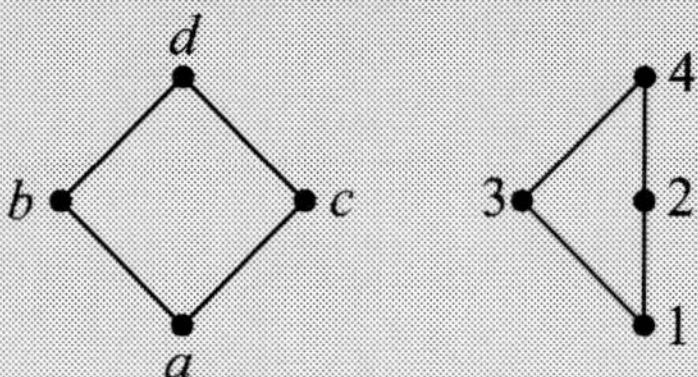


12. Determine whether the lattices shown in the figures are isomorphic or not.

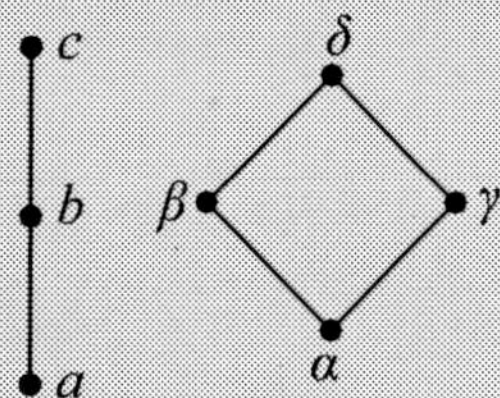


**Answers to Selected Problems**

1. (ii), (iii)
5. (i) Not a sublattice    (ii) Sublattice
   (iii) Sublattice    (iv) Not a sublattice
   (v) Sublattice.
7. $A \cup B$ is not a sublattice
11. Isomorphic
12. Not isomorphic.

## 8.13 BOUNDED LATTICE

A lattice $(L, \vee, \wedge)$ is called a bounded lattice if it has a greatest element and a least element. The greatest element and least element of a *bounded lattice* are denoted by I and $o$ and bounded lattice is denoted by $(L, \vee, \wedge, o, I)$.

If L is a bounded lattice, then for $x \in L$

(i) $x \vee o = x$    and    $x \wedge o = o$
(ii) $x \vee I = I$    and    $x \wedge I = x$

**Illustration**

1. Let S be a non-empty set and P(S), power set of S is a lattice under the relation '$\subseteq$'. Then the lattice $(P(S), \subseteq)$ is a bounded lattice where the greatest element is the set L itself and least element is $\phi$.
2. The lattice $(Z^+, \leq)$ with the relation '$\leq$' not a bounded lattice, as the lattice $(Z^+, \leq)$ has no greatest element.

**Example 13:** Show that the lattice shown below is a bounded lattice.

***Solution***

From the given Hasse diagram, let X = $\{a, b, c, d, e, f\}$

Since $x \leq a$ for all $x \in X$

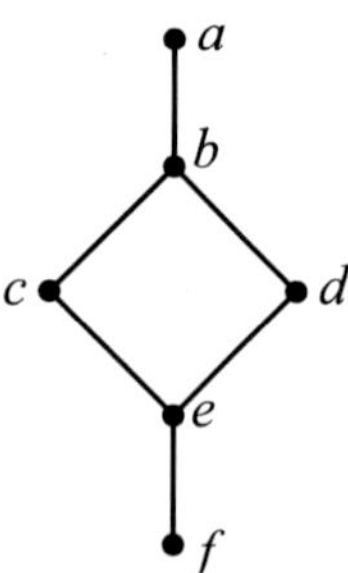

and $\quad f \leq x$ for all $x \in X$

$\therefore \quad$ The greatest element (I) of X is '$a$' and least element (0) of X is $f$.

Hence, the given lattice $(X, \leq)$ is a bounded lattice.

**Theorem 15** Prove that every finite lattice $L = \{a_1, a_2, a_3, \ldots, a_n\}$ is bounded.

***Solution***

We have $L = \{a_1, a_2, a_3, \ldots, a_n\}$.

Since L is a lattice.

$\therefore \quad a_1 \vee a_2 \vee a_3 \vee \ldots \vee a_n \in L \quad$ and $\quad a_1 \wedge a_2 \wedge a_3 \wedge \ldots \wedge a_n \in L$

Also, $a_i \leq a_1 \vee a_2 \vee a_3 \vee \ldots \vee a_n \quad$ and $\quad a_1 \wedge a_2 \wedge \ldots \wedge a_n \leq a_i \quad$ for all $a_i \in L$

$\therefore \quad a_1 \vee a_2 \vee a_3 \vee \ldots \vee a_n$ is the greatest element of L

and $\quad a_1 \wedge a_2 \wedge a_3 \wedge \ldots \wedge a_n$ is the least element of lattice L.

$\therefore \quad$ The greatest and least elements exist for every finite lattice L. Hence L is bounded.

## 8.14 JOIN-IRREDUCIBLE ELEMENT

Let $(L, \wedge, \vee)$ be a lattice. An element $a \in L$ is said to be join-irreducible if it cannot be expressible as the join of two distinct elements of L.

In other words $a \in L$ is join-irreducible if for any $b, c \in L$, $a = b \vee c$, then either $a = b$ or $a = c$.

For example,

(i) Every prime number under multiplication is a join-irreducible element, as $p$ is a prime number, then $p = ab \Rightarrow p = a$ or $p = b$.

(ii) In the given Hasse diagram, $a$ is not a join-irreducible element as exist $b_1$ and $b_2$ such that $a = b_1 \vee b_2$ i.e., $a$ has two immediate predecessors $b_1$ and $b_2$.

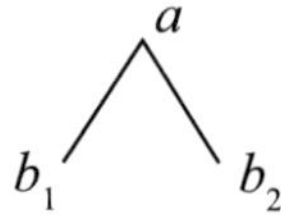

**Theorem 16** Let L be lattice with lower bound 0. Prove that $a(\neq 0) \in$ L is join-irreducible if and only if $a$ has a unique immediate predecessor.

### *Proof*

Here L is a lattice with lower bound 0. Let $a \in$ L, $a \neq 0$

Let $a$ be a join-irreducible element of L.

Since $a \neq 0$, $a$ has at least one immediate predecessor.

Let $b_1$ and $b_2$ be two immediate predecessors of $a$.

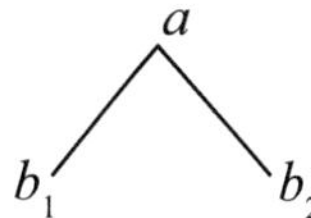

$\therefore \quad a = \sup. \{b_1, b_2\} = b_1 \vee b_2 \quad$ and $\quad a \neq b_1, a \neq b_2$.

$\Rightarrow \quad a$ is not join-irreducible element, which is a contradiction.

$\Rightarrow \quad a$ has a unique immediate predecessor.

*Conversely,* let $a$ has a unique immediate predecessor.

If $a$ is not a join-irreducible element, then there exist distinct elements $b, c$ in L such that $a = b \vee c$ and $a \neq b, a \neq c$.

$$a = b \vee c \quad \Rightarrow \quad b \leq a, c \leq a \quad \Rightarrow \quad b < a, c < a$$

Let $b', c'$ be such that $b \leq b' << a \quad$ and $\quad c \leq c' << a$.

$\Rightarrow \quad a$ has two immediate predecessors $b', c'$, which is not possible.

$\therefore \quad a$ must be join-irreducible element.

**Remark.** An element $a \in$ L is a join-irreducible element iff $a$ has a unique immediate predecessor.

In the Hasse diagram below, $a$ has unique immediate predecessor $c$.

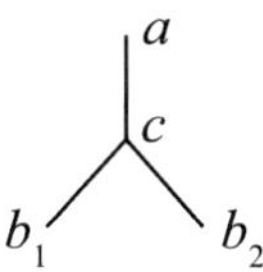

## 8.15 MEET-IRREDUCIBLE ELEMENT

Let $(L, \wedge, \vee)$ be a lattice. An element $a \in$ L is said to be meet-irreducible if it cannot be expressible as the meet of two distinct elements of L.

In other words $a \in$ L is meet-irreducible element if for any $b, c \in$ L, $a = b \wedge c$, then either $a = b$ or $a = c$.

**Example 14:** Determine the join-irreducible and meet-irreducible elements of the lattice shown in given figure.

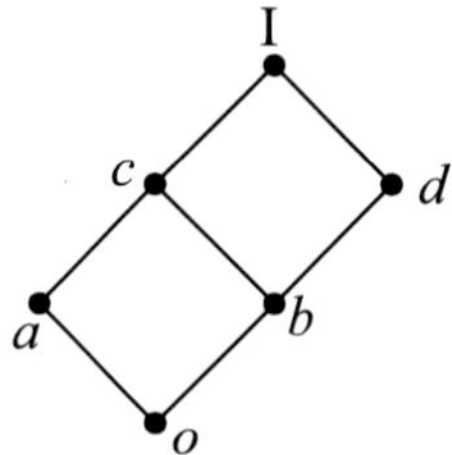

***Solution***

From the given figure, Let L = {$o$, $a$, $b$, $c$, $d$, I}

Clearly, $o$, $a$ and $b$ are join-irreducible elements of L.

Now, $c$ is not a join-irreducible element of L as $a$, $b \in$ L, $c = a \vee b$.

Also, $d$ is also a join-irreducible element of L and I is not a join-irreducible element of L as I $= c \vee d$.

∴    The join-irreducible elements of L are $a$, $b$ and $d$.

Similarly, $o$ is not a meet-irreducible element of L, as $o = a \wedge b$.

Also, $o$, $b$ is not a join-irreducible element of L as $b = c \wedge d$.

∴    The meet-irreducible elements of L are $a$, $c$, $d$ and I.

## Irredundant Join-irreducible Elements

The join-irreducible elements are said to be irredundant join-irreducible elements, if no join-irreducible element precedes any other join-irreducible element.

## 8.16 ATOMS

An element $a$ of a lattice L is called an atom if $a$ is an immediate successor of $o$, where $o$ is a lower bound of L.

If $a$ is an atom, therefore $a$ is the immediate successor of $o$ i.e., $a \gg o$ Þ $a^l o$

∴    $a$ has a unique predecessor $o$.

∴    $a$ is a join-irreducible.

Thus, every atom must be a join-irreducible element.

All join-irreducible elements need not be atoms.

For example, for the lattice in the given figure $e$ and $f$ are atoms.

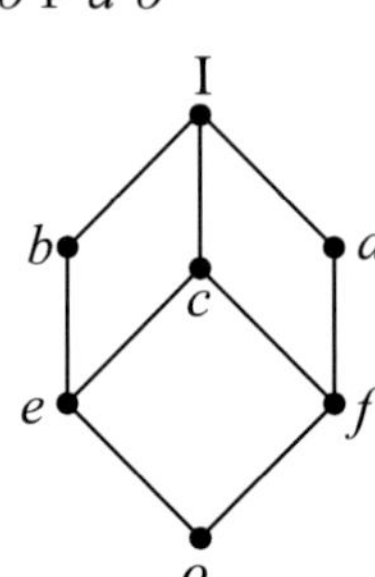

## 8.17 DISTRIBUTIVE LATTICES

A lattice L is said to be **distributive lattice**, if for any $x, y, z \in L$, the following conditions are held

(i) $x \vee (y \wedge z) = (x \vee y) \wedge (x \vee z)$

(ii) $x \wedge (y \vee z) = (x \wedge y) \vee (x \wedge z)$

By the principle of duality the condition (i) holds if and only if (ii) holds. Hence it is sufficient to verify any one of these two conditions.

If a lattice is not distributive, it is called non-distributive lattice.

For example, every chain is a distributive lattice.

### Illustration

The lattice $(P(X), \subseteq)$ is a distributive lattice as

For $A, B \in P(X)$,  $\quad A \cap (B \cup C) = (A \cap B) \cup (A \cap C)$

and  $\quad\quad\quad\quad\quad A \cup (B \cap C) = (A \cup B) \cap (A \cup C)$

**Example 15:** Give an example of a lattice which is not distributive.  [*KU 2007*]

*Solution*

Consider, a lattice $L = \{o, a, b, c, I\}$ with the given Hasse diagram.

For $a, b, c, \in L$

$\quad a \wedge (b \vee c) = a \wedge I = a$

and  $\quad (a \wedge b) \vee (a \wedge c) = o \vee o = o$

$\therefore \quad a \wedge (b \vee c) \neq (a \wedge b) \vee (a \wedge c)$, for all $a, b, c \in L$

$\therefore \quad$ L is not a distributive lattice.

$\therefore \quad$ L is a non-distributive lattice.

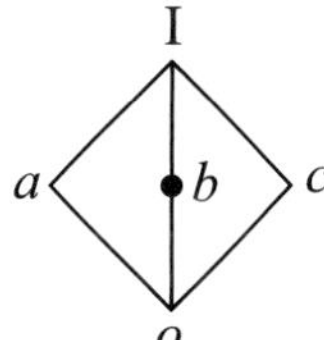


**Example 16:** Examine whether the given lattice in the figure is distributive or non-distributive.

*Solution*

From the given Hasse diagram, let $L = \{0, a, b, c, I\}$ be a lattice.

For $a, b, c, \in L$

$\quad a \vee (b \wedge c) = a \vee 0 = a$  $\quad\quad\quad [\because \; b \wedge c = o]$

But  $(a \vee b) \wedge (a \vee c) = I \wedge c = c$

$\therefore \quad a \vee (b \wedge c) \neq (a \vee b) \wedge (a \vee c)$

Hence, L is a non-distributive lattice.

**Remark.** A sublattice of a distributive lattice is also distributive.

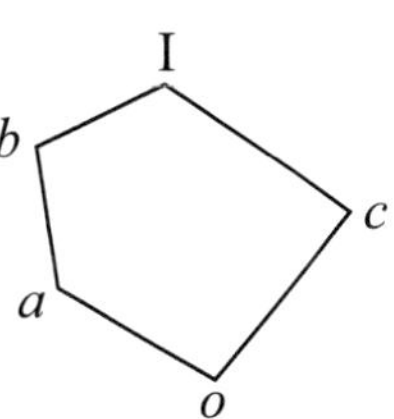

**Example 17:** Let L be a distributive lattice. If for $a, b, c \in$ L such that $a \wedge c = b \wedge c$ and $a \vee c = b \vee c$, then $a = b$.

### Solution

Here L is a distributive lattice.

Let $\quad a \wedge c = b \wedge c$ and $a \vee c = b \vee c$

Now, $\quad a = a \wedge (a \vee c)$ $\hfill$ [*By absorption*]

$\quad\quad = a \wedge (b \vee c) = (a \wedge b)(a \wedge c)$ $\hfill$ [$\because$ L *is distribution*]

$\quad\quad = (a \wedge b) \vee (b \wedge c) = b \wedge (a \vee c)$

$\quad\quad = b \wedge (b \vee c) = b.$

$\therefore \quad a \wedge c = b \wedge c$ and $a \vee c = b \vee c \Rightarrow a = b.$

**Example 18:** Show that a totally ordered set is always a distributive lattice.

### Solution

Let $(S, \leq)$ be a totally ordered set (toset).

$\therefore$ For $a, b \in$ S, either $a \leq b$ or $b \leq a$

$\hfill$ [*As every pair of elements in a toset are comparable*]

If $\quad a \leq b \quad \Rightarrow \quad$ sup. $\{a, b\} = b$, inf. $\{a, b\} = a$

If $\quad b \leq a \quad \Rightarrow \quad$ sup. $\{a, b\} = a$, inf. $\{a, b\} = b$

$\therefore \quad$ sup.$\{a, b\}$ and inf.$\{a, b\}$ both exist in S.

$\therefore \quad$ S is a lattice.

Again, for $a, b, c \in$ S, we have the following cases:

*Case (i):* When $a \leq b \leq c$

$\quad\quad a \wedge (b \vee c) = a \wedge c = a$ and $(a \wedge b) \vee (a \wedge c) = a \vee a = a$

$\therefore \quad\quad a \wedge (b \vee c) = (a \wedge b) \vee (a \wedge c) = a.$

*Case (ii):* When $a \leq c \leq b$

$\quad\quad a \wedge (b \vee c) = a \wedge b = a$ and $(a \wedge b) \vee (a \wedge c) = a \vee a = a$

$\therefore \quad\quad a \wedge (b \vee c) = (a \wedge b) \vee (a \wedge c) = a$

*Case (iii):* When $b \leq a \leq c$

$\quad\quad a \wedge (b \vee c) = a \wedge c = a$ and $(a \wedge b) \vee (a \wedge c) = b \vee a = a$

$\therefore \quad\quad a \wedge (b \vee c) = (a \wedge b) \vee (a \wedge c) = a$

*Case (iv)*: When $b \le c \le a$

$$a \wedge (b \vee c) = a \wedge c = c \quad \text{and} \quad (a \wedge b) \vee (a \wedge c) = b \vee c = c$$

$\therefore \qquad a \wedge (b \vee c) = (a \wedge b) \vee (a \wedge c) = c$

Similarly, we can show that for $c \le a \le b$ and $c \le b \le a$

$$a \wedge (b \vee c) = (a \wedge b) \vee (a \wedge c)$$

$\therefore \qquad a \wedge (b \vee c) = (a \wedge b) \vee (a \wedge c)$, for all $a, b, c \in S$ $\qquad (1)$

Also, by principle of duality, we can show that

$\therefore \qquad a \vee (b \wedge c) = (a \vee b) \wedge (a \vee c)$, for all $a, b, c \in S$

$\therefore$  The lattice S is distributive.

**Theorem 17** Let L be a finite distributive lattice. Prove that every element in L can be written uniquely (except for order) as the join of irredundant join-irreducible elements of L.

### *Proof*

Let $a \in L$ be any element. If $a$ is join-irreducible, then we have done.

If $a$ is not a join-irreducible element, then there exists elements $b_1, b_2 \in L$ in L such that $a = b_1 \vee b_2$ and $a \ne b_1, a \ne b_2$.

If both $b_1$ and $b_2$ are join-irreducible, we have done, otherwise we can write $b_1$ and $b_2$ as the join of other elements of L. Since the lattice L is a finite lattice, therefore this process must be terminated after a finite number of steps.

$\therefore$  $a$ can be expressed as the join of join-irreducible elements.

Let  $a = c_1 \vee c_2 \vee \ldots \vee c_n$,  where $c_1, c_2, \ldots, c_k$ are join-irreducible elements.

In the above expression if any $c_i$ precedes $c_j$ then we have $c_i \vee c_j = c_j$.

$\therefore$  $c_i$ can be removed from the above expression of $a$.

Thus, we can express $a$ as the join of join-irreducible elements where no $c$ precedes any other $c$.

Let $a = d_1 \vee d_2 \vee \ldots \vee d_m$ be an expression of $a$ as the join of join-irreducible elements and no $d$ precedes any other $d$.

$\therefore$  The elements $d_1, d_2, \ldots, d_m$ in the expression of $a$ are irredundant join-irreducible elements.

Now, we have to show that the expression

$$a = d_1 \vee d_2 \vee \ldots \vee d_m \qquad (1)$$

is unique, except for the order.

Let $\qquad\qquad\qquad a = e_1 \vee e_2 \vee, \ldots, \vee e_k \qquad (2)$

be any other representation of $a$ as the join of irredundant join-irreducible elements.

Now, we have $\qquad d_i \le d_1 \vee d_2 \vee \ldots \vee d_m$, for $1 \le i \le m$

$\Rightarrow \qquad\qquad d_i \le e_1 \vee e_2 \vee, \ldots, \vee e_k$

$\Rightarrow \qquad\qquad d_i = d_i \wedge (e_1 \vee e_2 \vee, \ldots, \vee e_k)$

Since L is distributive.

$\therefore \qquad\qquad d_i = (d_i \wedge e_1) \vee (d_i \wedge e_2) \vee (d_i \wedge e_3) \vee \ldots \vee (d_i \wedge e_k)$

Since $d_i$ is join-irreducible, therefore there exists $j$, $1 \le j \le k$ such that $d_i = d_i \wedge e_j$

$\Rightarrow \qquad\qquad d_i \le e_j \qquad\qquad\qquad (3)$

Similarly, we can show that $e_j \le d_r$ for some $r$ such that $1 \le r \le m \qquad\qquad (4)$

From (3) and (4), we have $d_i \le e_j \le d_r$

Since $d$'s are irredundant, we must have $d_i = e_j = d_r$.

$\therefore \qquad\qquad d_r \le e_j$

Thus, by selecting different values of $i$, the $d$'s and $e$'s may be paired off.

$\therefore$   The representation (1) of $a$ is unique except for the order.

**Note.**  The above theorem may not be true for non-distributive lattice.

Let L be the finite lattice given in the below figure.

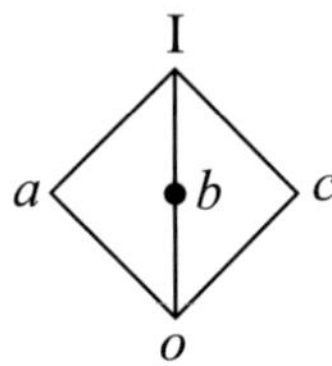


$\therefore$   L = $\{o, a, b, c, I\}$. Clearly, L is a non-distributive lattice.

Now, $I = a \vee b$ and $I = b \vee c$ are two distinct representations of I as the join or irredundant join-irreducible elements.

$\therefore$   Representation of I is not unique for non-distributive lattice.

**Theorem 18** Two lattices L and M are distributive lattices iff L × M is distributive.

***Proof***

Let L and M be distributive lattices.

Let $(a_1, b_1)$, $(a_2, b_2)$, $(a_3, b_3) \in$ L × M, for $a_1, a_2, a_3 \in$ L and $b_1, b_2, c_3 \in$ M

Now, $\qquad\qquad (a_1, b_1) \wedge [(a_2, b_2) \vee (a_3, b_3)]$

$\qquad\qquad\qquad\qquad = (a_1, b_1) \wedge (a_2 \vee a_3, a_2 \vee b_3)$

$\qquad\qquad\qquad\qquad = (a_1 \wedge (a_2 \vee a_3), b_1 \wedge (b_2 \vee b_3))$

$\qquad\qquad\qquad\qquad = ((a_1 \wedge a_2) \vee (a_1 \wedge a_3), (b_1 \wedge b_2) \vee (b_1 \wedge b_3))$

$\qquad\qquad\qquad\qquad\qquad\qquad [\because \quad$ L *and* M *are distributive*$]$

$$= (a_1 \wedge a_2, b_1 \wedge b_2) \vee (a_1 \wedge a_3, b_1 \wedge b_3)$$

$$= [(a_1, b_1) \wedge (a_2, b_2)] \vee [(a_1, b_1) \wedge (a_2, b_3)]$$

$\therefore \quad (a_1, b_1) \wedge [(a_2, b_2) \vee (a_3, b_3)] = [(a_1, b_1) \wedge (a_2, b_2)] \vee (a_1, b_1) \wedge (a_3, b_3)],$

for $\quad (a_1, b_1), (a_2, b_2), (a_3, b_3) \in L \times M.$

Similarly, we can show that

$(a_1, b_1) \vee [(a_2, b_2) \wedge (a_3, b_3)] = [(a_1, b_1) \vee (a_2, b_2)] \wedge [(a_1, b_1) \vee (a_3, b_3)],$

for $(a_1, b_1), (a_2, b_2), (a_3, b_3) \in L \times M.$

*Conversely,* let $L \times M$ be distributive lattice. To show L and M are distributive lattice $a_1, a_2,$ $a_3 \in L$ and $b_1, b_2, b_3 \in M$ be any elements

$\therefore \quad (a_1, b_1), (a_2, b_2), (a_3, b_3) \in L \times M$

$\Rightarrow \qquad (a_1, b_1) \wedge [(a_2, b_2) \vee (a_3, b_3)] = [(a_1, b_1) \wedge (a_2, b_2)] \vee (a_1, b_1) \wedge (a_3, b_3)]$

$\Rightarrow \qquad (a_1, b_1) \wedge (a_2, \vee a_3, b_2 \vee b_3) = (a_1 \wedge a_2, b_1 \wedge b_2) \vee (a_1 \wedge a_3, b_1 \wedge b_3)$

$\Rightarrow \quad (a_1 \wedge (a_2, \vee a_3), b_1 \wedge (b_2 \vee b_3)) = ((a_1 \wedge a_2) \vee (a_1 \wedge a_3), (b_1 \wedge b_2) \vee (b_1 \wedge b_3))$

$\Rightarrow \qquad\qquad a_1 \wedge (a_2, \vee a_3) = (a_1 \wedge a_2) \vee (a_1 \wedge a_3)$

and $\qquad\qquad b_1 \wedge (b_2 \vee b_3) = (b_1 \vee b_2) \vee (b_1 \wedge b_3)$

$\Rightarrow \quad$ L and M are distributive lattices.

## 8.18 MODULAR LATTICE

A lattice $(L, \leq)$ is called a modular lattice if for any $x, y, z$ in L, where $x \leq z.$

$$x \vee (y \wedge z) = (x \vee y) \wedge z$$

**Theorem 19** Prove that a distributive lattice is modular.

*Proof*

Let $(L, \leq)$ be a distributive lattice.

For $x, y, z \in L$ and $x \leq z$

$\therefore \qquad\qquad\qquad x \vee (y \wedge z) = (x \vee y) \wedge (x \vee z)$

$\qquad\qquad\qquad\qquad = (x \vee y) \wedge z \qquad\qquad [\because \quad x \vee z = z]$

$\therefore \quad$ The lattice L is a modular lattice.

**Theorem 20** A sublattice of a modular lattice is modular.

*Proof*

Let S be a sublattice of a modular lattice L.

For $x, y, z \in S$ with $x \in z$

$\Rightarrow \qquad\qquad\qquad\qquad\qquad x, y, z \in L$

$\therefore \qquad\qquad\qquad\qquad x \vee (y \wedge z) = (x \vee y) \wedge z$

Since S is closed under $\wedge$ and $\vee$, therefore result holds in S.

Hence, S is modular.

**Example 19:** Determine whether the lattice shown in the given figure is a modular lattice or not.

***Solution***

Let $L = \{o, u, v, w, I\}$ be the given lattice.

For $a = o, \quad b = v, \quad c = w$

Obviously, $a \leq w$

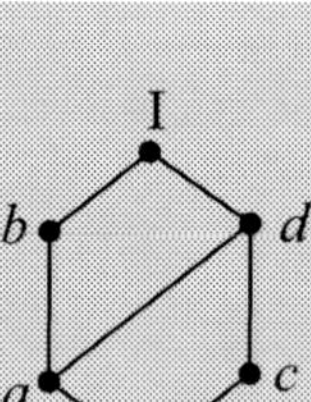

Now, $\qquad\qquad a \vee (b \wedge c) = o \vee (v \wedge w) = o \vee o = o$

and $\qquad\qquad (a \vee b) \wedge c = (o \vee v) \wedge w = v \vee w = o$

$\therefore \qquad\qquad a \vee (b \wedge c) = (a \vee b) \wedge c.$

$\therefore$ L is a modular lattice.

---

### Exercise 8.4

1. If L is a bounded lattice then show that

$$a \vee (b \wedge c) = (a \vee b) \wedge (a \vee c)$$

   when $a, b, c \in L$ and any one $a, b, c$ is either $o$ or $I$.

2. Show that if a bounded lattice has two or more elements then $o \neq I$.

3. Show that sublattice of a distributive lattice is distributive.

4. Let P be the power set of the set $\{1, 2, 3\}$. Show that the lattice P under the relation of set inclusion is distributive.

5. Let $L = \{o, a, b, c, d, I\}$ be a bounded lattice with the Hasse diagram given in the figure. Show that for any $x, y, z$ from the subset $\{a, b, d\}$ of L, we have

$$x \vee (y \wedge z) = (x \vee y) \wedge (x \vee z)$$

   and $x \wedge (y \vee z) = (x \wedge y) \vee (x \wedge z)$

6. Show that a totally ordered lattice is modular.

7. Show that two lattices $L_1$ and $L_2$ are modular iff $L_1 \times L_2$ is modular.

8. If $a, b, c$, are elements of a modular lattice L with greatest element $u$ and if

$$a \vee b = (a \wedge b) \vee c = u,$$ then show that

$$a \vee (b \wedge c) = b \vee (c \vee a) = c \vee (a \vee b) = u.$$

## 8.19 COMPLEMENT OF AN ELEMENT

If L is a bounded lattice with least element 0 and greatest element I and $a \in$ L be any. An element $a' \in$ L is called complement of a if $a \vee a' =$ I and $a \wedge a' = 0$.

Here, it can be noted that complements may not exist and may not be unique.

For example, in the bounded lattice $D_{20} = \{1, 2, 4, 5, 10, 20\}$. Elements 2 and 10 have no complements in $D_{20}$.

Since 1 is the least element and 20 is the greatest element of $D_{20}$.

$$\text{Now, for } 10 \in D_{20}, \quad \begin{array}{lll} 10' \neq 1, & \text{as} & 10 \vee 1 = 10 \neq 20 \\ 10' \neq 2, & \text{as} & 10 \vee 2 = 10 \neq 20 \\ 10' \neq 4, & \text{as} & 10 \vee 4 = 20 \quad \text{but} \quad 10 \wedge 4 = 2 \neq 1 \\ 10' \neq 5, & \text{as} & 10 \vee 5 = 10 \ 10 \neq 20 \\ 10' \neq 10, & \text{as} & 10 \vee 10 = 10 = 10 = \neq 20 \\ 10' \neq 20, & \text{as} & 10 \vee 20 = 20 \quad \text{but} \quad 10 \wedge 20 = 10 \neq 1 \end{array}$$

Similarly, $2 \in D_{20}$ has no complement in $D_{20}$.

We can see that $o' =$ I and $I' = o$ as $o \vee$ I $=$ I and $o \wedge$ I $= o$.


**Theorem 21** Prove that in a bounded distributive lattice, complement (if exists) of an element is unique.                                                       [*KU 2006*]

### *Proof*

Let L be a bounded distributive lattice with least element $o$ and greatest I.

Suppose that an element $a$ has two complements $b$ and $c$ in L.

$$\text{i.e.,} \qquad a \vee b = \text{I} \quad \text{and} \quad a \wedge b = 0 \tag{1}$$

$$a \vee c = \text{I} \quad \text{and} \quad a \wedge c = 0 \tag{2}$$

$$\begin{aligned} \text{Now,} \quad b &= b \wedge \text{I} \\ &= b \wedge (a \vee c) && [Using\ (2)] \\ &= (b \wedge a) \vee b \wedge c) && [\because \ \text{L } is\ distributive] \\ &= 0 \vee (b \wedge c) && [Using\ (1)] \\ &= (a \wedge c) \vee (b \wedge c) && [Using\ (2)] \\ &= (a \wedge b) \wedge c \\ &= \text{I} \wedge \text{c} \\ &= c. \end{aligned}$$

$$\Rightarrow \qquad\qquad b = c$$

Hence, if an element has a complement then the complement is unique.

**Example 20:** If $a$ and $b$ are elements in a bounded distributive lattice and $a$ has a complement $a'$, then show that:

(i) $a(a' \wedge b) = a \vee b$  (ii) $a \wedge (a' \vee b) = a \wedge b$.

### *Solution*

Let L be a bounded distributive lattice with least element $o$ and greatest element I. $a, b \in L$ be any elements.

Since $a$ has a complement $a'$

$\therefore \quad a \vee a' = I \quad$ and $\quad a \wedge a' = o$  (1)

(i) $\qquad a \vee (a' \wedge b) = (a \vee a') \wedge (a \vee b) \qquad [\because \quad$ L *is distributive*]

$\qquad\qquad = I \wedge (a \vee b) = a \vee b \qquad\qquad [Using\ (1)]$

(ii) $\qquad a \wedge (a' \vee b) = (a \wedge a') \vee (a \wedge b) \qquad [\because \quad$ L *is distributive*]

$\qquad\qquad = o \wedge (a \wedge b) = a \wedge b \qquad\qquad [Using\ (1)]$

**Example 21:** Find the complement of each element of the lattice $D_{35}$.

### *Solution*

Here $D_{35} = \{1, 5, 7, 35\}$ is a bounded lattice with 1 is the least element and 35 is the greatest elements. The Hasse diagram of the bounded lattice $D_{35}$ is shown in the adjoining figure.

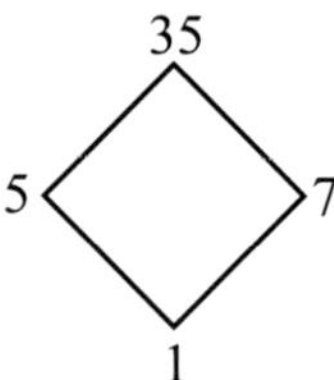


(i)  $1 \vee 35 = 35, \quad 1 \wedge 35 = 1$

$\therefore \quad 1' = 35 \quad$ and $\quad 35' = 1$

(ii)  $5 \vee 7 = 35, \quad 5 \wedge 7 = 1$

$\therefore \quad 5' = 7 \quad$ and $\quad 7' = 5$

$\therefore$  Complements of 1, 5, 7 and 35 are 35, 7, 5 and 1 respectively.

## 8.20 COMPLEMENTED LATTICES

Let L be a bounded lattice and every element $a \in L$ has a complement, then L is called complemented lattice.

**Illustration 1.** The lattice $D_{42}$ is complemented lattice as the lattice $D_{42}$ is bounded with least element 1 and greatest element 42.

Also, $\quad 1' = 42, \quad 2' = 21, \quad 3' = 14, \quad 6' = 7, \quad 7' = 6, \quad 14' = 3, \quad 21' = 2 \quad$ and $\quad 42' = 1$.

**Illustration 2.** The lattice in figure is a complemented lattice, where the complement of $a$ is $c$, the complement of $b$ is also $c$,

$\Rightarrow$  The complements of $c$ are $a$ and $b$. Also, 0 and 1 are complement of each other.

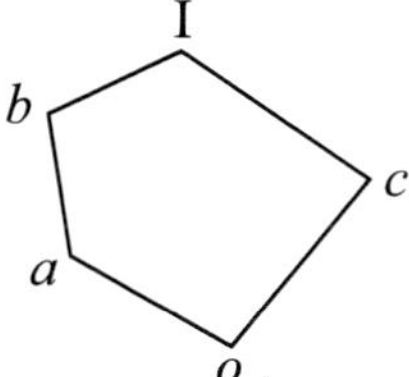

**Example 3:** Check whether the lattice, given in figure is complemented or not.

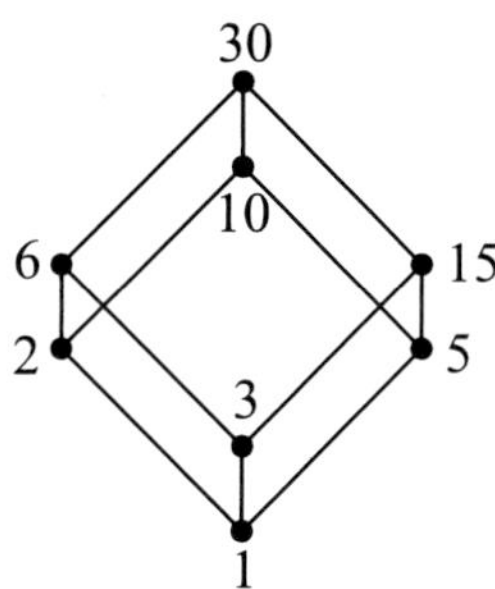

### Solution

Here $D_{30}$ {1, 2, 3, 5, 6, 10, 15, 30} is bounded with least element 1 and greatest element 30. 1 and 30 are complements of each other as $1 \vee 30 = 30$ and $1 \wedge 30 = 1$.

Also,     $2 \vee 15 = 30$   and   $2 \wedge 15 = 1$

∴    2 and 15 are complements of each other.

Similarly, 3 and 10 are complements of each other and 5 and 6 are complements of other.

∴    $D_{30}$ is a complemented lattice.


**Example 22:** Find whether the given bounded lattice is complemented or not.

### Solution

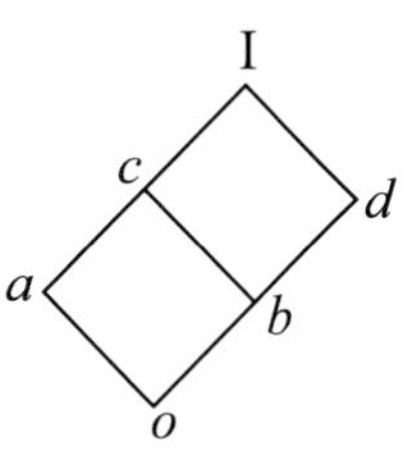

From the given Hasse diagram,

Let     $L = \{o, a, b, c, d, I\}$

Here    $o' = I$   and   $I' = o$

Now,    $a \vee d = I$   and   $a \wedge d = o$

∴    $a$ and $d$ are complements of each other.

Also,    $b \vee o = b$,   $b \vee a = c$,   $b \vee c = c$,   $b \vee d = d$   and   $b \vee I = I$

But,    $b \wedge I = b \neq o$

∴    $b$ has no complement.

∴    L is not a complemented lattice.

## Exercise 8.5

1. Let L be bounded a lattice shown in the figure. Show that each element of L has at least one complement.

2. Let L be a bounded lattice with at least two elements. Show that no element of L is its own-complement.

3. If L is a bounded distributive lattice then show that the set of complemented elements of L is sublattice of L.

4. Find whether the given bounded lattice is complemented or not.

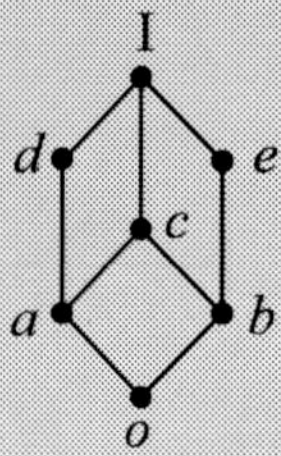

5. Let S be a non-empty set. Show that the lattice P(S), the power set of S, under the relation of set inclusion is a complemented lattice.

6. Let L be a finite complemented distributive lattice. Prove that every element in L can be written uniquely (except for order) as the join of irredundant atoms of L.

# 9

# Boolean Algebra

## 9.1 INTRODUCTION

George Boole (1815–1864), an English mathematician introduced a systematic treatment of logic and developed an algebra, now called "Boolean Algebra". In 1938, C.E. Shannon used a two-valued Boolean algebra called "Switching Algebra", to represent the properties of bistable electrical switching circuits which are used in designing the computer chips. In this chapter we shall study Boolean algebra as an algebraic structure Boolean algebra as a lattice and also discuss its applications to switching circuits and logic gates.

## 9.2 BINARY OPERATIONS

Let A be a non-empty set, then the function from $A \times A$ into A is called a **binary operation.**

Symbolically, a function '$*$' which is '$*$': $A \times A \to A$ is called a binary operation. The image of any $(a, b) \in A \times A$ under '$*$' is denoted by $a * b$.

**Illustrative Examples**

1. Addition on N is a binary operation, for example
$$2 \in N, 5 \in N \implies 2 + 5 = 7 \in N$$
2. Multiplication on N is a binary operation, for example
$$3 \in N, 4 \in N \implies 3 \times 4 = 12 \in N$$
3. Multiplication in the set of even integers is also a binary operation.
4. Subtraction on N is not a binary operation, since
$$2 \in N, 3 \in N \quad \text{but } (2 - 3) = -1 \notin N$$
5. Subtraction on the set I of all integers is a binary operation, since
$$a \in I, b \in I \implies a - b \in I \quad \text{for all } a, b \in I$$

## 9.3 ALGEBRAIC STRUCTURE

A non empty set B equipped with one or more operations defined on it is called an **algebraic structure.**

Suppose '$*$' is a binary operation on B, then (B, $*$) is an algebraic structure. Similarly, (N, +), (R, +, ×) etc. are all algebraic structures.

## 9.4 BOOLEAN ALGEBRA

A non-empty set B together with two binary operations '+' and '·' and one unary operation '$'$' and satisfying the following axioms is called **Boolean algebra**.

**1. Closure Laws:**

For all $a, b \in$ B,
(i) $a + b \in$ B
(ii) $a \cdot b \in$ B

**2. Commutative laws:**

For all $a, b \in$ B,
(i) $a + b = b + a$
(ii) $a \cdot b = b \cdot a$

**3. Associative laws:**

For all $a, b, c \in$ B,
(i) $a + (b + c) = (a + b) + c$
(ii) $a \cdot (b \cdot c) = (a \cdot b) \cdot c$

**4. Identity laws:**

There exist two distinct elements 0 and 1 in B (called the **zero element** and the **unit element** respectively) such that for all $a \in$ B,
(i) $a + 0 = a$
(ii) $a \cdot 1 = a$

**5. Complement laws:**

For every $a \in$ B, there exists an element $a' \in$ B such that
(i) $a + a' = 1$
(ii) $a \cdot a' = 0$
where $a'$ is complement or negation of $a$.

**6. Distributive laws:**

For all $a, b, c \in$ B, we have
(i) $a + (b \cdot c) = (a + b) \cdot (a + c)$
(ii) $a \cdot (b + c) = (a \cdot b) + (a \cdot c)$

**Note**

1. Each variable in Boolean algebra B can assume only two specific values 0 and 1.
2. + denotes $\vee$ (join or disjunction).
3. · denotes $\wedge$ (meet or conjunction).
4. $'$ denotes $\sim$ (negation).
5. 1 denotes true (T) and 0 denotes false (F).
6. Boolean algebra is generally denoted by (B, +, ·, $'$, 0, 1) in order to emphases its various parts i.e., the set B, two binary operations '+' and '·', the complement '$'$' and two special elements 0 and 1.

Given below is the table showing similarities between logic, set theory and Boolean algebra.

|  | **Set** | **Logic** | **Boolean algebra** |
|---|---|---|---|
| Statements | A, B, C, ... | $p, q, r, ...$ | $a, b, c, ...$ |
| Logical OR | $A \cup B$ | $p \vee q$ | $a + b$ |
| Logical AND | $A \cap B$ | $p \wedge q$ | $a \cdot b$ |
| Logical NOT | $A'$ | $\sim p$ | $a'$ |
| Tautology | Universal set U | $t$ | 1 (unit element) |
| Contradiction | $\phi$ | $c$ | 0 (zero element) |
| Identity laws | $A \cup \phi = A$ | $p \vee c = p$ | $a + 0 = a$ |
|  | $A \cap U = A$ | $p \wedge t = p$ | $a \cdot 1' = a$ |
| Complement laws | $A \cup A' = U$ | $p \vee \sim p = t$ | $a + a' = 1$ |
|  | $A \cap A' = \phi$ | $p \wedge \sim p = c$ | $a \cdot a' = 0$ |

**Example 1:**  Describe a Boolean algebra of sets?

***Solution***

Let B be a collection of sets closed under union, intersection and complement. Then B is a Boolean algebra with the union ($\cup$) as logical OR, intersection ($\cap$) as logical AND and complement ($'$) as logical NOT. Here the empty set $\phi$ is the zero element and the universal set U is the unit element of B.

**Example 2:**  Let A = $\{a, b\}$. Without assuming commutative laws and distributive laws of the set theory, show that $(P(A), \cup, \cap, ', \phi, A)$ is a Boolean algebra where P(A) is the power set of the set A.

***Solution***

We have P(A) = $\{\phi, \{a\}, \{b\}, \{a, b\}\}$.
The composition tables for the operations $\cup$ and $\cap$ are:

| $\cup$ | $\phi$ | $\{a\}$ | $\{b\}$ | $\{a, b\}$ |
|---|---|---|---|---|
| $\phi$ | $\phi$ | $\{a\}$ | $\{b\}$ | $\{a, b\}$ |
| $\{a\}$ | $\{a\}$ | $\{a\}$ | $\{a, b\}$ | $\{a, b\}$ |
| $\{b\}$ | $\{b\}$ | $\{a, b\}$ | $\{b\}$ | $\{a, b\}$ |
| $\{a, b\}$ | $\{a, b\}$ | $\{a, b\}$ | $\{a, b\}$ | $\{a, b\}$ |

| $\cap$ | $\phi$ | $\{a\}$ | $\{b\}$ | $\{a, b\}$ |
|---|---|---|---|---|
| $\phi$ | $\phi$ | $\phi$ | $\phi$ | $\phi$ |
| $\{a\}$ | $\phi$ | $\{a\}$ | $\phi$ | $\{a\}$ |
| $\{b\}$ | $\phi$ | $\phi$ | $\{b\}$ | $\{b\}$ |
| $\{a, b\}$ | $\phi$ | $\{a\}$ | $\{b\}$ | $\{a, b\}$ |

Here      $\phi' = \{a, b\}, \{a\}' = \{b\}, \{b\}' = (a), (a, b\}' = \phi$

1. **Closure laws:** As all the entries in the composition tables are the elements of set P(A)

   $\therefore$     Closure laws are satisfied.

2. **Commutative laws:** From the composition tables, we have for all $X, Y \in P(A)$

$$X \cup Y = Y \cup X \quad \text{and} \quad X \cap Y = Y \cap X$$

   Hence commutative laws are satisfied.

3. **Associative laws:** From the composition tables, we have, for all $X, Y, Z \in P(A)$

$$(X \cup Y) \cup Z = X \cup (Y \cup Z)$$

   and
$$(X \cap Y) \cap Z = X \cap (Y \cap Z)$$

   $\therefore$
$$(X + Y) + Z = X + (Y + Z)$$

   and
$$(X \cdot Y) \cdot Z = X \cdot (Y \cdot Z)$$

   Hence associative laws are satisfied.

4. **Identity laws:** From the composition tables, we have

$$\phi \cup \phi = \phi \qquad\qquad \phi \cap \{a, b\} = \phi$$
$$\{a\} \cup \phi = \{a\} \qquad\qquad \{a\} \cap \{a, b\} = \{a\}$$
$$\{b\} \cup \phi = \{b\} \qquad\qquad \{b\} \cap \{a, b\} = \{b\}$$
$$\{a, b\} \cup \phi = \{a, b\} \qquad \{a, b\} \cap \{a, b\} = \{a, b\}$$

   Thus, $\{a, b\}$ is the unit element and $\phi$ is the zero element.

5. **Complement laws:** From the composition tables, we have

$$\phi \cup \phi' = \{a, b\} \qquad\qquad \phi \cap \phi' = \phi \qquad\qquad [\because \quad \phi' = \{a, b\}]$$

   Similarly,    $\{a\} \cup \{a\}' = \{a, b\} \qquad \{a\} \cap \{a\}' = \phi$

   $\{b\} \cup \{b\}' = \{a, b\} \qquad \{b\} \cap \{b\}' = \phi$

   $\{a, b\} \cup \{a, b\}' = \{a, b\} \qquad \{a, b\} \cap \{a, b\}' = \phi$

   Hence complement laws are satisfied.

6. **Distributive laws:** From the composition tables, we have for all $X, Y, Z \in P(A)$

$$X \cup (Y \cap Z) = (X \cup Y) \cap (X \cup Z) \quad \text{and} \quad X \cap (Y \cup Z) = (X \cap Y) \cup (X \cap Z)$$

   For example     let    $X = \{a\}$,    $Y = \{b\}$   and    $Z = \{a, b\}$

   (i)
$$X \cup (Y \cap Z) = \{a\} \cup [\{b\} \cap \{a, b\}]$$
$$= \{a\} \cup [\{b\}]$$
$$= \{a\} \cup \{b\} = \{a, b\}$$

   and
$$(X \cup Y) \cap (X \cup Z) = [\{a\} \cup \{b\}] \cap [\{a\} \cup \{a, b\}]$$
$$= \{a, b\} \cap \{a, b\} = \{a, b\}$$

   $\therefore$
$$X \cup (Y \cap Z) = (X \cup Y) \cap (X \cup Z)$$

(ii) $$X \cap (Y \cup Z) = \{a\} \cap [\{b\} \cup \{a, b\}]$$
$$= \{a\} \cap \{a, b\} = \{a\}$$

and $$(X \cap Y) \cup (X \cap Z) = [\{a\} \cap \{b\}] \cup [\{a\} \cap \{a, b\}]$$
$$= \phi \cup \{a\} = \{a\}$$

∴ $$X \cap (Y \cup Z) = (X \cap Y) \cup (X \cap Z)$$

Hence $(P(A), \cup, \cap, ', \phi, A)$ is a Boolean algebra.

**Example 3:** Let $B = \{0, 1\}$. The binary operations '$+$' and '$\cdot$' on B are defined as follows:

| + | 0 | 1 |
|---|---|---|
| 0 | 0 | 1 |
| 1 | 1 | 1 |

| $\cdot$ | 0 | 1 |
|---|---|---|
| 0 | 0 | 0 |
| 1 | 0 | 1 |

A unary operation '$'$' is defined as $0' = 1$ and $1' = 0$. Show that $(B, +, \cdot, ', 0, 1)$ is a Boolean algebra.

***Solution***

1. **Closure laws:** As all the entries in the composition tables are the elements of set B

   ∴ Closure laws are satisfied.

2. **Commutative laws:**

| $x$ | $y$ | $x + y$ | $y + x$ |
|---|---|---|---|
| 0 | 0 | 0 | 0 |
| 1 | 0 | 1 | 1 |
| 0 | 1 | 1 | 1 |
| 1 | 1 | 1 | 1 |

| $x$ | $y$ | $x \cdot y$ | $y \cdot x$ |
|---|---|---|---|
| 0 | 0 | 0 | 0 |
| 0 | 1 | 0 | 0 |
| 1 | 0 | 0 | 0 |
| 1 | 1 | 1 | 1 |

From the above composition tables, we have

$$x + y = y + x \quad \text{and} \quad x \cdot y = y \cdot x \qquad \text{for all } x, y \in B$$

∴ Commutative laws are satisfied.

3. **Associative laws:**

| $x$ | $y$ | $z$ | $x + y$ | $y + z$ | $(x + y) + z$ | $x + (y + z)$ |
|---|---|---|---|---|---|---|
| 0 | 0 | 0 | 0 | 0 | 0 | 0 |
| 0 | 0 | 1 | 0 | 1 | 1 | 1 |
| 0 | 1 | 0 | 1 | 1 | 1 | 1 |
| 0 | 1 | 1 | 1 | 1 | 1 | 1 |
| 1 | 0 | 0 | 1 | 0 | 1 | 1 |
| 1 | 0 | 1 | 1 | 1 | 1 | 1 |
| 1 | 1 | 0 | 1 | 1 | 1 | 1 |
| 1 | 1 | 1 | 1 | 1 | 1 | 1 |

| $x$ | $y$ | $z$ | $x \cdot y$ | $y \cdot z$ | $(x \cdot y) \cdot z$ | $x \cdot (y \cdot z)$ |
|---|---|---|---|---|---|---|
| 0 | 0 | 0 | 0 | 0 | 0 | 0 |
| 0 | 0 | 1 | 0 | 0 | 0 | 0 |
| 0 | 1 | 0 | 0 | 0 | 0 | 0 |
| 0 | 1 | 1 | 0 | 1 | 0 | 0 |
| 1 | 0 | 0 | 0 | 0 | 0 | 0 |
| 1 | 0 | 1 | 0 | 0 | 0 | 0 |
| 1 | 1 | 0 | 1 | 0 | 0 | 0 |
| 1 | 1 | 1 | 1 | 1 | 1 | 1 |

From the above composition tables, we have for all $x, y, z \in$ B

$$(x + y) + z = x + (y + z) \quad \text{and} \quad (x \cdot y) \cdot z = x \cdot (y \cdot z)$$

$\therefore$    Associative laws are satisfied.

### 4. Identity laws:

| $x$ | 0 | $x + 0$ |
|---|---|---|
| 0 | 0 | 0 |
| 1 | 0 | 1 |

| $x$ | 1 | $x \cdot 1$ |
|---|---|---|
| 0 | 1 | 0 |
| 1 | 1 | 1 |

$\therefore$ $\qquad\qquad x + 0 = x \quad \text{and} \quad x \cdot 1 = x \quad \text{for all } x \in$ B

$\therefore$    $0 \in$ B is zero element of B and $1 \in$ B is unit element of B.

### 5. Complement laws:

| $x$ | $x'$ | $x + x'$ |
|---|---|---|
| 0 | 1 | 1 |
| 1 | 0 | 1 |

| $x$ | $x'$ | $x \cdot x'$ |
|---|---|---|
| 0 | 1 | 0 |
| 1 | 0 | 0 |

$\therefore$ $\qquad\qquad x + x' = 1 \quad \text{and} \quad x \cdot x' = 0 \quad \text{for all } x \in$ B

$\therefore$    Complement laws are satisfied.

### 6. Distributive laws:

| $x$ | $y$ | $z$ | $y + z$ | $x \cdot (y + z)$ | $x \cdot y$ | $x \cdot z$ | $x \cdot y + x \cdot z$ |
|---|---|---|---|---|---|---|---|
| 0 | 0 | 0 | 0 | 0 | 0 | 0 | 0 |
| 0 | 0 | 1 | 1 | 0 | 0 | 0 | 0 |
| 0 | 1 | 0 | 1 | 0 | 0 | 0 | 0 |
| 0 | 1 | 1 | 1 | 0 | 0 | 0 | 0 |
| 1 | 0 | 0 | 0 | 0 | 0 | 0 | 0 |
| 1 | 0 | 1 | 1 | 1 | 0 | 1 | 1 |
| 1 | 1 | 0 | 1 | 1 | 1 | 0 | 1 |
| 1 | 1 | 1 | 1 | 1 | 1 | 1 | 1 |

| $x$ | $y$ | $z$ | $y \cdot z$ | $x + (y \cdot z)$ | $x + y$ | $x + z$ | $(x + y) \cdot (x + z)$ |
|---|---|---|---|---|---|---|---|
| 0 | 0 | 0 | 0 | 0 | 0 | 0 | 0 |
| 0 | 0 | 1 | 0 | 0 | 0 | 1 | 0 |
| 0 | 1 | 0 | 0 | 0 | 1 | 0 | 0 |
| 0 | 1 | 1 | 1 | 1 | 1 | 1 | 1 |
| 1 | 0 | 0 | 0 | 1 | 1 | 1 | 1 |
| 1 | 0 | 1 | 0 | 1 | 1 | 1 | 1 |
| 1 | 1 | 0 | 0 | 1 | 1 | 1 | 1 |
| 1 | 1 | 1 | 1 | 1 | 1 | 1 | 1 |

From the above composition tables, we have for all $x, y, z \in B$

$$x \cdot (y + z) = x \cdot y + x \cdot z$$

and

$$x + (y \cdot z) = (x + y) \cdot (x + z)$$

$\therefore$   Distributive laws are satisfied.

Hence $(B, +, \cdot, ', 0, 1)$ is a Boolean algebra.


**Example 4:**  Let $B = \{a, b, c, d\}$. Two binary operations '+' and '$\cdot$' are defined as follows:

| + | $a$ | $b$ | $c$ | $d$ |
|---|---|---|---|---|
| $a$ | $a$ | $b$ | $b$ | $a$ |
| $b$ | $b$ | $b$ | $b$ | $b$ |
| $c$ | $b$ | $b$ | $c$ | $c$ |
| $d$ | $a$ | $b$ | $c$ | $d$ |

| $\cdot$ | $a$ | $b$ | $c$ | $d$ |
|---|---|---|---|---|
| $a$ | $a$ | $a$ | $d$ | $d$ |
| $b$ | $a$ | $b$ | $c$ | $d$ |
| $c$ | $d$ | $c$ | $c$ | $d$ |
| $d$ | $d$ | $d$ | $d$ | $d$ |

A binary operation ''' is defined as $a' = c$, $b' = d$, $c' = a$, $d' = b$. Show that $(b, +, \cdot, ')$ is a Boolean algebra.

### Solution

1. **Closure laws:**  As all the entries in the composition tables are the elements of set B.

   $\therefore$   Closure laws are satisfied.

2. **Commutative laws:**  From the composition tables, we have for all $x, y \in B$,

   $$x + y = y + x \quad \text{and} \quad x \cdot y = y \cdot x$$

   Hence commutative laws are satisfied.

3. **Associative laws:**  From the composition tables, we have for all $x, y, z \in B$,

   $$(x + y) + z = x + (y + z)$$

   and
   $$(x \cdot y) \cdot z = x \cdot (y \cdot z)$$

   Hence associative laws are satisfied.

4. **Identity laws:** From the composition tables, we have

$$a + b = a \qquad\qquad a \cdot b = a$$
$$b + d = b \qquad\qquad b \cdot b = b$$
$$c + d = c \qquad\qquad c \cdot b = c$$
$$d + d = d \qquad\qquad d \cdot b = d$$

Thus $b$ is the unit element and $d$ is the zero element.

5. **Complement laws:** From the composition tables, we have

$$a + d' = a + c = b \qquad\qquad a \cdot a' = a \cdot c = d$$
$$b + b' = b + d = b \qquad\qquad b \cdot b' = b \cdot d = d$$
$$c + c' = c + a = b \qquad\qquad c \cdot c' = c \cdot a = d$$
$$d + d' = d + b = b \qquad\qquad d \cdot d' = d \cdot b = d$$

Hence the complement laws are satisfied.

6. **Distributive laws:** From the composition table, we have for all $x, y, z \in B$

$$x + (y \cdot z) = (x + y) \cdot (x + z) \quad \text{and} \quad x \cdot (y + z) = (x \cdot y) + (x \cdot z)$$

For example, let $x = d$, $y = c$ and $z = b$

$$\therefore \qquad x + (y \cdot z) = d + c \cdot b \qquad \text{and} \qquad (x + y) \cdot (x + z) = (d + c) \cdot (d + b)$$
$$= d + c = c \qquad\qquad\qquad = c \cdot b = c$$

$$\text{Also} \qquad x \cdot (y + z) = d \cdot (c + b) \quad \text{and} \quad (x \cdot y) + (x \cdot z) = (d \cdot c) + (d \cdot b)$$
$$= d \cdot b = d \qquad\qquad\qquad = d + d = d$$

$$\therefore \qquad x + (y \cdot z) = (x + y) \cdot (x + z)$$

$$\text{and} \qquad x \cdot (y + z) = (x \cdot y) + (x \cdot z)$$

Hence $(B, +, \cdot, ')$ is a Boolean algebra.

**Example 5:** Let $D_{35}$ be the set of positive factors of 35. Two binary operations '+' and '·' are defined as follows:

$$a + b = \text{l.c.m.} \,(a, b) \quad \text{and} \quad a \cdot b = \text{g.c.d.} \,(a, b) \,\text{for all } a, b \in D_{35}$$

A unary operation '$'$' on $D_{35}$ is defined as $a' = \dfrac{35}{a}$ for all $a \in D_{35}$.

Show that $(D_{35} +, \cdot, ', 1, 35)$ is a Boolean algebra.

### *Solution*

We have $D_{35} = \{1, 5, 7, 35\}$. The composition tables for the binary operations '+' and '·' as defined in the given problem are

| + | 1 | 5 | 7 | 35 |
|---|---|---|---|----|
| 1 | 1 | 5 | 7 | 35 |
| 5 | 5 | 5 | 35 | 35 |
| 7 | 7 | 35 | 7 | 35 |
| 35 | 35 | 35 | 35 | 35 |

| · | 1 | 5 | 7 | 35 |
|---|---|---|---|----|
| 1 | 1 | 1 | 1 | 1 |
| 5 | 1 | 5 | 1 | 5 |
| 7 | 1 | 1 | 7 | 7 |
| 35 | 1 | 5 | 7 | 35 |

and $\quad 1' = \dfrac{35}{1} = 35, \quad 5' = \dfrac{35}{5} = 7, \quad 7' = \dfrac{35}{7} = 5 \quad$ and $\quad 35' = \dfrac{35}{35} = 1 \qquad \left[ \because \quad a' = \dfrac{35}{a} \right]$

1. **Closure laws:** As all the entries in the composition tables are the elements of set $D_{35}$

   $\therefore \quad$ Closure laws are satisfied.

2. **Commutative laws:** From the composition tables, we have for all $a, b \in D_{35}$

   $$a + b = b + a \quad \text{and} \quad a \cdot b = b \cdot a$$

   Hence, commutative laws are satisfied.

3. **Associative laws:** From the composition tables, we have for all $a, b, c \in D_{35}$

   $$(a + b) + c = a + (b + c) \quad \text{and} \quad (a \cdot b) \cdot c = a \cdot (b \cdot c)$$

   $\therefore \quad$ Associative laws are satisfied.

4. **Identity laws:** From the composition tables, we have

$$
\begin{array}{ll}
1 + 1 = 1 & \qquad 1 \cdot 35 = 1 \\
5 + 1 = 5 & \qquad 5 \cdot 35 = 5 \\
7 + 1 = 7 & \qquad 7 \cdot 35 = 7 \\
35 + 1 = 35 & \qquad 35 \cdot 35 = 35
\end{array}
$$

   Hence '1' is the zero element and '35' is the unit element.

5. **Complement laws:** From the composition tables, we have

$$
\begin{array}{ll}
1 + 1' = 1 + 35 = 35 & \qquad 1 \cdot 1' = 1 \cdot 35 = 1 \\
5 + 5' = 5 + 7 = 35 & \qquad 5 \cdot 5' = 5 \cdot 7 = 1 \\
7 + 7' = 7 + 5 = 35 & \qquad 7 \cdot 7' = 7 \cdot 5 = 1 \\
35 + 35' = 35 + 1 = 35 & \qquad 35 \cdot 35' = 35 \cdot 1 = 1
\end{array}
$$

   Hence the complement laws are satisfied.

6. **Distributive laws:** From the composition tables, we have for all $a, b, c \in D_{35}$

$$a + (b \cdot c) = (a + b) \cdot (a + c) \quad \text{and} \quad a \cdot (b + c) = (a \cdot b) + (a \cdot c)$$

   For example let $\quad a = 5, \quad b = 7 \quad$ and $\quad c = 35$

$$
\begin{array}{lll}
\therefore & a + (b \cdot c) = 5 + (7 \cdot 35) & \text{and} \quad (a + b) \cdot (a + c) = (5 + 7) \cdot (5 + 35) \\
& \qquad\qquad = 5 + 7 = 35 & \qquad\qquad\qquad\quad = 35 \cdot 35 = 35
\end{array}
$$

$$
\begin{array}{lll}
\text{Also} & a \cdot (b + c) = 5 \cdot (7 + 35) & \text{and} \quad (a \cdot b) + (a \cdot c) = (5 \cdot 7) + (5 \cdot 35) \\
& \qquad\qquad = 5 \cdot 35 = 5 & \qquad\qquad\qquad\quad = 1 + 5 = 5
\end{array}
$$

$$\therefore \quad a + (b \cdot c) = (a + b) \cdot (a + c) \quad \text{and} \quad a \cdot (b + c) = (a \cdot b) + (a \cdot c)$$

   Thus the distributive laws are satisfied.

   Hence $(D_{35}, +, \cdot, ', 1, 35)$ is a Boolean algebra.

**Example 6:** Let B = {1, 2, 4, 5, 10, 20} be the set of positive factors of 20. Two binary operations '+' and '·' are defined as follows:

$$a + b = \text{l.c.m } (a, b) \quad and \quad a \cdot b = \text{g.c.d. } (a, b) \text{ for all } a, b \in B$$

A unary operation '′' on B is defined as $a' = \dfrac{20}{a}$ for all $a \in B$. Show that (B, +, ·, ′, 1, 20) is not a Boolean algebra.

### Solution

Here B = {1, 2, 4, 5, 10, 20}. The composition tables for the binary operations '+' and '.' as defined in the given problem are

| +  | 1  | 2  | 4  | 5  | 10 | 20 |
|----|----|----|----|----|----|----|
| 1  | 1  | 2  | 4  | 5  | 10 | 20 |
| 2  | 2  | 2  | 4  | 10 | 10 | 20 |
| 4  | 4  | 4  | 4  | 20 | 20 | 20 |
| 5  | 5  | 10 | 20 | 5  | 10 | 20 |
| 10 | 10 | 10 | 20 | 10 | 10 | 20 |
| 20 | 20 | 20 | 20 | 20 | 20 | 20 |

| ·  | 1  | 2  | 4  | 5  | 10 | 20 |
|----|----|----|----|----|----|----|
| 1  | 1  | 1  | 1  | 1  | 1  | 1  |
| 2  | 1  | 2  | 2  | 1  | 2  | 2  |
| 4  | 1  | 2  | 4  | 1  | 1  | 4  |
| 5  | 1  | 1  | 1  | 5  | 5  | 5  |
| 10 | 1  | 2  | 1  | 5  | 10 | 10 |
| 20 | 1  | 2  | 4  | 5  | 10 | 20 |

From the above composition tables, we have for each $a \in B$

$$a + 1 = a \quad \text{and} \quad a \cdot 20 = a$$

∴  1 is the zero element and 20 is the unit element.

Now,

| $a$ | $a' = \dfrac{20}{a}$ | $a + a' = \text{l.c.m. } (a, a')$ |
|-----|----------------------|-----------------------------------|
| 1   | 20                   | 20                                |
| 2   | 10                   | 10                                |
| 5   | 5                    | 20                                |
| 4   | 4                    | 20                                |
| 10  | 2                    | 10                                |
| 20  | 1                    | 20                                |

Thus for each $a \in B$, $a + a' = 20$ is not true.

∴   Complement laws are not satisfied.

Hence (B, +, ., ′, 1, 20) is not a Boolean algebra

**Note.** If $m$ can be expressed as the product of distinct prime numbers, then the set $D_m$ of all positive factors of $m$ is a Boolean algebra with the operations defined in example 6 and if $m$ cannot be expressed as the product of distinct prime numbers, then $D_m$ is not a Boolean algebra.

**Theorem 1** Product of two Boolean algebras is also a Boolean algebra i.e. If A and B are Boolean algebras, then $A \times B$ is also a Boolean algebra.

***Proof***

Here $A \times B = \{(a, b): a \in A$ and $b \in B\}$. Let $(a_1, b_1), (a_2, b_2) \in A \times B$

Let '+' and '·' be the binary operations said '′' be the unary operation on $A \times B$, defined by

$$(a_1, b_1) + (a_2, b_2) = (a_1 + a_2, b_1 + b_2)$$
$$(a_1, b_1) \cdot (a_2, b_2) = (a_1 \cdot a_2, b_1 \cdot b_2)$$

and
$$(a_1, b_1)' = (a_1', b_1')$$

1. **Closure laws:**

   For any $(a_1, b_1), (a_2, b_2) \in A \times B$

   $$(a_1 + a_2, b_1 + b_2) \in A \times B \quad \text{and} \quad (a_1 \cdot a_2, b_1 \cdot b_2) \in A \times B$$

   ∴   Closure laws are satisfied.

2. **Commutative laws:**

   For any $(a_1, b_1), (a_2, b_2) \in A \times B$

   $$(a_1, b_1) + (a_2, b_2) = (a_1 + a_2, b_1 + b_2) = (a_2 + a_1, b_2 + b_1)$$
   $$= (a_2, b_2) + (a_1, b_1)$$

   and
   $$(a_1, b_1) \cdot (a_2, b_2) = (a_1 \cdot a_2, b_1 \cdot b_2) = (a_2 \cdot a_1, b_2 \cdot b_1)$$
   $$= (a_2, b_2) \cdot (a_1, b_1)$$

   Hence, commutative laws are satisfied.

3. **Associative laws:**

   For any $(a_1, b_1), (a_2, b_2), (a_3, b_3) \in A \times B$

   $$((a_1, b_1) + (a_2, b_2)) + (a_3, b_3) = (a_1 + a_2, b_1 + b_2) + (a_3, b_3)$$
   $$= ((a_1 + a_2) + a_3, (b_1 + b_2) + b_3)$$
   $$= (a_1 + (a_2 + a_3), b_1 + (b_2 + b_3))$$
   $$= (a_1, b_1) + (a_2 + a_3, b_2 + b_3)$$
   $$= (a_1, b_1) + ((a_2, b_2) + (a_3, b_3))$$

   and
   $$((a_1, b_1) \cdot (a_2, b_2)) \cdot (a_3, b_3) = (a_1 \cdot a_2, b_1 \cdot b_2) \cdot (a_3, b_3)$$
   $$= ((a_1 \cdot a_2) \cdot a_3, (b_1 \cdot b_2) \cdot b_3)$$
   $$= (a_1 \cdot (a_2 \cdot a_3), b_1 \cdot (b_2 \cdot b_3))$$
   $$= (a_1, b_1) \cdot (a_2 \cdot a_3, b_2 \cdot b_3)$$
   $$= (a_1, b_1) \cdot ((a_2, b_2) \cdot (a_3, b_3))$$

   Hence, associative laws are satisfied.

### 4. Identity laws:

Let $0_1$ and $0_2$ be zero elements of A and B respectively.

$\therefore \quad (0_1, 0_2) \in A \times B$

For any $(a_1, b_1) \in A \times B$

$$(a_1, b_1) + (0_1, 0_2) = (a_1 + 0_1, b_1 + 0_2) = (a_1, b_1)$$

$\therefore \quad (0_1, 0_2)$ is a zero element of $A \times B$.

Let $1_1$ and $1_2$ be unit elements of A and B respectively.

$\therefore \qquad\qquad\qquad (1_1, 1_2) \in A \times B$

$$(a_1, b_1) \cdot (1_1, 1_2) = (a_1 \cdot 1_1, b_1 \cdot 1_2) = (a_1, b_1)$$

$\therefore \quad (1_1, 1_2)$ is a unit element of $A \times B$.

Hence, identity laws are satisfied.

### 5. Complement laws:

Let $(a, b) \in A \times B$

Since A and B are Boolean algebras

$\therefore \quad$ There exist $a' \in A$ and $b' \in B$ such that

$$a + a' = 1_1, \quad b + b' = 1_2$$

and $\qquad\qquad\qquad a \cdot a' = 0_1, \quad a \cdot a' = 0_2$

$\therefore \qquad\qquad\qquad (a', b') \in A \times B$

$$(a, b) + (a', b') = (a + a'b + b') = (1_1, 1_2)$$

and $\qquad\qquad (a, b) \cdot (a', b') = (a \cdot a', b \cdot b') = (0_1, 0_2)$

Hence, complement laws are satisfied.

### 6. Distributive laws:

For any $(a_1, b_1), (a_2, b_2), (a_3, b_3) \in A \times B$

$$\begin{aligned}
(a_1, b_1) + ((a_2, b_2) \cdot (a_3, b_3)) &= (a_1, b_1) + (a_2 \cdot a_3, b_2 \cdot b_3) \\
&= ((a_1 + (a_2 \cdot a_3), b_1 + (b_2 \cdot b_3)) \\
&= ((a_1 + a_2) \cdot (a_1 + a_3), (b_1 + b_2) \cdot (b_1 + b_3)) \\
&= (a_1 + a_2, b_1 + b_2) \cdot (a_1 + a_3, b_1 + b_3) \\
&= ((a_1, b_1) + (a_2, b_2)) \cdot ((a_1, b_1) + (a_3, b_3))
\end{aligned}$$

Also, $\quad$
$$\begin{aligned}
(a_1, b_1) \cdot ((a_2, b_2) + (a_3, b_3)) &= (a_1, b_1) \cdot (a_2 + a_3, b_2 + b_3) \\
&= ((a_1 \cdot (a_2 + a_3), b_1 \cdot (b_2 + b_3)) \\
&= (a_1 \cdot a_2 + a_1 \cdot a_3, b_1 \cdot b_2 + b_1 \cdot b_3) \\
&= (a_1 \cdot a_2, b_1 \cdot b_2) + (a_1 \cdot a_3, b_1 \cdot b_3) \\
&= (a_1, b_1) \cdot (a_2, b_2) + (a_1, b_1) \cdot (a_3, b_3)
\end{aligned}$$

Hence, distributive laws are satisfied.

$\therefore \quad A \times B$ is a Boolean algebra.

## Exercise 9.1

1. Let $B = \{\phi, \{a\}, \{b\}, \{c\}, \{a, b\}, \{a, c\}, \{b, c\}, \{a, b, c\}\}$; show that $(B, \cup, \cap, ')$ is a Boolean algebra.

2. Let $S$ be any non-empty set. Let $B$ be the set of all subsets of the set $S$. Define operations '+' and '·' by $X + Y = X \cup Y$, $X \cdot Y = X \cap Y$ and $X' = S - X$, where $X \in B$. Show that $(B, +, \cdot, ')$ is a Boolean algebra.

3. Let $B = \{\alpha, \beta, \gamma, \delta\}$. Two binary operations '+' and '·' are defined as follows

| + | $\alpha$ | $\beta$ | $\gamma$ | $\delta$ |
|---|---|---|---|---|
| $\alpha$ | $\alpha$ | $\beta$ | $\beta$ | $\alpha$ |
| $\beta$ | $\beta$ | $\beta$ | $\beta$ | $\beta$ |
| $\gamma$ | $\beta$ | $\beta$ | $\gamma$ | $\gamma$ |
| $\delta$ | $\alpha$ | $\beta$ | $\gamma$ | $\delta$ |

| · | $\alpha$ | $\beta$ | $\gamma$ | $\delta$ |
|---|---|---|---|---|
| $\alpha$ | $\alpha$ | $\alpha$ | $\delta$ | $\delta$ |
| $\beta$ | $\alpha$ | $\beta$ | $\gamma$ | $\delta$ |
| $\gamma$ | $\delta$ | $\gamma$ | $\gamma$ | $\delta$ |
| $\delta$ | $\delta$ | $\delta$ | $\delta$ | $\delta$ |

The unary operations '′' is defined as $\alpha' = \gamma$, $\beta' = \delta$, $\gamma' = \alpha$ and $\delta' = \beta$. Show that $(B, +, \cdot, ')$ is a Boolean algebra.

4. Let $L$ be the set of all logical statements. Define operations '+', '·' and '′' by $p + q = p \vee q$; $p \cdot q = p \wedge q$; $p' = \sim p$ for all $p, q \in L$ where $\vee, \wedge$ and $\sim$ have usual meanings in mathematical logic. Show that $(L, +, \cdot, ')$ is a Boolean algebra.

5. Let $D_6$ be the set of positive factors of 6. Two binary operations '+', and '·' on $D_6$ are defined as follows: $a + b = $ l.c.m. $(a, b)$ and $a \cdot b = $ g.c.d $(a, b)$ for all $a, b \in D_6$

A unary operation '′' on $D_6$ is defined as $a' = \dfrac{6}{a}$ for all $a \in D_6$.

Show that $(D_6, +, \cdot, ', 1, 6)$ is a Boolean algebra.

6. Let $B = \{1, 2, 4\}$. Define '+', '·' and complement '′' in $B$ by $a + b = $ l.c.m $(a, b)$ i.e., least common multiple of $a$ and $b$ and $a \cdot b = $ g.c.d $(a, b)$, i.e., greatest common divisor of $a$ and $b$; and $a' = \dfrac{4}{a}$, where $a, b \in B$. Show that $(B, +, \cdot, ', 1, 4)$ is not a Boolean algebra.

7. Let $B = \{1, 2, 3, 4, 6, 12\}$ be the set of positive factors of 12. Two binary operations '+' and '·' are defined as follows: $a + b = $ l.c.m $(a, b)$ and $a \cdot b = $ g.c.d $(a, b)$ for all $a, b \in B$.
   A unary operation '′' is defined on $B$ as $a' = \dfrac{12}{a}$ for all $a \in B$.
   Show that $(B, +, \cdot, ', 1, 12)$ is not a Boolean algebra.

8. Let $n$ be a positive integer and is divisible by $a^2$ where $a$ is a prime number. Let $B$ be the set positive factors of $n$. Two binary operations '+' and '·' are defined as follows: $a + b = $ l.c.m. $(a, b)$ and $a \cdot b = $ g.c.d. $(a, b)$ for all $a, b \in B$.
   A unary operation on $B$ is defined as $a' = \dfrac{n}{a}$ for all $a \in B$.
   Show that $(B, +, \cdot, ')$ is not a Boolean algebra.

9. Let $S_n = \{1, 2, 3, \ldots, n\}$. Two binary operations '+' and '·' are defined as follows: $a + b = \max\{a, b\}$ and $a \cdot b = \min\{a, b\}$ for all $a, b \in S_n$.
   Show that $S_n$ is not a Boolean algebra w.r.t. binary operations '+' and '·' and any unary operation defined on $S_n$.

10. Let $B = \{0, 1\}$, the set of binary digits with operations '+' '·' and '′' defined as follows:

| + | 0 | 1 |
|---|---|---|
| 0 | 0 | 1 |
| 1 | 1 | 1 |

| · | 0 | 1 |
|---|---|---|
| 0 | 0 | 0 |
| 1 | 0 | 1 |

| $a$ | $a'$ |
|---|---|
| 0 | 1 |
| 1 | 0 |

$\{B, +, ., ', 0,1\}$ is a Boolean algebra. Show that $B^n = B \times B \times \ldots \times B$ ($n$ factors) is a Boolean algebra with component wise addition, multiplication and complement.

**Answers to Selected Problems**

4. Let $t$ and $f$ represent a tautology and a contradiction respectively in the set of logical statements. Then $t$ can be used as a unit element for $\wedge$ (multiplicative) identity and $f$ can be used as a zero element for $\vee$ (additive) identity.

8. Complement laws are not satisfied.

## 9.5 GENERAL PROPERTIES OF BOOLEAN ALGEBRA

**(1) The zero element 0 and unit element 1 of Boolean algebra B are unique.**

*Proof*

(a) Suppose there exist two zero elements $0_1$ and $0_2$

Then, $\qquad 0_1 + 0_2 = 0_2 \qquad\qquad$ (i) $\quad [\because \quad 0_1 \text{ is zero element}]$

$\qquad\qquad\qquad 0_1 + 0_2 = 0_1 \qquad\qquad$ (ii) $\quad [\because \quad 0_2 \text{ is zero element}]$

From (i) and (ii), we have $0_1 = 0_2$

Hence zero element of Boolean algebra B is unique.

(b) Suppose there exist two unit elements $1_1$ and $1_2$

Then, $\qquad 1_1 \cdot 1_2 = 1_2 \qquad\qquad$ (i) $\quad [\because \quad 1_1 \text{ is unit element}]$

$\qquad\qquad\qquad 1_1 \cdot 1_2 = 1_1 \qquad\qquad$ (ii) $\quad [\because \quad 1_2 \text{ is unit element}]$

From (i) and (ii), we have $1_1 = 1_2$

Hence unit element of Boolean algebra B is unique.

**(2) The complement of each element in Boolean algebra B is unique.**

*Proof*

Suppose $a'$ and $a''$ are two complements of $a$ in Boolean algebra B.

Then, $\qquad\qquad a + a' = 1, \quad a \cdot a' = 0$

$\qquad\qquad\qquad a + a'' = 1, \quad a \cdot a'' = 0$

Now, $\qquad\qquad a' = a' \cdot 1 = a' \cdot (a + a'')$

$\qquad\qquad\qquad\qquad = a' \cdot a + a' \cdot a'' = 0 + a' \cdot a''$

$$= a \cdot a'' + a' \cdot a'' = a'' \cdot (a + a')$$
$$= a'' \cdot (1) = a''$$
$$\therefore \qquad a' = a''$$

Hence complement $a'$ of $a$ is unique.

**(3) The identity elements 0 and 1 in Boolean algebra B are complement of each other.**

### *Proof*

The two identity elements in Boolean algebra B are 0 and 1

| | | |
|---|---|---|
| (a) | $0' + 0 = 0'$ | $[\because \quad 0 \ is \ zero \ element]$ |
| Also, | $0 + 0' = 1$ | $[\because \quad a + a' = 1]$ |
| $\therefore$ | $0' = 1$ | |
| (b) | $1 + 1' = 1$ | $[\because \quad a + a' = 1]$ |
| Also, | $1 + 0 = 1$ | $[\because \quad a + 0 = a]$ |
| $\therefore$ | $1 + 1' = 1 + 0$ | |
| or | $1' = 0$ | |

Hence identity elements 0 and 1 are complements of each other.

**(4) A Boolean algebra cannot have three elements.**

### *Proof*

Suppose $a$ is the third element other than 0 and 1 in Boolean algebra B.

| | | |
|---|---|---|
| Since | $0' = 1 \quad$ and $\quad 1' = 0$ | |
| Therefore | $a' = a$ | |
| Now | $a \cdot a' = 0$ | |
| $\Rightarrow$ | $a \cdot a = 0$ | $[\because \quad a' = a]$ |
| $\Rightarrow$ | $a = 0$ | |

which is a contradiction since $a \neq 0$

$\therefore \quad$ Our supposition is wrong.

Hence, there cannot be a Boolean algebra containing three elements.

**(5) Involution law: For each $a$ belonging to Boolean algebra B, $(a')' = a$**

### *Proof*

| | | |
|---|---|---|
| | $a + a' = 1$ | |
| and | $a \cdot a' = 0$ | |
| Now, | $a' + a = 1$ | $[\because \quad a + a' = a' + a]$ |
| and | $a' \cdot a = 0$ | $[\because \quad a \cdot a' = a' \cdot a]$ |

Hence by the definition of complement $(a')' = a$.

**(6) Idempotent laws: For each $a$ belonging to Boolean algebra B,**
     **(i) $a + a = a$   (ii) $a \cdot a = a$**

*Proof*

(i) For each $a$ belonging to Boolean algebra B, we have

$$a + a = (a + a) \cdot 1$$
$$= (a + a) \cdot (a + a') \qquad [\because \; a + a' = 1]$$
$$= a + a' \cdot a \qquad [Distributive\ law]$$
$$= a + 0 = a$$

Hence, $a + a = a$ for each $a$ belonging to Boolean algebra B.

(ii) For each $a$ belonging to Boolean algebra B, we have

$$a \cdot a = a \cdot a + 0 \qquad [\because \; a + 0 = a]$$
$$= a \cdot a + a \cdot a' \qquad [\because \; a \cdot a' = 0]$$
$$= a \cdot (a + a') = a \cdot 1 = a \qquad [\because \; a + a' = 1]$$

Hence, $a \cdot a = a$ for each $a$ belonging to Boolean algebra B.

**(7) Boundedness laws: For each $a$ belonging to Boolean algebra B,**
     **(i) $a + 1 = 1$   (ii) $a \cdot 0 = 0$**

*Proof*

(i) For each $a$ belonging Boolean algebra B, we have

$$a + 1 = a + (a + a') \qquad [\because \; a + a' = 1]$$
$$= a + a + a'$$
$$= a + a' = 1 \quad [\because \; a + a = a\ by\ idempotent\ law]$$

Hence, $\qquad a + 1 = 1.$

(ii) For each $a$ belonging to Boolean algebra B, we have

$$a \cdot 0 = a \cdot 0 + 0$$
$$= 0 + (a \cdot 0)$$
$$= (a \cdot a') + (a \cdot 0)$$
$$= a \cdot (a' + 0)$$
$$= a \cdot a' = 0$$

Hence, $\qquad a \cdot 0 = 0.$

**(8) Associative laws: For all $a, b, c$ belonging to Boolean algebra B,**
     **(i) $a + (b + c) = (a + b) + c$   (ii) $a \cdot (b \cdot c) = (a \cdot b) \cdot c$**

*Proof*

(i) For all $a, b, c$ belonging to Boolean algebra B, let

$$L = a + (b + c) \quad \text{and} \quad R = (a + b) + c$$

$$a \cdot L = a \cdot [a + (b + c)] = a \cdot a + a \cdot (b + c)$$
$$= a + a \cdot b + a \cdot c \qquad [\because \quad a \cdot a = a, \textit{by idempotent law}]$$
$$= a \cdot (1 + b + c) = a \cdot 1 = a$$
$$[\because \quad (1 + b + c) = 1, \textit{by boundedness law}]$$

and
$$a \cdot R = a \cdot [(a + b) + c] = a \cdot (a + b) + a \cdot c$$
$$= a \cdot a + a \cdot b + a \cdot c$$
$$= a + a \cdot b + a \cdot c = a \cdot (1 + b + c) = a \cdot 1 = a$$

Hence
$$a \cdot L = a \cdot R \qquad\qquad (1)$$

Also,
$$a' \cdot L = a' \cdot [a + (b + c)] = a' \cdot a + a' \cdot (b + c)$$
$$= 0 + a' \cdot (b + c) = a' \cdot (b + c) \qquad [\because \quad a \cdot a' = 0]$$

and
$$a' \cdot R = a' \cdot [(a + b) + c] = a' \cdot (a + b) + a' \cdot c$$
$$= a' \cdot a + a' \cdot b + a' \cdot c$$
$$= 0 + a' \cdot (b + c) = a' \cdot (b + c)$$

Hence
$$a' \cdot L = a' \cdot R \qquad\qquad (2)$$

Now,
$$L = L \cdot 1 = L \cdot (a + a') = L \cdot a + L \cdot a'$$
$$= a \cdot L + a' \cdot L \qquad [\textit{Commutative law}]$$
$$= a \cdot R + a' \cdot R \qquad [\textit{Using (1) and (2)}]$$
$$= (a + a') \cdot R = 1 \cdot R = R$$

Hence,
$$a + (b + c) = (a + b) + c.$$

For all $a$, $b$, $c$ belonging to Boolean algebra B, let

$$L = a \cdot (b \cdot c) \quad \text{and} \quad R = (a \cdot b) \cdot c$$

$$\therefore \quad a + L = a + a \cdot (b \cdot c) = (a + a) \cdot (a + b \cdot c) \qquad [\textit{Distributive law}]$$
$$= a \cdot (a + b \cdot c) \qquad [\because \quad a + a = a]$$
$$= a \cdot a + a \cdot b \cdot c$$
$$= a + a \cdot b \cdot c \qquad [\because \quad a \cdot a = a]$$
$$= a \cdot (1 + b \cdot c) = a \cdot 1 = a$$
$$[\because \quad 1 + b \cdot c = 1, \textit{by idempotent law}]$$

and
$$a + R = a + (a \cdot b) \cdot c = (a + a \cdot b) \cdot (a + c)$$
$$= [a (1 + b)] \cdot (a + c) = a \cdot (a + c)] \qquad [\because \quad 1 + b = 1]$$
$$= a \cdot a + a \cdot c = a + a \cdot c \qquad [\because \quad a \cdot a = a]$$
$$= a \cdot (1 + c) = a \cdot 1 = a \qquad [\because \quad 1 + c = 1]$$

Hence,
$$a + L = a + R \qquad\qquad (1)$$

Also,
$$a' + L = a' + a \cdot (b \cdot c) = (a' + a) \cdot (a' + b \cdot c) \qquad [\textit{Distributive law}]$$
$$= a' + b \cdot c \qquad [\because \quad a' + a = 1]$$

and
$$a' + R = a' + (a \cdot b) \cdot c = (a' + a \cdot b) \cdot (a' \cdot c)$$
$$= (a' + a) \cdot (a' + b) \cdot (a' + c)$$
$$= 1 \cdot (a' + b) \cdot (a' + c) \qquad [\because \ a' + a = 1]$$
$$= a' + b \cdot c \qquad [\textit{Distributive law}]$$

Hence, $\qquad a' + L = a' + R \qquad\qquad (2)$

Now, $\qquad L = L + 0 \qquad\qquad [\because \ a + 0 = a]$
$$= L + a \cdot a'$$
$$= (L + a) \cdot (L + a')$$
$$= (a + L) \cdot (a' + L) \qquad [\textit{Commutative law}]$$
$$= (a + R) \cdot (a' + R) \qquad [\textit{Using } (1) \textit{ and } (2)]$$
$$= R + a \cdot a' = R + 0 = R$$

Hence, $\qquad a \cdot (b \cdot c) = (a \cdot b) \cdot c \qquad\qquad [\textit{Distributive law}]$

**(9) Absorption laws: For all $a$, $b$ belonging to Boolean algebra B,**

      **(i) $a + (a \cdot b) = a$    (ii) $a \cdot (a + b) = a$**

***Proof***

(i) For all $a$, $b$ belonging to Boolean algebra B, we have
$$a + (a \cdot b) = (a \cdot b) + a \qquad\qquad [\textit{Commutative law}]$$
$$= (a \cdot b) + (a \cdot 1) \qquad\qquad [\because \ a = a \cdot 1]$$
$$= a \cdot (b + 1)$$
$$= a \cdot 1 = a \qquad [\because \ b + 1 = 1, \textit{ by idempotent law}]$$

Hence, $a + (a \cdot b) = a$ for all $a$, $b$ belonging to Boolean algebra B.

(ii) For all $a$, $b$ belonging to Boolean algebra B, we have
$$a \cdot (a + b) = (a + b) \cdot a$$
$$= (a + b) \cdot (a + 0)$$
$$= a + (b \cdot 0) = a + 0 = a \qquad [\textit{Distributive law}]$$

Hence, $a \cdot (a + b) = a$ for all $a$, $b$ belonging to Boolean algebra B.

**(10) De-Morgan's laws: For all $a$, $b$ belonging to Boolean algebra B,**

      **(i) $(a + b)' = a' \cdot b'$    (ii) $(a \cdot b)' = a' + b'$**

***Proof***

(i) For all $a$, $b$ belonging to Boolean algebra B,
$$(a + b) + a' \cdot b' = (a + b + a') \cdot (a + b + b') \qquad [\textit{Distributive law}]$$
$$= (a + a' + b) \cdot (a + b + b')$$
$$= (1 + b) \cdot (a + 1) \qquad\qquad [\because \ a + a' = 1]$$
$$= 1 \cdot 1 = 1 \qquad\qquad [\because \ a + 1 = 1]$$

Also,
$$(a + b) \cdot (a' \cdot b') = a \cdot (a' \cdot b') + b \cdot (a' \cdot b')$$
$$= (a \cdot a') \cdot b' + (b \cdot b') \cdot a'$$
$$= 0 \cdot b' + 0 \cdot a' \qquad [\because \ a \cdot a' = 0]$$
$$= 0 + 0 = 0$$

Hence, by the definition of complement, $(a + b)' = a' \cdot b'$.

(ii) For all $a$, $b$ belonging to Boolean algebra B,
$$(a \cdot b) + (a' \cdot b') = a \cdot b + a' + b'$$
$$= a' + a \cdot b + b' \qquad [Commutative\ law]$$
$$= (a' + a) \cdot (a' + b) + b' \qquad [Distributive\ law]$$
$$= 1 \cdot (a' + b) + b' \qquad [\because \ a' + a = 1]$$
$$= a' + b + b'$$
$$= a' + (b + b') = a' + 1 = 1$$
$$[\because \ a' + 1 = 1,\ by\ boundedness\ law]$$

Also,
$$(a \cdot b) \cdot (a' + b') = (a \cdot b) \cdot a' + (a \cdot b) \cdot b' \qquad [Distributive\ law]$$
$$= (a \cdot a') \cdot b + a \cdot (b \cdot b') \qquad [Associative\ law]$$
$$= 0 \cdot b + a \cdot 0 = 0 + 0 = 0$$

Hence by the definition of complement $(a \cdot b)' = a' + b'$.

**(11) Cancellation laws: For all $a$, $b$, $c$ belonging to Boolean algebra B,**

    (i) $a + b = a + c$ **and** $a' + b = a' + c$ $\Rightarrow$ $b = c$

    (ii) $a \cdot b = a \cdot c$ **and** $a' \cdot b = a' \cdot c$ $\Rightarrow$ $b = c$

***Proof***

(i) We have
$$b = b + 0$$
$$= b + (a \cdot a') \qquad [\because \ a \cdot a' = 0]$$
$$= (b + a) \cdot (b + a')$$
$$= (a + b) \cdot (a' + b) \qquad [Commutative\ law]$$
$$= (a + c) \cdot (a' + c) \qquad [Given]$$
$$= (a \cdot a') + c = 0 + c = c.$$

Hence, $b = c$ for all $a$, $b$, $c$ belonging to Boolean algebra B.

(ii) We have
$$b = b \cdot 1$$
$$= b \cdot (a + a') \qquad [\because \ a + a' = 1]$$
$$= b \cdot a + b \cdot a'$$
$$= a \cdot b + a' \cdot b \qquad [Commutative\ law]$$
$$= a \cdot c + a' \cdot c \qquad [Given]$$
$$= (a \cdot a') \cdot c = 1 \cdot c = c$$

Hence $b = c$ for all $a$, $b$, $c$ belonging to Boolean algebra B.

## 9.6 BOOLEAN EXPRESSION

Let $(B, +, \cdot, ')$ be a Boolean algebra. All elements of B are **_constants_** in B. A symbol representing an arbitrary element of Boolean algebra B is called a **_variable_** in B and is usually denoted by letters $a, b, c, ..., p, q, r, ...; x, y, z$.

An expression formed with binary variables combined by binary operators AND and OR and the unary operator NOT is called a **Boolean expression.** It is usually denoted by E.

In other words, A Boolean expression is an algebraic statement containing finite number of Boolean variables and operations '+', '·', and '′'.

For example,
$$E_1 = x + y + z, \qquad E_2 = (x' + yz)' + xyz,$$
$$E_3 = (xy'\, z' + y + x'z')', \quad E_4 = (xy + z') + x'\, yz + y'$$

are Boolean expressions in the variables $x$, $y$, and $z$.

Also, the sum and product of two Boolean expressions is also a Boolean expression over the same Boolean Algebra. i.e., if $E_1$ and $E_2$ are Boolean expressions, then $E_1 + E_2$, $E_1 \cdot E_2$ are also Boolean expressions. A Boolean expression containing variables $x_1, x_2, x_3, ..., x_n$ is denoted by $E(x_1, x_2, ..., x_n)$.

Let $E(x_1, x_2, ..., x_n)$ is a Boolean expression in Boolean algebra $(B, +, \cdot, ')$, then we can evaluate the expression $E(x_1, x_2, x_3 ..., x_n)$ by substituting the elements of B as the values of the variables in the expression.

For example, let $E(x, y) = xy' + x (x' + y)$ be a Boolean expression over the Boolean algebra $(\{0,1\}, +, \cdot, ')$ of binary digits 0 and 1. For $x = 1$, $y = 0$, we have

$$E(1,0) = 1 \cdot 0' + 1(1' + 0) = 1 \cdot 1 + 1(0 + 0) = 1' + 0 = 1$$

Further, two Boolean expression of $n$ variables over same Boolean algebra are called **_equivalent_** if they have same values for any assignment of values to the $n$ variables and if two Boolean expressions are equivalent then one of them is a simplified form of the other.

Thus, if we say to simplify a Boolean expression, then our aim is to simplify it into an equivalent form. Since the elements of Boolean algebra will be assigned as values of the variables in the Boolean expression, therefore all the laws of Boolean algebra can be applied to simplify Boolean expression.

The rule for simplification of a Boolean expression is "An expression within a bracket is simplified by observing precedence of '′' over '·' and precedence '·' over '+'."

Thus, an expression is simplified by the usual convention that NOT (′) has precedence over AND (·) and AND (·) has precedence over OR (+) unless changed by the parentheses.

For example:

    (i)   $a + b \cdot c$ means $a + (b \cdot c)$ and not $(a + b) \cdot c$

   (ii)   $a \cdot b'$ means $a \cdot (b')$ and not $(a \cdot b')$

  (iii)   $a \cdot (b + c)'$ means $a \cdot (b' \cdot c')$ and not $(a \cdot b') + (a \cdot c')$.

**Example 7:** *S*how that if $a$, $b$, $c$ belong to Boolean algebra B, then $[(a \cdot b) + c]'$ is a Boolean expression in $a$, $b$ and $c$.

### Solution

Here $a$, $b$, $c$ belong to Boolean algebra B.

$\therefore$  $a$, $b$, $c$ are all Boolean expressions

$\Rightarrow$  $a \cdot b$ is a Boolean expression

Now, $a \cdot b$ and $c$ are Boolean expressions

$\therefore$  $(a \cdot b) + c$ is a Boolean expression

Hence, $[(a \cdot b) + c]'$ is a Boolean expression.

**Example 8:** Find the value of the Boolean expression $(a \cdot b) + c$ if $a = 0$, $b = 0$ and $c = 1$.

### Solution

$$(a \cdot b) + c = (0 \cdot 0) + 1 = 0 + 1 = 1.$$

Hence, the value of the given Boolean expression is 1.

**Example 9:** In the Boolean algebra (B, $+$, $\cdot$, $'$) show that (i) $(a' + b')' = a \cdot b$ for all $a, b \in$ B (ii) $(a' \cdot b')' = a + b$ for all $a, b \in$ B.

### Solution

(i) $\qquad\qquad (a' + b')' = [(a'') \cdot (b'')]$ $\qquad\qquad$ [*De-Morgan's law*]

$\qquad\qquad\qquad\qquad = (a) \cdot (b) = a \cdot b$ $\qquad\qquad\qquad$ [$\because\ a'' = a$]

$\therefore \qquad\qquad (a' + b')' = a \cdot b$ for all $a, b \in$ B.

(ii) $\qquad\qquad (a' \cdot b')' = [(a')' + (b')']$ $\qquad\qquad$ [*De-Morgan's law*]

$\qquad\qquad\qquad\qquad = a + b$ $\qquad\qquad\qquad\qquad\qquad$ [$\because\ (a')' = a$]

$\therefore \qquad\qquad (a' \cdot b')' = a + b$ for all $a, b \in$ B.

**Example 10:** For each $x$ in Boolean algebra B, prove that $\left.\begin{array}{l} x + y = 1 \\ x \cdot y = 0 \end{array}\right\} \Rightarrow y = x'.$

### Solution

| Given | $x + y = 1$ | (1) |
|---|---|---|
| and | $x \cdot y = 0$ | (2) |

Now, $\qquad\qquad\quad x' = x' + 0$ $\qquad\qquad\qquad\qquad$ [$\because\ x = x + 0$]

$\qquad\qquad\qquad\quad = x' + x \cdot y$ $\qquad\qquad\qquad\qquad$ [*Using* (2)]

$\qquad\qquad\qquad\quad = (x' + x) \cdot (x' + y)$ $\qquad\qquad$ [*Distributive law*]

$$= 1 \cdot (x' + y)$$
$$= (x + y) \cdot (x' + y) \qquad [Using\ (1)]$$
$$= (y + x) \cdot (y + x') \qquad [Commutative\ law]$$
$$= y + (x \cdot x') \qquad [Distributive\ law]$$
$$= y + 0 \qquad [\because\ x \cdot x' = 0]$$
$$x' = y.$$

**Example 11:** If B is $a$ Boolean algebra and $x, y \in$ B, then show that $(x + y) + (x' \cdot y') = 1$.

**Solution**

L.H.S
$$= (x + y) \cdot (x' \cdot y')$$
$$= x + [y + (x' \cdot y')]$$
$$= x + [(y + x') \cdot (y + y')] \qquad [Distributive\ law]$$
$$= x + (y + x') \cdot 1 = x + (y + x') \qquad [\because\ y + y' = 1]$$
$$= x + (x' + y) = (x + x') + y$$
$$= 1 + y = y + 1 = 1 = R.H.S.$$

**Example 12:** In the Boolean algebra (B, $+, \cdot, '$), show that (i) $a + b + c \cdot a' = a + b + c$ for all $a, b, c \in$ B.   (ii) $a \cdot b + c \cdot (a' + b') = a \cdot b + c$ for all $a, b, c \in$ B.

**Solution**

(i)
$$a + b + c \cdot a' = a + c \cdot a' + b \qquad [Commutative\ law]$$
$$= (a + c) \cdot (a + a') + b \qquad [Distributive\ law]$$
$$= (a + c) \cdot 1 + b \qquad [\because\ a + a' = 1]$$
$$= (a + c) + b = a + c + b = a + b + c$$
$$\therefore \quad a + b + c \cdot a' = a + b + c \text{ for all } a, b, c \in B$$

(ii)
$$a \cdot b + c \cdot (a' + b') = a \cdot b + c \cdot (a + b)' \qquad [Using\ De\text{-}Morgan's\ law]$$
$$= (a \cdot b + c) \cdot [a \cdot b + (a \cdot b')] \qquad [Distributive\ law]$$
$$= (a \cdot b + c) \cdot (1) \qquad [\because\ a + a' = 1]$$
$$\therefore a \cdot b + c \cdot (a' + b') = a \cdot b + c \text{ for all } a, b, c \in B.$$

**Example 13:** In the Boolean algebra (B, $+, \cdot, '$), show that (i) $(a + a' \cdot b) \cdot (a' + a \cdot b) = b$ for all $a, b \in$ B.   (ii) $(a + b) \cdot [a \cdot b' + b]' = 0$ for all $a, b \in$ B.

**Solution**

(i) $\quad (a + a' \cdot b) \cdot (a' + a \cdot b) = [(a + a') \cdot (a + b)] \cdot (a' + a \cdot b) \qquad [Distributive\ law]$
$$= [1 \cdot (a + b)] \cdot [(a' + a) \cdot (a' + b)] \qquad [\because\ a + a' = 1]$$
$$= (a + b) \cdot [1 \cdot (a' + b)]$$

$$= (a + b) \cdot (a' + b)$$
$$= b + a \cdot a' \qquad\qquad [\textit{Distributive law}]$$
$$= b + 0 = b \qquad\qquad [a \cdot a' = 0]$$

$\therefore \qquad (a + d' \cdot b) \cdot (a' + a \cdot b) = b$ for all $a, b \in$ B.

(ii)
$$(a + b) \cdot [a \cdot b' + b]' = (a + b) \cdot [b + a \cdot b']' \qquad [\textit{Commutative law}]$$
$$= (a + b) \cdot [(b + a) \cdot (b + b')]' \qquad [\textit{Distributive law}]$$
$$= (a + b) \cdot [(a + b) \cdot (b + b')]' \qquad [\textit{Commutative law}]$$
$$= (a + b) \cdot [(a + b) \cdot 1]' \qquad [\because \ a + a' = 1]$$
$$= (a + b) \cdot (a + b)'$$
$$= 0 \qquad\qquad [\because \ a \cdot a' = 0]$$

$\therefore \qquad (a + b) \cdot [a \cdot b' + b]' = 0$ for all $a, b \in$ B.

**Example 14:** In the Boolean algebra (B, +, $\cdot$, $'$), show that (i) $(a + b + c)' = a' \cdot b' \cdot c'$ for all $a$, $b, c \in$ B.  (ii) $(a \cdot b \cdot c)' = a' + b' + c'$ for all $a, b \in$ B.

***Solution***

(i)
$$(a + b + c)' = [(a + b) + c]'$$
$$= (a + b)' \cdot c' \qquad [\textit{De-Morgan's law}]$$
$$= (a' \cdot b') \cdot c' = a' \cdot b' \cdot c'$$

$\therefore \qquad (a + b + c)' = a' \cdot b' \cdot c'$ for all $a, b, c \in$ B.

(ii)
$$(a \cdot b \cdot c)' = [(a \cdot b) \cdot c]'$$
$$= (a \cdot b)' + c' \qquad [\textit{De-Morgan's law}]$$
$$= (a' + b') + c' = a' + b' + c'$$

$\therefore \qquad (a \cdot b \cdot c)' = a' + b' + c'$ for all $a, b, c \in$ B.

**Example 15:** Let (B, +, $\cdot$, $'$) be a Boolean algebra. Show that $a \cdot b' = 0$ if and only if $a' + b = 1$.

***Solution***

Let $\qquad\qquad a' + b = 1$

Now $\qquad\qquad (a' + b)' = 1'$

$\Rightarrow \qquad\qquad [(a')' \cdot b'] = 1' \qquad [\textit{De-Morgan's law}]$

$\Rightarrow \qquad\qquad a \cdot b' = 0 \qquad [\because \ (a')' = a]$

*Conversely, let* $\qquad a \cdot b' = 0$

$\Rightarrow \qquad\qquad (a \cdot b')' = 0'$

$\Rightarrow \qquad\qquad a' + (b')' = 1 \qquad [\textit{De-Morgan's law}]$

$\Rightarrow \qquad\qquad a' + b = 1$

Hence, $a \cdot b' = 0$ if and only if $a' + b = 1$.

**Example 16:** Show that in Boolean algebra, B,

$$\text{(i) } a \vee (a' \wedge b) = a \vee b \quad \text{(ii) } a \wedge (a' \vee b) = a \wedge b \quad \text{where, } a, a', b \in \text{B}.$$

### *Solution*

(i) For all $a, b$ belonging to Boolean algebra B, we have

$$a \vee (a' \wedge b) = (a \vee a') \wedge (a \vee b) \qquad [Distributive\ law]$$
$$= 1 \wedge (a \vee b) \qquad [Complement\ law]$$
$$= (a \vee b) \qquad [Identity\ law]$$

Hence, $\qquad a \vee (a' \wedge b) = a \vee b$

(ii) For all $a, b$ belonging to Boolean algebra B, we have

$$a \wedge (a' \vee b) = (a \wedge a') \vee (a \wedge b) \qquad [Distributive\ law]$$
$$= 0 \vee (a \wedge b) \qquad [Complement\ law]$$
$$= (a \wedge b) \qquad [Identity\ law]$$

Hence, $a \wedge (a' \vee b) = a \wedge b$ for all $a, b$ belonging to Boolean algebra B.

**Example 17:** In the Boolean algebra (B, $+$, $\cdot$, $'$,) simplify the Boolean expression $x \cdot (x + y) + [(y' + x) \cdot y]'$.

### *Solution*

The given Boolean expression is
$$= x \cdot (x + y) + [(y' + x) \cdot y]'$$
$$= x \cdot x + x \cdot y + [(y' + x) \cdot y]'$$
$$= x + x \cdot y + (y \cdot y' + y \cdot x]'$$
$$= x \cdot (1 + y) + (0 + y \cdot x)'$$
$$= x \cdot 1 + (y \cdot x)'$$
$$= x + (x' + y')$$
$$= 1 + y' = 1.$$

**Example 18:** Simplify the following Boolean expression: $a\,[b + c\,(ab + ac)']$.

### *Solution*

The given Boolean expression is
$$= a\,[b + c\,(ab + ac)']$$
$$= a\,[b + c\,((ab)'\,(ac))'] \qquad [De\text{-}Morgan's\ law]$$
$$= ab + ac\,[(ab)'\,(ac)']$$
$$= ab + [(ab)'\,(ac)\,(ac)'] \qquad [Associative\ law]$$
$$= ab + (ab)' \cdot 0$$
$$= ab + 0$$
$$= ab$$

$$\therefore \qquad a\,[b + c\,(ab + ac)'] = ab$$

**Example 19:** Simplify the following Boolean expression: $[a (a + b) + (b' + a)b]'$.

***Solution***

The given Boolean expression is $\quad [a \cdot (a + b) + (b' + a) \cdot b]'$

$$
\begin{aligned}
\text{Now,} \quad [a (a + b) + (b' + a) \cdot b] &= [a \cdot a + a \cdot b) + (b' \cdot b + a \cdot b) \quad [\textit{Distributive law}] \\
&= (a + a \cdot b) + (0 + a \cdot b) \\
&\qquad\qquad [\because \quad a \cdot a = 1 \text{ and } b' \cdot b = 0] \\
&= (a + a \cdot b) + (a \cdot b + 0) \qquad [\textit{Commutative law}] \\
&= a \cdot 1 + a \cdot b) + a \cdot b \\
&= a (1 + b) + a \cdot b \qquad\qquad [\textit{Distributive law}] \\
&= a \cdot 1 + a \cdot b \qquad\qquad\quad [\because \quad 1 + b = b] \\
&= a \cdot (1 + b) \\
&= a \cdot 1 \\
&= a \qquad\qquad\qquad\qquad [\because \quad a \cdot 1 = a]
\end{aligned}
$$

$\therefore \qquad [a (a + b) + (b' + a)b]' = a'$.

---

### Exercise 9.2

1. State and prove the following laws of a Boolean algebra:

   (i) Idempotent laws
   (ii) Involution laws
   (iii) Cancellation laws
   (iv) Absorption laws.

2. State and prove the following laws in a Boolean algebra:

   (i) Associative law of addition
   (ii) Associative law of multiplication.

3. Prove that in a Boolean algebra:

   (i) Zero element is unique
   (ii) Unit element is unique.

4. Let $(B, +, \cdot, ')$ be a Boolean algebra. For $a \in B$, let $x \in B$ such that $a + x = 1$ and $a \cdot x = 0$. Prove that $x$ is equal to the complement of $a$.

5. If $a, b$ and $c$ are elements of the Boolean algebra B, then find the complement of the following expressions:

   (i) $a + b'$   (ii) $(a')' \cdot b'$
   (iii) $a \cdot b \cdot c'$
   (iv) $b + c \cdot a'$   (v) $(a + b) \cdot (a + c)'$
   (vi) $(b + c') \cdot (a + bc)'$

6. In the Boolean algebra $(B, +, \cdot, ')$; show that

   (i) $a \cdot b + a \cdot b' + a' \cdot b + a' b' = 1$ for all $a, b \in B$
   (ii) $(a + b) \cdot (a' \cdot b') = 0$
   (iii) $(a + b)' + (a + b')' = a'$
   (iv) $a \cdot b + [(a + b') \cdot b]' = 1$
   (v) $[a + a' \cdot b'] \cdot [a' + a \cdot b] = a' \cdot b' + a \cdot b$

7. In the Boolean algebra $(B, +, \cdot, ')$ show that

   (i) $(a + b) \cdot (c + d) = a \cdot c + a \cdot d + b \cdot c + b \cdot d$ for all $a, b, c, d \in B$
   (ii) $(a \cdot b \cdot c) + (a \cdot b \cdot c') + (a' \cdot b \cdot c) = b \cdot (a + c)$ for all $a, b, c \in B$
   (iii) $(a \cdot b \cdot c) + (a \cdot b \cdot c') + (a \cdot b' \cdot c) + (a' \cdot b \cdot c) = a \cdot b + b \cdot c + c \cdot a$, for all $a, b, c \in B$.

8. In the Boolean algebra $(B, +, \cdot, ')$ show that

   (i) $[a' \cdot (a + b)]' + [b \cdot (b + a')]' + [b \cdot (b' + a)]' = 1$ for all $a, b \in B$

   (ii) $(a \cdot b') + [a \cdot (b \cdot c)'] + c = a + c$ for all $a, b, c \in B$

9. In the Boolean algebra $(B, +, \cdot, ')$ simplify the Boolean expression $(x + x') \cdot (y + t) \cdot (z + x) \cdot (z + t)$.

10. Prove the following statement in Boolean algebra, $x + x \cdot (y + 1) = x$.

11. Simplify each of the following Boolean expression:

   (i) $ab + ac + bd + cd$

   (ii) $(a + b)(a + c)(a' \cdot b)'$

   (iii) $[a + (b(c + a'))]'$

   (iv) $a + a'(a + b) + bc$

12. In the Boolean algebra $(B, +, \cdot, ')$ show that for all $a, b \in B$

   (i) $a + b = 0$ if and only if $a = 0, b = 0$

   (ii) $a' + b = 1$ if and only if $a + b = b$

   (iii) $a \cdot b' = 0$ if and only if $a \cdot b = a$.

13. Let $(B, +, \cdot, ')$ be a Boolean algebra. If $x \cdot y = x \cdot z$ and $x' \cdot y = x' \cdot z$, then prove that $y = z$.

14. Let $(B, +, \cdot, ')$ be a Boolean algebra. If $x + y = x + z$ and $x' + y = x' + z$, then prove that $y = z$.

15. Let $(B, +, \cdot, ')$ be a Boolean algebra. If for $x, y, z \in B$,

$$x + y = x + z \quad \text{and} \quad x \cdot y = x \cdot z, \quad \text{then}$$

prove that $y = z$.

**Answers to Selected Problems**

5.   (i) $a' \cdot b$
  (ii) $a' + b$
  (iii) $a' + b' + c$
  (iv) $b' \cdot (c' + a)$
  (v) $a' \cdot b' + a' \cdot c$
  (vi) $b' \cdot c + a' \cdot b' + a' \cdot c$

9. $t \cdot (x + z) + y \cdot z$

11.   (i) $(a + d)(b + c)$
  (ii) $a$
  (iii) $(a + b)'$
  (iv) $a + b$

## 9.7 PRINCIPLE OF DUALITY

The principle of duality states that if in any law or theorem of Boolean algebra, the operators ($+$ and $\cdot$) and identity elements (0 and 1) are interchanged simultaneously then the new law or theorem is also valid.

Since in the Boolean algebra $(B, +, \cdot, ', 0, 1)$, the dual of each axion is also an axiom, therefore the dual of any theorem in a Boolean algebra is also a theorem and the dual theorem can be proved by using the dual of each step in the proof of the original theorem.

Thus the **dual** of a given Boolean expression can be obtained by:

(a) Interchanging the binary operators '$+$' and '$\cdot$' and

(b) Interchanging the element 0 and 1.

These interchanges are done simultaneously in the whole expression to obtain the dual expression.

For example, let $a, b, c \in B$, then the dual of $a + (b \cdot c)$ is $a \cdot (b + c)$.

**Illustration**

| Original equation | Dual |
|---|---|
| $a + 0 = a$ | $a \cdot 1 = a$ |
| $a + 1 = 1$ | $a \cdot 0 = 0$ |
| $a + a' = 1$ | $a \cdot a' = 0$ |

**Example 20:** Write the dual of the following Boolean statements?

$$\text{(i) } (a \cdot 1) \cdot (0 + a') = 0 \quad \text{(ii) } a + a' \cdot b = a + b$$

*Solution*

To obtain the dual statement, we interchange the operators '+' and '·' and simultaneously interchange 0 and 1 appearing in the original statement.

| Equation | Dual equation |
|---|---|
| (i) $(a \cdot 1) \cdot (0 + a') = 0$ | $(a + 0) + (1 \cdot a') = 1$ |
| (ii) $a + a' \cdot b = a + b$ | $a \cdot (a' + b) = a \cdot b$ |

**Example 21:** Write the dual of each of the following Boolean expressions:

$$\text{(i) } a + a' \cdot b = a \cdot b \quad \text{(ii) } a \cdot 0 + a \cdot 1 = a \quad \text{(iii) } a \cdot b + b \cdot c = (a + c) \cdot b$$

*Solution*

For obtaining the dual expression of the given Boolean expression, we interchange the operators '+' and '· ' and simultaneously interchange 0 and 1 everywhere in the expression.

| Equation | Dual |
|---|---|
| (i) $a + a' \cdot b = a \cdot b$ | $a \cdot (a' + b) = a + b$ |
| (ii) $a \cdot 0 + a \cdot 1 = a$ | $(a + 1) \cdot (a + 0) = a$ |
| (iii) $a \cdot b + b \cdot c = (a + c) \cdot b$ | $(a + b) \cdot (b + c) = (a \cdot c) + b$ |

**Example 22:** In the Boolean algebra $(B, +, \cdot, ')$, find the truth value of the dual statement of the statement $a \cdot b + a \cdot b' + a' \cdot b + a' \cdot b' = 1$ *for all a, b* $\in$ B.

*Solution*

The given statement is $a \cdot b + a \cdot b' + a' \cdot b + a' \cdot b' = 1$.

The dual of this statement is obtained by interchanging '+' and '·' and interchanging 0 and 1.

Therefore, the dual statement is $(a + b) \cdot (a + b') \cdot (a' + b) \cdot (a' + b') = 0$ \hfill (1)

$$\begin{aligned}
\text{L.H.S. of } (1) &= (a + b) \cdot (a + b') \cdot (a' + b) \cdot (a' + b') \\
&= (a + bb') \cdot (a' + b) \cdot (a' + b') \quad [\textit{Distributive law}] \\
&= (a + 0) \cdot (a' + bb') \quad\quad\quad\quad [\because \ b \cdot b' = 0]
\end{aligned}$$

$$= a \cdot (a' + 0)$$
$$= a \cdot a' = 0 = \text{R.H.S. of (1)}$$

Hence the truth value of the dual statement (1) is 'T'.

**Example 23:** In the Boolean algebra B, show that the dual statement of the statement $a' \cdot b' + a \cdot b = (a' + b) \cdot (a + b')$ for all $a, b \in$ B is true.

### *Solution*

The given statement is $a' \cdot b' + a \cdot b = (a' + b) \cdot (a + b')$

The dual of the given statement is

$$(a' + b') \cdot (a + b) = (a' \cdot b) \cdot (a \cdot b')$$

Now, $\quad (a' + b') \cdot (a + b) = a' \cdot a + a' \cdot b + b' \cdot a + b' \cdot b \qquad\qquad (1)$

$$= 0 + a' \cdot b + b' \cdot a = 0 \qquad\qquad [\because \ a' \cdot a = 0]$$

$$= a' \cdot b + a \cdot b'$$

$\therefore \qquad (a' + b') \cdot (a + b) = (a' \cdot b) + (a \cdot b')$ for all $a, b \in$ B

Hence, the dual statement (1) is true.

**Example 24:** If $a, b, c$ are any elements in a Boolean algebra B, then for the following laws verify that each is dual of other.

(i) **Idempotent laws:**
    (a) $a + a = a$      (b) $a \cdot a = a$

(ii) **Boundedness laws:**
    (a) $a + 1 = 1$      (b) $a \cdot 0 = 0$

(iii) **Absorption laws:**
    (a) $a + (a \cdot b) = a$      (b) $a \cdot (a + b) = a$

(iv) **Associative laws:**
    (a) $(a + b) + c = a + (b + c)$      (b) $(a \cdot b) \cdot c = a \cdot (b \cdot c)$

(v) **De-Morgan's laws:**
    (a) $(a + b)' = a' \cdot b'$

### *Solution*

(i) (a) Here $a + a = a$

The dual statement of $a + a = a$ is $a \cdot a = a$.

This is the statement (i) (b).

The dual statement of $a \cdot a = a$ is $a + a = a$

This is the statement (i) (a).

∴    Statement (i) (a) and (i) (b) are duals of each other.

(ii)  (a) Here $a + 1 = 1$

The dual statement of $a + 1 = 1$ is $a \cdot 0 = 0$

This is the statement (ii) (b)

The dual statement of $a \cdot 0 = 0$ is $a + 1 = 1$

This is the statement (ii) (a)

∴    Statements (ii) (a) and (ii) (b) are duals of each other.

**Note.** Other laws can be proved similarly.

**Exercise 9.3**

**Write the dual of each of the following statements in $(B, +, \cdot, \,')$:**

1. $(a' + b')' = a \cdot b$

2. $a + [a \cdot (b + 1)] = a$

3. $(a + b) \cdot (a + 1) = a + a \cdot b + b$

4. $(a + b)' = a' \cdot b'$

5. $(a' + b) \cdot (a + b') = a' \cdot b' + a \cdot b$

6. $(a \cdot b)' = a' + b'$

7. $[(a' + b) \cdot (b' + c)] \cdot (a' + c') = 0$

8. $a + (b \cdot a) = a$

9. $a + [(b' + a) \cdot b]' = 1$

10. $a \cdot b' + b = a + b$

11. If $a + b = 0$, then $a = b = 0$

12. $a = 0$, if and only if $b = a \cdot b' + a' \cdot b$ for all $b$.

13. $a \cdot b' = 0$ if and only if $a \cdot b = a$.

14. $(1 + a) \cdot (0 + b) = b$

15. $(1 \cdot a)' = 0 + a'$

16. In the Boolean algebra $(B, +, \cdot, \,')$ show that the dual statement of the statement $a \cdot b \cdot c + a' + b' + c' = 1$ for all $a, b, c \in B$ is true.

17. In the Boolean algebra $(B, +, \cdot, \,')$, find the truth value of the dual statement of
$$(a + b) \cdot (c + d) = a \cdot c + a \cdot d + b \cdot c + b \cdot d$$
for all $a, b, c, d \in B$

**Answers to Selected Problems**

1. $(a' \cdot b')' = a + b$

2. $a \cdot [a + (b \cdot 0)] = a$

3. $(a \cdot b) + (a \cdot 0) = a \cdot (a + b) \cdot b$

4. $(a \cdot b)' = a' + b'$

5. $(a' \cdot b) + (a \cdot b') = (a' + b') \cdot (a + b)$

6. $(a + b)' = a' \cdot b'$

7. $[(a' \cdot b) + (b' \cdot c)] + (a' \cdot c') = 1$

8. $a \cdot (b + a) = a$

9. $a \cdot [(b' \cdot a) + b]' = 0$

10. $(a + b') \cdot b = a \cdot b$

11. If $a \cdot b = 1$, then $a = b = 1$

12. $a = 1$ if and only if $b = (a + b') \cdot (a' + b)$ for all $b$

13. $a + b' = 1$ if and only if $a + b = a$

14. $(0 \cdot a) + (1 \cdot b) = b$

15. $(0 + a)' = 1 \cdot a'$

17. T.

## 9.8 BOOLEAN ALGEBRA AS A LATTICE

In the previous chapter, we have studied that in a distributive lattice, if complement of an element exist then it is unique. Also, if every element of bounded lattice has complement then it is complemented. Now in this section we will define a Boolean lattice and Boolean algebra.

*A lattice is called a Boolean lattice if it is distributive and complemented.* Since complements are unique in a Boolean lattice, therefore we can consider a Boolean lattice as an algebra with two binary operations '+' and '·' and one unary operation ''. Boolean lattices so considered are called **Boolean algebras**.

**Theorem 2** If $(B, +, \cdot, ', 0, 1)$ is a Boolean algebra, then prove that it is also a Boolean algebra as a lattice.

### *Proof*

Here $(B, +, \cdot, ', 0, 1)$ is a Boolean algebra.

$\therefore$ It satisfies commutative laws, associative laws and absorption laws w.r.t. binary operations '+ ' and '·'and unary operation ''. Let '$\leq$' be a relation on B defined as:

$$x \leq y \text{ if } x + y = y \quad \text{and} \quad x \cdot y = x, \quad \text{for } x, y \in B \tag{1}$$

Now, if $x + y = y$, then $x \cdot y = x \cdot (x + y) = x$          *[By absorption law]*

and if $x \cdot y = x$, then $x + y = x \cdot y + y = y + (y \cdot x) = y$     *[By absorption law]*

$\therefore$   $x + y = y$   iff   $x \cdot y = x$

$\therefore$   From (1), $x \leq y$ if $x + y = y$ (or $x \cdot y = x$)

The relation '$\leq$' is a partial order relation on B.       *[Refer theorem on page]*

Also, we can show that

$$\sup.\{x, y\} = x + y \quad \text{and} \quad \inf.\{x, y\} = x \cdot y \quad \text{for all } x, y \in B$$

$\therefore$   supremum and infimum are exist for all $x$ and $y$ in B.

$\therefore$   $(B, \leq)$ is a lattice.         *[Refer theorem in lattice on page]*

Since $(B, +, \cdot, ')$ is a Boolean algebra.

$\therefore$   Distributive laws satisfies in $(B, +, \cdot, ')$.

$\therefore$   The lattice $(B, \leq)$ is a distributive.

Again, Boolean algebra satisfies the boundedness laws

i.e.,                 $x + 1 = 1$    and    $x \cdot 0 = 0$    for $x \in B$

$\therefore$                 $x \leq 1$    and    $0 \leq x$    for all $x \in B$

$\therefore$   0 and 1 are the least and greatest elements of B respectively.

$\therefore$   The lattice $(B, \leq)$ is bounded.

Now, by complement laws in B, for $x \in B$, there exists $x' \in B$ such that

$$x + x' = 1 \quad \text{and} \quad x \cdot x' = 0$$

$\therefore$   $x'$ is complement of $x$ in the lattice $(B, \leq)$.

$\therefore$   The lattice $(B, \leq)$ is also complemented.

Thus, the lattice $(B, \leq)$ being distributive and complemented, is a Boolean algebra.

Hence, the Boolean algebra $(B, +, \cdot, ', 0, 1)$ is also a Boolean algebra as a lattice.

**Theorem 3** If the lattice $(B, \leq)$ is a Boolean algebra, then prove that it is also a Boolean algebra as an algebraic structure.

### *Proof*

Since $(B, \leq)$ is a lattice.

$\therefore$ Supremum and infimum exist for any $x, y \in B$.

The binary operations '+' and '$\cdot$' in B are define as:

$$x + y = \text{sup.} \ \{x, y\} \quad \text{and} \quad x \cdot y = \text{inf.} \ \{x, y\}$$

1. **Closure laws:**

   For $x, y \in B$

   $$x + y = \text{sup.} \ \{x, y\} \in B \quad \text{and} \quad x \cdot y = \text{inf.} \ \{x, y\} \in B$$

   $\therefore$   Closure laws are satisfied.

2. **Commutative laws:**

   For $x, y \in B$

   $$x + y = \text{sup.} \ \{x, y\} = \text{sup.} \ \{y, x\} = y + x$$

   and $\qquad\qquad\qquad x \cdot y = \text{inf.} \ \{x, y\} = \text{inf.} \ \{y, x\} = y \cdot x$

   $\therefore$   Commutative laws are satisfied.

3. **Distributive laws:**

   Since, the lattice $(B, \leq)$ is a Boolean algebra, it is distributive and complemented.

   $\therefore$   For $x, y, z \in B$

   $$x + (y \cdot z) = (x + y) \cdot (x + z)$$

   and $\qquad\qquad\qquad x \cdot (y + z) = x \cdot y + x \cdot z$

   $\therefore$   Distributive laws are satisfied.

4. **Identity laws:**

   The lattice $(B, \leq)$ being complemented, is bounded.

   Let 0 and 1 be the least and greatest elements respectively in lattice $(B, \leq)$.

   $\therefore$   $0 \leq x$   and   $x \leq 1$ for any $x \in B$

   $$x + 0 = \text{sup.} \ \{x, 0\} = x$$

   and $\qquad\qquad\qquad x \cdot 1 = \text{inf.} \ \{x, 1\} = x$

$\therefore \quad x + 0 = x \quad$ and $\quad x \cdot 1 = x$, for $x \in$ B

$\therefore \quad$ Identity laws are satisfied.

Here 1 and 0 are unit element are zero element respectively.

### 5. Complemented laws:

Since the lattice (B, $\leq$) is distributive and complemented, it has a unique complement for each element of B.

$\therefore \quad$ For each $x \in$ B, there exists a unique complement $x'$, which defines a unary operation '' on B.

Also, by definition of complement, $x + x' = 1 \quad$ and $\quad x \cdot x' = 0$

$\therefore \quad$ Complement laws are also satisfied.

$\therefore \quad$ (B, +, $\cdot$, ', 0, 1) is a Boolean algebra.

Hence, the lattice (B, $\leq$) is a Boolean algebra as an algebraic structure.

**Example 25:** Let S be a finite set. The power set of S, i.e., P(S) is a lattice under the partial order relation '$\subseteq$' of set inclusion. Show that P(S) is a Boolean algebra.

### *Solution*

Here (P(S), $\subseteq$) is a lattice.

For A, B $\in$ P(S), we have

$$A \vee B = A \cup B \quad \text{and} \quad A \wedge B = A \cap B$$

Let A, B, C $\in$ P(S)

$\therefore \qquad A \vee (B \wedge C) = A \cup (B \cap C) = (A \cup B) \cap (A \cup C) = (A \vee B) \wedge (A \vee C)$

and $\qquad A \wedge (B \vee C) = A \cap (B \cup C) = (A \cap B) \cup (A \cap C) = (A \wedge B) \vee (A \wedge C)$

$\therefore \quad$ The lattice (P(S), $\subseteq$) is distributive.

Also (P(S), $\subseteq$) is a bounded lattice, in which the universal upper bound is S, and the universal lower bound is $\phi$, as $\phi \subseteq$ S for X $\in$ P(S).

Hence the lattice is bounded.

Now, the complement of any set T in P(S) is the set $T' = S - T$

Hence the lattice is complemented

and $\qquad\qquad T \vee T' = T \cup T' = S, \quad T \wedge T' = T \cap T' = \phi$

$\therefore \quad$ The lattice (P(S), $\subseteq$) is a Boolean algebra.

**Example 26:** Let $D_m$ be the set of all positive factors of $m$ ($m > 2$). If $m$ is the product of distinct primes, the prove that $D_m$ is a Boolean algebra.

### Solution

We already proved in lattice, that the set $D_m$ is a lattice with the relation divisibility.

Also, 1 and $m$ are the least and greatest elements of $D_m$ respectively. This lattice is also a distributive lattice.

Also, $m$ is the product of distinct primes,

For $x \in D_m$. Let $x' = \dfrac{m}{x}$

$$\therefore \qquad x + x' = \sup. \{x, y] = m \quad \text{and} \quad x \cdot x' = \inf. \{x \cdot x'\} = 1$$

$\therefore$  $x'$ is the complement of $x$ in $D_m$.

$\therefore$  The lattice $D_m$ is distributive and complemented.

Here, $D_m$ is a Boolean algebra.

**Example 27:** If $D_m$ is a Boolean algebra, find the atoms of $D_m$.

### Solution

Here $D_m$ is a Boolean algebra

$\therefore$  $m$ must be a product of distinct prime numbers.

Let $m = p_1 p_2 \ldots p_k$, where $p_1, p_2, \ldots p_k$, are distinct prime numbers. We know that $D_m$ is a lattice with lower bound 1. The immediate successors of 1 are the atoms of $D_m$.

Also, $1 << p_1$, $1 << p_2$, ..., $1 << p_n$,

$\therefore$  The atoms of the Boolean algebra B are the prime numbers $p_1, p_2, \ldots, p_n$.

## 9.9 ISOMORPHIC BOOLEAN ALGEBRA

Let B and B$'$ be two Boolean algebras, then B and B$'$ are said to be *isomorphic Boolean algebras*, if there exists a one-one function $f$: B $\rightarrow$ B$'$ which preserves the three operations '+', '·' and '$'$' such that

$$f(x + y) = f(x) + f(y), \quad f(x \cdot y) = f(x) \cdot f(y)$$

and $\qquad\qquad f(x') = (f(x))'$ for any $x, y \in$ B.

## 9.10 REPRESENTATION THEOREM

The representation theorem gives the number of elements in a finite Boolean algebra. It states that, **'if (B, +, ·, $'$, 0, 1) is a finite Boolean algebra and A is the set of all atoms of B. If A contains $n$ elements then B has $2^n$ elements'.**

### Proof

Here B is a finite Boolean algebra and A is the set of all atoms of B.

$\therefore$  B is a finite distributive and complemented lattice.

$\therefore$ Every element in B can be expressible uniquely (except for order) as the sum of irredundant atoms of B.

Let P(A) be the set of all subsets of A under the operations '$\cup$', '$\cap$' and '$'$'

$\therefore$ P(A) is a Boolean algebra.

If $x \in$ B, then there exists unique atoms $a_1, \ldots, a_n$ of B such that $x = a_1 + \ldots + a_n$.

We define a mapping $f: B \to P(A)$ such that

$$f(x) = \{a_1, \ldots, a_n\} \in P(A)$$

Let $\qquad x = a_1 + \cdots + a_r + b_1 + \cdots + b_s$

and $\qquad y = b_1 + \cdots + b_s + c_1 + \cdots + c_t$ be any elements of B and

$$A = \{a_1, \ldots, a_r, b_1, \ldots, b_s, c_1, \ldots, c_t, d_1, \ldots, d_k\}$$

Clearly, we can see that in $x$ and $y$, the common atoms are $b_1 + \cdots + b_s$.

$d_1, \ldots, d_k$ are the atoms of A which are not the expression of in $x$ and $y$.

If $e_i \in$ A then $e_i + e_i = e_i$ and $e_i \cdot e_i = e_i$

If $e_i, e_j \in$ A then $e_i \cdot e_j = 0$, whenever $i \neq j$ as $e_i$ and $e_j$ are both immediate successors of 0.

$\therefore \qquad x + y = a_1 + \cdots + a_r + b_1 + \cdots + b_s + c_1 + \cdots + c_t$

and

$$xy = b_1 b_1 + \cdots + b_s b_s = b_1 + \cdots + b_s$$

$\therefore \qquad f(x + y) = \{a_1, \ldots, a_r, b_1, \ldots, b_s, c_1, \ldots, c_t\}$

$$= \{a_1, \ldots, a_r, b_1, \ldots, b_s\} \cup \{b_1, \ldots, b_s, c_1, \ldots, c_t\}$$

$$= f(x) \cup f(y)$$

and $\qquad f(xy) = \{b_1, \ldots, b_s\}$

$$= \{a_1, \cdots, a_r, b_1, \cdots, b_s\} \cap \{b_1, \cdots, b_s, c_1, \cdots, c_t\}$$

$$= f(x) \cap f(y)$$

Let $\qquad x' = c_1 + \cdots + c_t + d_1 + \cdots + d_k$

$$x + x' = a_1 + \cdots + a_r + b_1 + \cdots + b_s + c_1 + \cdots + c_t + d_1 + \cdots + d_k$$

The element '1' of B is also expressible as the sum of irredundant atoms of B.

Let $\quad 1 = e_1 + \cdots + e_m \quad$ and $\quad A = \{e_1, \ldots, e_m, e_{m+1}, \ldots, e_p\}$

$$e_1 + \cdots + e_m \leq (e_1 + \cdots + e_m) + (e_{m+1} + \cdots + e_p)$$

$\Rightarrow \qquad 1 \leq e_1 + \cdots + e_p \qquad\qquad (1) [\because \text{ Each } e_i \leq 1]$

Also, $\qquad e_1 + \cdots + e_p \leq 1 \qquad\qquad (2)$

From (1), and (2), $\qquad e_1 + \ldots + e_p = 1$

$\therefore$ Sum of all atoms of B is 1

$\therefore \qquad a_1 + \cdots + e_r + b_1 + \cdots + b_s + c_1 + \cdots + c_t + d_1 + \cdots + d_k = 1$

$$\Rightarrow \qquad x + x' = 1 \qquad\qquad [\because \ e_i \cdot e_j = 0 \ for \ i \neq j]$$

Also, $\qquad\qquad x \cdot x' = 0$

$\therefore$ $x'$ is the complement of $x$ in B.

Also, $\qquad\qquad f(x') = f(c_1 + \cdots + c_t + d_1 + \cdots + d_k)$
$$= \{c_1, ..., c_t, d_1, ..., d_k\}$$
$$= \{a_1, ..., a_r, b_1, ..., b_s\}' = (f(x))'$$

The mapping $f$ is also one-one and onto.

$\therefore$ The mapping $f \colon B \to P(A)$ is a Boolean algebra is isomorphism.

$\therefore$ The number of elements in B = The number of elements in $P(A) = 2^n$.

## 9.11 SUBALGEBRA

Let $(B, +, \cdot, ', 0, 1)$ be a Boolean algebra. A non-empty subset S of B is said to be Boolean subalgebra of B if S is a Boolean algebra w.r.t. the operation of B.

**Example 28:** Show that $D_{30}$ is a Boolean algebra. Draw Hasse diagram of $D_{30}$. Find complements, atoms and subalgebras of $D_{30}$.

*Solution*

Here $\qquad D_{30} = \{1, 2, 3, 5, 6, 10, 15, 30\}$

Now, $\qquad 30 = 2 \times 3 \times 5$

$\therefore$ 30 is the product of distinct prime numbers

$\therefore$ $D_{30}$ is a Boolean algebra.

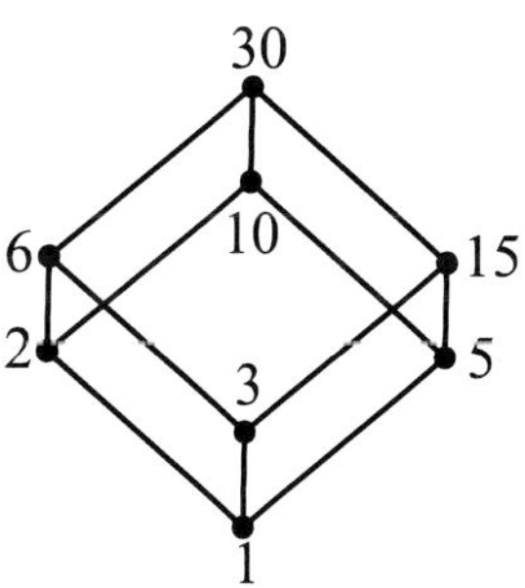


The Hasse diagram of $D_{30}$ is shown in the adjoining figure.

Now, $\qquad 1' = 30, \quad 2' = 15, \quad 3' = 10$

$\qquad\qquad 5' = 6, \quad 6' = 5, \quad 10'\,3, \quad 15' = 2 \quad$ and $\quad 30' = 1$

The immediate successors of the lower bound 1 are 2, 3 and 5.

$\therefore$ The **atoms** of $D_{30}$ are 2, 3 and 5.

$\therefore$ Number of elements in $D_{30} = 2^3 = 8$.

$\therefore$ A subalgebra of $D_{30}$ can have 2 or 4 or 8 elements

Also, $\{1, 30\}$ and $D_{30}$ are subalgebras of order 2 and 8 respectively.

Let $\{1, x, x', 30\}$ be a subalgebra of $D_{30}$ of order 4.

$\therefore$ $x$ can take either 2 or 3 or 5

$\therefore$ The subalgebras of $D_{30}$ of order 4 are $\{1, 2, 15, 30\}$, $\{1, 3, 10, 30\}$ and $\{1, 5, 6, 30\}$.

$\therefore$ The subalgebras of $D_{30}$ are $\{1, 30\}$, $\{1, 2, 15, 30\}$, $\{1, 3, 10, 30\}$, $\{1, 5, 6, 30\}$ and $D_{30}$.

### Exercise 9.4

1. Which of the lattices:
   (i) $D_6$ (ii) $D_{12}$ (iii) $D_{20}$ (iv) $D_{210}$ are Boolean algebras and what are their atoms?

2. Show that $D_{105}$ is a Boolean algebra, Draw the Hasse diagram of $D_{105}$. Find the complements, atoms and sub algebras of $D_{105}$.

3. If B is a Boolean algebra, show that:
   (i) $0 \leq x \leq 1$ for $x \in$ B
   (ii) $a < b \iff b' < a'$ for $a, b \in$ B.

4. Determine whether the lattice $D_{55}$ is a Boolean algebra? Draw the Hasse diagram of $D_{55}$. If $D_{55}$ is a Boolean algebra, then find the complements and atoms.

5. Show that the subset $\{1, 2, 3, 6\}$ of $D_{30}$ is not a subalgebra of Boolean algebra $D_{30}$, but it is a Boolean algebra in itself.

### Answers to Selected Problems

1. (i) $D_6$ is a Boolean algebra; Atoms are 2 and 3
   (ii) $D_{12}$ is not a Boolean algebra
   (iii) $D_{20}$ is not a Boolean algebra
   (iv) $D_{210}$ is a Boolean algebra; Atoms are 2, 3, 5 and 7.

2. Complements: $1' = 105$, $3' = 35$, $5' = 21$, $7' = 15$, $15' = 7$, $21' = 5$, $35' = 3$ and $105' = 1$
   Subalgebras: $\{1, 105\}$, $\{1, 3, 35, 105\}$, $\{1, 5, 21, 105\}$, $\{1, 7, 15, 105\}$ and $D_{105}$.

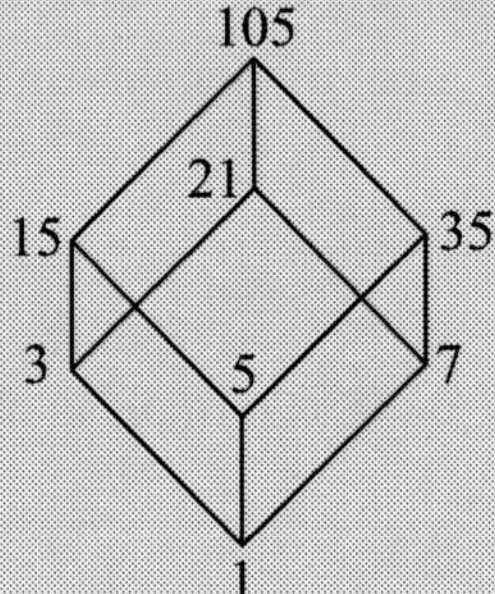


4. Yes, complements $1' = 55$, $5' = 11$, $11' = 5$, $55' = 1$. Atoms are 5 and 11.

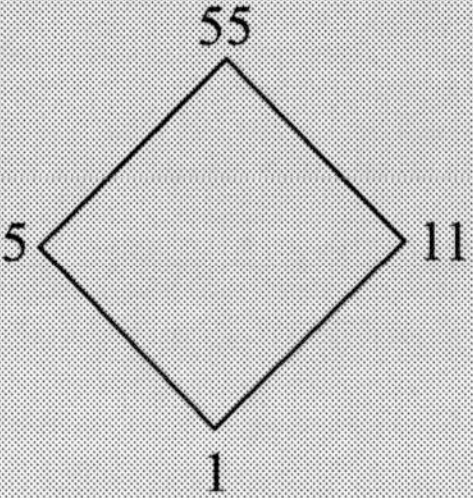


## 9.12 FUNDAMENTAL PRODUCT

Let $(B, +, \cdot, ')$ be a Boolean algebra with binary operations '+' and '·' and a many operation '·'.

A *literal* is a variable or a complemented variable *i.e.*, if $x$ is any variable, then the variable $x$ and complemented variable $x'$ are called *literals* corresponding to $x$.

A **fundamental product** is a literal or a product of two or more literals in which no two literals involve the same variable.

For example, $x \cdot z'$, $xy'z$, $x'yz'$ and $x'y'z'$ are the fundamental products, whereas $xyx'$ is not a fundamental product as the literal $x$ and $x'$ involve the same variable.

Also, $xyzy'$ and $x'yxz'$ are not fundamental products.

**Note.** Any product of literals can be reduced to either 0 or to a fundamental product.
For example

(i) $xyzy = xyyz = xyz$                                    *[By idempotent law, $y \cdot y = y$]*

(ii) $xyx'z = xx'yz = 0 \ \ yz = 0$                *[By complement law, $x \cdot x' = 0$]*

**Example 29:** In Boolean algebra (B, +, ·, ′) reduce the following products to either 0 or to a fundamental product (i) $xyzz'$   (ii) $x'yyx$   (iii) $xyz'z'xyyy$   (iv) $xyy'zz'$   (v) $xxyzz$.

***Solution***

(i) $xyzz' = (xy)(zz') = xy \cdot 0 = 0.$

(ii) $x'yyx = x'(yy)x = x'yx = (x'x)y = 0 \cdot y = 0.$

(iii) $xyz'z'xyyy = xy(z'x')x(yyy) = xyz'xy = (xy)(xy)z' = xyz'$

(iv) $xyy'zz' = x(yy')(zz') = x \cdot 0 \cdot 0 = 0.$

(v) $xxyzz = (xx)y(zz) = xyz.$

## 9.13 'CONTAINED-IN' FUNDAMENTAL PRODUCT

Let $E_1$ and $E_2$ be two fundamental products. If the literals of $E_1$ are also the literals of $E_2$, then the fundamental product $E_1$ is **contained-in** the fundamental product $E_2$.

For example, the fundamental product $x'y$ is contained-in the fundamental product $x'yz$, whereas the fundamental product $x'y$ is not contained in the fundamental product $xyz$ as $x'$ is not a literal of $xyz$.

**Illustrations**

(i) $xy'z$ is contained in $xy'zt$.

(ii) $x'y'$ is contained-in $x'y'z'$

(iii) $xy'z'$ is not contained-in $xyz'$

If fundamental product $E_1$ is contained-in the fundamental product $E_2$, then the literals of $E_1$ are also the literals of $E_2$.

∴ We can write $E_2 = E_1F$, where F is the product of literals of $E_2$ which are not literals of $E_1$.

∴                       $E_1 + E_2 = E_1 + E_1F = E_1$            *[By Absorption law]*

For example, Let $E_1 = x'y'$ and $E_2 = x'y'z'$ be two fundamental products.
Here fundamental product, $E_1$ is contained-in fundamental product $E_2$.

∴                       $E_1 + E_2 = x'y' + x'y'z' = x'y' + (x'y')z'$

$$= x'y' \quad \textit{[By Absorption law } a + (a \cdot b) = a]$$

## 9.14 SUM OF PRODUCT EXPRESSION

A Boolean expression E is called a Sum-of-product' (SOP) expression, if E is a fundamental product or the sum (join) of two or more fundamental products, none of which is contained-in another.

For example, in the Boolean algebra $(B, +, \cdot, ')$, the sum of fundamental products $xy + xz + x'yz$ is a SOP expression but $xz + xyz$ is not a SOP expression as $xz$ is contained in $xyz$.

### Illustration

(i) $xy + xy'z + x'yz + xy'z'$ is a SOP expression, as none of the term is contained-in other.
(ii) $x'z + x'y + x'y'z$ is not a SOP expression, as $x'z$ is contained in $x'y'z$.

## 9.15 SUM OF PRODUCT FORM

Let E be a Boolean expression, a *sum of product (SOP) form or disjunctive normal form (DNF)* of E is an equivalent Boolean SOP expression, where two Boolean expressions are said to be equivalent if one can be obtained from the other by a finite number of applications of identities of a Boolean algebra.

For example, let $E(x, y) = xy' + xy'z + x'y$ be a Boolean expression in the Boolean algebra $(B, +, \cdot, ')$,

Since $xy'$ is contained-in $xy'z$.

$$\therefore \qquad\qquad xy' + (xy')z + x'y = xy' + x'y \qquad\qquad [By\ absorption\ law]$$

$\therefore$ $xy' + x'y$ is a SOP expression.

$\therefore$ $xy' + x'y$ is a sum of product (SOP) form of the Boolean expression $xy' + xy'z + x'y$

**Steps for finding the SOP form of a Boolean expression:**

Let $E(x, y, z) = ((xy)'z)' ((x' + z) (y' + z'))'$ be a Boolean expression in Boolean algebra $(B, +, \cdot, ¢)$.

**Step 1.** Since complement occurs outside the bracket, therefore by using De-Morgan laws, we have

$$E(x, y, z) = ((xy)'' + z') ((x' + z)' + (y' + z')')$$
$$= ((xy)'' + z') (x'' \cdot z' + y'' \cdot z'')$$

**Step 2.** Since resultant of Step 1 contains double complement, therefore by using, *involution laws*, we have

$$E(x, y, z) = (xy + z') (xz' + yz) \qquad\qquad (1)$$

The Boolean expression E in (1) now consists of only sum and products of literals.

**Step 3.** Using *distributive law* in (1), we have

$$E(x, y, z) = xyxz' + xyyz + z'xz' + z'yz$$
$$= (xx)yz' + x(yy)z + xz'z' + yz'z \qquad\qquad (2)$$

**Step 4.** Using Commutative laws, complement laws and idempotent laws, in (2), we have

$$E(x, y, z) = xyz' + xyz + xz' + 0 \tag{3}$$

Each term in Boolean expression (3) is a fundamental product or 0.

**Step 5.** Since $xz'$ is contained-in $xyz'$, therefore by *absorption law* and *identity laws* in (3),

$$E(x, y, z) = xz' + xyz$$

The resultant of Step 5 is a sum of product (SOP) expression.

**Example 30:** Express the following Boolean expressions in sum-of product (SOP) form:

(i) $(xy')'z$   (ii) $(x + y)'(x' + y)$   (iii) $(x'y' + z)(x + y')'$

   *Solution*

$$
\begin{aligned}
\text{(i)} \quad (x\,y')'z &= (x' + y'')z \\
&= (x' + y)z && [By\ involution\ law] \\
&= x'z + yz && [By\ De\text{-}Morgan's\ law] \\
\text{(ii)} \quad (x + y)'(x' + y') &= (x' \cdot y') \cdot (x' + y') && [By\ De\text{-}Morgan's\ law] \\
&= x'x'y' + x'y'y' && [By\ distributive\ law] \\
&= x'y' + x'y' = x'y' && [By\ idempotent\ law] \\
\text{(iii)} \quad (x'y' + z)'(x + y')' &= (x'y' + z)(x' \cdot y'') && [By\ De\text{-}Morgan's\ law] \\
&= (x'y' + z)(x' \cdot y) && [By\ involution\ law] \\
&= x'y'z'y + zx'y && [By\ distributive\ law] \\
&= (x'x')(y'y) + x'yz && [By\ commutative\ law] \\
&= x' \cdot 0 + x'yz = 0 + xyz \\
& && [By\ idempotent\ and\ complement\ laws] \\
&= x'yz. && [By\ identity\ law]
\end{aligned}
$$

## 9.16 COMPLETE SUM OF PRODUCT FORM

Mintum. If a fundamental product involves all the variables under consideration, then the fundamental product is called a mintum. For example, if the variables under consideration are $x$ and $y$ then the mintums are $xy$, $xy'$, $x'y$, $x'y'$.

Let E be a Boolean expression. A sum of product (SOP) form called a *complete SOP form* of E is equivalent to the sum of product (SOP) form and each product in this sum is mintum. Complete sum of product form is also called *full disjunctive normal form* or *disjunctive canonical form*.

**Remark.** If $n$ variables are under consideration, then the number of mintums is $2^n$ and the complete sum of product (SOP) form of a Boolean expression can have atmost $2^n$ terms.

**Steps for finding the complete sum of product form of a Boolean expression:**

**Step 1.** Consider a Boolean expression E $(x_1, x_2, ..., x_n)$ and find the sum of product form of the Boolean expression E.

**Step 2.** If ever a product in the sum of product form of E involves all the variables $x_1, x_2, ..., x_n$, then the sum product form of E is the complete sum of product form of the Boolean expression E.

If any product P in the sum of product form of E does not involve the variable $x_i$, then multiply P by $x_i + x'_i$ and then use distributive law and idempotent law to simplify the sum of product form of E.

**Step 3.** Repeat step-2 until every product P in E involves all the variables i.e., a mintum.

**Example 31:** In the Boolean algebra (B, +, ·, '), express the Boolean expression E(x, y, z) given by $xy'z + (x' + y)'$ in its complete sum of product form.

### *Solution*

Here
$$E(x, y, z) = xy'z + (x' + y)' \qquad [By \ De\text{-}Morgan's \ law]$$
$$= xy'z + x'' \cdot y' \qquad [By \ involution \ law]$$
$$= xy'z + xy'$$
$$= xy'z + xy' \qquad (1) \ [By \ identity \ law]$$
$$= xy'z + xy'(z + z')$$
$$= xy'z + xy'z + xy'z'$$

which is the required complete sum of product form the given expression in variable $x$, $y$ and $z$

### 9.17 BOOLEAN FUNCTIONS

Let (B, +, ·, ') be a Boolean Algebra and $E(x_1, x_2, ..., x_n)$ be a Boolean expression of $n$ variables $x_1, x_2, ..., x_n$ over B.

Define function $f$: $B^n \rightarrow B$ such that $f(x_1, x_2, ..., x_n) = E(x_1, x_2, ..., x_n)$, then $f$ is a called a Boolean function.

Thus, the functions which can be represented by Boolean expressions are called **Boolean functions.**

For example, $f(x_1, x_2) = x_1 \cdot x_2' + x_1' \cdot x_2$ is a Boolean function. In this function if $x_1$ is true and $x_2$ is false then $x_1'$ false and $x_2'$ is true. $x_1 \cdot x_2'$ is therefore true and $x_1' \cdot x_2$ is false therefore $f(x_1, x_2)$ is true. The function is also true when $x_1$ is false and $x_2$ is true.

**Boole's Expansion Theorem**

If (B, +, ..., 0, 1) be a Boolean Algebra $f(x_1, x_2, ..., x_n)$ be a Boolean function in $n$ variables $x_1, x_2, ..., x_n$, then

$$f(x_1, x_2, ..., x_n) = f(1, 1, ..., 1) x_1 \cdot x_2, ..., x_n + \cdots + f(1, 0, ..., 1) x_1 \cdot x_2, ..., x_n + \cdots +$$
$$f(0, 1, ..., 1) x_1 \cdot x_2, ..., x_n + \cdots + f(0, 0, ..., 1) x_1 \cdot x_2, ..., x_n + \cdots +$$
$$f(0, 0, ...) x_1 \cdot x_2, ..., x_n.$$

**Example 32:** Construct an input / output table for the Boolean function defined by $f(a, b) = (a \cdot b)' + b$.

***Solution***

The given function is $f(a, b) = (a \cdot b)' + b$       (1)

The Boolean expression on the R.H.S. contains two variables, so the given function is 2-place Boolean function and thus the domain is $(\{0, 1\})^2$ where

$$(\{0, 1\})^2 = \{0, 1\} \times \{0, 1\} = \{(0, 0), (0, 1), (1, 0), (1, 1)\}.$$

From (1), we have

$$f(0, 0) = (0 \cdot 0)' + 0 = 0' + 0 = 1 + 0 = 1$$
$$f(0, 1) = (0 \cdot 1)' + 1 = 0' + 1 = 1 + 1 = 1$$
$$f(1, 0) = (1 \cdot 0)' + 0 = 0' + 0 = 1 + 0 = 1$$
$$f(1, 1) = (1 \cdot 1)' + 1 = 1' + 1 = 0 + 1 = 1$$

Thus the required input-output table is as given below:

| Input | | Output |
|---|---|---|
| $a$ | $b$ | $(a \cdot b)' + b$ |
| 0 | 0 | 1 |
| 0 | 1 | 1 |
| 1 | 0 | 1 |
| 1 | 1 | 1 |

**Example 33:** Construct an input / output table for the Boolean function $f$ defined by $f(x_1, x_2) = x_1' \cdot x_2$.

***Solution***

The number of variables in the Boolean expression given on the right-hand side of (1) are 2, so the given function is a 2-place Boolean function and thus the domain given by

$$(\{0, 1\})^2 = \{0, 1\} \times \{0, 1\} = \{(0, 0), (0, 1), (1, 0), (1, 1)\}$$

From (1), we have

$$f(0, 0) = 0' \cdot 0 = 1 \cdot 0 = 0$$
$$f(0, 1) = 0' \cdot 1 = 1 \cdot 1 = 1$$
$$f(1, 0) = 1' \cdot 0 = 0 \cdot 0 = 0$$
$$f(1, 1) = 1' \cdot 1 = 0 \cdot 1 = 0$$

Thus, the required input-output table is as given below:

| Input | | Output |
|---|---|---|
| $x_1$ | $x_2$ | $x_1' \cdot x_2$ |
| 0 | 0 | 0 |
| 0 | 1 | 1 |
| 1 | 0 | 0 |
| 1 | 1 | 0 |

**Example 34:** Show that the function $f$ given by the following table is a Boolean function:

| Input | | | Output |
|:---:|:---:|:---:|:---:|
| $x_1$ | $x_2$ | $x_3$ | $f(x_1, x_2, x_3)$ |
| 1 | 1 | 1 | 1 |
| 1 | 1 | 0 | 1 |
| 1 | 0 | 1 | 1 |
| 1 | 0 | 0 | 0 |
| 0 | 1 | 1 | 0 |
| 0 | 1 | 0 | 0 |
| 0 | 0 | 1 | 0 |
| 0 | 0 | 0 | 1 |

*Solution*

*Construction of Boolean function:*

1. Mark all the rows of the table having output 1. Here 1st, 2nd, 3rd and 8th rows have output 1.

2. In the above identified rows, replace 1 by $x_i$ and 0 $x_i'$ according to their respective positions where $i = 1, 2, 3$.

   Thus, following combinations are obtained:

$$x_1 \cdot x_2 \cdot x_3 \qquad \text{for the 1st row}$$
$$x_1 \cdot x_2 \cdot x_3' \qquad \text{for the 2nd row}$$
$$x_1 \cdot x_2' \cdot x_3 \qquad \text{for the 3rd row}$$
$$x_1' \cdot x_2' \cdot x_3' \qquad \text{for the 8th row}$$

3. By applying OR to all the above combinations, we get the following Boolean expression

$$(x_1 \cdot x_2 \cdot x_3) + (x_1 \cdot x_2 \cdot x_3') + (x_1 \cdot x_2' \cdot x_3) + (x_1' \cdot x_2' \cdot x_3')$$

The required function is

$$f(x_1, x_2, \cdot x_3) = (x_1 \cdot x_2 \cdot x_3) + (x_1 \cdot x_2 . x_3') + (x_1 \cdot x_2' . x_3) + (x_1' \cdot x_2' \cdot x_3')$$

As the function is represented by a Boolean expression, therefore it is a Boolean function.

## 9.18 SWITCHING CIRCUITS

In practice, we often require making (switching on) and breaking (switching off) of an electrical circuit. It is also desirable and sometimes essential that the switching operations (making and breaking) of an electrical circuit is very fast and without sparking. For this purpose, we need a **switch** which is a device that can turn the current **on** or **off** in an electrical circuit. The circuit that can turn the current *on* or *off* in an electrical circuit is called as **switching circuit**.

Most of the electronic switching circuits are binary in nature and are known as binary switching circuits or logic circuits. This is because a switch can either be in 'on' or in 'off' position.

It cannot be '*on*' as well as '*off*' at the same time. This property of switches has an analogy (resemblances) with those of statements which can be either true or false but cannot be both at the same time.

Switches can be inter-connected in various ways to perform complex control functions. For example, an air-conditioning system might be controlled by the thermostats in two rooms. If the temperature in either room exceeds 30º C, thermostat in that room will close a switch to turn on the air-conditioning system.

The two most fundamental ways of interconnecting switches are **series** connection and **parallel** connection.

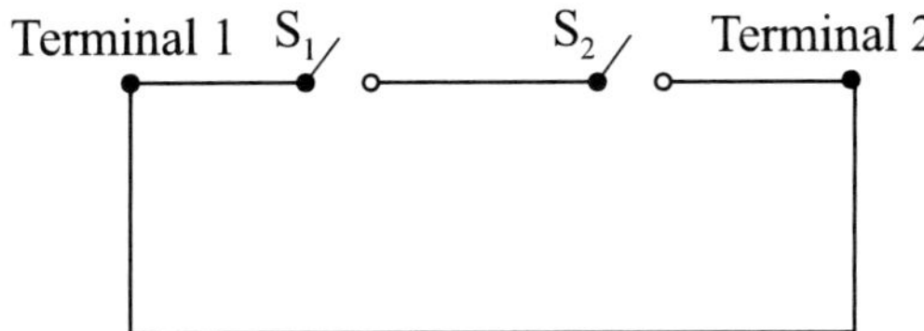


In a series connection of two switches, the circuit is open between the terminals 1 and 2 if even one or both of the switches are in the off position and the circuit is closed if both the switches are in the on position.

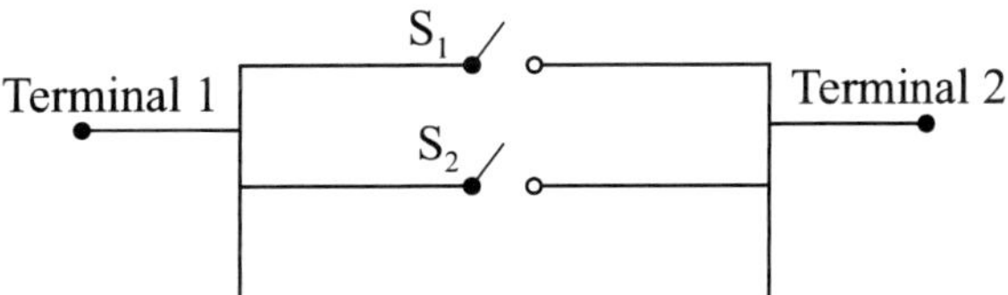


In parallel connection of two switches, the circuit is open between terminals 1 and 2 if both switches $S_1$ and $S_2$ are in the off position and the circuit is closed if even one or both the switches are in the on position.

The tables for the circuits in series and parallel are given below:

<table>
<tr><td colspan="3">Switches in Series</td><td colspan="3">Switches in Parallel</td></tr>
<tr><td colspan="2">Input</td><td>Output</td><td colspan="2">Input</td><td>Output</td></tr>
<tr><td>$S_1$</td><td>$S_2$</td><td>$S_1 \cdot S_2$</td><td>$S_1$</td><td>$S_2$</td><td>$S_1 + S_2$</td></tr>
<tr><td>off</td><td>off</td><td>off</td><td>off</td><td>off</td><td>off</td></tr>
<tr><td>on</td><td>off</td><td>off</td><td>on</td><td>off</td><td>on</td></tr>
<tr><td>off</td><td>on</td><td>off</td><td>off</td><td>on</td><td>on</td></tr>
<tr><td>on</td><td>on</td><td>on</td><td>on</td><td>on</td><td>on</td></tr>
</table>

It can be easily observed as shown in the table given above that the parallel connection behaves like an OR (disjunction) operation while the series connection behaves as an AND (conjunction) operation. A switch which is 'off' when the switch $S_1$ is 'on' and is 'on' when the switch $S_1$ is 'off' is called the **complement** of the switch $S_1$ and is denoted by $S_1'$.

Thus all the results of Boolean algebra can be applied to the analysis and synthesis of the switching networks.

To study the behaviour of complex switching circuits, let us define the following algebraic system.

Let $(B, +, \cdot, ', 0, 1)$ be an algebraic system where '+' and '$\cdot$' are two binary operations which represent connection in *parallel and series* respectively (i.e., equivalent to logical OR and logical AND) and '$'$' be the unary operation of taking complement of switches. The switch is a variable that can either assume the value 0 (i.e., off position) of the value 1 (i.e., on position).

## 9.19 SWITCHING ALGEBRA

A switching circuit consisting of parallel and series connection of switches can be described algebraically in our algebraic system.

### 1. Commutative laws

For every two switches $x$ and $y$ belonging to the algebraic system B, we have

| Input | | Output | |
|---|---|---|---|
| $x$ | $y$ | $x + y$ | $y + x$ |
| 0 | 0 | 0 | 0 |
| 0 | 1 | 1 | 1 |
| 1 | 0 | 1 | 1 |
| 1 | 1 | 1 | 1 |

| Input | | Output | |
|---|---|---|---|
| $x$ | $y$ | $x \cdot y$ | $y \cdot x$ |
| 0 | 0 | 0 | 0 |
| 0 | 1 | 0 | 0 |
| 1 | 0 | 0 | 0 |
| 1 | 1 | 1 | 1 |

$\therefore \qquad x + y = y + x$ and $x \cdot y = y \cdot x$ for all $x, y \in B$.

Hence the commutative laws are satisfied.

### 2. Identity laws

For every switch $x$ belonging to the algebraic system B, we have

| Input | | Output |
|---|---|---|
| $x$ | 0 | $x + 0$ |
| 0 | 0 | 0 |
| 1 | 0 | 1 |

| Input | | Output |
|---|---|---|
| $x$ | 1 | $x \cdot 1$ |
| 0 | 1 | 0 |
| 1 | 1 | 1 |

$\therefore \qquad x + 0 = x$ and $x \cdot 1 = x$ for all $x \in B$.

Hence the identity laws are satisfied.

### 3. Complement laws

For every switch $x$ belonging to the algebraic system B, we have

| Input | | Output |
|---|---|---|
| $x$ | $x'$ | $x + x'$ |
| 0 | 1 | 1 |
| 1 | 0 | 1 |

| Input | | Output |
|---|---|---|
| $x$ | $x'$ | $x \cdot x'$ |
| 0 | 1 | 0 |
| 1 | 0 | 0 |

$\therefore \qquad\qquad x + x' = 1 \quad \text{and} \quad x \cdot x' = 0 \quad \text{for all } x \in \text{B}$

Hence the complement laws are satisfied.

## 4. Distributive laws

For every three switches $x$, $y$ and $z$ belonging to the algebraic system B, we have

| Input | | | | | | Output | |
|---|---|---|---|---|---|---|---|
| $x$ | $y$ | $z$ | $y \cdot z$ | $x + y$ | $x + z$ | $x + y \cdot z$ | $(x + y) \cdot (x + z)$ |
| 0 | 0 | 0 | 0 | 0 | 0 | 0 | 0 |
| 0 | 0 | 1 | 0 | 0 | 1 | 0 | 0 |
| 0 | 1 | 0 | 0 | 1 | 0 | 0 | 0 |
| 0 | 1 | 1 | 1 | 1 | 1 | 1 | 1 |
| 1 | 0 | 0 | 0 | 1 | 1 | 1 | 1 |
| 1 | 0 | 1 | 0 | 1 | 1 | 1 | 1 |
| 1 | 1 | 0 | 0 | 1 | 1 | 1 | 1 |
| 1 | 1 | 1 | 1 | 1 | 1 | 1 | 1 |

$\therefore \qquad\qquad x + y \cdot z = (x + y) \cdot (x + z) \quad \text{for all } x, y, z \in \text{B.}$

| Input | | | | | | Output | |
|---|---|---|---|---|---|---|---|
| $x$ | $y$ | $z$ | $y + z$ | $x \cdot y$ | $x \cdot z$ | $x \cdot (y + z)$ | $x \cdot y + x \cdot z$ |
| 0 | 0 | 0 | 0 | 0 | 0 | 0 | 0 |
| 0 | 0 | 1 | 1 | 0 | 0 | 0 | 0 |
| 0 | 1 | 0 | 1 | 0 | 0 | 0 | 0 |
| 0 | 1 | 1 | 1 | 0 | 0 | 0 | 0 |
| 1 | 0 | 0 | 0 | 0 | 0 | 0 | 0 |
| 1 | 0 | 1 | 1 | 0 | 1 | 1 | 1 |
| 1 | 1 | 0 | 1 | 1 | 0 | 1 | 1 |
| 1 | 1 | 1 | 1 | 1 | 1 | 1 | 1 |

$\therefore \qquad\qquad x \cdot (y + z) = x \cdot y + x \cdot z \quad \text{for all } x, y, z \in \text{B}$

Hence the distributive laws are satisfied.

Therefore (B, +, ., $'$, 0, 1) is a Boolean algebra and is usually referred to as ***Switching algebra***. Thus, it is possible to simplify the switching circuits using Boolean algebra. The simplification of switching circuit is equivalent to the simplification of the corresponding Boolean function. Boolean algebra thus can be used to design switching circuits for performing the required operations.

The design of a digital system consists of the following steps:

    (i) To convert the given problem into the input-output relationship given by truth tables.

    (ii) Construction of switching function from the truth tables.

    (iii) Simplification of switching functions by the theorems and axioms of Boolean algebra.

    (iv) Realization of simplified switching functions by means of logic-gates.

**Example 35:** Express the following circuit as Boolean function in the switch algebra.

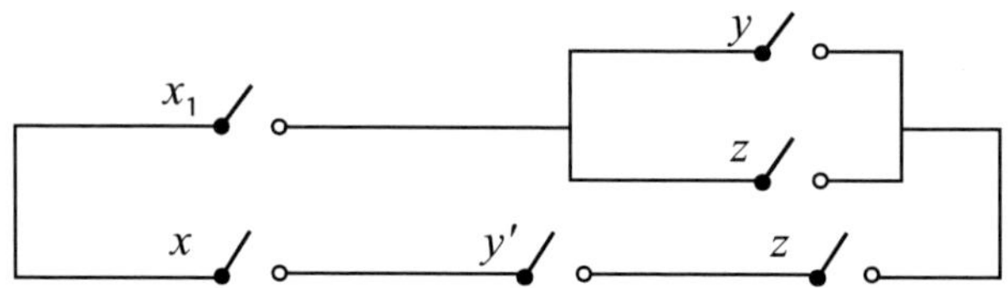

***Solution***

In the given circuit there are two series connections and one parallel connection.

1. The inputs to the first series connection are $x_1$ and $x_2$ and so the output is $x_1 \cdot x_2$
2. The inputs to the second series connection are $x_1'$ and $x_2$ and so the output is $x_1' \cdot x_2$
3. The inputs to the parallel connection are $(x_1 \cdot x_2)$ and $(x_1' \cdot x_2)$ and so the output is $x_1 \cdot x_2 + x_1' \cdot x_2$.

Hence the required Boolean function is $\quad x_1 \cdot x_2 + x_1' \cdot x_2$.

**Example 36:** Write the Boolean function for the following circuit.

***Solution***

In the given circuit there are two series connections and two parallel connections.

1. The inputs to the first parallel connection are $y$ and $z$ and so the output is $(y + z)$.
2. The inputs to the first series connection are $x'$ and $(y + z)$ and so the output is $x' \cdot (y + z)$.
3. The inputs to the second series connection are $x$, $y'$ and $z$ and so the output is $x \cdot y' \cdot z$.
4. The inputs to the second parallel connection are $x' \cdot (y + z)$ and $x \cdot y' \cdot z$ and so the output is $x' \cdot (y + z) + x \cdot y' \cdot z$.

Hence the required Boolean expression is $\quad x' \cdot (y + z) + x \cdot y' \cdot z$.

**Example 37:** Construct the circuit corresponding to the Boolean function $x \cdot y + x \cdot y' + x' \cdot y$ in the switching algebra.

***Solution***

Let $f(x, y) = x \cdot y + xy' + x'y$ be the given Boolean function.

In, the function $f(x, y)$, the circuits corresponding to $x \cdot y, x \cdot y'$ and $x' \cdot y$ are connected in parallel.

Also, the circuits corresponding to $x$ and $y$, $x$ and $y'$, $x'$ and $y$ are connected in series respectively.

$\therefore$ The circuit of the given Boolean function is given by the following figure.

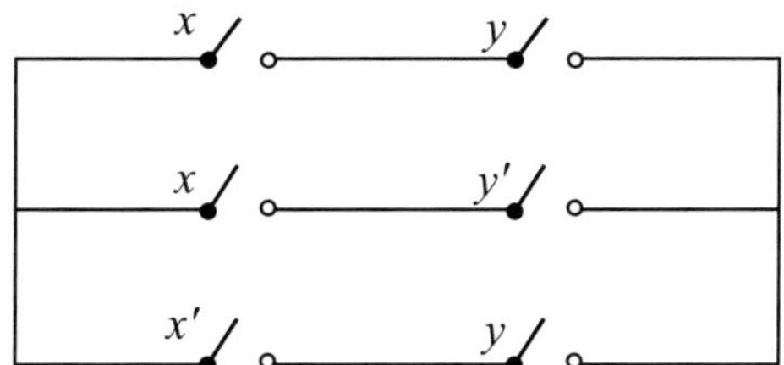

**Example 38:** Write the Boolean function for the following circuit. Simplify the function and construct the switching circuit for the simplified Boolean function.

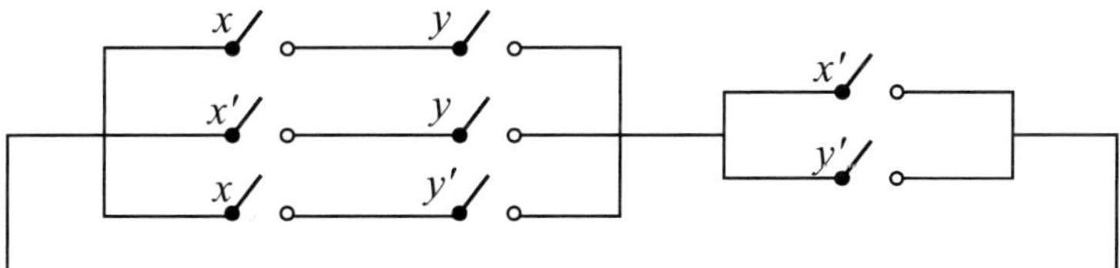

***Solution***

The Boolean expression for the given circuit can be obtained, proceeding as in previous examples, which is

$$
\begin{aligned}
f(x, y) &= (x \cdot y + x' \cdot y + x \cdot y') \cdot (x' + y') \\
&= [(x + x') \cdot y + x \cdot y'] \cdot (x' + y') \\
&= (y + x \cdot y') \cdot (x' + y') \\
&= (y + y') \cdot (y + x) \cdot (x' + y') \\
&= (y + x) \cdot (x' + y') \\
&= y \cdot x' + y \cdot y' + x \cdot x' + x + y' \\
&= x' \cdot y + x \cdot y' + 0 + 0 \\
&= x' \cdot y + x \cdot y'
\end{aligned}
$$

The circuit representing the simplified Boolean function is:

**Example 39:** Write the Boolean function for the following circuit. Simplify the function and construct the switching circuit for the simplified Boolean function.

***Solution***

The Boolean function for the given circuit can be obtained, proceeding as in previous examples, which is

$$f(x, y, z) = x \cdot y' + x \cdot y \cdot z + x' \cdot z$$
$$= x \cdot (y' + y \cdot z) + x' \cdot z$$
$$= x \cdot (y + y') \cdot (y' + z) + x' \cdot z$$
$$= x \cdot (y' + z) + x' \cdot z$$
$$= x \cdot y' + x \cdot z + x' \cdot z$$
$$= x \cdot y' + z \cdot (x \cdot x')$$
$$= x \cdot y' + z$$

The circuit representing the simplified Boolean function is

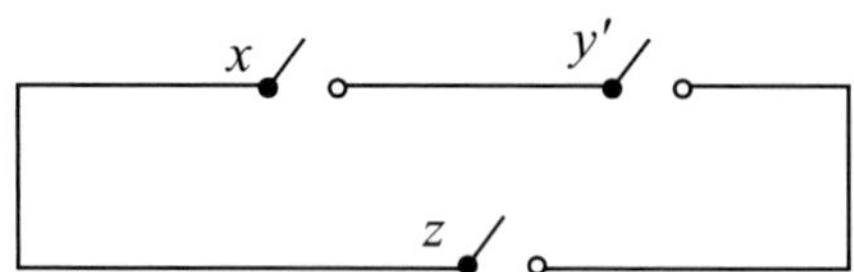

**Example 40:** Write the Boolean function for the following circuit. Simplify the function and construct the switching circuit for the simplified Boolean function.

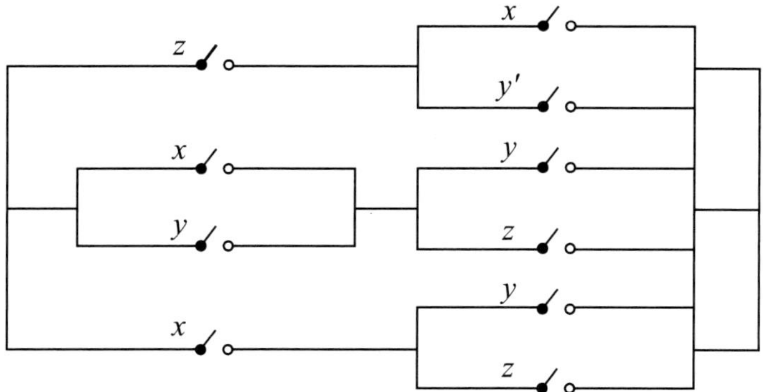

### *Solution*

The Boolean function for the given circuit can be obtained, proceeding as in previous examples, which is

$$f(x, y, z) = z \cdot (x + y) + (x + y) \cdot (y + z) + x \cdot (y + z)$$
$$= z \cdot (x + y) + y + x \cdot z + x \cdot (y + z)$$
$$= z \cdot x + z \cdot y + y + x \cdot z + x \cdot y + x \cdot z$$
$$= z \cdot x + z \cdot y + y + x \cdot y$$
$$= x \cdot z + y \cdot (1 + z) + x \cdot y$$
$$= x \cdot z + y + x \cdot y$$
$$= x \cdot z + y \cdot (1 + x)$$
$$= x \cdot z + y$$

The circuit representing the simplified circuit is

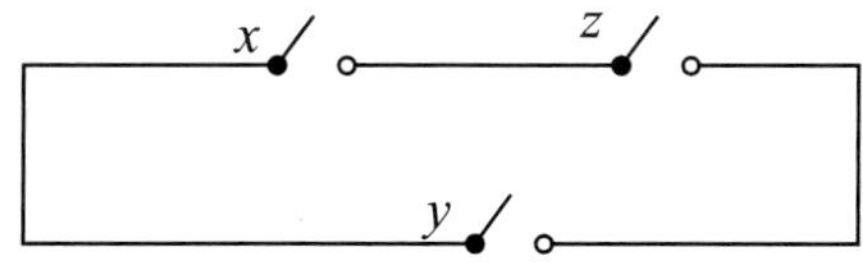

**Example 41:** Write the Boolean expression for the following circuit. Simplify the expression and construct the switching circuit for the simplified expression.

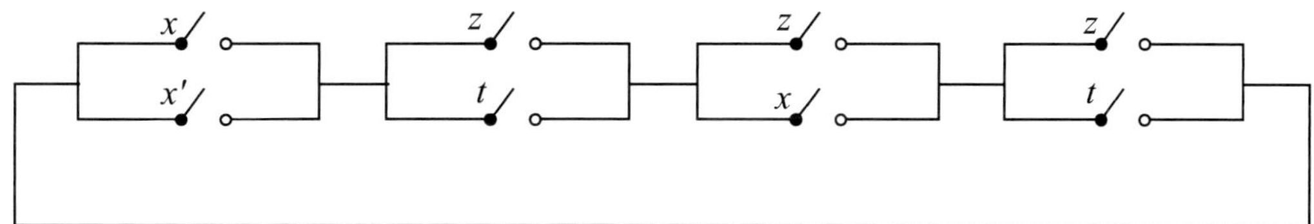

### *Solution*

The Boolean expression for the given circuit can be obtained, proceeding as in previous examples, which is

$$f(x, y, z, t) = (x + x') \cdot (y + t) \cdot (z + x) \cdot (z + t)$$
$$= 1 \cdot (y + t) \cdot (z + x) \cdot (z + t)$$
$$= (y + t) \cdot (z + t \cdot x)$$
$$= y \cdot z + y \cdot t \cdot x + t \cdot z + t \cdot t \cdot x$$
$$= y \cdot z + y \cdot t \cdot x + t \cdot z + t \cdot x$$
$$= z \cdot (y + t) + t \cdot x \cdot (y + t)$$
$$= z \cdot (y + t) + t \cdot x$$

The circuit representing the simplified circuit is

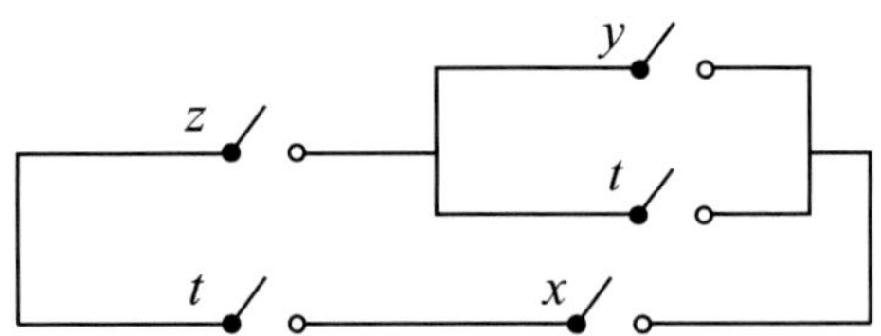

### Exercise 9.5

1. Write the Boolean function for the following circuit in switching algebra.

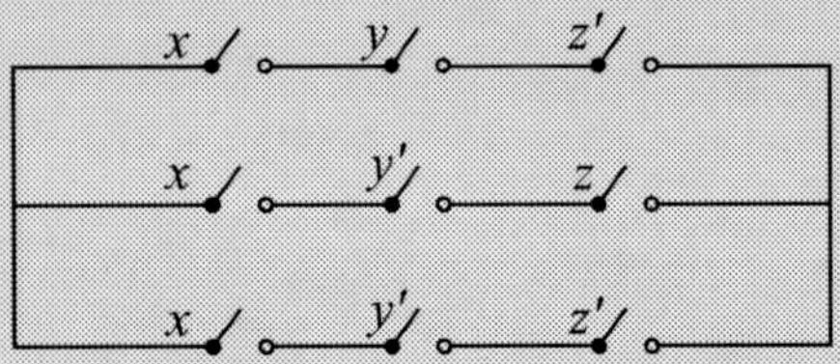

2. Construct the circuit corresponding to the Boolean function $(x \cdot y + z) \cdot (x' + y \cdot z') \cdot (x' + z)$ in the switching algebra.

3. Construct the circuit corresponding to the Boolean function $(x \cdot y \cdot z') + x'(y + z')$ in the switching algebra.

4. Construct the circuit corresponding to the Boolean function $x \cdot y' \cdot z + (y + z) \cdot x'$ in the switching algebra.

5. Simplify the circuit given below:

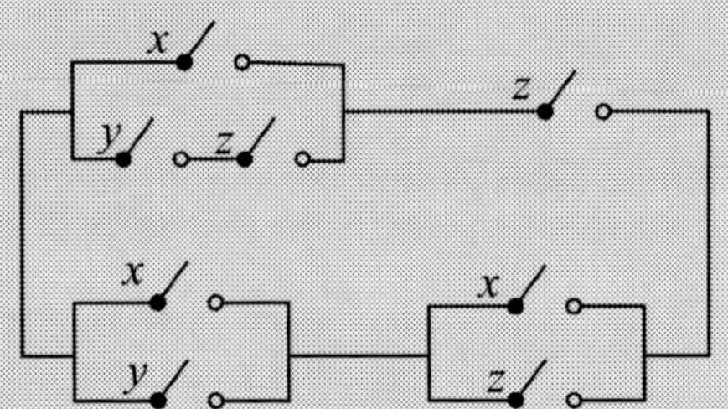

6. Simplify the circuit given below:

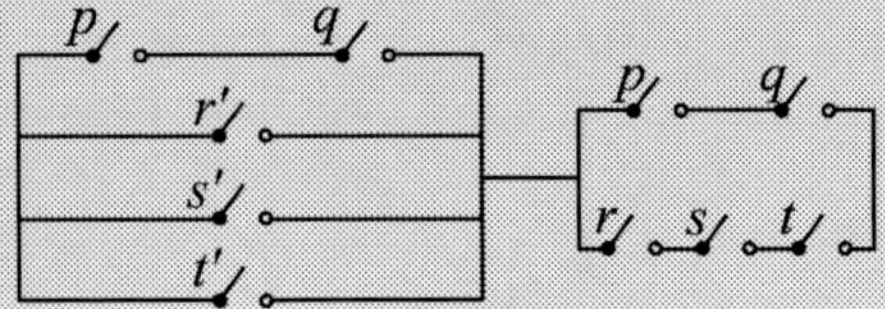

7. Simplify the circuit given below:

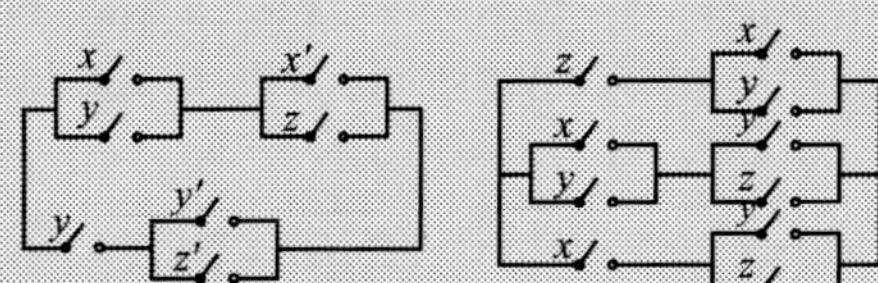

***Write the Boolean function for the following circuits. Simplify the function and construct the switching circuit for the simplified Boolean function.***

8.

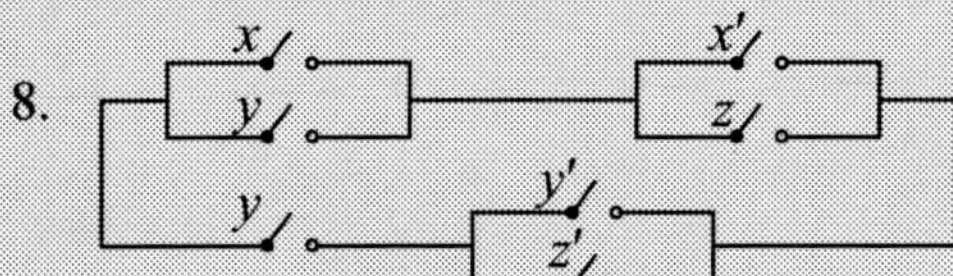

9.

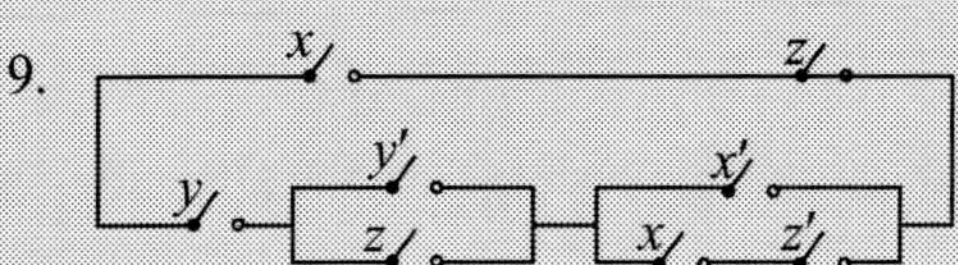

10.

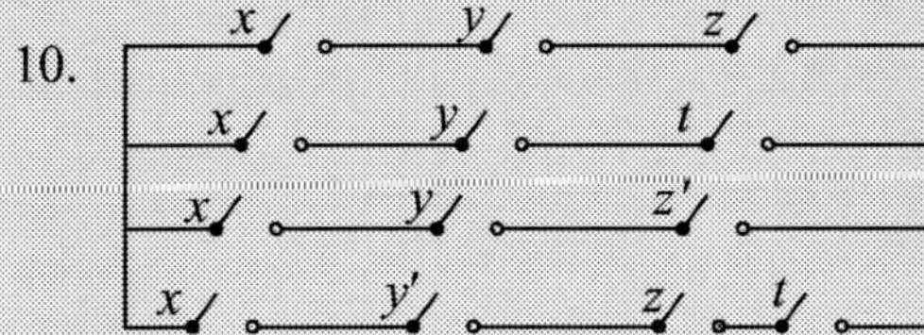

**Answers**

1. $x(y' + z')$

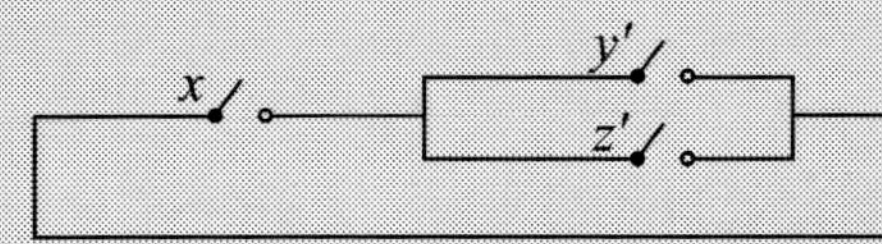

2.

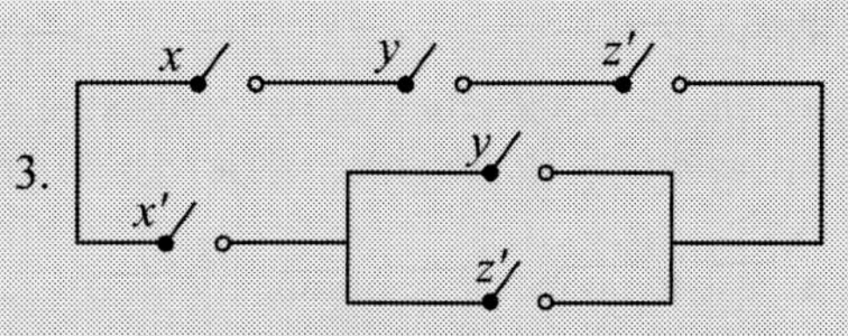

3.

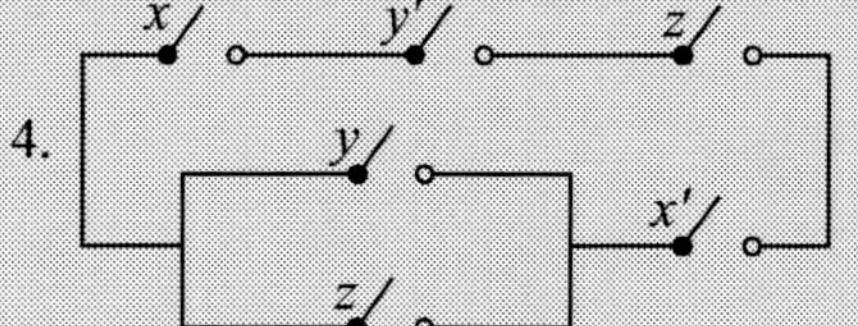

4.

5. $x + y = z$;

6. $p \cdot q$;

7. $x \cdot z + y$;

8. $x \cdot z + y$;

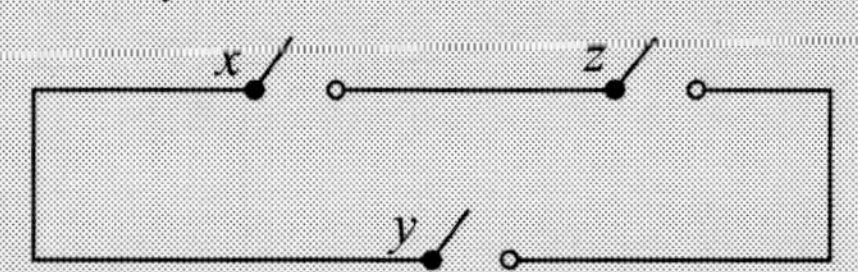

9. $z \cdot (x + y)$;

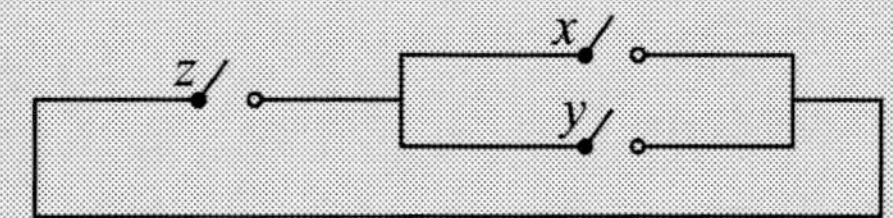

10. $x \cdot (y + z \cdot t)$;

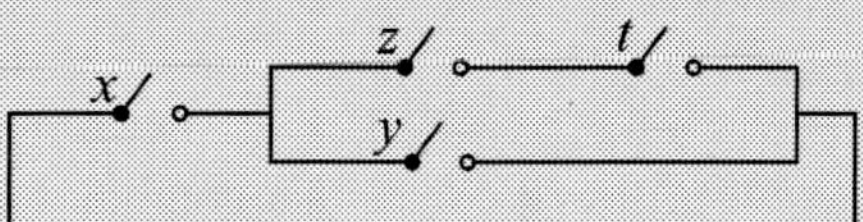

## 9.20 LOGIC GATES

A digital circuit which either allows a signal to pass through it or stops it, is called a gate. These gates are said to be **logic gates** if they allow the signal to pass when some logical conditions are satisfied.

The basic logic gates are of three types:

**1. OR gate     2. AND gate     3. NOT gate**

### 9.20.1 OR Gate

An OR gate is the Boolean function defined by

$$f(a, b) = a + b, \quad a, b \in \{0, 1\}.$$

The symbol (+) is referred to as OR in Boolean algebra. The Boolean expression $x = a + b$ indicates that $x$ is equal to $a$ OR $b$.

The **OR gate** is a device that combines $a$ with $b$ to give as the result. Thus the OR gate is a device which has two or more inputs and only one output.

The logic symbol of OR gate with inputs $a$ and $b$ and output $x$ is shown in above figure.

The input / output table for OR gate is given below:

| Input | | Output |
|---|---|---|
| $a$ | $b$ | $a + b$ |
| 1 | 1 | 1 |
| 1 | 0 | 1 |
| 0 | 1 | 1 |
| 0 | 0 | 0 |

### 9.20.2 AND Gate

An AND gate is the Boolean function defined by

$$f(a, b) = a \cdot b, \quad a, b \in \{0, 1\}.$$

The sign '$\cdot$' is referred to as AND in Boolean algebra. The Boolean expression $x = a \cdot b$ indicates that $x$ is equal to $a$ AND $b$.

The **AND gate** is a device that combines $a$ and $b$ to give $x$ as the result. Thus the AND gate is a device which has two or more inputs and only one output.

The logic symbol of AND gate with inputs $a$ and $b$ and output $x$ is shown in the below figure.

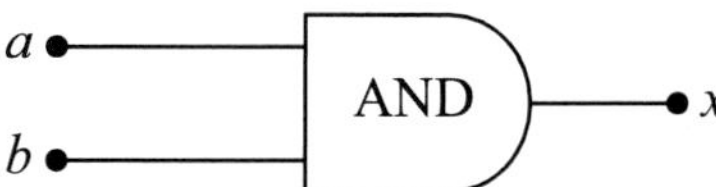

The input / output table for AND gate is given below:

| Input | | Output |
|:---:|:---:|:---:|
| $a$ | $b$ | $a \cdot b$ |
| 1 | 1 | 1 |
| 1 | 0 | 0 |
| 0 | 1 | 0 |
| 0 | 0 | 0 |

### 9.20.3 NOT Gate

A NOT gate is the Boolean function defined by $f(a) = a'$, $a \in \{0, 1\}$. The symbol '$'$' is referred to as NOT in Boolean algebra. The Boolean expression $a' = x$ indicates that $x$ equals NOT $a$.

The **NOT gate** is a device which inverts the input. If input is 1, then output is 0 and if input is 0, then output is 1. The NOT gate is one input and one output device. The logic symbol of NOT gate is shown in the figure below.

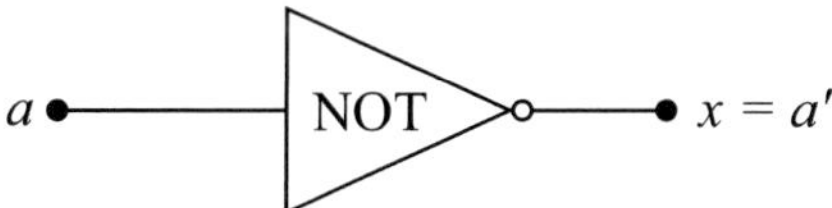

The input-output table for NOT gate is given below:

| Input | Output |
|:---:|:---:|
| $a$ | $a'$ |
| 0 | 1 |
| 1 | 0 |

**Note.** Performing operation OR is the same as taking the maximum of the bits $x_1$ and $x_2$. Similarly performing operation AND is the same as taking the minimum of the bits $x_1$ and $x_2$.

**Example 42:** Find the value of Boolean expression $[(a \cdot b') + c] \cdot a'$, given that

    (i) $a = 1$, $b = 0$, $c = 0$
    (ii) $a = 0$, $b = 1$, $c = 0$
    (iii) $a = 0$, $b = 0$, $c = 1$.

***Solution***

(i) Here    $a = 1$,  $b = 0$,  $c = 0$        *[Given]*

$\therefore \quad a' = 0, \quad b' = 0, \quad c' = 1$

$\Rightarrow$
$$a \cdot b' = 1 \cdot 1 = 1$$
$$(a \cdot b') + c = 1 + 0 = 1$$
$$[a \cdot b' + c] \cdot a' = 1 \cdot 0 = 0$$

Hence the value of Boolean expression $[a \cdot b' + c] \cdot a'$ is **0**.

(ii) Here $\quad a = 0, \quad b = 1, \quad c = 0$ $\hfill$ [*Given*]

$\therefore \quad a' = 1, \quad b' = 0, \quad c' = 1$

$\Rightarrow$
$$a \cdot b' = 0 \cdot 0 = 0$$
$$(a \cdot b') + c = 0 + 0 = 0$$
$$[a \cdot b' + c] \cdot a' = 0 \cdot 1 = 0$$

Hence the value of Boolean expression $[(a \cdot b') + c] \cdot a'$ is **0**.

(iii) Here $\quad a = 0, \quad b = 0, \quad c = 1$

$\therefore \quad a' = 1, \quad b' = 1, \quad c' = 0$

$\Rightarrow$
$$a \cdot b' = 0 \cdot 1 = 0$$
$$(a \cdot b') + c = 0 + 1 = 1$$
$$[a \cdot b' + c] \cdot a' = 1 \cdot 1 = 1$$

Hence the value of Boolean expression $[(a \cdot b') + c] \cdot a'$ is **1**.

**Example 43:** Find the output for a given input from the following circuit:

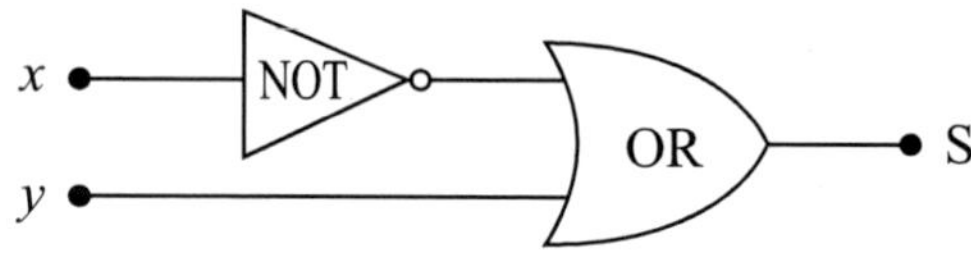


(i) Inputs: $x = 0, y = 1$.
(ii) Inputs: $x = 1, y = 0$.

### Solution

(i) In the given circuit, the NOT gate changes the input $x = 0$ to 1 and so both the inputs to the OR gate are 1 and 1. Hence the output is S = 1 as shown in figure below.

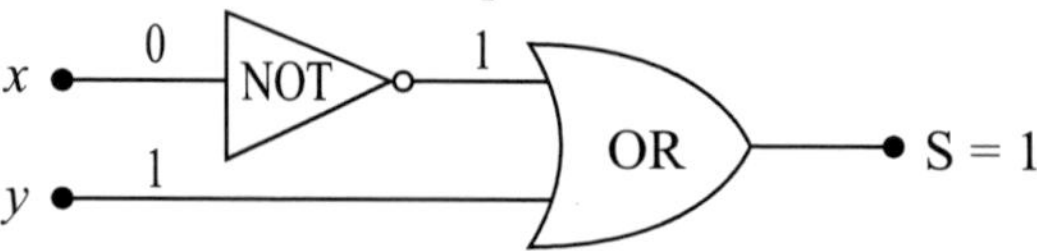


(ii) In the given circuit the NOT gate changes the input $x = 1$ to 0 and so both the inputs to the OR gate are 0 and 0. Hence the output is S = 0 as shown in figure below.

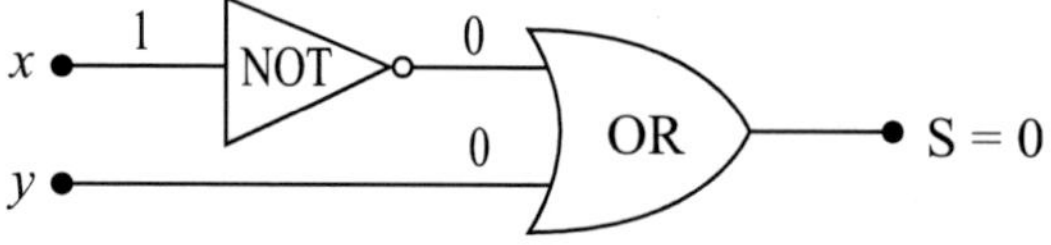

**Example 44:** Find the output for a given input from the following circuit:

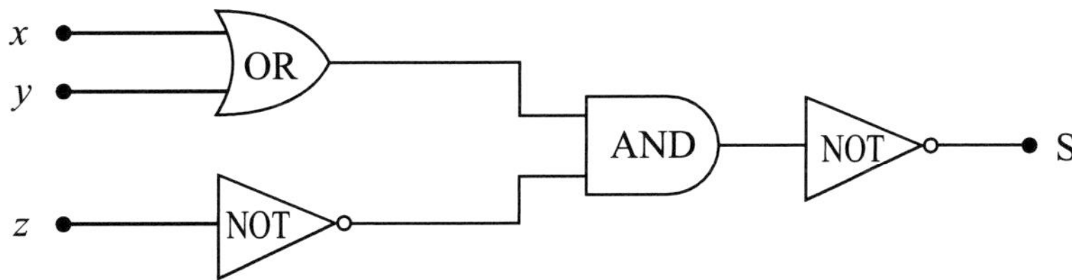

(i)  Inputs: $x = 1$, $y = 0$ and $z = 1$.
(ii)  Inputs: $x = 0$, $y = 1$ and $z = 0$.

*Solution*

(i)  In the given circuit for the inputs $x = 1$, $y = 0$, the output of the OR gate is 1 which NOT gate changes the input $z = 1$ to 0.

Now for the inputs 1 and 0, the output of the AND gate is 0. The NOT gate further changes the input 0 to 1. Thus the output of the circuit is S = 1 as shown in figure below.

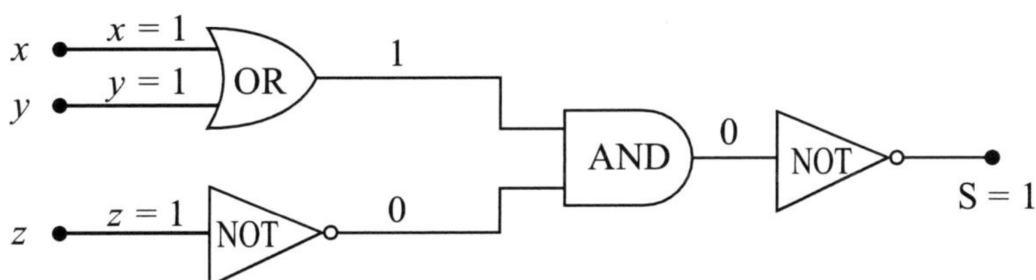

(ii)  In the given circuit for the inputs $x = 0$, $y = 1$ the output of the OR gate is 1 while NOT gate changes the input $z = 0$ to 1.

Now for the inputs 1 and 1, the output of the AND gate is 1. The NOT gate further changes the input 1 to 0. Thus the output of the circuit is S = 0 as shown in figure below.

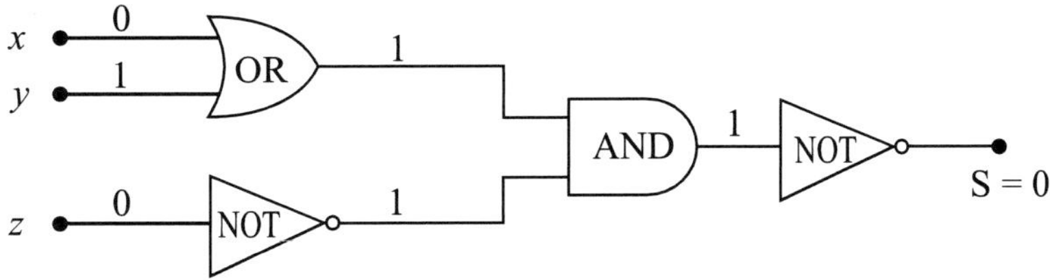

**Example 45:** Construct the input-output table from the following circuit:

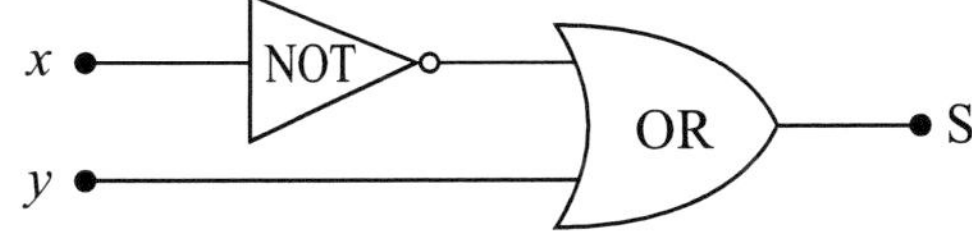

### *Solution*

There are 4 possible cases:

**Case I.** Inputs $x = 1$ and $y = 1$:

In the given circuit, the NOT gate changes the input $x = 1$ to 0. Therefore the two inputs to the OR gate are 0 and 1.

Hence the output is S = 1 as shown in figure below.

**Case II.** Inputs $x = 1$ and $y = 0$:

In the given circuit, the NOT gate changes the input $x = 1$ to 0. Therefore, the two inputs the OR gate are 0 and 0. Hence the output of the circuit is S = 0 as shown in figure below.

**Case III.** Inputs $x = 1$ and $y = 1$:

In the given circuit, the NOT gate changes the input $x = 0$ to 1. Therefore, both the inputs to the OR gate are 1 and 1. Hence the output is S = 1 as shown in figure below.

**Case IV.** Inputs $x = 0$ and $y = 0$:

In the given circuit, the NOT gate changes the input $x = 0$ to 1. Therefore, the two inputs to the OR gate are 1 and 0. Hence the output is S = 1 as shown in figure below.

Hence the required input-output table is as given below:

| Input | | Output |
|---|---|---|
| $x$ | $y$ | S |
| 1 | 1 | 1 |
| 1 | 0 | 0 |
| 0 | 1 | 1 |
| 0 | 0 | 1 |

**Example 46:** Show that the following circuits are equivalent*:

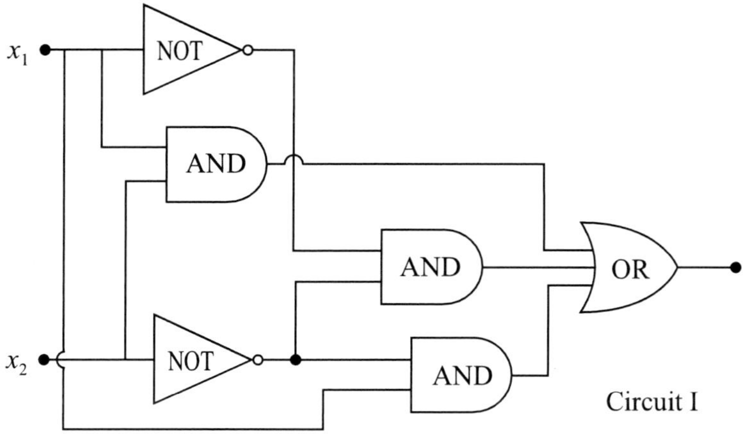


* The circuits having the identical input-output tables are said to be **equivalent.**

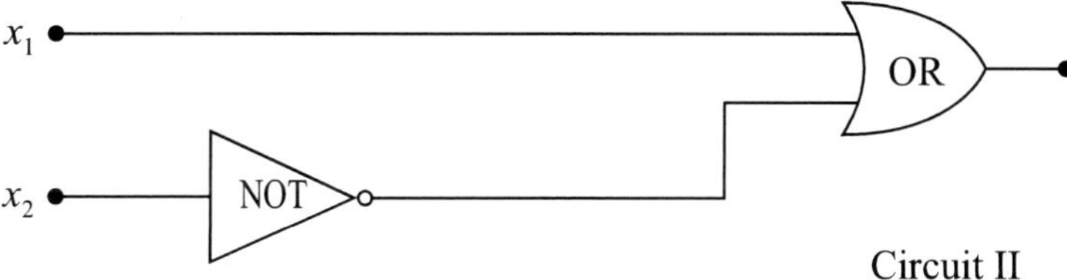


### Solution

The input-output tables for the two circuits can be obtained as in previous example which are as under:

| Input | | Output | | Input | | Output |
|---|---|---|---|---|---|---|
| $x_1$ | $x_2$ | $f(x_1, x_2)$ | | $x_1$ | $x_2$ | $f(x_1, x_2)$ |
| 0 | 0 | 1 | | 0 | 0 | 1 |
| 0 | 1 | 0 | | 0 | 1 | 0 |
| 1 | 0 | 1 | | 1 | 0 | 1 |
| 1 | 1 | 1 | | 1 | 1 | 1 |

|  Circuit - I  |  Circuit - II  |

Since the input-output tables are identical, hence the two circuits are equivalent.

**Example 47:** Find the combinatorial circuit corresponding to the Boolean expression $\{x_1 + (x_2' \cdot x_3)\} + x_3$ and write the input-output table for the circuit obtained.

### *Solution*

The given Boolean expression is $\{x_1 + (x_2' \cdot x_3)\} + x_3$           (1)

    Here there is one NOT gate corresponding to input $x_2$.

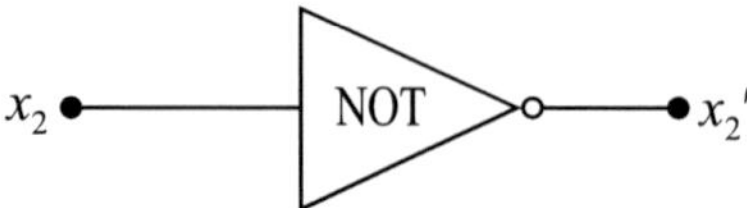


In the innermost parenthesis, the expression is $x_2' \cdot x_3$. This is AND of $x_2'$ and $x_3$. Therefore we add an AND gate to the above circuit with inputs $x_2'$ and $x_3$.

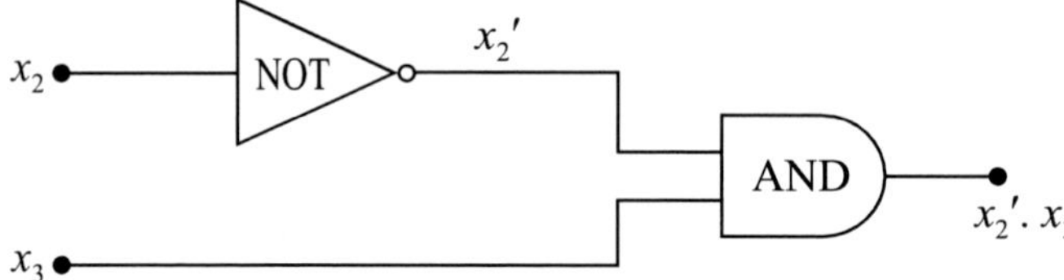


    Now for expression $x_1 + (x_2' \cdot x_3)$, which is an OR of $x_1$ and $x_2' \cdot x_3$ we add an OR gate to the above circuit with inputs $x_1$ and $x_2' \cdot x_3$

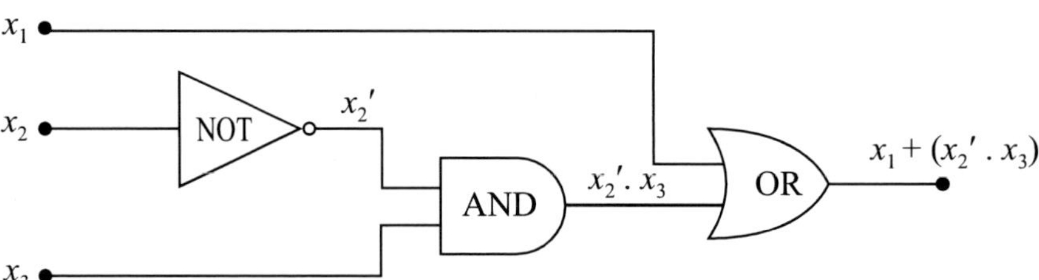


    Finally for the given expression $\{x_1 + (x_2' \cdot x_3)\} + x_3$, which is an OR of $x_1 + (x_2' \cdot x_3)$ and $x_3$, we add an OR gate to the above circuit with inputs $x_1 + (x_2 \cdot x_3)$ and $x_3$.

    Hence the required combinatorial circuit for the given expression is as shown in figure below.

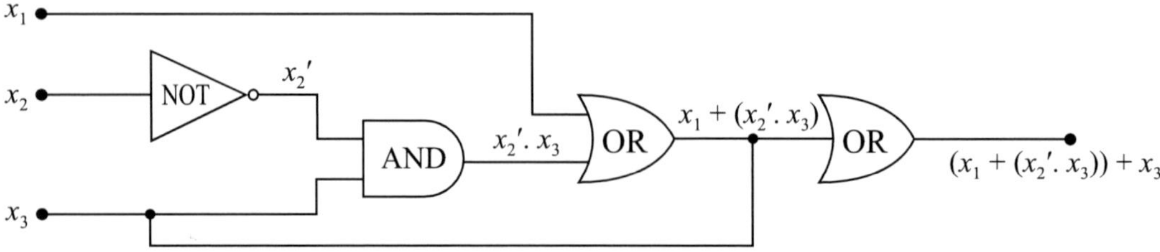

Now,

$$f(x_1, x_2, x_3) = \{x_1 + (x_2' \cdot x_3)\} + x_3$$
$$\therefore \ f(1, 1, 1) = \{1 + (1' \cdot 1)\} + 1$$
$$= \{1 + (0 \cdot 1)\} + 1 = (1 + 0) + 1 = 1 + 1 = 1$$
$$f(1, 1, 0) = \{1 + (1' \cdot 0)\} + 0 = \{1 + (0 \cdot 0)\} + 0$$
$$= (1 + 0) + 0 = 1 + 0 = 1$$
$$f(1, 0, 1) = \{1 + (0' \cdot 1)\} + 1 = \{1 + (1 \cdot 1)\} + 1$$
$$= (1 + 1) + 1 = 1 + 1 = 1$$
$$f(1, 0, 0) = \{1 + (0' \cdot 0)\} + 0 = \{1 + (1 \cdot 0)\} + 0$$
$$= (1 + 0) + 0 = 1 + 0 = 1$$
$$f(0, 1, 1) = \{0 + (1' \cdot 1)\} + 1 = \{0 + (0 \cdot 1)\} + 1$$
$$= (0 + 0) + 1 = 0 + 1 = 1$$
$$f(0, 1, 0) = \{0 + (1' \cdot 0)\} + 0 = \{0 + (0 \cdot 0)\} + 0$$
$$= (0 + 0) + 0 = 0 + 0 = 0$$
$$f(0, 0, 1) = \{0 + (0' \cdot 1)\} + 1 = \{0 + (1 \cdot 1)\} + 1$$
$$= (0 + 1) + 1 = 1 + 1 = 1$$
$$f(0, 0, 0) = \{0 + (0' \cdot 0)\} + 0 = \{0 + (1 \cdot 0)\} + 0$$
$$= (0 + 0) + 0 = 0 + 0 = 0$$

Hence the required input-output table is:

| Input | | | Output |
|:---:|:---:|:---:|:---:|
| $x_1$ | $x_2$ | $x_3$ | S |
| 1 | 1 | 1 | 1 |
| 1 | 1 | 0 | 1 |
| 1 | 0 | 1 | 1 |
| 1 | 0 | 0 | 1 |
| 0 | 1 | 1 | 1 |
| 0 | 1 | 0 | 0 |
| 0 | 0 | 1 | 1 |
| 0 | 0 | 0 | 0 |

**Example 48:** Construct a combinatorial circuit from the following input/output table:

| Input | | Output |
|---|---|---|
| $x_1$ | $x_2$ | S |
| 1 | 1 | 1 |
| 1 | 0 | 0 |
| 0 | 1 | 1 |
| 0 | 0 | 0 |

***Solution***

First we construct the Boolean function from the given input-output table.

1. In the output column we mark all those rows having output 1.
2. In the above specified rows, replace 1 by $x_1$ or $x_2$ and 0 by $x_1'$ or $x_2'$ according to their respective positions.

Thus the following combinations are formed:

$$x_1 \cdot x_2$$
$$x_1' \cdot x_2$$

3. By applying OR to the above combinations, the Boolean expression is

$$x_1 \cdot x_2 + x_1' \cdot x_2 \tag{1}$$

*Construction of circuit:*

Here there is one NOT gate corresponding to input $x_1$.

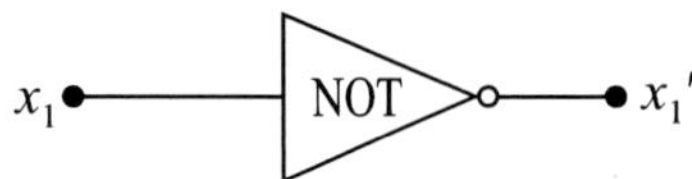


The expression $x_1' \cdot x_2$ is the AND of $x_1'$ and $x_2$.

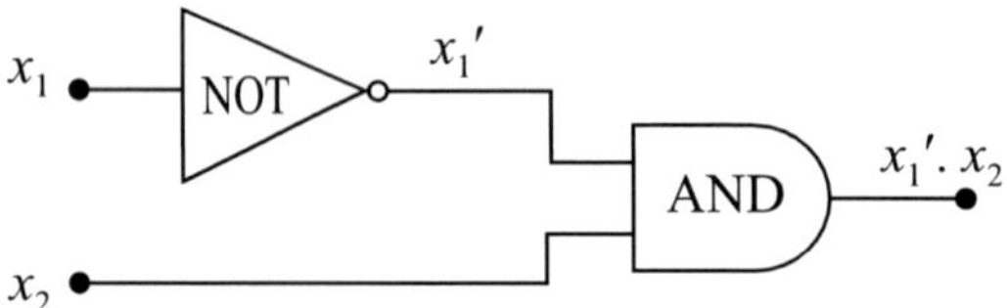


Therefore we add an AND gate to the above circuit with inputs $x_1'$ and $x_2$.

The expression $x_1 \cdot x_2$ is the AND of $x_1$ and $x_2$

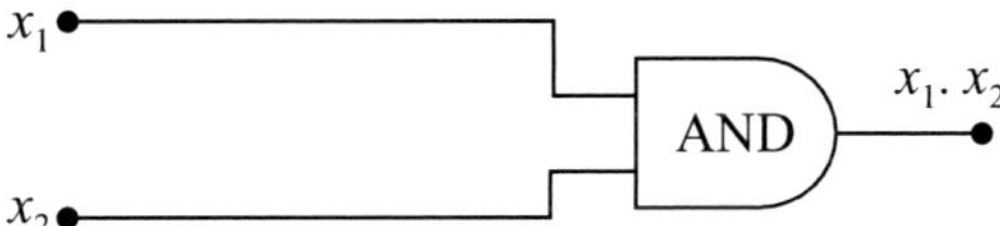

Now for the expression $x_1 \cdot x_2 + x_1' \cdot x_2$ which is an OR of $x_1 \cdot x_2$ and $x_1' \cdot x_2$, we add an OR gate to the above circuits with inputs $x_1 \cdot x_2$ and $x_1' \cdot x_2$.

Hence the required combinatorial circuit is as shown in figure below.

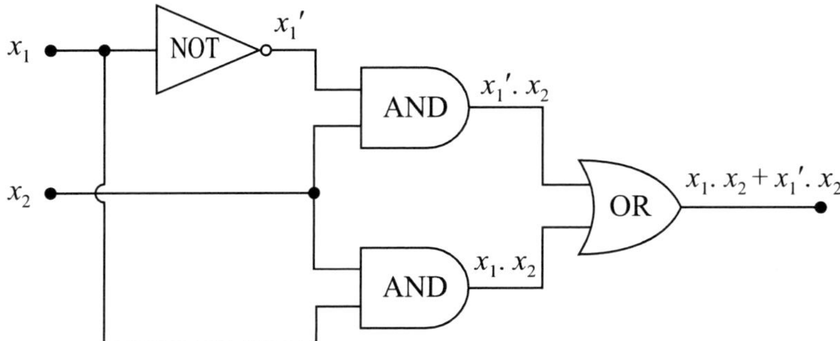

**Example 49:** Express the following circuit as a Boolean algebra of switching circuits.

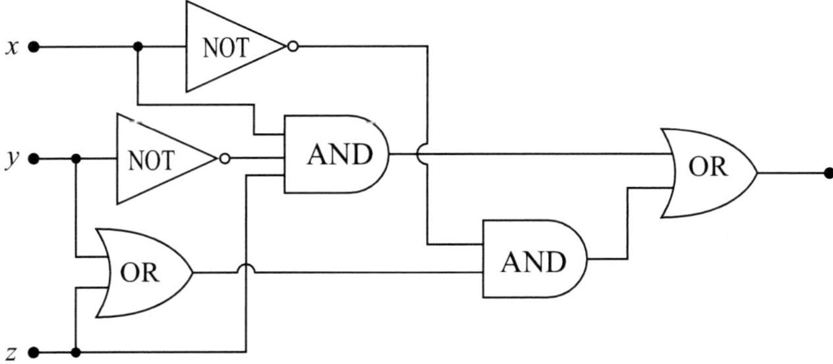

***Solution***

In the given circuit there are two OR gates, two AND gates and two NOT gates

1. The inputs to the first OR gate are $y$ and $z$ and so output is $(y + z)$.
2. The inputs to the first AND gate are $x$, $y'$ and $z$ and so output is $(x \cdot y' \cdot z)$.
3. The inputs to the second AND gate are $x'$ and $(y + z)$ and so the output is $x' \cdot (y + z)$.
4. The inputs to the last OR gate are $x' \cdot (y + z)$ and $(x \cdot y' \cdot z)$ and so output is

$$x' \cdot (y + z) + (x \cdot y' \cdot z)$$

Hence the required Boolean function is $f(x, y, z) = x' \cdot (y + z) + (x \cdot y' \cdot z)$.

**Example 50:** Simplify the combinatorial circuit given below:

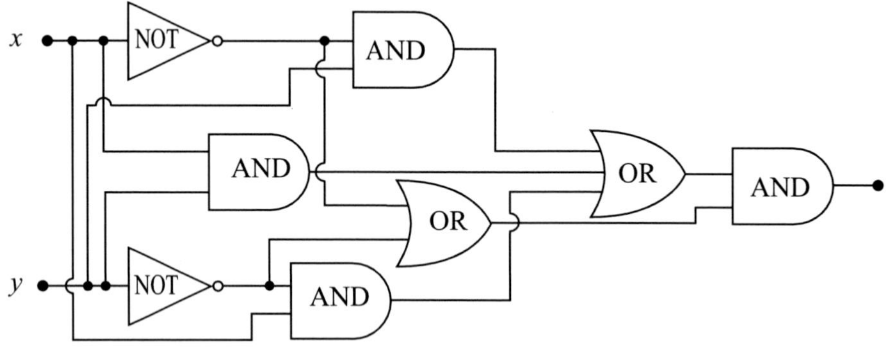

### Solution

Let $f(x, y)$ be the Boolean function of the given circuit in the Boolean algebra of switching circuits. Proceeding as in previous example, the equivalent Boolean function is

$$
\begin{aligned}
f(x, y) &= (x \cdot y + x' \cdot y + x \cdot y') \cdot (x' + y') \\
&= [y \cdot (x + x') + x \cdot y'] \cdot (x' + y') \\
&= (y + x \cdot y') \cdot (x' + y') && [\because \quad a + a' = 1] \\
&= y \cdot x' + y \cdot y' + x \cdot y' \cdot x' + x \cdot y' \cdot y' \\
&= x' \cdot y + 0 + 0 + x \cdot y' && [\because \quad a \cdot a' = \text{and } a \cdot a = a] \\
&= x' \cdot y + x \cdot y'
\end{aligned}
$$

The simplified circuit is as shown in figure below.

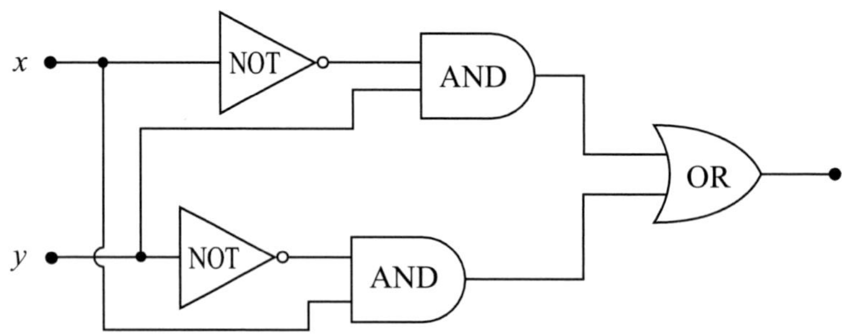

**Example 51:** Simplify the combinatorial circuit given below:

### Solution

Let $f(x, y, z)$ be the Boolean function of the given circuit in the Boolean algebra of switching circuits. Proceeding as in example 13, the equivalent Boolean function is

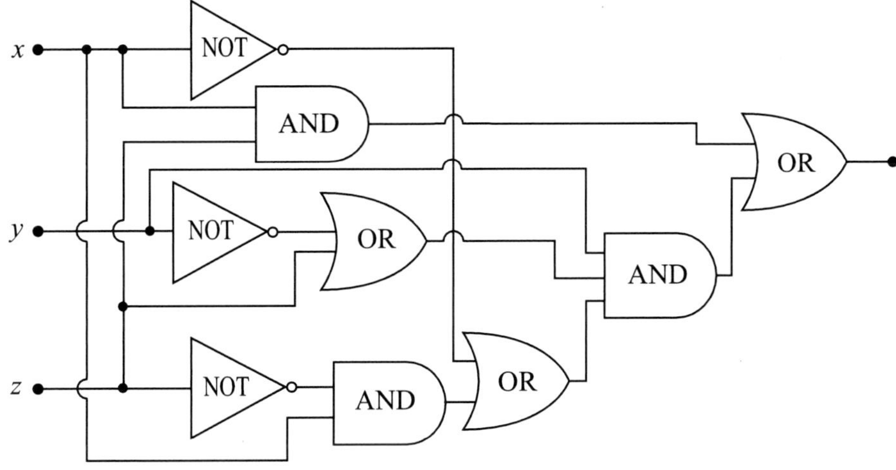

$$f(x, y, z) = x \cdot z + y \cdot (y' + z) \cdot (x' + x \cdot z')$$
$$= x \cdot z + (y \cdot y' + y \cdot z) \cdot (x' + x) \cdot (x' + z') \quad [\textit{Distributive law}]$$
$$= x \cdot z + (0 + y \cdot z) \cdot 1 \cdot (x' + z') \quad\quad [\because \quad a + a' = 1]$$
$$= x \cdot z + y \cdot z \cdot (x' + z')$$
$$= x \cdot z + y \cdot z \cdot x' + y \cdot z \cdot z'$$
$$= x \cdot z + y \cdot z \cdot x' + 0$$
$$= z \cdot (x + y\,x')$$
$$= z \cdot (x + y)\,(x + x')$$
$$= z \cdot (x \mid y)$$

The simplified circuit is as shown in figure below.

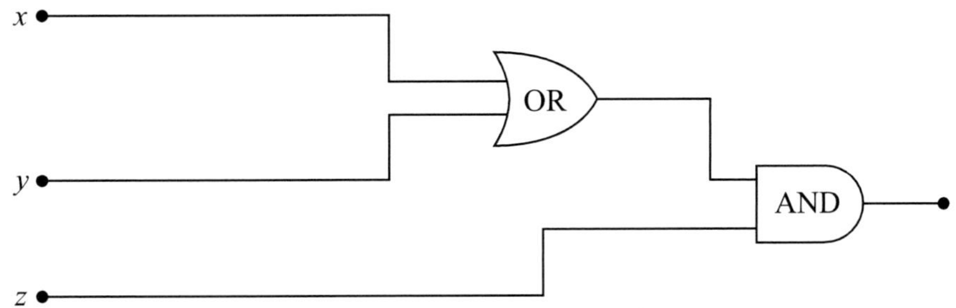

## Exercise 9.6

1. Construct an input-output table for the Boolean functions defined by:

   (i) $f(x_1, x_2) = x_1 \cdot x_2'$
   (ii) $f(x_1, x_2, x_3) = x_1 \cdot x_2' + x_2 \cdot x_3'$
   (iii) $f(a, b, c) = (a \cdot b') + c$
   (iv) $f(a, b, c) = (a \cdot b') \cdot c$

2. Show that the function $f$ given by the following table is a Boolean function:

   | Input | | Output |
   |---|---|---|
   | $x_1$ | $x_2$ | S |
   | 1 | 1 | 1 |
   | 1 | 0 | 0 |
   | 0 | 1 | 0 |
   | 0 | 0 | 0 |

3. Write the Boolean function given by the following input/output table:

   | Input | | | Output |
   |---|---|---|---|
   | $x_1$ | $x_2$ | $x_3$ | S |
   | 1 | 1 | 1 | 1 |
   | 1 | 1 | 0 | 1 |
   | 1 | 0 | 1 | 1 |
   | 1 | 0 | 0 | 0 |
   | 0 | 1 | 1 | 0 |
   | 0 | 1 | 0 | 0 |
   | 0 | 0 | 1 | 0 |
   | 0 | 0 | 0 | 1 |

4. Find the output for a given input from the following circuits:

(i) 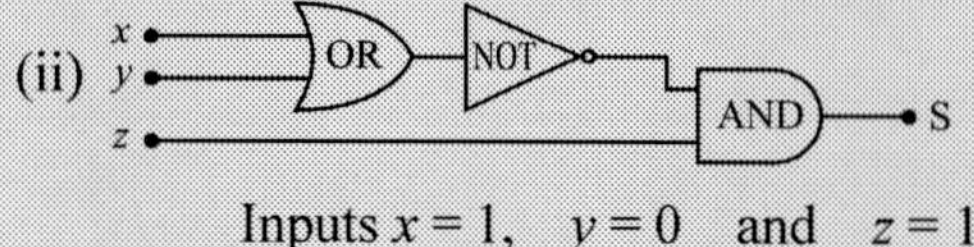

Inputs $x = 0$, $y = 1$

(ii) 

Inputs $x = 1$, $y = 0$ and $z = 1$

5. Construct the input-output table for the circuits given in Q4 (i) and Q4 (ii).

6. Show that the given circuits are equivalent

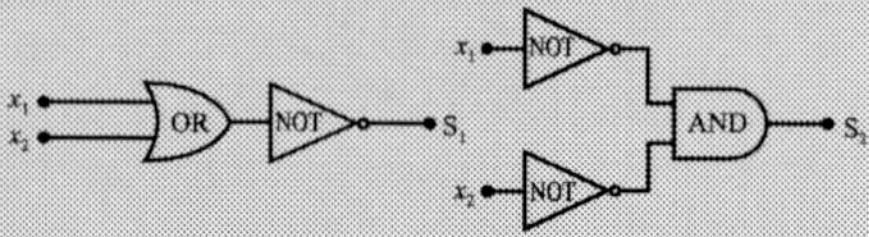

7. Find the combinatorial circuits corresponding to the following Boolean expression:

   (i) $f(x, y) = x + x \cdot y$
   (ii) $(x \cdot y \cdot z') + x' \cdot (y \cdot z')$
   (iii) $(x \cdot y \cdot z) \cdot (x' + y \cdot z') \cdot (x' + z')$
   (iv) $x \cdot y + x' \cdot (x + y + y')$
   (v) $[(x_1 \cdot x_2) \cdot (x_1 \cdot x_2')] \cdot x_2$

8. Construct a combinatorial circuit from the following input/output table:

   | Input | | Output |
   |---|---|---|
   | $x_1$ | $x_2$ | S |
   | 1 | 1 | 1 |
   | 1 | 0 | 0 |
   | 0 | 1 | 0 |
   | 0 | 0 | 1 |

*Express the following circuits as a Boolean function of the Boolean algebra of switching circuits:*

9. 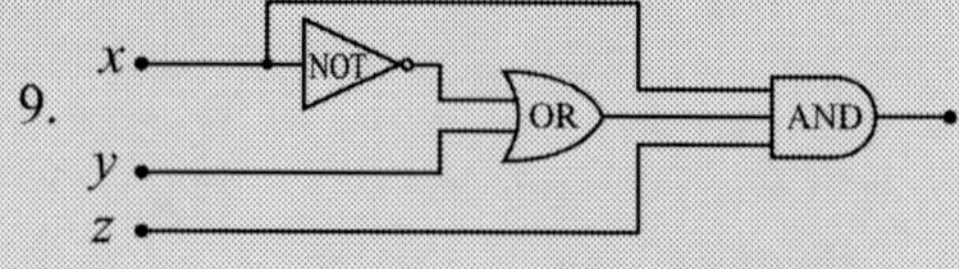

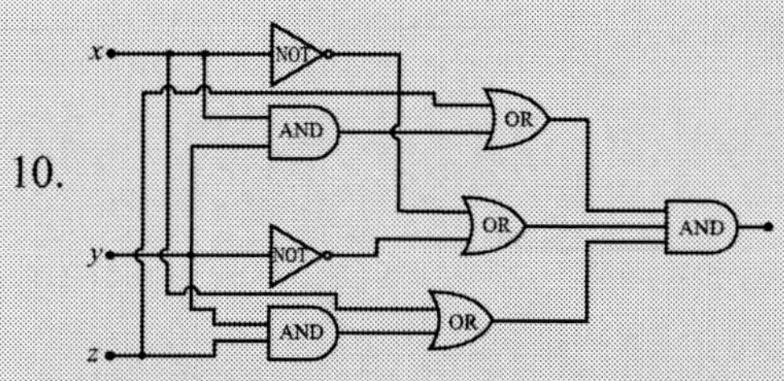

10.

***Simplify the combinatorial circuits given below:***

11.

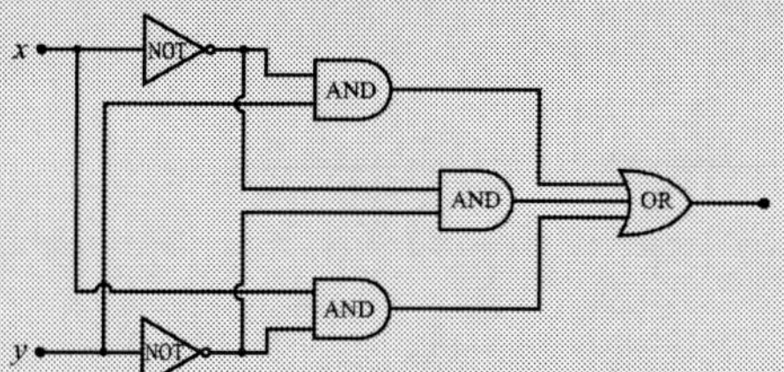

12.

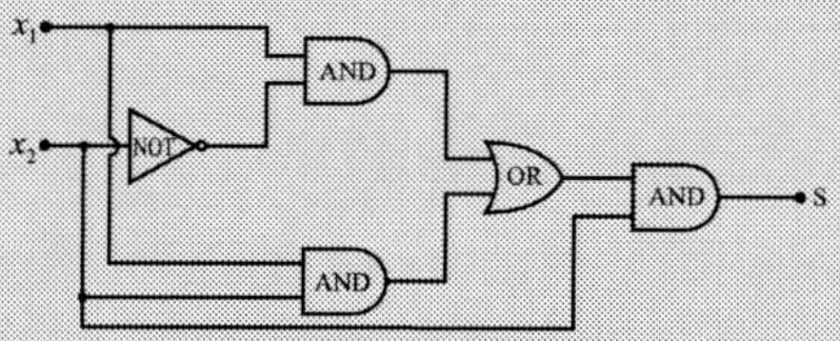

## Answers to Selected Problems

1. (i)

| Input | | Output |
|---|---|---|
| $x_1$ | $x_2$ | $x_1 \cdot x_2'$ |
| 0 | 0 | 0 |
| 0 | 1 | 0 |
| 1 | 0 | 1 |
| 1 | 1 | 0 |

(ii)

| Input | | | Output |
|---|---|---|---|
| $x_1$ | $x_2$ | $x_3$ | $x_1 \cdot x_2 + x_2 \cdot x_3'$ |
| 0 | 0 | 0 | 0 |
| 0 | 0 | 1 | 0 |
| 0 | 1 | 0 | 1 |
| 0 | 1 | 1 | 0 |
| 1 | 0 | 0 | 0 |
| 1 | 0 | 1 | 0 |
| 1 | 1 | 0 | 1 |
| 1 | 1 | 1 | 1 |

(iii)

| Input | | | Output |
|---|---|---|---|
| $a$ | $b$ | $c$ | $(a \cdot b') + c$ |
| 1 | 1 | 1 | 1 |
| 1 | 1 | 0 | 0 |
| 1 | 0 | 1 | 1 |
| 1 | 0 | 0 | 1 |
| 0 | 1 | 1 | 1 |
| 0 | 1 | 0 | 0 |
| 0 | 0 | 1 | 1 |
| 0 | 0 | 0 | 0 |

(iv)

| Input | | | Output |
|---|---|---|---|
| $a$ | $b$ | $c$ | $(a \cdot b') \cdot c$ |
| 1 | 1 | 1 | 0 |
| 1 | 1 | 0 | 0 |
| 1 | 0 | 1 | 1 |
| 1 | 0 | 0 | 0 |
| 0 | 1 | 1 | 0 |
| 0 | 1 | 0 | 0 |
| 0 | 0 | 1 | 0 |
| 0 | 0 | 0 | 0 |

2. $f(x_1, x_2) = x_1 \cdot x_2$

3. $f(x_1, x_2, x_3) = (x_1 \cdot x_2 \cdot x_3) + (x_1 \cdot x_2 \cdot x_3')$
$\qquad\qquad + (x_1 \cdot x_2' \cdot x_3)$
$\qquad\qquad + (x_1' \cdot x_2' \cdot x_3')$

4. (i)

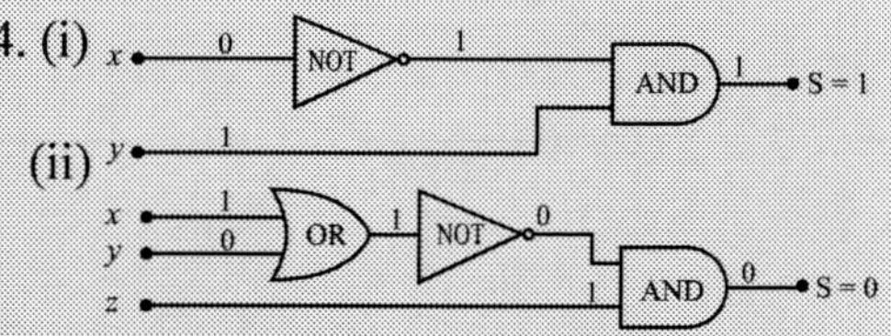


(ii)



5. (i)

| Input | | Output |
|---|---|---|
| $x$ | $y$ | S |
| 0 | 0 | 0 |
| 0 | 1 | 1 |
| 1 | 0 | 0 |
| 1 | 1 | 0 |

(ii)

| Input | | | Output |
|---|---|---|---|
| $x$ | $y$ | $z$ | S |
| 0 | 0 | 0 | 0 |
| 0 | 0 | 1 | 1 |
| 0 | 1 | 0 | 0 |
| 0 | 1 | 1 | 0 |
| 1 | 0 | 0 | 0 |
| 1 | 0 | 1 | 0 |
| 1 | 1 | 0 | 0 |
| 1 | 1 | 1 | 0 |

7. (i)

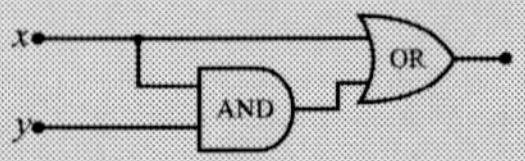


(ii)

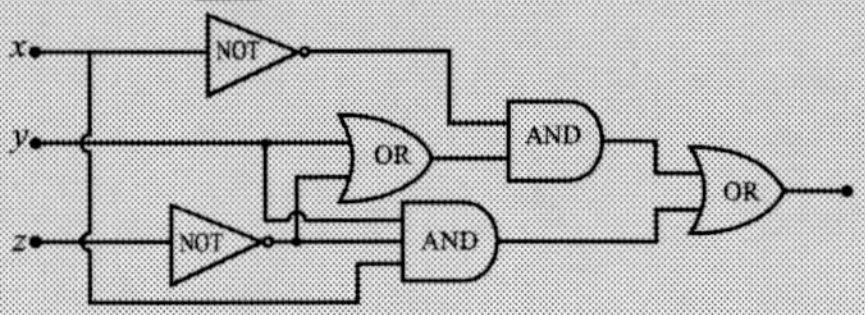


(iii)

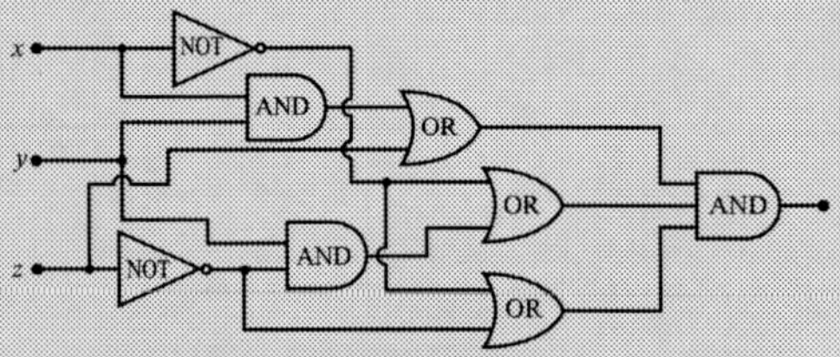


(iv)

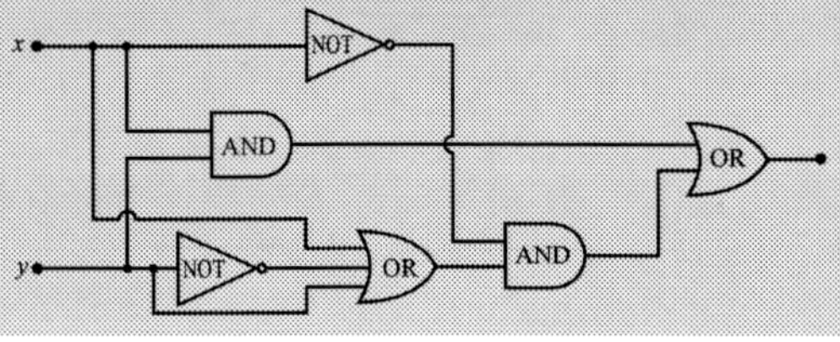


8.

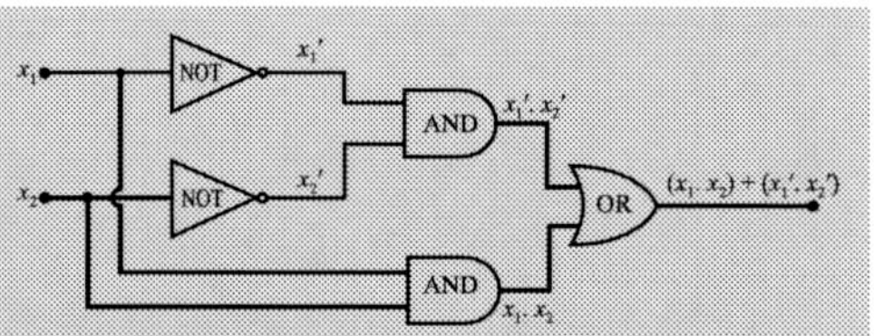


9. $x \cdot (y + z') \cdot z$

10. $(x \cdot y + z) \cdot (x + y \cdot z) \cdot (x' + y')$

11.

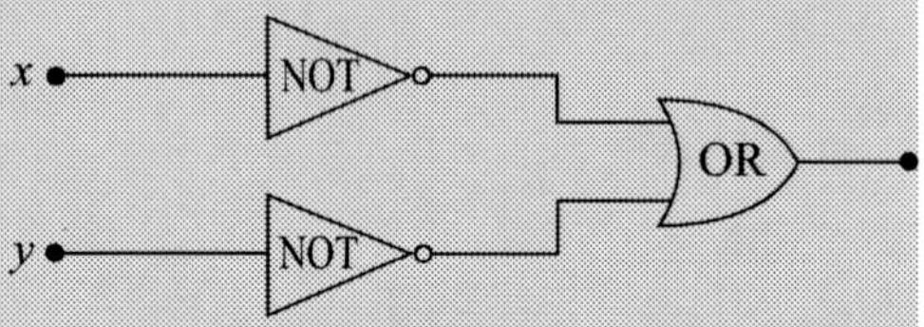


12.

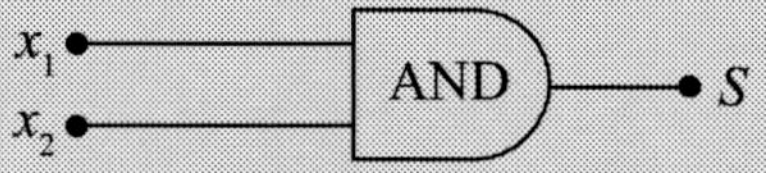

# Section 4

# 10

# Graph Theory

## 10.1 INTRODUCTION

This part of the course introduces some mathematical ideas which can be used to model certain classes of problem which generally involve, in one sense or another, connections between entities. The basic idea of graph theory was introduced by Euler. Graphs provide valuable information about a physical problem. The word graph refers to a specific mathematical structure usually represented as a diagram consisting of points joined by lines. In applications the points may, for instance, correspond to chemical atoms, towns, electrical terminals or anything that can be connected in pairs. The lines may be chemical bonds, roads, wires or other connections. Applications of graph theory are found in communications, structures and mechanisms, electrical networks, transport systems, social networks and computer science.

## 10.2 DEFINITION OF GRAPH

A graph denoted by, $G = (V, E)$ consists of two things:

(i)  A set $V = V(G)$ whose elements are known as points, nodes or vertices of G.
(ii)  A set $E = E(G)$ of pairs of distinct vertices called edges of G.

If $(u, v) \in E$ then, we can say that $u$ and $v$ are connected by an edge where $u$ and $v$ are vertices in the set V. Graphs are pictured by diagrams in the plane. Each vertex $v$ in set V is represented by a dot or a small circle and each edge $e = (u, v)$ is represented by a curve/line which connects these two vertices $u$ and $v$.

For example, Fig. 1 represents the graph, G(V, E) where

(i)  V consists of vertices A, B, C, D and E.
(ii)  E consists of edges $e_1 = (A, B)$, $e_2 = (B, D)$, $e_4 = (E, C)$, $e_5 = (C, A)$, $e_6 = (B, E)$ and $e_7 = (C, D)$.

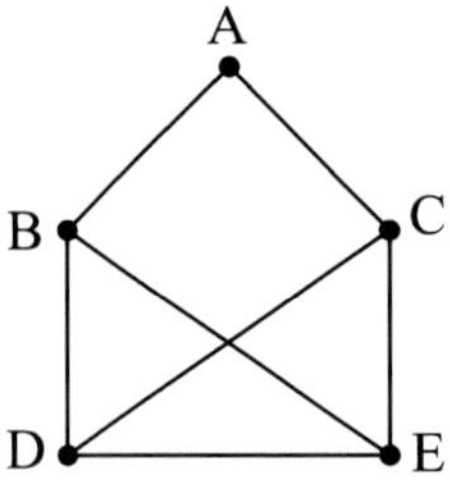

**Fig. 1**

In fact, a graph is usually denoted by drawing its diagram rather than explicitly listing its vertices and edges. If the edge set E = E(G) consists of unordered pairs of distinct vertices, then the graph G = (V, E) is known as undirected graph. For example, Fig. 2 represents the undirected graph.

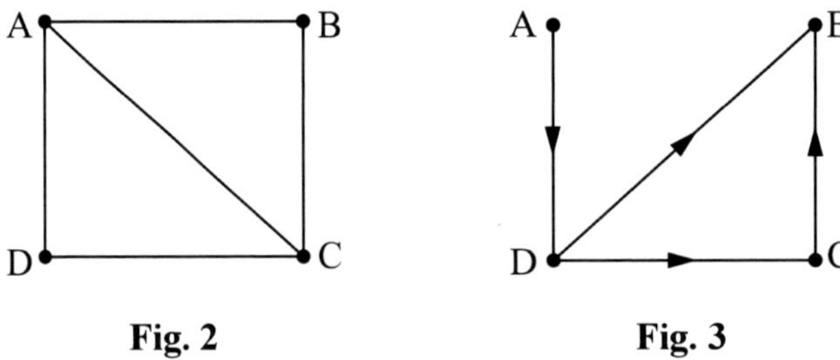

**Fig. 2**          **Fig. 3**

If the edge set E = E(G) consists of ordered pairs of distinct vertices, then the graph G = (V, E) is known as **directed graph.** In this graph each edge is assigned a direction.

For example, Fig. 3 represents the directed graph.

## 10.3 GRAPH TERMINOLOGY

**(i) Adjacent Vertices**

Two vertices $u$ and $v$ are said to be adjacent if they are connected by an edge $e = (u, v)$. i.e., if there is an edge $e = (u, v)$, then we can say that the vertex $u$ is adjacent to $v$ and vice-versa. In such a case $u$ and $v$ are called endpoints of $u$ and $v$ and the edge $e$ is said to connect $u$ and $v$. Also, the edge is said to be incident on each of its endpoints $u$ and $v$. For example, an Fig. 4, then adjacent vertices are $(v_1, v_2)$, $(v_2, v_s)$, $(v_1, v_4)$, $(v_2, v_4)$, $(v_2, v_5)$, $(v_s, v_5)$, $(v_4, v_5)$, $(v_5, v_6)$.

**(ii) Isolated Vertex**

A vertex of degree zero i.e., a vertex that has no edge, incident with it is called as isolated vertex. For example, in Fig. 4, the pendent vertex is $v_7$.

**(iii) Odd Vertex**

A vertex is said to be odd if its degree is an odd number. Thus $v_4$ and $v_6$ are odd vertices in graph G is Fig. 4.

**(iv) Even Vertex**

A vertex is said to be even if its degree is an odd number, thus $v_1$, $v_2$, $v_3$ and $v_5$ are even vertices in graph G of Fig. 4 (vi).

**(v) Pendent Vertex**

A vertex of degree one is called as pendent vertex. For example, in Fig. 4, the pendent vertex is $v_6$.

**(vi) Degree**

The degree of a vertex $v$ in graph G(V, E) is equal to the number of edges that are incident on vertex $v$ and is denoted by deg($v$). The self-loop is counted twice where self-loop is defined as the edge $e$ whose endpoints are the same vertex. For example, in Fig. 4, the degree of each vertex is as follows:

$\deg(v_1) = 2$,   $\deg(v_2) = 4$,   $\deg(v_3) = 2$,   $\deg(v_4) = 3$,   $\deg(v_5) = 4$,   $\deg(v_6) = 1$, $\deg(v_7) = 0$

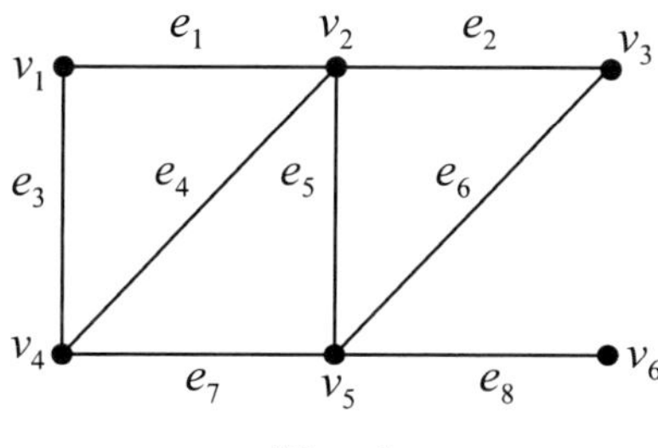


**Fig. 4**

**(vii) Incident**

An edge is said to be incident with the vertices it joins.

For example, the edge $e_1$ is incident with the vertices $v_1$ and $v_2$.

**(viii) Pendent Edge**

The only edge, which is incident on pendent vertex, is called a pendent edge.

For example, in Fig. 4, the pendent edge is $e_8$.

**Theorem 1** Show that the sum of degree of all the vertices in a graph G is even and is twice the number of edges in the graph.

***Proof***

Each edge contributes one degree to each of the vertices on which it is incident. So, each edge contributes two degrees in a graph as an edge is in between two vertices. Hence, if there are M edges in a graph, then

$$2M = d(v_1) + d(v_2) + d(v_3) + \ldots + d(V_{M-1}) + d(V_M) \text{ and } 2M \text{ is always even.}$$

Hence proved.

**Theorem 2** Prove that in any graph G, there are even numbers of vertices of odd degree.

### *Proof*

Consider a graph G having even and odd vertices i.e., vertices with even and odd degree. Now, divide these vertices into two categories, one of even vertices, say $V_1, V_2, \ldots, v_m$. And suppose,

$$X = d(V_1) + d(V_2) + \ldots + d(V_m)$$
$$Y = d(U_1) + d(U_2) + \ldots + d(U_n)$$

Now, as we know, sum of degrees of all the vertices is even (From Theorem 1). So, $X + Y$ is even.

Since, X is an even number as it is the sum of $m$ even numbers. But, Y is an odd number as it is the sum of $n$ odd numbers. So, to make Y an even number, the number $n$ must be even.

Hence proved.

**Example 1:** Consider the Graph G in Fig. 5 and find the following:

  (i) The vertex set V(G)

 (ii) The edge set E(G)

(iii) Degree of each vertex in V(G)

### *Solution*

  (i) There are six vertices so V(G) = (A, B, C, D, E, F)

 (ii) There are ten pairs $(u, v)$ of vertices where the vertex $u$ is connected to vertex $v$. Hence, E(G) = {(A, B), (B, D), (D, F), (F, E), (E, C), (C, A), (A, F), (B, E), (B, F), (C, D)}

(iii) The degree of a vertex is number of edges to which it belongs.

    e.g. degree of vertex B is 4 since B belongs to the edges (B, A), (B, D), (B, E), (B, F). Similarly, deg(A) = 3, deg(C) = 3, deg(D) = 3, deg(E) = 3, deg(F) = 4.

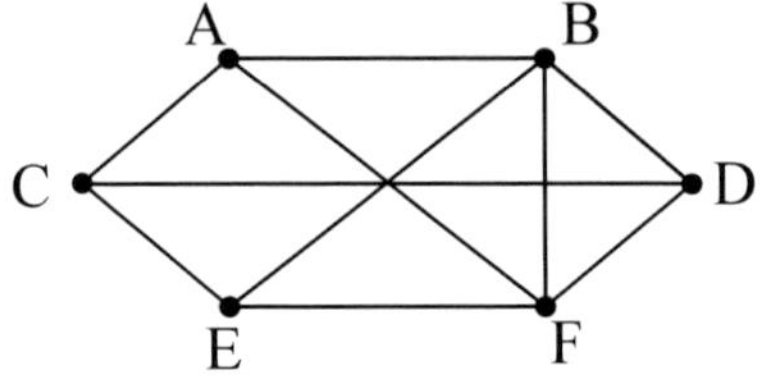


**Fig. 5**

## 10.4 TYPES OF GRAPHS

### 10.4.1 Directed Graph

A directed graph G = (V, E) or simply diagraph consists of two things:

(i) A set V = V(G) whose elements are known as points, nodes or vertices of G.
(ii) A set E = E(G) of ordered pairs of distinct vertices known as directed edges or simply edges of G.

The following *terminology* is used:

(a) An edge is said to be incident with the vertices it joins. For example, the edge $(u, v)$ is incident with the vertices $u$ and $v$. More specifically, we can say that the edge $(u, v)$ is incident from $u$ and is incident into $v$.
(b) The vertex $u$ is called initial the vertex and vertex $v$ is called the terminal vertex of the edge $(u, v)$.
(c) A vertex that is incident from and into the same vertex, e.g. $(b, b)$ is called a loop.
(d) Two vertices are said to be adjacent if they are joined by an edge. For example, if $(u, v)$ is an edge, then the vertex $u$ is said to be adjacent *to* the vertex $v$ and the vertex $v$ is said to be *adjacent from* the vertex $u$.
(e) A vertex is said to be an isolated vertex if there is no edge incident with it.
(f) If $(u, v)$ is an edge, then $v$ is called a *successor* of $u$ and the set of all the successor of vertex $u$ is called as *successor list* or *adjacency list* of $u$.

Directed Graphs are pictured by diagrams in the plane.

Each vertex $v$ in set V is represented by a dot or a small circle and each directed edge $e = (u, v)$ is represented by an arrow or directed curve from the initial point $u$ of $e$ to the terminal point $v$.

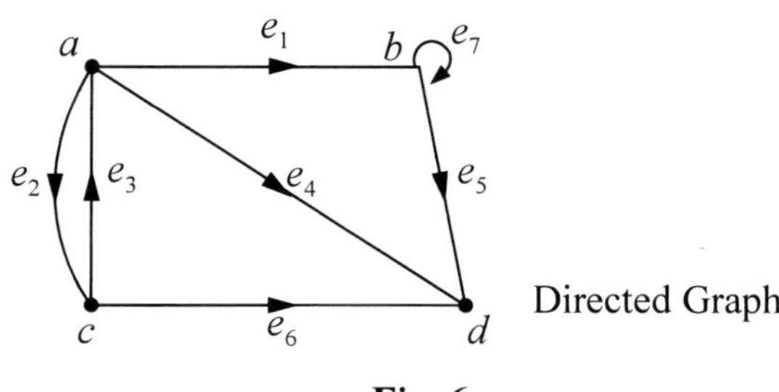


**Fig. 6**

For example, Fig. 6 represents the directed graph G(V, E) where:

(i) V consists of vertices $a$, $b$, $c$, $d$ and $e$
(ii) E consists of edges $e_1 = (a, b)$, $e_2 = (a, c)$, $e_3 = (c, a)$, $e_4 = (a, d)$, $e_5 = (b, d)$, $e_6 = (c, d)$ and $e_7 = (b, b)$.

### 10.4.2 Null Graph

A graph, which consists of only isolated vertex, is called as null graph. For example, the graph shown in Fig. 7(a) is a null graph.

### 10.4.3 Trivial Graph

A graph, which consists of only a single vertex and no edge i.e., a single point, is called a trivial graph.

### 10.4.4 Multigraphs

A multigraph is defined as the one which contains multiple edges and self-loop where *self-loop* is defined as the edge $e$ whose endpoints are the same vertex i.e., an edge that is incident from and into the same vertex and *multiple edges* are defined as two or more edges, which connect the same endpoints. The example multigraph is as in Fig. 7(b).

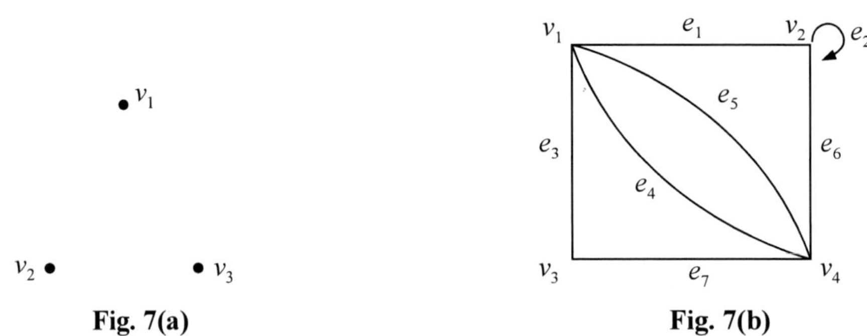


<table>
<tr><td align="center">**Fig. 7(a)**</td><td align="center">**Fig. 7(b)**</td></tr>
</table>

Here, the edges $e_4$ and $e_5$ are multiple edges since they connect the same end points and the edge $e_2$ is loop since its endpoints are the same vertex.

## 10.5 PATH

A path in a graph G is an alternating sequence of vertices and edges of the form

$v_0, e_1, v_1, e_2, v_2, e_3, \ldots e_{n-1}, v_{n-1}, e_n, v_n$

where each edge $e_i$ connects the vertices $v_i$ and $v_{i-1}$

The number $n$ of edges is called the length of the path. A path is generally denoted by listing its sequence of vertices $v_1, v_2, \ldots, v_n$.

For example, in Fig. 8, the sequence K = $(a, b, c, f, e, c, d, g)$ is a path from $a$ to $g$.

### 10.5.1 Simple Path

A path is a simple path in which all the vertices are distinct.

For example, in Fig. 8, the sequence, L = $(a, b, c, e, f, g)$ is a simple path from $a$ to $g$.

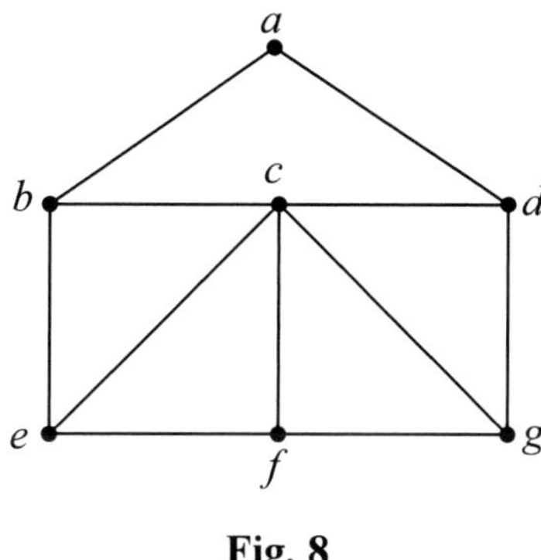

**Fig. 8**

### 10.6 TRAIL

A path is a trail in which all the edges are distinct.

For example, in Fig. 8, the sequence M = $(a, b, c, f, e, c, d, g)$ is a trial as no edge is repeated.

### 10.7 CIRCUIT OR CLOSED PATH

A circuit or closed path is a path which starts and ends at the same vertex i.e., $v_0 = v_n$.

For example, in Fig. 8, the sequence N = $(b, c, d, g, f, c, d, e, b)$ is a closed path.

### 10.8 CYCLE

A cycle is a closed path in which all the vertices are distinct except $v_0 = v_n$.

A cycle of length $k$ is known as $k$-cycle. In a graph, a cycle must have a length of three or more.

For example, in Fig. 8, the sequence X = $(b, c, d, g, f, e, b)$ is a cycle.

**Example 2:** Consider the graph in Fig. 8 and consider the following sequences:

$A_1 = (b, c, f, e, c, f, g)$    $A_2 = (a, b, c, f, b, c, g)$
$A_3 = (b, c, f, e, c, d, g)$    $A_4 = (b, c, g, f, c, e, b)$

The sequence $A_1$ is a path from $b$ to $g$ but it is not a trail as the edge $(c, f)$ is used twice.

Also, it is not a simple path as vertex $c$ and $f$ are used twice.

The sequence $A_2$ is a not path as there is no edge $(f, b)$.

The sequence $A_3$ is a trail as no edge is repeated but it is not a simple path as vertex $c$ is used twice.

The sequence $A_4$ is a closed path but it is not a cycle as the vertex $c$ is used twice.

**Theorem 3** If a graph G has two distinct paths from a vertex $u$ to a vertex $v$, then show that graph G has a cycle.

### *Proof*

Consider two distinct paths from $u$ to $v$ be $P_1 = (e_1, e_2, ..., e_n)$ and $P_2 = (e_1, e_2, ..., e_n)$.

Now, delete from paths $P_1$ and $P_2$ all the initial edge, which are common i.e., of we have.

$$e_1 = e_1, e_2 = e_2, ..., e_k = e_k, \quad \text{but } e_{k+1} \neq e_{k+1}.$$

We will delete all the first $k$ edges of both the paths $P_1$ and $P_2$.

Now, after deleting the $k$-edges, both paths start from the same vertex, say, $u_1$ and end at $v$.

Now, to construct a cycle, start from vertex $u_1$ and follow the left over path of $P_1$ until we first meet any vertex of the leftover path of $P_2$.

If this vertex is $u_2$ then the remaining cycle is computed by following the leftover path of $P_2$ that starts from $u_2$ and ends at $v$.

**Example 3:** Consider the Graph in figure and find the following:

(i) All simple paths from A to F.
(ii) All the trails from A to F.
(iii) All cycles which include vertex B.

### *Solution*

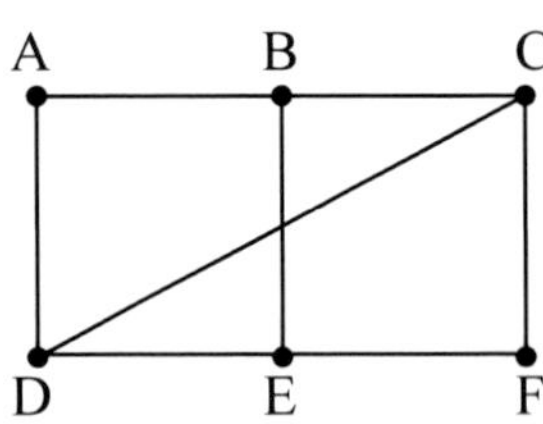

(i) A simple path from A to F is a path that has no vertex and hence no edge is repeated. There are eight such paths and are given as

(A, B, C, F), (A, B, E, F), (A, B, C, D, E, F), (A, B, E, D, C, F), (A, D, E, F), (A, D, C, F), (A, D, E, C, B, F), (A, D, C, B, E, F)

(ii) A trail from A to F is a path such that no edge is repeated. There are eight such trails and are given as:

(A, B, C, F), (A, B, E, F), (A, B, C, D, E, F), (A, B, E, D, C, F), (A, D, E, F),
(A, D, C, F), (A, D, E, C, B, F), (A, D, C, B, E, F)

(iii) A cycle is a closed path in which no vertex is repeated (except the first and last vertex). There are 4 cycles which include vertex B and are as:

(B, E, F, C, B), (B, E, D, C, B), (B, E, D, A, B), (B, A, D, C, B), (B, A, D, E, F, C, B)

## 10.9 COMPLEMENTARY GRAPH

The complement of a graph G is defined as a graph G′ which has same number of vertices as in graph G and has two vertices connected if and only if they are not connected in graph G.

For example, the complement of a graph G in Fig. 9 is a graph G′ in Fig. 9(a).

## 10.10 SUBGRAPHS

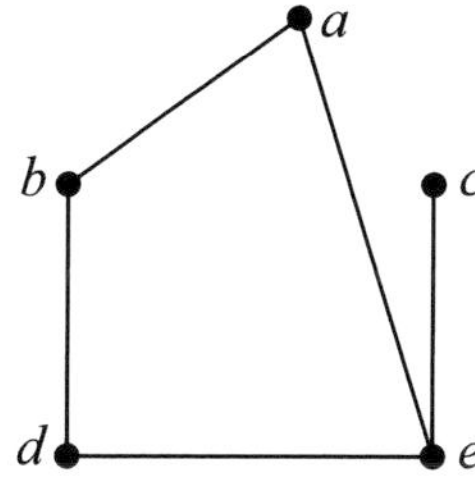

**Fig. 9**  Graph G

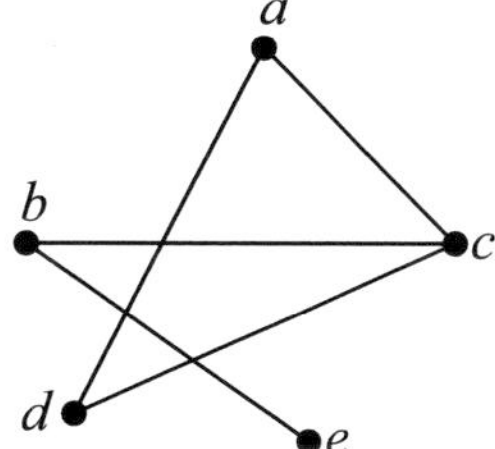

**Fig. 9(a)**  Graph G′

A subgraph H = H (V′, E′) of a graph G = G (V, E) is a graph in which V′ ⊆ V and E′ ⊆ E i.e., the vertices and edges of graph H are contained in vertices and edges of graph G.

**Example 4:**  Consider the graph G in Fig. 10. Show its different subgraph.

*Solution*

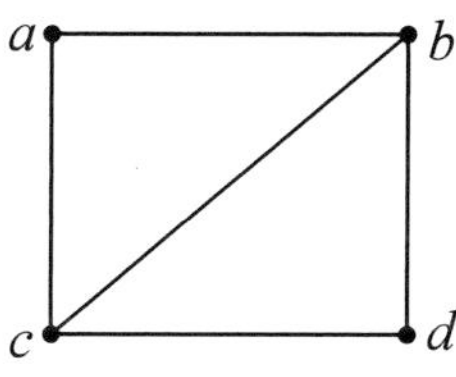

**Fig. 10**

The following are some subgraph, as shown in Fig. 11(a)–11(d) of the given graph G.

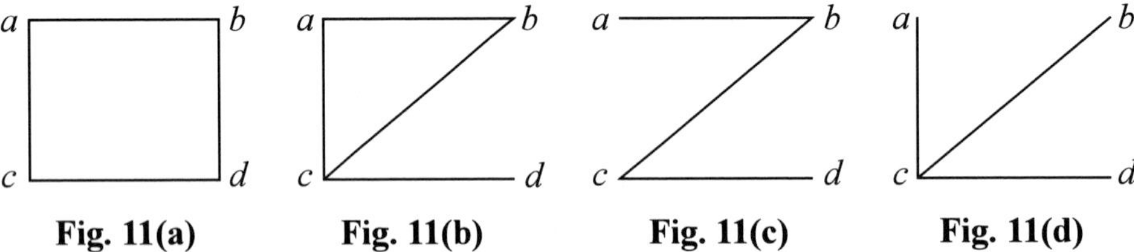

Fig. 11(a)    Fig. 11(b)    Fig. 11(c)    Fig. 11(d)

## 10.11 SPANNING SUBGRAPH

A subgraph H of G is said to be a spanning subgraph if H contains all the vertices of G.

For example the graph H shown in Fig. 12 is a spanning subgraph of G shown in Fig. 11(a).

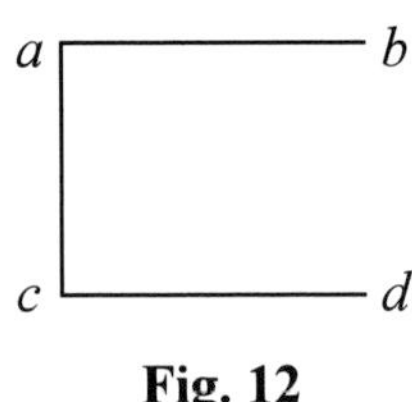

Fig. 12

## 10.12 ISOMORPHIC GRAPH

Two graphs G and G′ are said to be isomorphic graphs if there is one-to-one correspondence (i.e., one-one and onto) between their vertices and between their edges such that incidence is preserved. In other words. Suppose that an edge $e$ is incident on vertices $v_1$ and $v_2$ in graph G then the corresponding edge $e'$ must be incident on vertices $v_1'$ and $v_2'$ in graph G′.

For example:

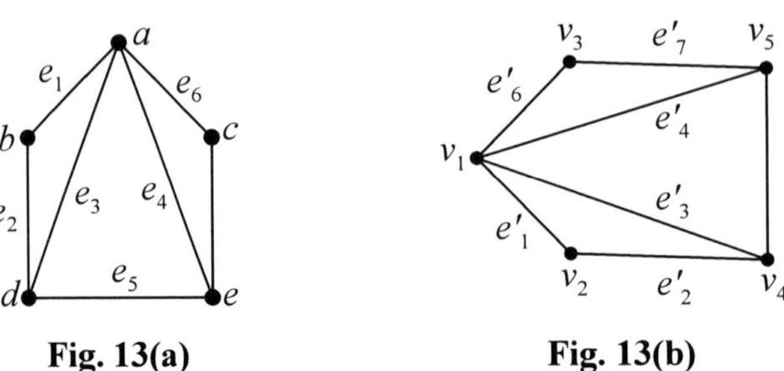

Fig. 13(a)                Fig. 13(b)

In Fig. 13(a)–(b), the vertices of Fig. 13(a) are corresponding to the vertices of Fig. 13(b) i.e., the vertices $a$, $b$, $c$, $d$, $e$ in Fig. 13(a) are corresponding to the vertices $v_1$, $v_2$, $v_3$, $v_4$, $v_5$ in Fig. 13(b). Also the incidence relationship is preserved by the edges i.e., the edges $e_1$, $e_2$, $e_3$, $e_4$, $e_5$, $e_6$, $e_7$ of Fig. 13(a) are corresponding to edges $e_1'$, $e_2'$, $e_3'$, $e_4'$, $e_5'$, $e_6'$, $e_7'$ of Fig. 13(b).

Therefore for a graph to be isomorphic we must have the following:

1. The same number of vertices.
2. The same number of edges.
3. An equal number of vertices with a given degree.

If anyone of these conditions is violated then the given graphs are not isomorphic.

For example:

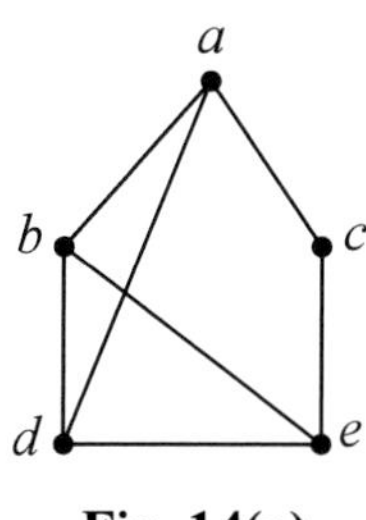

**Fig. 14(a)**

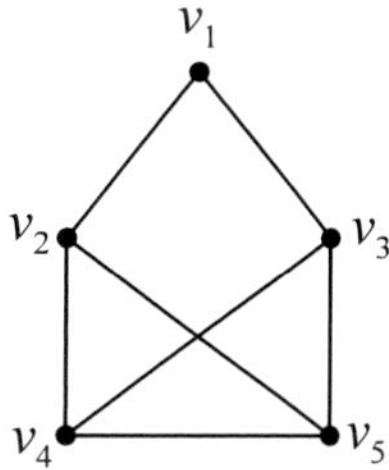

**Fig. 14(b)**

Here, the number of vertices are same i.e., 5
and the number of edges are also same i.e., 7

and $\deg(a) = \deg(v_2)$, $\deg(b) = \deg(v_4)$, $\deg(c) = \deg(v_1)$, $\deg(d) - \deg(v_5)$, $\deg(e) = \deg(v_3)$.

So, these two graphs are isomorphic as these satisfy all the above three properties.

**Definition:** The two graph $G = (V, E)$ and $G' = (V', E')$ are said to be isomorphic if there is a one-to-one and onto function $f$ from V to V' such that if two vertices $a$ and $b$ adjacent in G, then $f(a)$ and $f(b)$ are adjacent in G', for all $a$ and $b$ in V of graph G. Such a function $f$ is called isomorphic function.

For example:

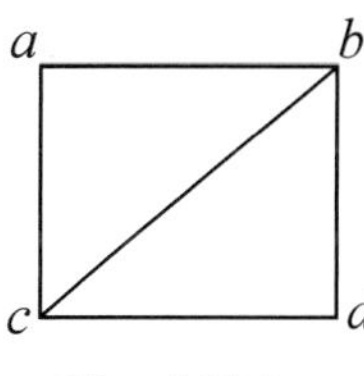

**Fig. 15(a)**

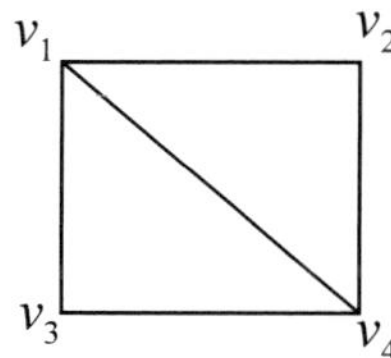

**Fig. 15(b)**

The above two graphs are isomorphic and isomorphism is as given below:

$$f(a) = v_2 \qquad f(b) = v_1 \qquad f(c) = v_4 \qquad f(d) = v_3$$

## 10.13 HOMEOMORPHIC GRAPH

Two graph G and G′ are said to be homeomorphic graph if graph G′ can be obtained from G by dividing the edges of graph G with additional vertices.

For example, the graphs (b) and (c) in Fig. 16(a)–16(c) are homeomorphic graph since they can be obtained from graph (a) by adding appropriate vertices. In graph (b), we introduce two vertices $v_6$ and $v_7$ on two edges $(v_2, v_5)$ and $(v_5, v_3)$ respectively. Similarly in graph (c), we introduce two vertices $v_6$ and $v_7$ on two edges $(v_4, v_5)$ and $(v_5, v_1)$ respectively.

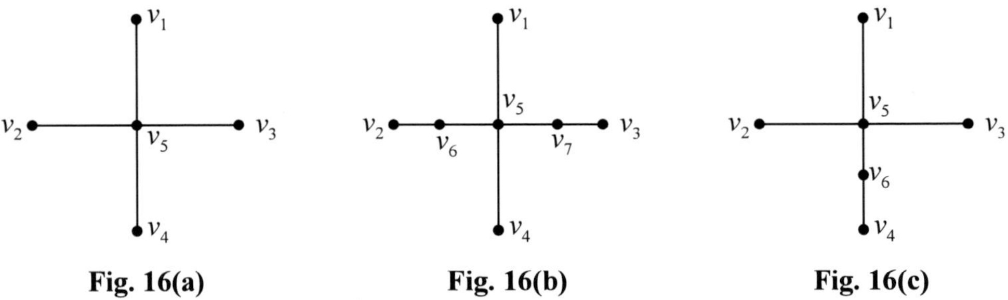

| Fig. 16(a) | Fig. 16(b) | Fig. 16(c) |

Another example of homeomorphic graph is as in Fig. 17(a)–17(b).

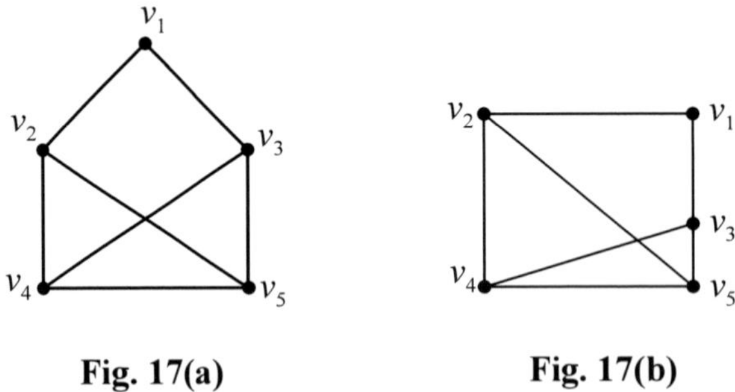

**Fig. 17(a)**  **Fig. 17(b)**

## 10.14 CONNECTED GRAPH

A graph G is connected if there is a path between any two of its vertices i.e., if there is a path from any vertex $u$ to any vertex $v$ or vice-versa.

For example, the graph in Fig. 18 is a connected graph.

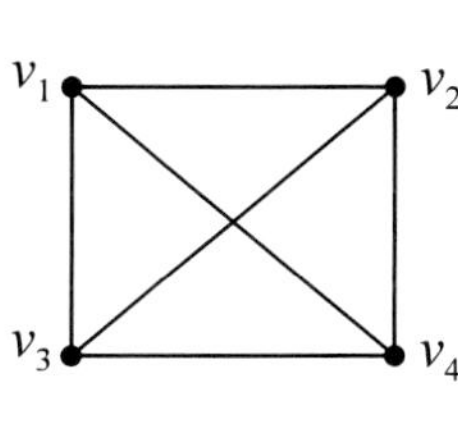

**Fig. 18**

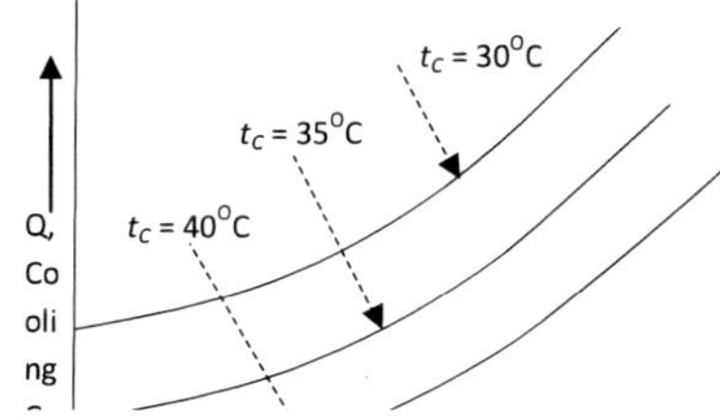


**Fig. 19**

## 10.15 DISCONNECTED GRAPH

A graph G is disconnected if there is a no path between any two of its vertices.

For example, the graph in Fig. 19 is a disconnected graph as there is no path between vertex $v_1$ and $v_5$.

## 10.16 CONNECTED COMPONENT

A connected subgraph H of a graph G is called a connected component if H is not contained in any larger connected subgraph of graph G. So, a given graph G can be partitioned into its connected components.

For example, the graph G is Fig. 19 has four connected components given by $(v_1, v_2, v_3, v_4)$, $(v_5, v_6, v_7, v_8)$, $(v_9, v_{10})$ and $(v_{11}, v_{12}, v_{13})$.

## 10.17 CUT SET

A cut set for a given connected graph G is the smallest set of edges such that removal of this set will increase the number of connected components in the remaining subgraph, whereas the removal of any proper subset of it will not. It follows that in a connected graph, the removal of cut-set will separate the graph in two parts.

For example, cut set for the graph G in Fig. 20 is $\{(v_1, v_3), (v_3, v_4)\}$ or $\{(v_3, v_5), (v_3, v_7)\}$.

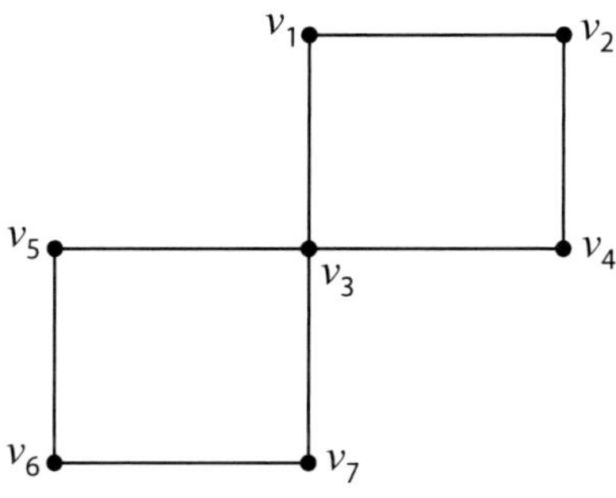

**Fig. 20**

## 10.18 CUT POINTS OR CUT VERTICES

A cut point for a given connected graph G is a vertex $v$ in G if G-$v$ is disconnected i.e., G-$v$ has more connected components than graph G. The subgraph G-$v$ is obtained by removing vertex $v$ and the edges that are incident on $v$ from graph G.

For example, in Fig. 21 the cut point for graph G is $v_5$.

G-$v_5$ is as in Fig. 22.

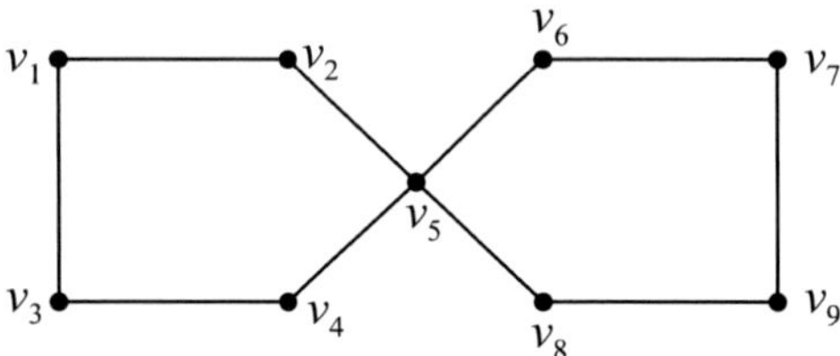

**Fig. 21**

## 10.19 CUT EDGES OR BRIDGES

A bridge for a connected graph G is an edge $e$ if G-$e$ is disconnected i.e., G-$e$ has more connected components than graph G. The subgraph G-$e$ obtained by deleting an edge $e$ from G.

For example, in Fig. 22, the bridges for graph g are $e_1$, $e_2$ and $e_5$.

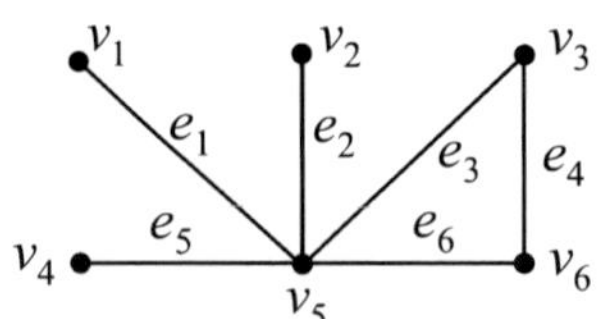

**Fig. 22**

**Example 5:** Consider the graph G in Fig. 23.1 and 23.2. Find the following:

   (i) All cut-points.

  (ii) All bridges.

 (iii) Subgraph H of G generated by V' = {A, B, E, C}.

***Solution***

   (i) A vertex $v$ is cut-point if G-$v$ is disconnected. Thus there are 2 cut-points B and E.

  (ii) An edge $e$ is bridges if G-$e$ is disconnected. Thus there are 3 bridges:

       (B, C), (B, E) and (B, F).

 (iii) The subgraph H is as in Fig. 23.2. Subgraph H consists of the vertices V' and the edge set E' of all the edges whose end-points belongs to V' i.e.

       E' = {(B, E), (B, C), (A, E)}

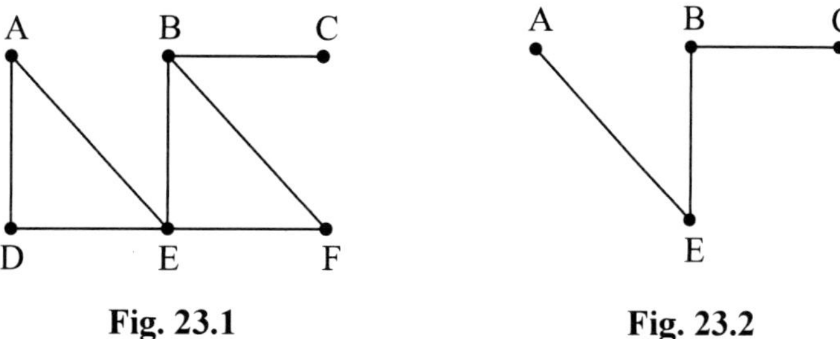

Fig. 23.1                    Fig. 23.2

**Example 6:** Consider the Graph G in Fig. 24.1. Find all the subgraphs obtained when each edge is deleted. Does G have any bridges?

*Solution*

The seven graphs obtained by deleting each of the edges of graph G are as in Fig. 24.2(a)–24.2(g).

As all the seven graphs so obtained are connected, so no edge is a bridge.

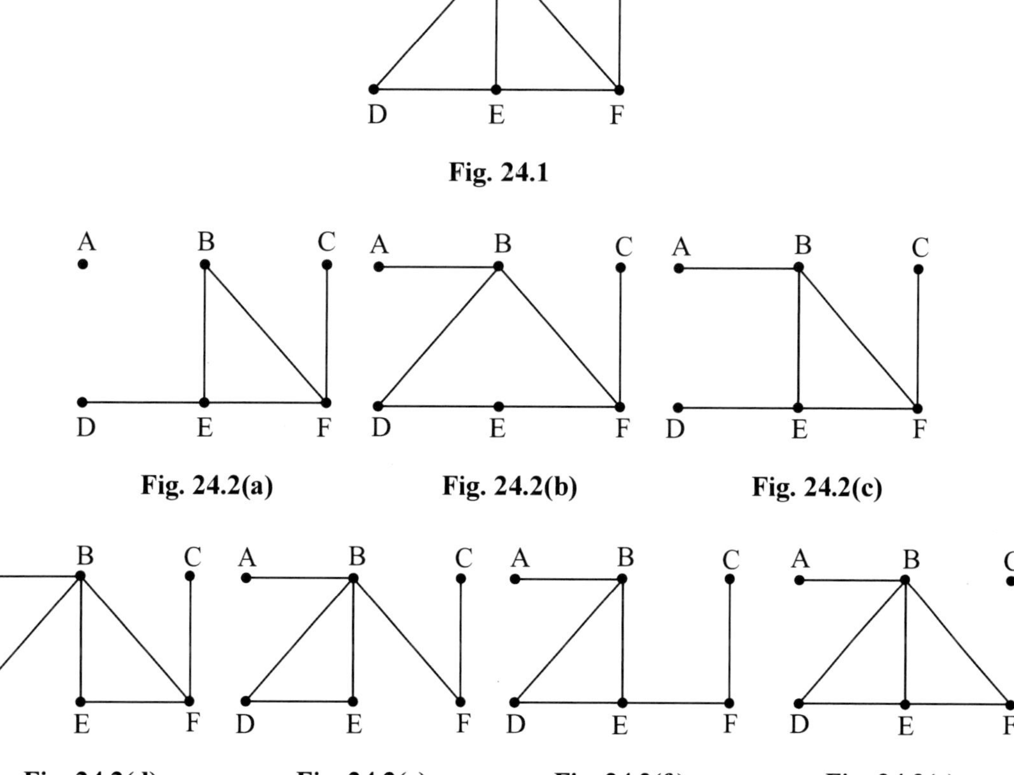

Fig. 24.1

Fig. 24.2(a)                Fig. 24.2(b)                Fig. 24.2(c)

Fig. 24.2(d)            Fig. 24.2(e)            Fig. 24.2(f )            Fig. 24.2(g)

**Example 7:** Consider the multigraph in Fig. 25(a)–25(c) and find the following:

  (i)  Which are simple graphs?

 (ii)  Which of them are connected? If they are not, then find the number of connected components in each case.

(iii)  Which are cycle-free? If they are not, then find the number of cycles in each case.

(iv)  Which are loop-free (i.e., without loops)?

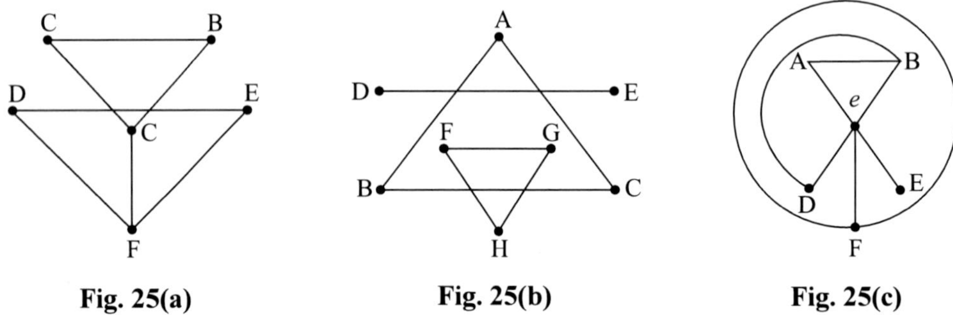

     **Fig. 25(a)**        **Fig. 25(b)**        **Fig. 25(c)**

***Solution***

  (i)  Only graphs in Fig. 25(a) and 25(b) are simple graphs and graph in Fig. 25(c) is a multigraph as it has a loop (F, F).

 (ii)  The graphs in Fig. 25(a) and 25(b) are connected and the graph in Fig. 25(c) is disconnected and its connected components are (A, B, C), (D, E, F), (G, H).

(iii)  No graph is cycle free.

     Graph in Fig. 25(a) has cycle (A, B, C, A) and (D, E, F, D).

     Graph in Fig. 25(b) has cycle (A, B, C, A) and (F, G, H, F).

     Graph in Fig. 25(c) has cycle (A, B, C, A), (B, C, D, B) and (A, C, D, B, A).

(iv)  Only graph in Fig. 25(c) has a loop (F, F).

**Example 8:** Show that the graph in Fig. 26(b) and Fig. 26(c) are homeomorphic.

***Solution***

The graph in Fig. 26(b) and 26(c) are homeomorphic since graph in Fig. 26(a) can be obtained from graph in Fig. 26(a) by dividing the edge (A, B) and graph in Fig. 26(c) can be obtained from graph in Fig. 26(a) by dividing the edge (A, D).

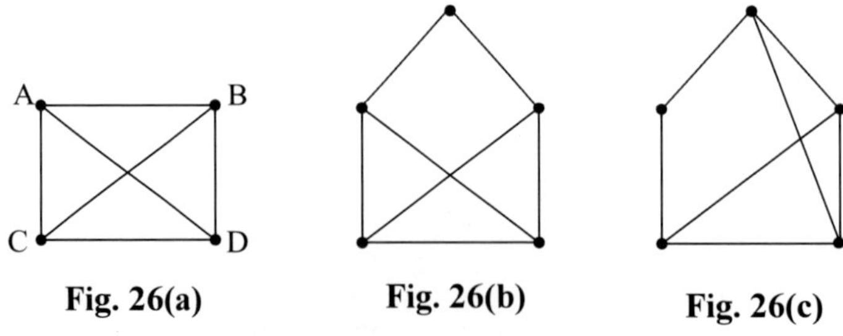

     **Fig. 26(a)**        **Fig. 26(b)**        **Fig. 26(c)**

**Example 9:** Which of the following graphs in Fig. 27(a)–27(b) are isomorphic graphs?

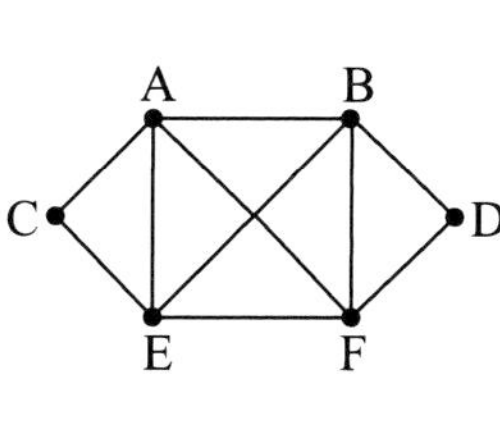
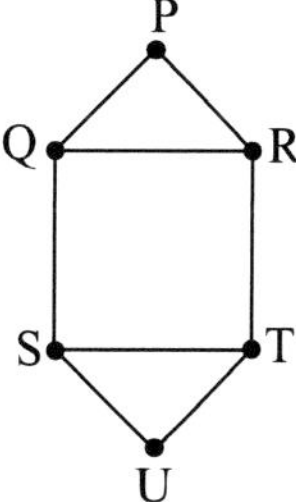

**Fig. 27(a)**

**Fig. 27(b)**

*Solution*

The two graphs in Fig. 27(a) and 27(b) are isomorphic since we have:

$$\deg(A) = \deg(Q), \deg(B) = \deg(R), \deg(C) = \deg(P)$$
$$\deg(D) = \deg(U), \deg(E) = \deg(S), \deg(F) = \deg(T)$$

### Exercise 10.1

1. Consider the graph G in Fig. Ex. 1 and
   (a) find the following:
   (i) The vertex set V(G)
   (ii) The edge set E(G)
   (iii) Degree of each vertex in V(G)
   (b) Show that the sum of degrees of all the vertices is even.

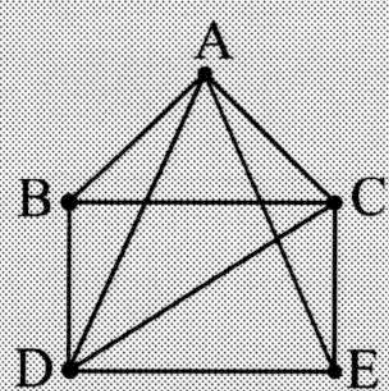

**Fig. Ex. 1**

2. Consider the graph G in Fig. Ex. 1 and find the followings:
   (i) All simple paths from A to E.
   (ii) All the trails from A to E.
   (iii) All cycles which include vertex C.
   (iv) All the cycles in G.

3. Consider the graph G in Fig. Ex. 3 and find the followings:

(i) All simple paths from B to F.
(ii) All the cycles.
(iii) All cut-points.
(iv) All bridges.
(v) Subgraph H of G generated by
   (a) V′ = {B, C, D, E}
   (b) V′ = {A, C, D, E}
   (c) V′ = {A, C, E, F}

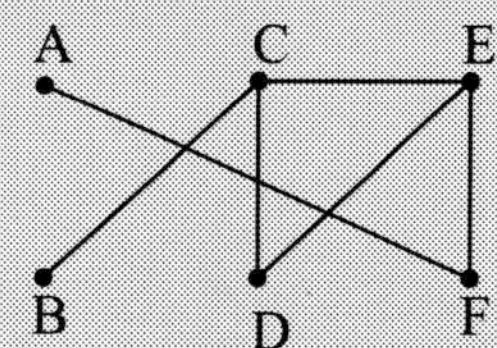

**Fig. Ex. 3**

4. Consider the graph G in Fig. Ex. 3. Find all the subgraphs obtained when each vertex is deleted. Does G have any cut-points?

5. Consider the graph G in Fig. Ex. 3. Find all the subgraphs obtained when each edge is deleted. Does G have any bridges?

6. Show that the graph in Fig. Ex. 6(a) and Fig. Ex. 6(b) are homeomorphic.

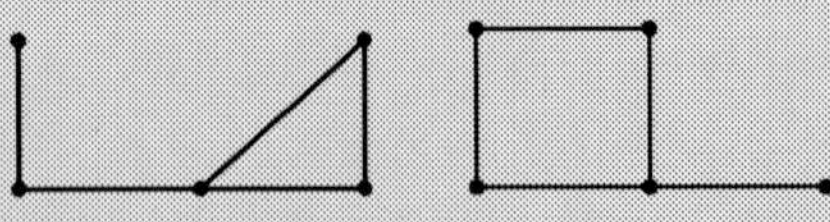

**Fig. Ex. 6(a)**          **Fig. Ex. 6(b)**

7. Consider the multigraph in Fig. Ex. 7(a)–Ex. 7(c) and find the following:

   (i) Which of them are connected? If they are not, then find the number of connected component in each case.

   (ii) Which are cycle-free? If they are not, then find the number of cycles in each case.

   (iii) Which are loop-free (i.e., without loops)?

   (iv) Which are simple graphs?

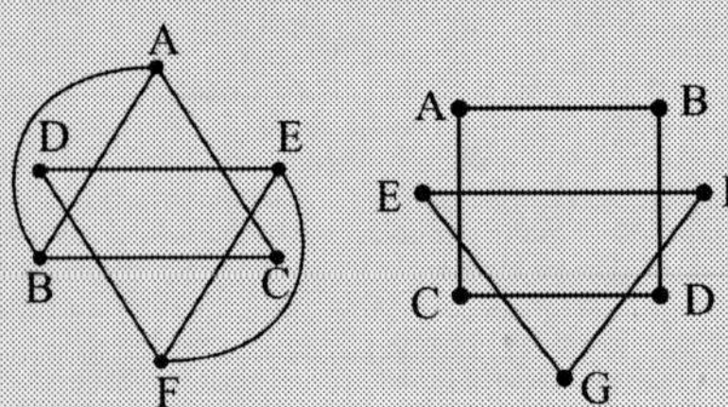

**Fig. Ex. 7(a)**          **Fig. Ex. 7(b)**

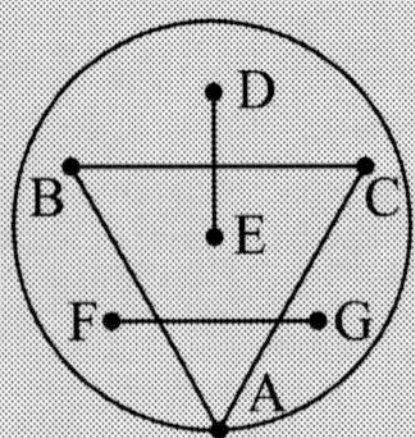

**Fig. Ex. 7(c)**

8. Which of the followings graphs in Fig. Ex. 8(a)–Ex. 8(c) are isomorphic graphs?

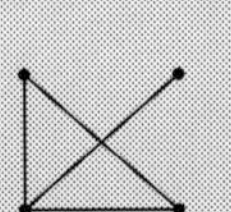    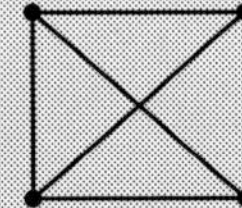

**Fig. Ex. 8(a)**   **Fig. Ex. 8(b)**   **Fig. Ex. 8(c)**

**Answers to Selected Problems**

1(a). (i) $V(G) = \{A, B, C, D, E\}$

   (ii) $E(G) = \{(A, B), (A, C), (A, D), (A, E), (B, C), (B, D), (C, D), (C, E), (D, E)\}$

   (iii) Degree of vertex

   $A = 4; B = 3; C = 4; D = 4; E = 3$

2. (i) ABDE, ABCE, ACE, ABCDE, AE, ACDE

   (ii) ABDE, ABCE, ACE, ABCDE, AE, ACDE, ABCDAE

   (iii) BCD, ACE, ABC, CDE

   (iv) ABC, ACE, CDE, BCD, ADE

3. (i) BCEF, BCDEF

   (ii) CDE

   (iii) C, E, F

   (iv) AF, EF, BC

   (v)

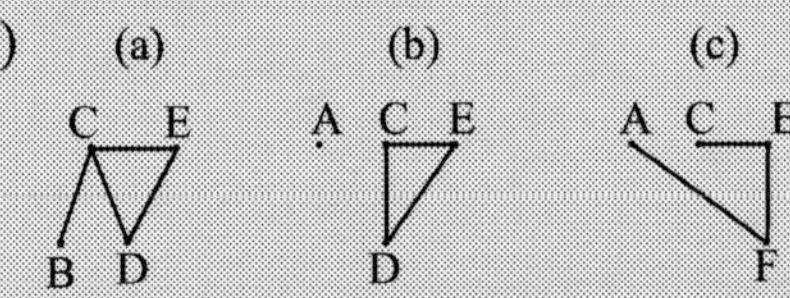

7. (i) Fig. 7(a)–7(c) are not connecters. Connected components for each figure are:

   7(a)   DEF and ABC

   7(b)   EFG and ABCD

   7(c)   ABC, FG, DE

   (ii) Each graph has cycles in 7(a) there are 4 cycles, in 7(b) there are 2 cycles, in 7(c) there are 2 cycles.

   (iii) 7(a) and 7(b) are loop-free

   (iv) Fig. 7(b) is simple graph.

8. Graph shown by 8(a) and 8(b) are isomorphic.

## 10.20 COMPLETE GRAPH

A complete graph $G = (V, E)$ of $n$ vertices is a graph in which every vertex is connected to every other vertex i.e., there exists an edge between every pair of distinct vertices. It is denoted by $K_n$. A complete graph with $n$ vertices have $n(n-1)/2$ edges.

For example, the complete graph $K_4$ and $K_5$ are as follows:

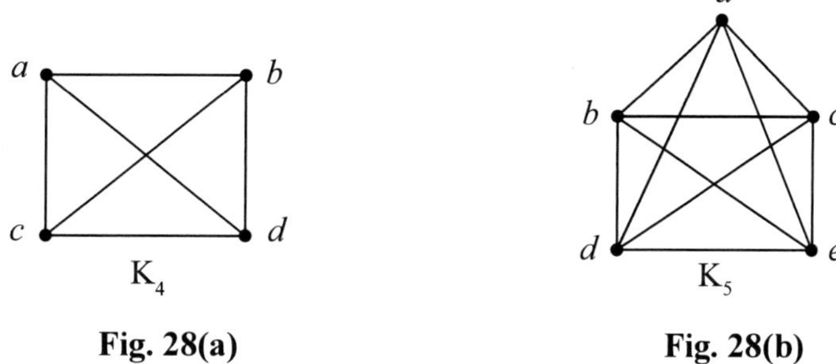

**Fig. 28(a)**                    **Fig. 28(b)**

## 10.21 LABELED GRAPH

A graph G is called a labeled graph if its edges and/or edges are assigned some name or data. Then, we can use these labels in the edge set.

For example, the graph in Fig. 29 is a labeled graph.

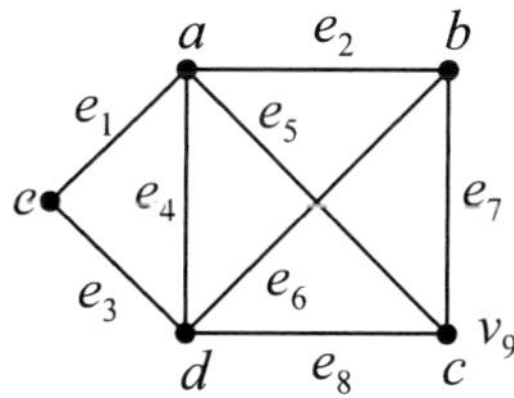

**Fig. 29**

## 10.22 WEIGHTED GRAPH

A graph G is called a weighted graph if each $e$ of G is assigned a non-negative number $w$ called as weight/length of edge $e$. The weight/length of a path in a weighted graph G is the sum of weights of edges in the path.

For example, the graph G in Fig. 30 is a weighted graph.

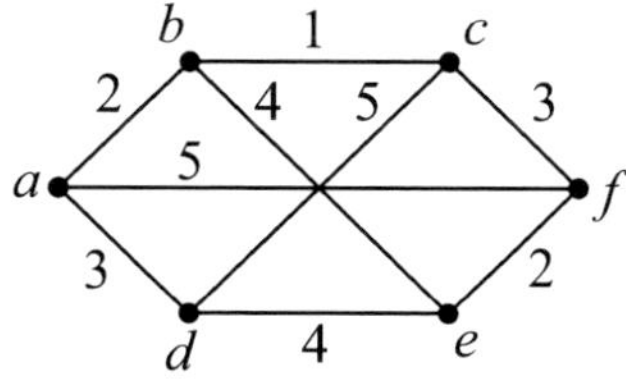

**Fig. 30**

## 10.23 REGULAR GRAPH

A graph G is called a regular graph of degree $k$ or $k$-regular if every vertex of G has the degree $k$ i.e., all the vertex of G are of same degree $k$. A complete graph $K_n$ of $n$ vertices is a regular graph of degree $n - 1$.

For example, the regular graphs of degree 2 are as in Fig. 31(a)–31(c).

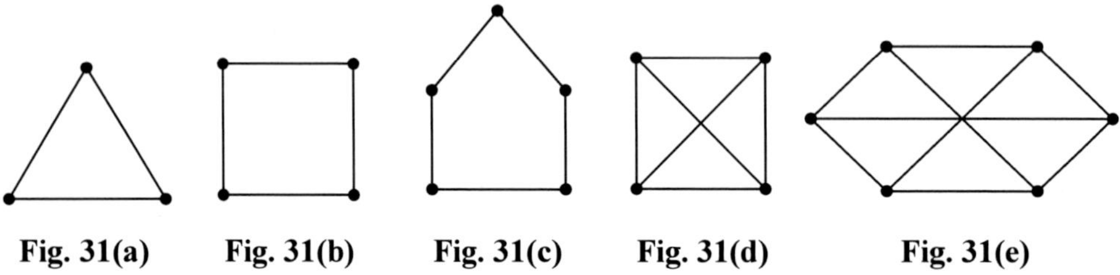

|  |  |  |  |  |
|---|---|---|---|---|
| **Fig. 31(a)** | **Fig. 31(b)** | **Fig. 31(c)** | **Fig. 31(d)** | **Fig. 31(e)** |

The odd degree regular graph (e.g., 3-regular graph) must have even number of vertices, as sum of degrees of all the vertices is always even number (Theorem).

For example, 3-regular graph is as in Fig. 31(d) and 31(e).

**Theorem 4** Prove that K-regular graph must have even number of vertices when the value of K is odd.

### *Proof*

Consider a graph with $n$ vertices. Let S be the sum of degree of all the $n$ vertices of a K-regular graph. Then, we have

$$S = K * n$$

The sum S is always even as the sum of degree of all the vertices of a given graph is even (from theorem 1).

Now, suppose that K is odd, so for S to be even, $n$ must be an even number.

Hence proved.

## 10.24 BIPARTITE GRAPH

A graph G is said to be a bipartite graph if its vertex set V can be partitioned into subsets P and Q such that each edge of G connects a vertex of P to a vertex of Q. It is denoted by $K_{m,n}$ where $m$ and $n$ are the number of vertices in P and Q respectively. For standardization, we will assume that $m \le n$.

For example, the graph in Fig. 32(a) is a bipartite $K_{2,3}$ graph.

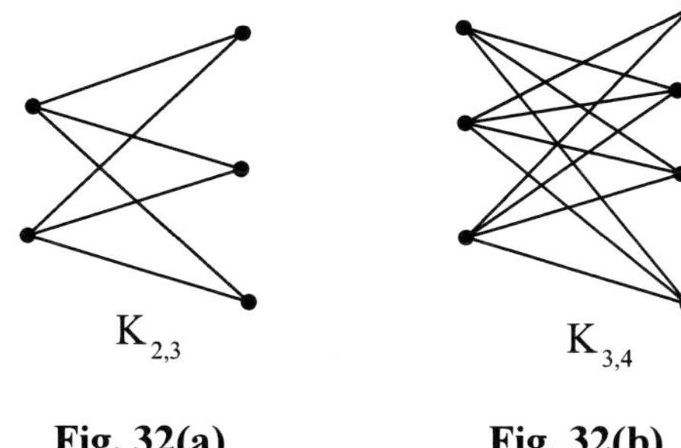



**Fig. 32(a)**          **Fig. 32(b)**

## 10.25 COMPLETE BIPARTITE GRAPH

A graph G is said to be a complete bipartite graph if its vertex set V can be partitioned into subsets P and Q such that each vertex of P is connected to each vertex of Q. It is denoted by $K_{m,n}$ where $m$ and $n$ are the number of vertices in P and Q respectively. For standardization, we will assume that $m \leq n$. Clearly, the graph $K_{m,n}$ has $m * n$ edges.

For example, the graph in Fig. 32(b) is a complete bipartite $K_{3,4}$ graph.

## 10.26 REPRESENTATION OF GRAPHS

There are two standard ways to represent a graph G with the matrices i.e., adjacency matrix and incidence matrix representation.

### (a) Representation of Undirected Graph

#### (i) Adjacency matrix representation

Suppose that an undirected graph G consists of $n$ vertices, say, $v_1, v_2, \ldots, v_n$. Then the *adjacency* matrix $A = [a_{ij}]$ of undirected graph G is an $n \times n$ matrix defined by

$$a_{ij} = \begin{cases} 1 \text{ if } (v_i, v_j) \text{ is an edge i.e., } v_i \text{ is adjacent to } v_j \\ 0 \text{ if there is no edge between } v_i \text{ and } v_j \end{cases}$$

If there exists an edge between vertex $v_i$ and $v_j$ where $i$ is a row and $j$ is a column, then the value of $a_{ij} = 1$.

If there is no edge between vertex $v_i$ and $v_j$ then the value of $a_{ij} = 0$.

For example, the adjacency matrix $M_A$ of graph G shown in Fig. 33(a) is

$$M_A = \begin{array}{c|cccc} & A & B & C & D \\ \hline A & 0 & 1 & 0 & 1 \\ B & 1 & 0 & 1 & 0 \\ C & 0 & 1 & 0 & 1 \\ D & 1 & 0 & 1 & 0 \end{array}$$

Since the graph $g$ consists of four vertices. Therefore, the adjacency matrix is $4 \times 4$ matrix

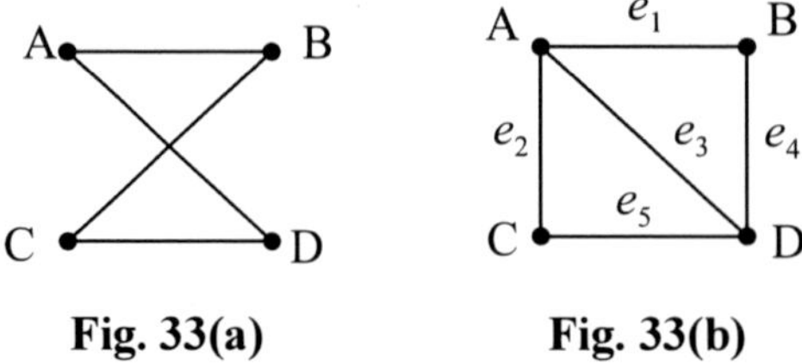

Fig. 33(a)          Fig. 33(b)

### (ii) Incidence matrix representation

Suppose that an undirected graph G consists of $n$ vertices and $m$ edge, then the incidence matrix

$C = [C_{ij}]$ of undirected graph G is an $n \times m$ matrix defined by

$$C_{ij} = \begin{cases} 1 \text{ if the vertex } v_i \text{ is incident by edge } e_j \\ 0 \text{ otherwise} \end{cases}$$

There is a row for every vertex and a column for every edge in the incidence matrix.

The number of 1's in an incidence matrix of an undirected graph (without loops) is equal to the sum of degrees of all the vertices of the graph.

For example, consider the undirected graph G as shown in Fig. 33(b) and its incidence matrix $M_1$ is as shown below:

$$M_1 = \begin{array}{c|ccccc} & 1 & 2 & 3 & 4 & 5 \\ \hline A & 1 & 1 & 1 & 0 & 0 \\ B & 1 & 0 & 0 & 1 & 0 \\ C & 0 & 1 & 0 & 0 & 1 \\ D & 0 & 0 & 1 & 1 & 1 \end{array}$$

Since the undirected graph consists of four vertices and five edges. Therefore, the incidence matrix is an $4 \times 5$ matrix.

## (b) Representation of Directed Graph

### (i) Adjacency matrix representation

Suppose that a directed graph G consists of $n$ vertices, say $v_1, v_2, ...., v_n$. Then the *adjacency* matrix A = $[a_{ij}]$ of directed graph G is an $n \times n$ matrix defined by

$$a_{ij} = \begin{cases} \text{if } (v_i, v_j) \text{ is edge i.e., if } v_i \text{ is the initial vertex and } v_j \text{ is the final vertex} \\ 0 \text{ if there is no edge between } v_i \text{ and } v_j \end{cases}$$

If there exists an edge between vertex $v_i$ and $v_j$ as the initial vertex and $v_j$ as the final vertex, then value of $a_{ij} = 0$.

If there exists no edge between vertex $v_i$ and $v_j = 0$.

The number one's in the *adjacency matrix* of a directed graph is equal to the number of edges.

For example, consider the directed graph shown in Fig. 34(a) and its adjacency matrix $M_A$ is given by

$$M_A = \begin{array}{c|ccccc} & A & B & C & D & E \\ \hline A & 0 & 0 & 1 & 1 & 0 \\ B & 1 & 0 & 0 & 0 & 0 \\ C & 0 & 0 & 0 & 1 & 1 \\ D & 0 & 1 & 0 & 0 & 1 \\ E & 0 & 0 & 0 & 0 & 0 \end{array}$$

Since the directed graph G consists of five vertices. Therefore, the adjacency matrix will be a $5 \times 5$ matrix.

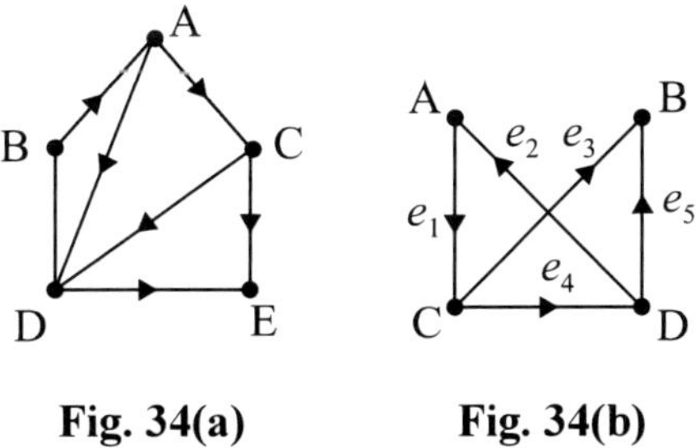


**Fig. 34(a)**          **Fig. 34(b)**

## (ii) Incidence matrix representation

Suppose that a directed graph G consists of $n$ vertices and $m$ edges, then the *incidence matrix* $C = [C_{ij}]$ of directed graph G is an $n \times m$ matrix define by

$$C_{ij} = \begin{cases} 1 & \text{if } v_i \text{ is initial vertex of edge } e_j \\ -1 & \text{if } v_i \text{ is final vertex of edge } e_j \\ 0 & \text{if } v_i \text{ is not incident on edge } e_j \end{cases}$$

The number of 1's in the incidence matrix is equal to the number of edges in the graph.

For example, consider the directed graph G shown in Fig. 34(b) and its incidence matrix $M_1$ is as:

$$M_1 = \begin{array}{c|ccccc} & 1 & 2 & 3 & 4 & 5 \\ \hline A & 0 & -1 & 0 & 0 & 0 \\ B & 0 & 0 & -1 & 0 & -1 \\ C & -1 & 0 & 1 & 1 & 0 \\ D & 0 & 1 & 0 & -1 & 1 \end{array}$$

Since, the directed graph consists of four vertices and five edges. Therefore, the incidence matrix is $4 \times 5$.

## (c) Representation of Multigraph

Multigraphs can be represented only by adjacency matrix representation.

### (i) Adjacency matrix representation

Suppose that a multigraph G consists of $n$ vertices, say $v_1, v_2, ...., v_n$. Then the *adjacency* matrix $A = [a_{ij}]$ of multigraph G is an $n \times n$ matrix defined by

$$a_{ij} = \begin{cases} n & \text{if there is one or more than one edge between vertex } v_i \text{ and } v_j \\ & \text{where } n \text{ is the number of edge} \\ 0 & \text{otherwise} \end{cases}$$

If there exists one or more than one edge between vertex $v_i$ and $v_j$, then $a_{ij} = n$, where $n$ is the number of edges between $V_i$ and $V_j$.

If there is an edge between vertex $v_i$ and $v_j$, then value of $a_{ij} = 0$.

For example, consider the multigraph of Fig. 35 and its adjacency matrix $M_A$ is given as:

$$M_A = \begin{array}{c|ccccc} & A & B & C & D & E \\ \hline A & 0 & 1 & 3 & 0 & 0 \\ B & 1 & 0 & 1 & 1 & 1 \\ C & 3 & 1 & 0 & 1 & 2 \\ D & 0 & 1 & 0 & 0 & 1 \\ E & 0 & 1 & 2 & 1 & 1 \end{array}$$

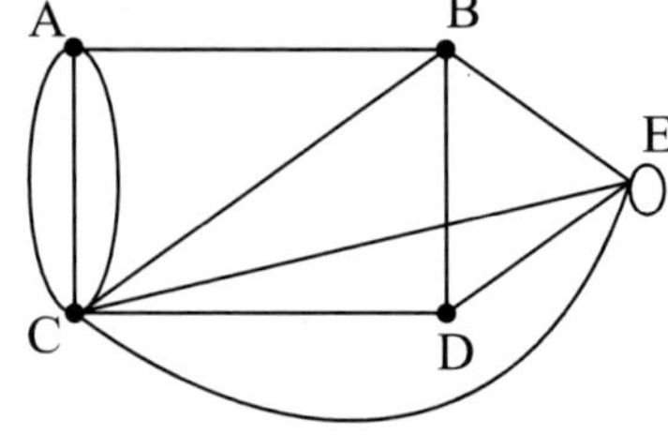


**Fig. 35**

Since, the multigraph consists of five vertices, so, the adjacency matrix is a $5 \times 5$ matrix.

**Example 10:** Find the adjacency matrix M = $[m_{ij}]$ of graph in Fig. 36(a)–36(b).

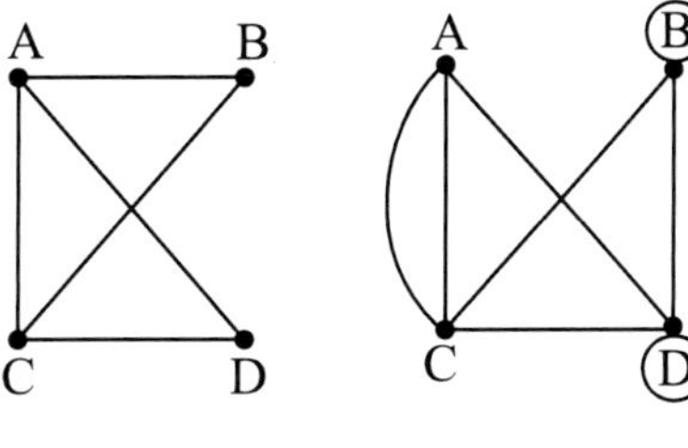

**Fig. 36(a)**　　　**Fig. 36(b)**

### Solution

Set $m_{ij} = n$ if there are $n$ edges $\{v_i, vj\}$ and
Set $m_{ij} = 0$ otherwise. Hence:

(a) $M =$

|   | A | B | C | D |
|---|---|---|---|---|
| A | 0 | 1 | 1 | 1 |
| B | 1 | 0 | 1 | 0 |
| C | 1 | 1 | 0 | 1 |
| D | 1 | 0 | 1 | 0 |

(b) $M =$

|   | A | B | C | D |
|---|---|---|---|---|
| A | 0 | 0 | 2 | 1 |
| B | 0 | 1 | 1 | 1 |
| C | 2 | 1 | 0 | 1 |
| D | 1 | 1 | 1 | 1 |

**Example 11:** Draw the graph G according to each adjacency matrix:

(a) $M = \begin{bmatrix} 0 & 1 & 0 & 0 \\ 1 & 0 & 1 & 1 \\ 0 & 1 & 0 & 0 \\ 0 & 1 & 0 & 0 \end{bmatrix}$
(b) $M = \begin{bmatrix} 1 & 0 & 2 & 1 \\ 2 & 1 & 0 & 2 \\ 1 & 1 & 0 & 1 \\ 1 & 2 & 1 & 0 \end{bmatrix}$

### Solution

(a) Since M is a 4-square matrix, so, G has 4 vertices, say A, B, C, D. Draw an edge from $A_i$ to $A_j$ when M = 1 and zero otherwise and the graph appears as in Fig. 37(a).

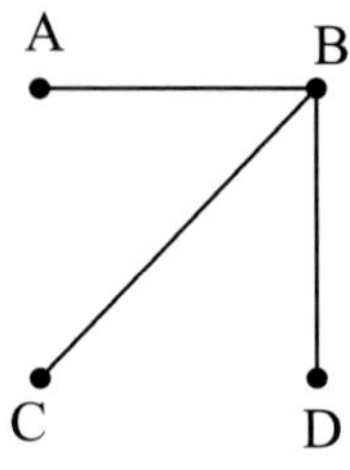

**Fig. 37(a)**

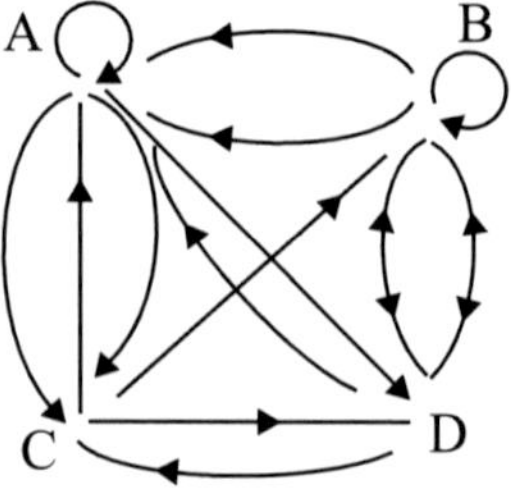

**Fig. 37(b)**

(b) Since M is a 4-square matrix, so, G has 4 vertices, say A, B, C, D. Draw $n$ edge from $A_i$ to $A_j$ when $M_{ij} = n$ and zero otherwise. Also draw $n$ loops at $A_i$ when $M_{ij} = n$ and the graph appears as in Fig. 37(b).

**Example 12:**  Draw a directed weighted graph whose weight matrix W is as given below:

$$W = \begin{bmatrix} 0 & 2 & 1 & 5 \\ 0 & 0 & 0 & 1 \\ 4 & 0 & 0 & 3 \\ 0 & 3 & 2 & 0 \end{bmatrix}$$

*Solution*

Since W is a 4-square matrix, so, G has 4 vertices, say $V_1$, $V_2$, $V_3$, $V_4$. Draw an edge from $V_i$ to $V_j$ with weight $w$ when $W_{ij} = w$ and zero otherwise, and the graph appears as in Fig. 38.

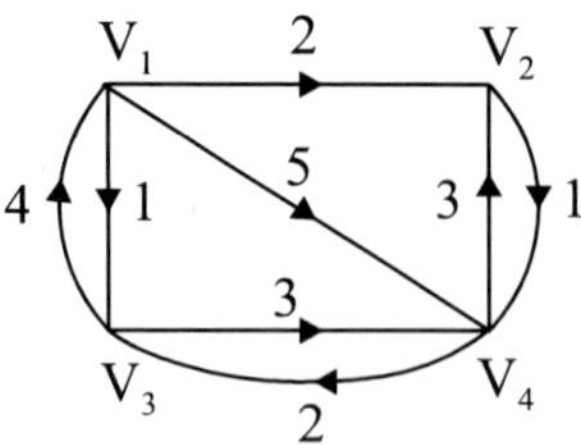

**Fig. 38**

## 10.27 TRAVERSABLE MULTIGRAPH

A multigraph G is said to be traversable if it "can be drawn without any breaks in the curve and without repeating any edges" i.e., if there is a path which includes all the vertices and uses each edge exactly once. Since no edge is used twice, such a path must be a trail and is called as **traversable trail**.

For example, the Fig. 39(b) shows the traversable trail of the multigraph in Fig. 39(a).

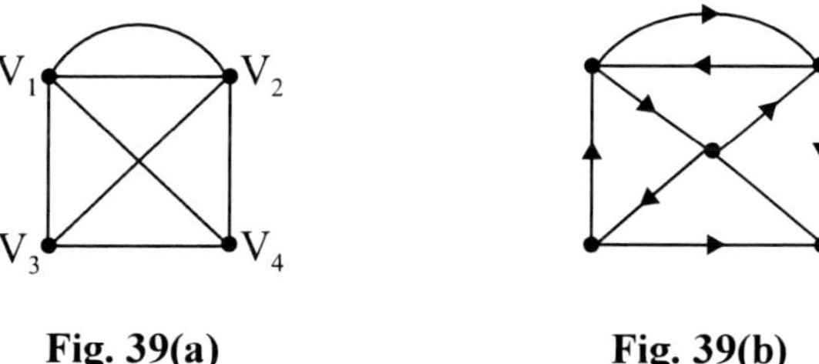

**Fig. 39(a)**  **Fig. 39(b)**

**Theorem 5** The *necessary and sufficient condition* for a multigraph to be traversable is that it should be connected and has either zero or two odd vertices (vertex with odd degree). If a multigraph has two vertices of odd degree, then the traversable trail may begin at either odd vertex and will end at the other odd vertex.

### *Proof*

Suppose that a multigraph is traversable and the traversable trail does not begin or end at even vertex M. Then whenever the traversable trail enters M by an edge, there must always be an edge which is not previously used and by which the traversable trail can leave M. Thus the edges in trail incident with M must appear in pair and so M is an even vertex. Therefore, if N is a odd vertex, the traversable trail must begin or end at N. Consequently, a multigraph with more than two odd vertices cannot be traversable.

For example the graph in Fig. 39(a) has two even vertex ($V_1$, $V_2$) and two odd vertex ($V_3$, $V_4$). So, the multigraph is traversable and the traversable trail begins at odd vertex $V_3$ and end at other odd vertex $V_4$.

## 10.28 THE BRIDGES OF KONIGSBERG [MDU 2008]

The eighteenth century East Prussian town of Konigsberg included two islands and seven bridges as shown in Fig. 40(a). The problem was that beginning anywhere and ending anywhere, can a person walk through town crossing all the seven bridges but not crossing any bridge twice? This problem was solved by Euler. Euler in 1736 proved that such a walk is impossible. He replaced the two islands (A and B) and the two banks of the rivers (C and D) by points and the bridges with curves as in Fig. 40(b). Now, it is clear that the problem of crossing each of Konigsberg bridges once and only once is equivalent to finding a path in the multigraph in Fig. 40(b) that traverses each of the edges once and only once.

Euler proves that the multigraph in Fig. 40(b) is not traversable and hence the walk in Konigsberg is impossible. Recall that a multigraph is traversable if it has either zero or two vertices of odd degree (Theorem). But the multigraph corresponding to Konigsberg bridge problem has four odd vertices. Thus one cannot walk through Konigsberg bridge so that each bridge is crossed exactly once.

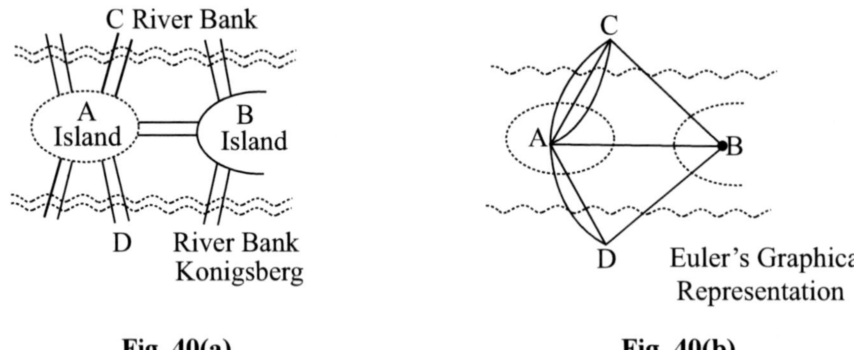



**Fig. 40(a)**　　　　　　　　　　**Fig. 40(b)**

## 10.29 EULER PATH

An Euler path in a graph G is a path whose edge list contains each edge of the graph exactly once i.e. path that traverses each edge in the graph once and only once.

## 10.30 EULER CIRCUIT

An Euler circuit in a graph G is a *circuit* that traverses each edge in the graph once and only once. In other words, an Euler circuit is a path that contains each edge of the graph exactly once and is closed known as closed traversable trail. In Euler circuit, each edge is traversed exactly once, but vertices may be repeated (or some vertices may not be included).

## 10.31 EULER GRAPH

An Euler graph G is a graph that possess an Euler circuit i.e., a graph G is Eulerian if their exists a closed traversable trail called Eulerian trail.

　　For example, the graph in Fig. 41($a$) is an Euler graph and the Euler circuit for this graph in Fig. 41($b$) is as follows:

　　$V_1, V_2, V_5, V_4, V_2, V_4, V_3, V_1, V_3, V_5, V_1$

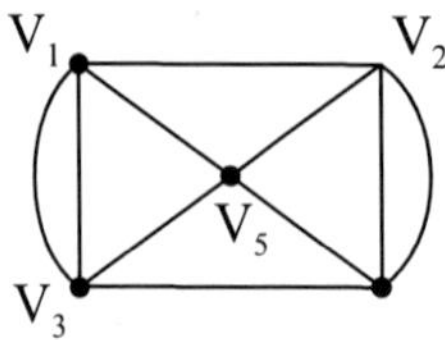



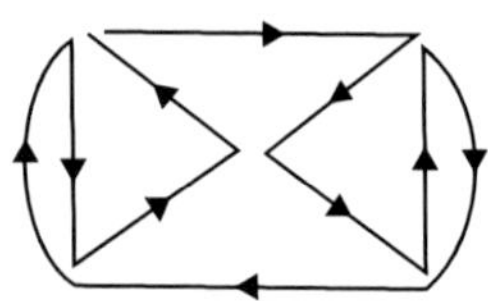

**Fig. 41(a)**　　　　　　　　　　**Fig. 41(b)**

A **simple criteria** to find out whether a graph is Eulerian or not is that a connected undirected graph is Eulerian if and only if each vertex has even degree (**Theorem**).

**Theorem 6** An undirected graph possesses an Eulerian path if and only if it is connected and has either zero or two vertices of odd degree.

### *Solution*

Suppose that the graph possesses a Eulerian path, that the graph must be connected is obvious. When the Eulerian path is traced, we observe that every time the path meets a vertex, it goes through two edges, which are incident with the vertex and have not been traced before. Thus, except for the two vertices at the two ends of the path, the degree of any vertex in the graph must be even. If the two vertices at the two ends of the Eulerian path are distinct, then they are the only two vertices with odd degree. If they coincide, all vertices have even degree, and the Eulerian path becomes a Eulerian circuit. Thus, the necessity of the stated condition is proved.

To prove the sufficiency of the stated condition, we construct an Eulerian path by starting at one of the two vertices that are of odd degree and going through the edges of the graph in such a way that no edge will be traced more than once. For a vertex of even degree, whenever the path enters the vertex through an edge, it can always leave the vertex through another edge that has not been traced before. Therefore when the construction eventually comes to an end, we must have reached the other vertex of odd degree. If all of the edges in the graph were traced this way, clearly, we would have an Eulerian path. If not all of the edges in the graph were traced, clearly, we shall remove those edges that have been traced and obtain a subgraph formed by the remaining edges. The degrees of vertices of this subgraph are all even. Moreover, this subgraph must touch the path that we have traced at one or more vertices since the original path is connected. Starting from one of these vertices, we can again construct a path that passes through the edges. Because the degrees of all the vertices are all even this path must return eventually to the vertex at which it starts. We can combine this path with the path we have constructed to obtain one which starts and ends at the two vertices of odd degree. If necessary, the argument is repeated until we obtain a path that transverses all the edges in the graph.

**Theorem 7** An undirected graph possesses a Eulerian circuit if and only if it is connected and each vertex has even degree.

### *Proof*

Suppose G is Eulerian and T is the closed Eulerian trail. For any vertex $v$ of G, the trial T enters and leaves $v$ the same number of times without repeating any edge. Hence, $v$ has even number.

Suppose conversely that each vertex of G has even degree. We construct the Eulerian trail. We begin a trail $T_1$ at any edge $e$. We extend $T_1$ by adding one edge after the other. If $T_1$ is not closed at any step, say, $T_1$ begins at $u$ but ends at $v \neq u$, then only an odd number of edges incident on $v$ appear in $T_1$; hence we can extend $T_1$ by adding another edge incident on $v$. Thus, we can continue to extent $T_1$ until $T_1$ returns to its initial vertex $u$, i.e., until $T_1$ is closed. If $T_1$ includes all the edges of G, then $T_1$ is our Eulerian trail.

Suppose $T_1$ doesn't include all the edges of G. Consider the graph H obtained by deleting all edges of $T_1$ from G. H may not be connected, but each vertex of H has even degree since $T_1$ contains an even number of the edges incident on any vertex. Since G is connected, there is an edge $e'$ of H which has an endpoint $u'$ in $T_1$, we construct trail $T_2$ in H beginning at $u'$ and using $e'$. Since all vertices in H has even degree, we can continue to extend $T_2$ in H until $T_2$ returns to $u'$ as shown in Fig. 42. We can clearly put $T_1$ and $T_2$ together to form a large closed trail in G. We continue to this process until all the edges of G are used. We finally obtain an Eulerian trail and so G is Eulerian.

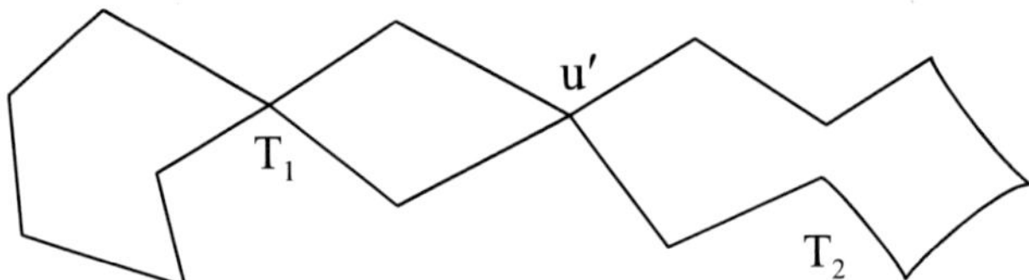


**Fig. 42**

**Theorem 8** A directed graph possesses an Eulerian circuit if and only if it is connected and the incoming degree of every vertex is equal to its outgoing degree.

### *Proof*

A directed graph possesses an Eulerian path if and only if it is connected and the incoming degree of every vertex is equal to its outgoing degree with the possible exception of two vertices. For these two vertices, the incoming degree of one is one larger than its outgoing degree and the incoming degree of the other is one less than its outgoing degree where incoming degree of a vertex is the number of edges that are incident into it and outgoing degree of a vertex is the number of edges that are incident from it.

## 10.32 HAMILTONIAN PATH

A Hamiltonian path in a graph G is a path whose vertex list contains each vertex of the graph exactly once i.e., a path that traverses each vertex exactly once.

## 10.33 HAMILTONIAN CIRCUIT

A Hamiltonian circuit in a graph G is a Hamiltonian path in which initial (starting) vertex is same as final (terminating) vertex *i.e.*, a circuit which passes through each of the vertices in a graph exactly once. In Hamiltonian circuit, each vertex is traversed exactly once, but edges may be repeated (or some edges may not be included).

## 10.34 HAMILTONIAN GRAPH

A Hamiltonian graph G is a graph that possesses a Hamiltonian circuit.

For example, the graph in Fig. 43(a) is a Hamiltonian graph and the Hamiltonian circuit is as follows:

$$V_1, V_2, V_5, V_6, V_3, V_8, V_7, V_4, V_1.$$

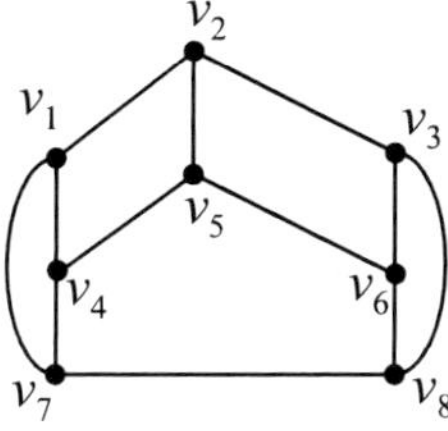

Hamiltonian Graph

**Fig. 43(a)**

Note that a Eulerian circuit traverses each edge exactly once, but it may repeat vertices, while a Hamiltonian circuit traverses each edge exactly once, but it may repeat edges Fig. 43(b) given an example of graph that is Eulerian but not Hamiltonian and Fig. 43(c) given an example of graph that is Hamiltonian but not Eulerian.

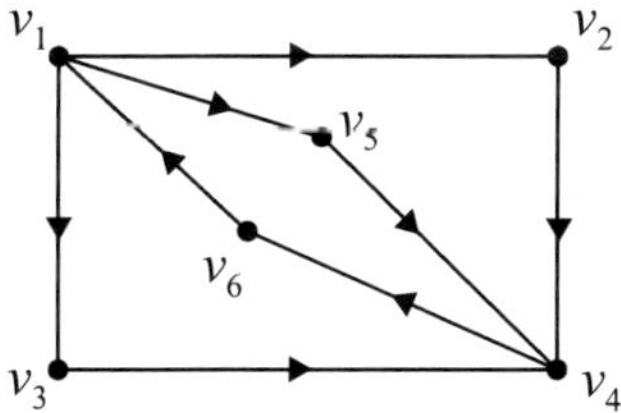

Eulerian but not Hamiltonian

**Fig. 43(b)**

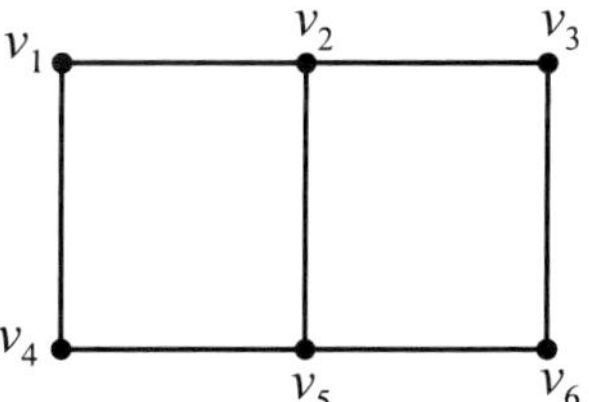

Hamiltonian but not Eulerian

**Fig. 43(c)**

There is no simple criterion to find out whether or not a graph is Hamiltonian as is for Euler graph. The only sufficient condition to find out whether or not a graph is Hamiltonian is as follows (given by G.A. Dirac.)

Let G be a connected graph with $n$ vertices. Then G is Hamiltonian if $n \geq 3$ and $n \leq \deg(v)$ for each vertex $v$ in G (**Theorem**).

**Example 13:** Consider the graph in Fig. 44 and check if the graph:

    (i) Is traversable i.e., has an Euler path?

    (ii) Is Eulerian graph i.e., has Euler circuit?

(iii) Have a Hamiltonian path?

(iv) Have a Hamiltonian circuit?

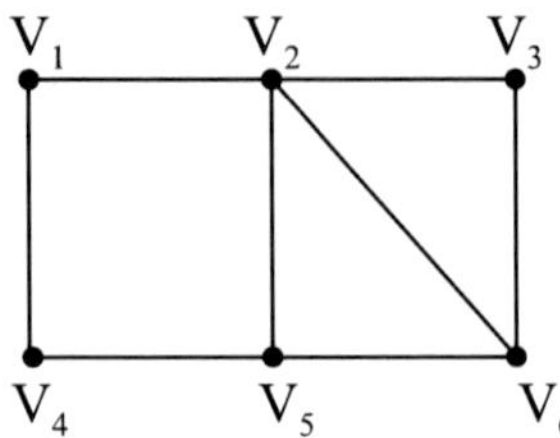

**Fig. 44**

***Solution***

(i) A graph G is traversable i.e., has an Euler path if either 0 or 2 vertices have odd degree. As there are only two vertices $V_5$ and $V_6$ with odd degree. So, the graph has an Euler path. In case of Euler path, if the graph has two odd vertices, then the Euler path begins from one odd vertex and ends at another odd vertex. One such Euler path starting from vertex $V_5$ is given by

$$V_5, V_4, V_1, V_2, V_3, V_6, V_2, V_5, V_6$$

(ii) A graph is Eulerian i.e., has an Euler circuit if all the vertices are of odd degree. As the given graph contains two odd vertices as stated above in part (i), so the given graph doesn't contain an Euler circuit.

(iii) Hamiltonian path is given by

$$V_1, V_2, V_3, V_6, V_5, V_4$$

(iv) Hamiltonian circuit is given by

$$V_1, V_2, V_3, V_6, V_5, V_4, V_1$$

**Example 14:** Which of the graphs in Fig. 45(a)–45(b)

(i) Are not traversable i.e. don't have an Euler path?

(ii) Are Eulerian graph i.e. have Euler circuit?

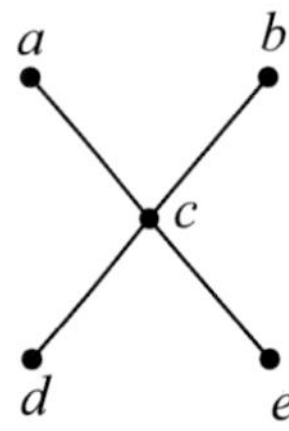

**Fig. 45(a)**

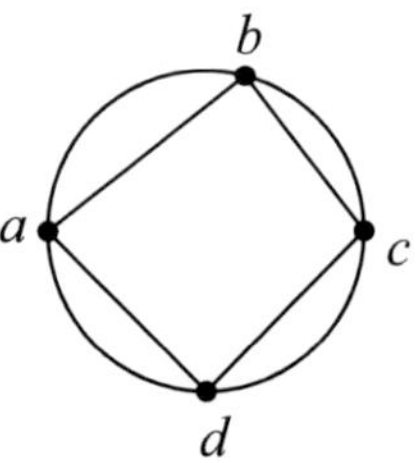

**Fig. 45(b)**

### *Solution*

(i) A graph G is traversable i.e., has an Euler path if either 0 or 2 vertices have odd degree.
   (a) As the graph in Fig. 45(a) contains all four odd vertices, so the given graph dosen't contain an Euler path.
   (b) As the graph in Fig. 45(b) contains all four even vertices, so the given graph contains an Euler circuit and is given by *a, b, c, d, a, b, c, d, a.*

(ii) A graph is Eulerian i.e., has an Euler circuit if all the vertices are of odd degree,
   (a) As the graph in Fig. 45(a) contains all four odd vertices, so the given graph dosen't contain an Euler circuit.
   (b) As the graph in Fig. 45(b) contains all four even vertices, so the given graph contains an Euler circuit and is given by *a, b, c, d, a, b, c, d, a.*

---

**Exercise 10.2**

1. Which of the graph in Fig. Ex. 1(a)–Ex. 1(d)
   (i) Are traversable i.e., don't have a Euler path?
   (ii) Are not traversable i.e., don't have an Euler path?
   (iii) Are Eulerian graph

[*Hint.* A graph is traversable i.e., Euler path if either 0 or 2 vertices have odd degree and graph is Eulerian i.e., have Euler circuit if all the vertices are of odd degree.]

2. Show that Euler formula holds true for the following connected planar graphs as shown in Fig. Ex. 2(a)–Ex. 2(c).

3. Give an example of the graph that has

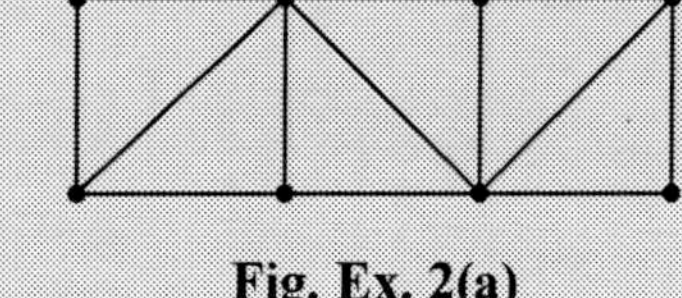

**Fig. Ex. 2(a)**

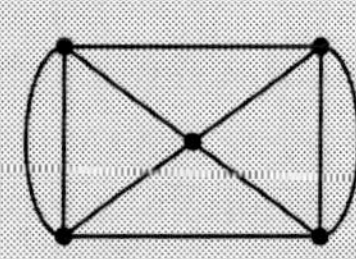

**Fig. Ex. 2(b)**    **Fig. Ex. 2(c)**

   (i) both an Eulerian circuit and a Hamiltonian circuit.
   (ii) an Eulerian circuit but no Hamiltonian circuit.
   (iii) no Eulerian circuit but has Hamiltonian circuit.
   (iv) no Eulerian circuit and no Hamiltonian circuit.

4. A complete bipartite graph is denoted by $K_{m,n}$.
   (i) is there a Hamiltonian path in $K_{4,4}$, $K_{4,5}$ and $K_{4,6}$.
   (ii) is there a Hamiltonian circuit in $K_{4,4}$, $K_{4,5}$ and $K_{4,6}$.

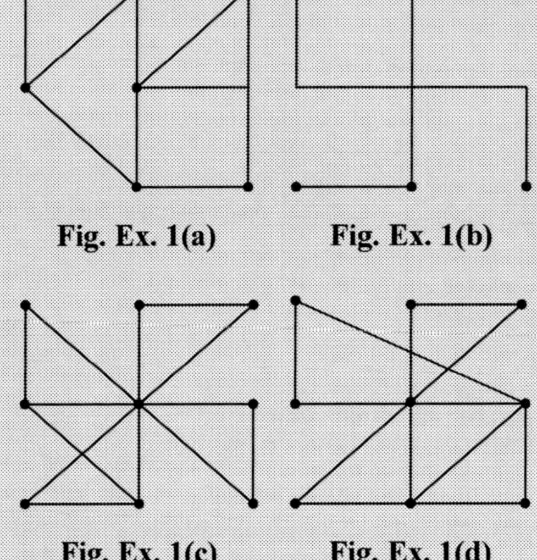

**Fig. Ex. 1(a)**    **Fig. Ex. 1(b)**

**Fig. Ex. 1(c)**    **Fig. Ex. 1(d)**

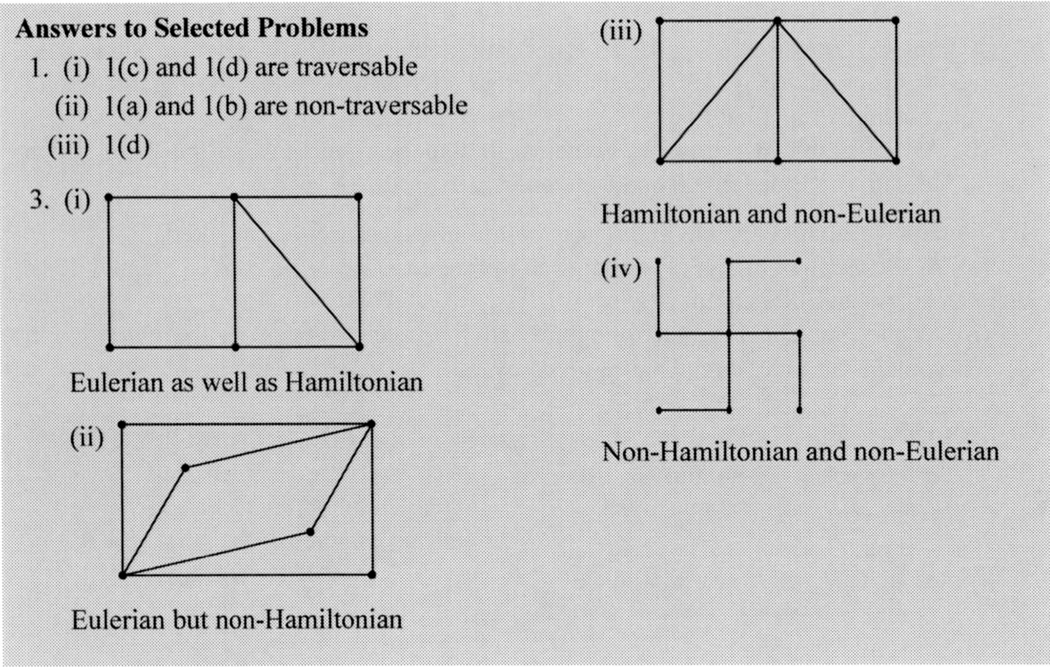

**Answers to Selected Problems**

1. (i) 1(c) and 1(d) are traversable
   (ii) 1(a) and 1(b) are non-traversable
   (iii) 1(d)

3. (i)

Eulerian as well as Hamiltonian

(ii)

Eulerian but non-Hamiltonian

(iii)

Hamiltonian and non-Eulerian

(iv)

Non-Hamiltonian and non-Eulerian

## 10.35 PLANAR GRAPH

A graph or multigraph that can be drawn on plane in such a way that no edges cross one another is called a planar graph.

For example, a complete graph $K_4$ in Fig. 46(a) is a planar graph.

Another example of planar graph is as in Fig. 46(b).

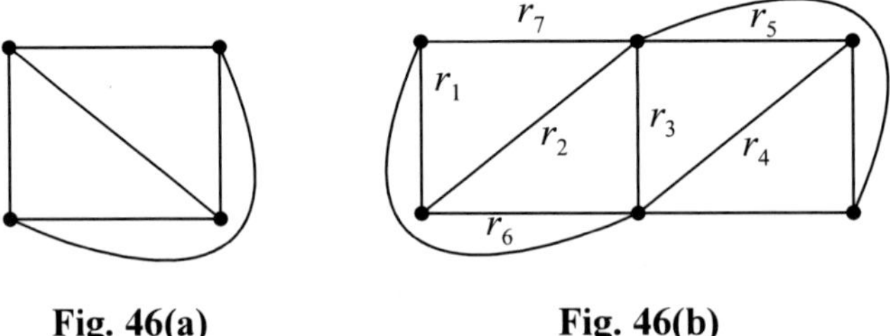

**Fig. 46(a)**          **Fig. 46(b)**

**(a) Region of a Planar Graph**

A region of a planar graph G is defined as an area of plane that is bounded by edges and is not further divided into subareas. A planar graph divides the plane into one or more finite regions and one infinite region where finite region is defined as a region whose area is finite and *infinite region* is defined as a region whose area is infinite. A planar graph has always only one infinite region.

For example, the planar graph G is Fig. 46(b) has a total of seven regions with six finite regions, namely, $r_1$, $r_2$, $r_3$, $r_4$, $r_5$, $r_6$ and one infinite region, namely, $r_7$.

### (b) Properties of Planar Graph

1. If a connected planar graph G has $v$ vertices, $e$ edges and $r$ regions, then $v - e + r = 2$ (Euler Formula).
2. If a connected planar graph G has $e$ edges and $r$ regions, then $r \leq 2/3\, e$.
3. If a connected planar graph G has $v$ vertices and $e$ edges, then $3v - e \geq 6$.

**Theorem (Euler Formula) 9** For any connected planar graph, $V - E + R = 2$ where V, E and R are the number of vertices, edges and regions of the graph respectively.

*Proof*

We will prove this theorem by induction on the number of edges.

**Basis of Induction.** Assume that the edges $e = 1$. Then we have two cases, graphs of which are shown in Fig. 45 and Fig. 46.

In Fig. 47(a) we have $V = 2$ and $R = 1$. Thus $2 - 1 + 1 = 2$.

In Fig. 47(b) we have $V = 1$ and $R = 2$. Thus $1 - 1 + 2 = 2$.

Hence, the basis of induction is verified.

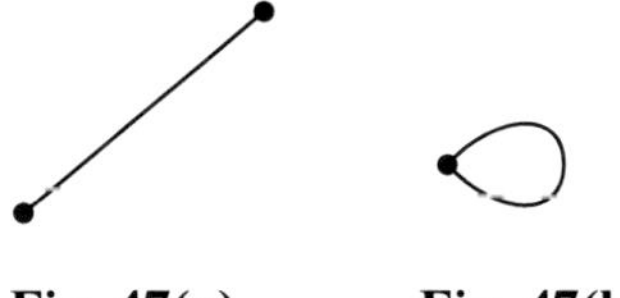

**Fig. 47(a)**          **Fig. 47(b)**

**Induction Step.** Let us assume that the formula holds for connected planar graph with K edges.

Let G be a graph with K + 1 edges.

Firstly, we suppose that G contains no circuits. Now, take a vertex $v$ and find a path starting at $v$. Since G is circuit free, so, whenever we find an edge, we have a new vertex. At least we will reach a vertex $u$ with degree 1. So, we cannot move further as shown in Fig. 48(a).

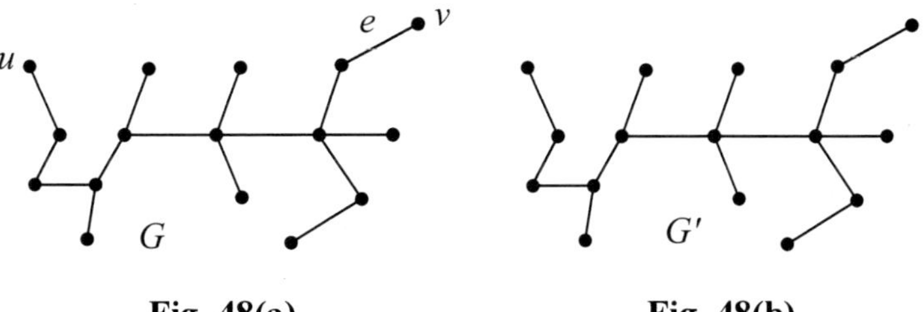

**Fig. 48(a)**                         **Fig. 48(b)**

Now remove vertex $v$ and the corresponding edges incident on $v$. So, we are left with a graph G′ having K edges as shown in Fig. 48(b).

Hence, by inductive assumption, Euler's formula holds for G′.

Now, since G has one more edge than G′, one more vertex than G′ with same number of regions as in G′. Hence, the formula also holds for G.

Secondary, we assume that G contains a circuit and $e$ is an edge in the circuit shown in Fig. 49.

Now, as $e$ is the part of a boundary for two regions. So, we only remove the edge and we are left with graph G′ having K edges (Fig. 50).

Hence, by inductive assumption, Euler's formula holds for G′.

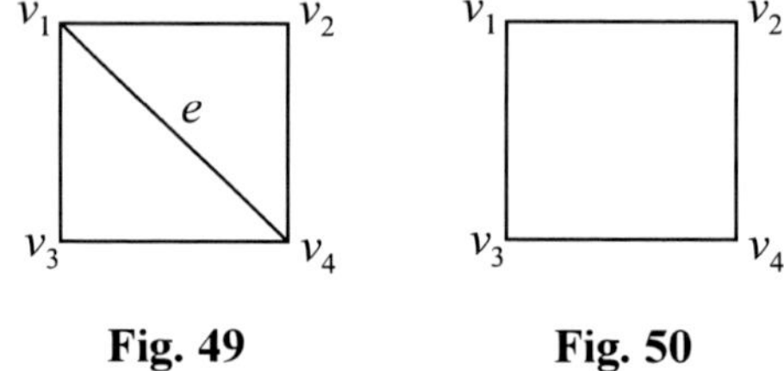

**Fig. 49**          **Fig. 50**

Now, since G has one more edge than G′, one more region than G′ with the same number of vertices as G′. Hence the formula also holds for G′. Hence the formula also holds for G, which, verifies the inductive step and hence proves the theorem.


## 10.36 NON-PLANAR GRAPH

A graph is called a non-planar graph if it cannot be drawn on plane in such a way that no edges cross one another.

For example, the graph in Fig. 51 is a non-planar graph.

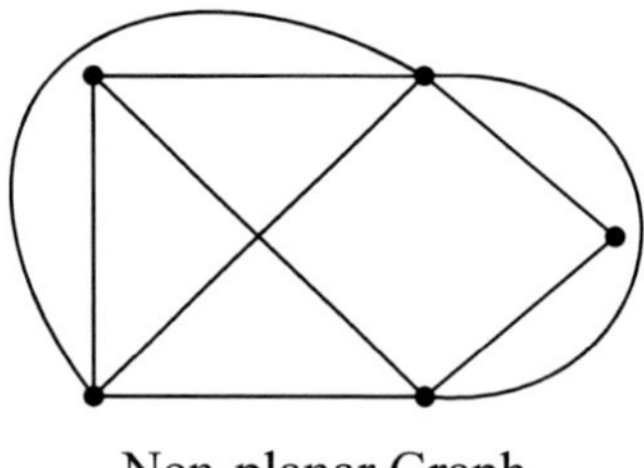

Non-planar Graph

**Fig. 51**


**Theorem (Kuratowski's) 10** A graph is non-planar if and only if it contains a subgraph homeomorphic to $K_{3,3}$ or $K_5$. These two graph $K_5$ and $K_{3,3}$ are called Kuratowski's graphs and are shown in Fig. 52(a) and 52(b).

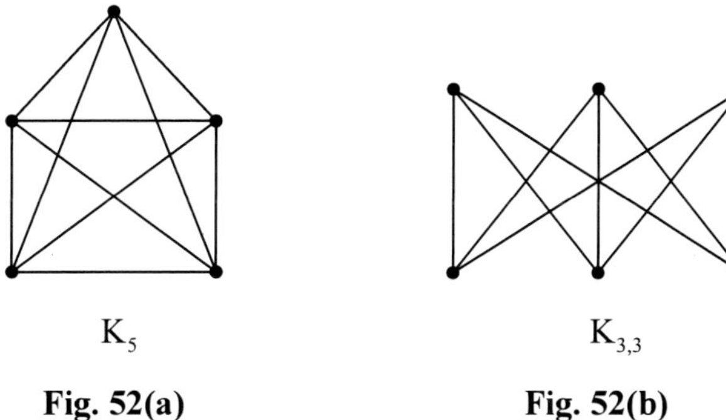

$$K_5 \qquad\qquad K_{3,3}$$

**Fig. 52(a)**  **Fig. 52(b)**

**Theorem 11** Let G be a finite connected planar graph with at least three vertices. Show that G has at least one vertex of degree 5 or less.

### *Proof*

Consider a graph G with all of its vertices of degree 6 or more. Then the sum of degrees of all the vertices would be greater than or equal to $6v$. But, as we know that the sum of degrees of all the vertices in a graph G is twice the number of edges. So, we have

$$6v \leq 2e$$

or $\qquad v \leq e/3$ $\hspace{6cm}$ (1)

But, for any Planar graph, we have

$$r \leq 2e/3$$ $\hspace{6cm}$ (2)

From Euler formula's we have

$$v - e + r = 2$$ $\hspace{6cm}$ (3)

Now, putting value of $v$ and $r$ from equation (1) and (2) in equation (3), we have

$$e/3 - e + 3e/3 \geq 2$$
$$\text{i.e.} \quad 0 \geq 2$$

which is not true. Hence we conclude that there exists at least one vertex in G with degree 5 or less.

**Example 15:** Draw the graph $K_{2,4}$.

### *Solution*

$K_{2,4}$ consists of 6 vertices partitioned into a set X of 2 vertices, say, $v_1$, $v_2$ and a set Y of 4 vertices, say, $u_1$, $u_2$, $u_3$, $u_4$ and all possible edges from a vertex $v_i$ in set X to a vertex $u_j$ in set Y. Thus there are 8 edges and the graph is as shown in Fig. 53.

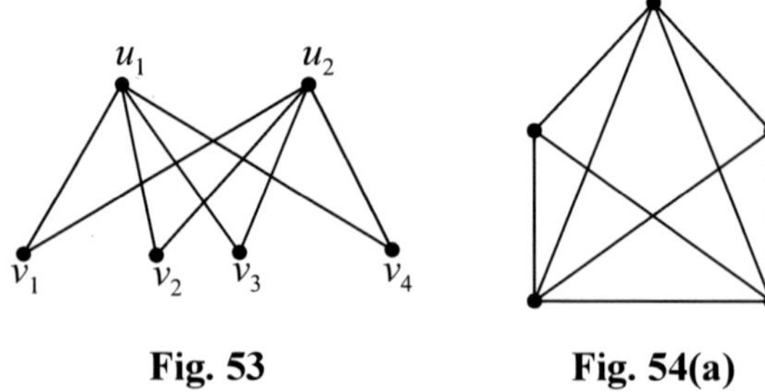

**Fig. 53**                    **Fig. 54(a)**

**Example 16:**  Draw a planar representation of the graph G in Fig. 54(a), if possible.

*Solution*

The planar representation of the graph is as in Fig. 54(b).

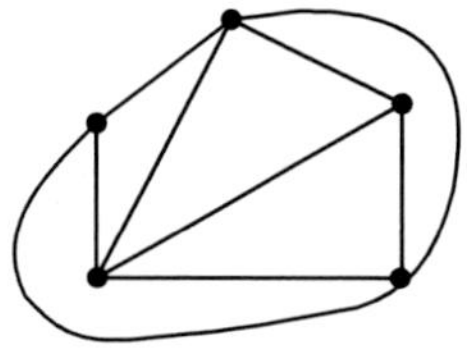

**Fig. 54(b)**

**Example 17:**  Show that Euler formula holds true for the connected planar graphs as shown in Fig. 55.

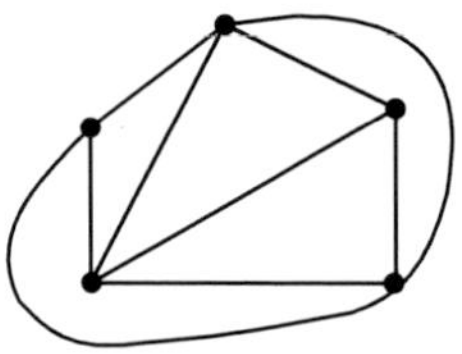

**Fig. 55**

*Solution*

Here V = 5, E = 9, R = 6

So,   V − E + R = 5 − 9 + 6 = 2,        Hence Proved.


## 10.37 GRAPH COLORING

Suppose that G = (V, E) is a graph with no multiple edges. A vertex colouring or simply colouring of G is an assignment of colours to the vertices of G in such a way that adjacent vertices have different colours. A graph G is $k$-colourable if there exists a colouring of G that uses only $k$-colours.

**Proper/improper Coloring.** A colouring is proper if any two adjacent vertices $u$ and $v$ have different colours, otherwise, it is called improper colouring.

## Exercise 10.3

1. Draw a planar representation of each graph G in Fig. Ex. 1(a)–Ex. 1(d), if possible, otherwise show that it has a subgraph homeomorphic to $K_5$ or $K_{3,3}$.

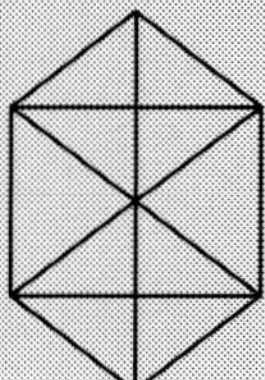

**Fig. Ex. 1(a)**

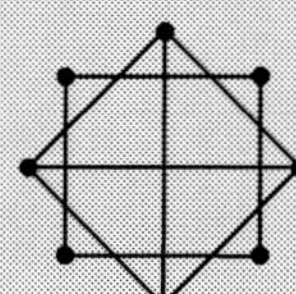

**Fig. Ex. 1(b)**

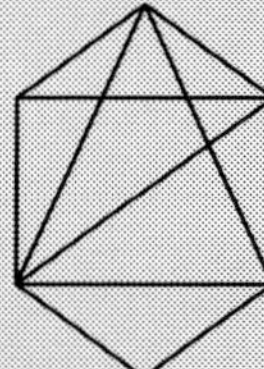

**Fig. Ex. 1(c)**

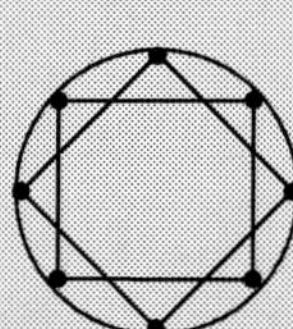

**Fig. Ex. 1(d)**

2. Draw the complete Bipartite graph $K_{2,3}$, $K_{3,4}$, $K_{2,5}$ and $K_{4,6}$.

3. Draw the undirected graph $K_4$ and $K_6$.

4. Give the planar representation of the graph as in Fig. Ex. 4(a)–Ex. 4(c), if possible

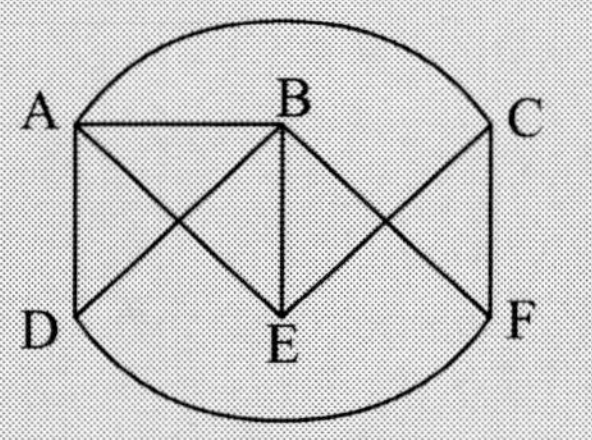


**Fig. Ex. 4(a)**

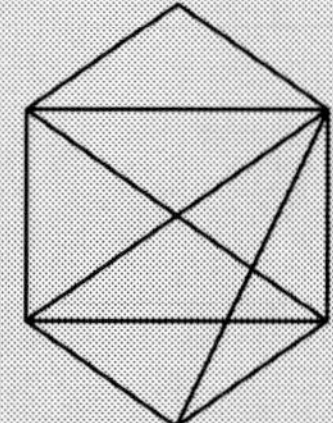

**Fig. Ex. 4(b)**

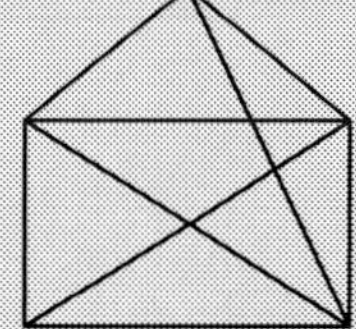

**Fig. Ex. 4(c)**

5. Which connected graph can be both regular and bipartite?

   [**Hint**. $K_{m,m}$ bipartite graph]

### Answers to Selected Problems

1. (a) Non-planar
   (b) Non-planar
   (c) Planar representation doesn't exist
   (d)

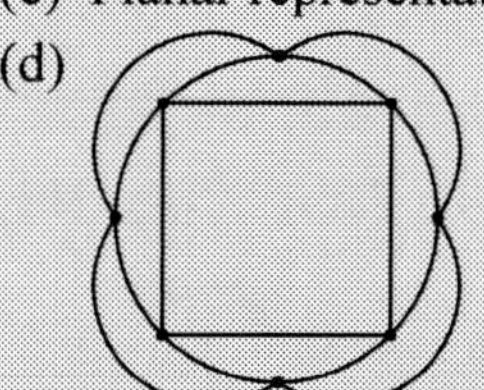

2.

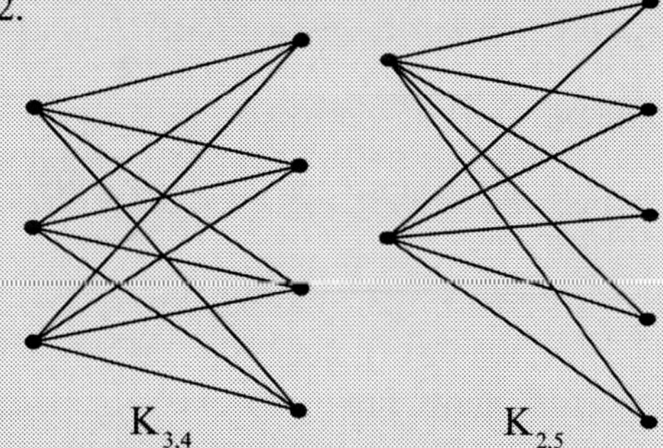


Similarly $K_{2,3}$ and $K_{4,6}$ may be drawn.

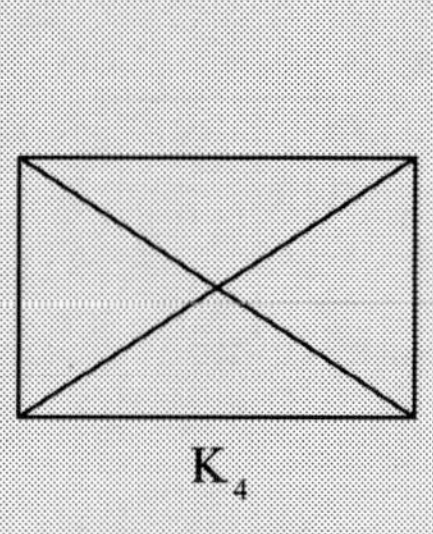


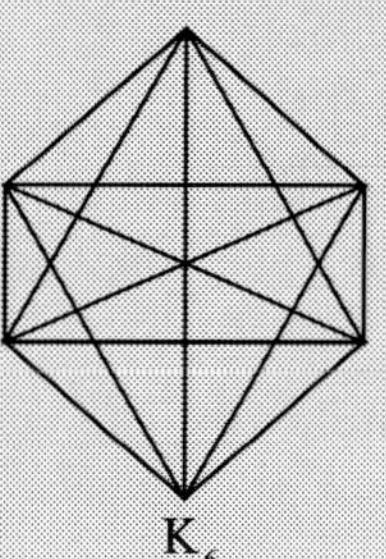

**Chromatic number.** A graph can be coloured by assigning a different colour to each of its vertices. However, for most graphs a colouring can be found that uses fewer colours than the number of vertices in the graph. The minimum number of colours needed to produce a proper colouring of a graph G is called the chromatic number of G and is denoted by $\chi$ (G).

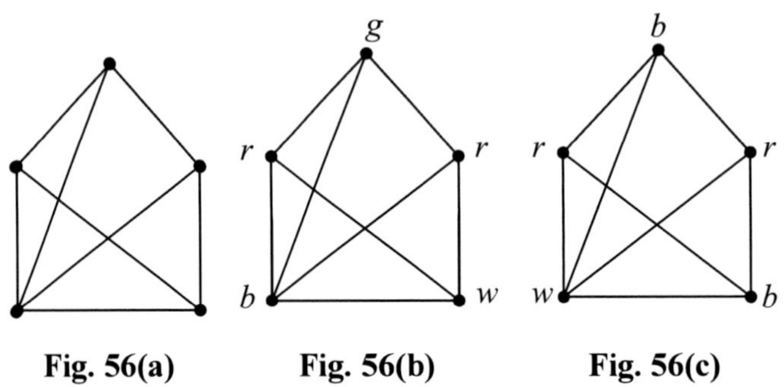

**Fig. 56(a)**          **Fig. 56(b)**          **Fig. 56(c)**

**Example 18:**  Consider the following graph G in Fig. 56(a) and colour C = {r, g, b, w}. Color the properly:

(i) Using all colours
(ii) Using fewer colours (if possible)

*Solution*

(i) Fig. 56(b) shows the graph properly coloured with all the four given colours, i.e. *r, g, b* and *w.*
(ii) Fig. 56(c) shows the graph properly coloured with three colours, i.e. *r, b,* and *w.*

It is not possible to colour this graph properly with two colours.
The graph shown in Fig. 56(c) is minimum 3-colourable, hence, $\chi$ (G) = 3.

**Example 19:**  The chromatic number of complete graph $k_n$ *d* with *n* vertices is *n.*

*Solution*

Since every vertex is adjacent to every other vertex, so, no two vertices can be assigned the same colour. Thus a colouring of $k_n$ can be constructed using *n* colours by assigning a different colour to each vertex. Hence the chromatic number of $k_n$ i.e. $\chi$ $(k_n) = n.$

**Example 20:**  The chromatic number of complete bipartite graph $K_{m,n}$ where *m* and *n* are positive integers is two.

*Solution*

The number of colours needed does not depend upon *m* and *n.* However, only two colours are needed. Firstly, colour the set of *m* vertices with one colour and, then, the set of *n* vertices with a second colour. Since, edges connect only a vertex from the set of *m* vertices and a vertex from the set of $n$ vertices, no two adjacent vertices have the same colour. Hence, $c(k_m, n) = 2$.

***Note:*** There is no simple way to actually determine whether an arbitrary graph is $k$-colourable. However, the following theorem gives a simple characterization of 2-colourable graphs.

**Theorem 12** Prove that the following are equivalent or a graph G:
  (i)  G is 2-colourable
 (ii)  G is bipartite
(iii)  Every cycle of G has even Length.

***Proof***

  (i)  Suppose that every cycle of G has even length. We pick a vertex in each connected component and paint it with first colour, say blue. We then successively paint all the vertices as follows; if a vertex is painted blue, then any vertex adjacent to it will be painted the second colour, say green. If a vertex is painted green, then any vertex adjacent to it will be painted the second blue. Since every cycle has even length, no adjacent vertices will be painted the same colour. Hence G is 2-colourable.
 (ii)  Suppose $g$ is 2-colourable. Let X be the set of vertices painted the first colour, and let Y be set of vertices painted the second colour. The X and Y form a bipartite partition of the vertices of G since neither the vertices of X nor the vertices of Y can be adjacent to each other since they are of the same colour. So, G is a bipartite graph.
(iii)  Lastly, suppose that G is bipartite and X and Y form the bipartite partition of the vertices of G. If a cycle begins at a vertex $u$ of say, X, then it will go to a vertex of Y, and then to a vertex of X and then to Y and so on. Hence, when the cycle returns to $u$, it must be of even length. So, every cycle of G has even length.

**Theorem 13** The vertices of every planar graph can be properly coloured with 5 colours.

***Proof***

We will prove this theorem by induction on the number $n$ of vertices of G.

If $n \leq 5$, then the theorem obviously holds. Now, suppose $n > 5$ and the theorem holds true for the graph with $n - 1$ vertices (Induction Hypothesis).

As we know that every planar graph G has a vertex such that deg $(v) \leq$ (Theorem). So, by induction G-$v$ is 5-colourable (by Induction Hypothesis). Now, we have 2 cases:

  (i)  The vertices adjacent to $v$ use less than 5 colours. Then, we can simply paint $v$ with one of the remaining colours and obtain a 5-colouring of G.
 (ii)  The vertex $v$ is adjacent to the five vertices, which are painted different colours. Say, the vertices, moving clockwise about $v$ and $v_1$, $v_2$, $v_3$, $v_4$, $v_5$ and are painted respectively with colours $c_1$, $c_2$, $c_3$, $c_4$, $c_5$ respectively as shown in Fig. 57.

Consider the subgraph H of G generated by the vertices painted colours $c_1$ and $c_3$. Suppose that there is a path P in H from $v_1$ to $v_3$ whose vertices are painted either colour $c_1$ or $c_3$. The path P together with the edges $(v, v_1)$ and $(v, v_3)$ form a cycle $c$ that encloses either $v_2$ or $v_4$. Then, a similar path between vertices $v_2$ and $v_4$ painted with colours $c_2$ and $c_4$ cannot exist, otherwise, it will intersect with the cycle C that encloses the vertex $v_2$ or $v_4$. So, there is no path between vertices $v_2$ and $v_4$, which is painted with colours $c_2$ and $c_4$ . So, we can paint $v_2$ and $v_4$ with the same colour $c_2$ and we can choose colour $c_4$ to paint $v$ and obtain a 5-colouring of G. Thus G is 5-colourable and the theorem is proved.

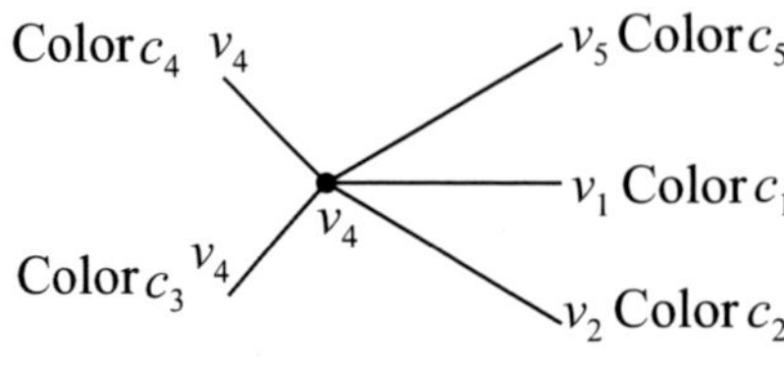


**Fig. 57**

## 10.38 SHORTEST PATH IN WEIGHTED GRAPH

Let G = (V, E, $w$) be a weighted graph where $w$ is function from E to the set of positive real numbers. Such a graph can be used to represent highways connecting different cities. Suppose that V represent a set of cities, E represent a set of highways connecting these cities and the weight of edge, $w\,(i,\,j)$, represent the length of the edge i.e., distance between cities. The length/distance of the path in G is defined as the sum of length of edges in the path. One common problem is to determine the shortest path from one vertex to another vertex in V. There are several well-known procedures for solving this problem. We represent here one discovered by E.W. Dijkstra which is applicable to both undirected graph as will as directed graph and is as follows:

**Dijkstra's Algorithm.** This algorithm maintains a set of vertices whose shortest path from the source is already known. The graph is represented by its cost adjacency matrix, where cost being the weight of the edge. In the cost adjacency matrix of the graph, all the diagonal values are zero. If there is no path from source vertex $V_5$ to any other vertex $V_i$ then it is represented by $+\infty$. In this algorithm, we have assumed that all the weights are positive.

1. Initially there is no vertex in sets.
2. Include the source vertex $V_s$ in S. Determine all the paths from $V_s$ to all the vertices without going through any other vertex.
3. Now, include that vertex in S, which is nearest to $V_s$ and find shortest paths to all the vertices through this vertex and update the values.
4. Repeat the step 3 until $n - 1$ vertices are not included in S if there are $n$ vertices in the graph.

After completion of the process, we got the shortest paths to all other vertices from the source vertex.

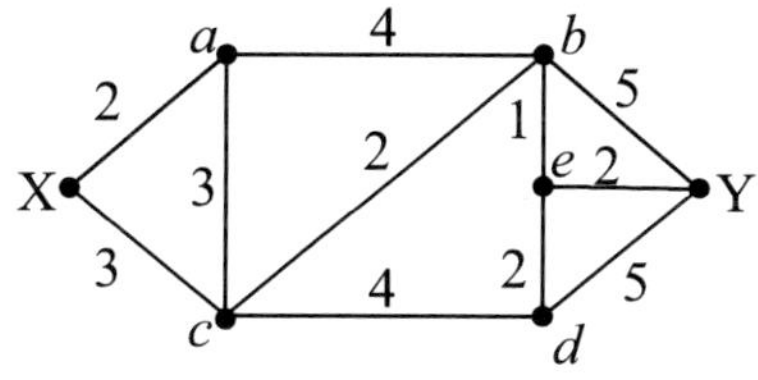

**Fig. 58(a)**

**Example 21:** Find the shortest path between X and Y in the graph shown in Fig. 58(a) by using Dijkstra's Algorithm.

*Solution*

**Step I.** Include the vertex X in set S and determine all the direct paths from X to all other vertices without going through any other vertex.

| S | Distance to all other vertices | | | | | | |
|---|---|---|---|---|---|---|---|
| | X | $a$ | $b$ | $c$ | $d$ | $e$ | Y |
| X | 0 | 2(X) | $\infty$ | 3(X) | $\infty$ | $\infty$ | $\infty$ |

**Step II.** Include the vertex in S that is nearest to X and determine shortest paths to all the vertices going through this vertex and update the values. The 1st nearest vertex is '$a$' and is included in S.

| S | Distance to all other vertices | | | | | | |
|---|---|---|---|---|---|---|---|
| | X | $a$ | $b$ | $c$ | $d$ | $e$ | Y |
| X, $a$ | 0 | 2(X) | 6(X, $a$) | 3(X) | $\infty$ | $\infty$ | $\infty$ |

**Step III.** The vertex, which is 2nd nearest to X is '$c$' and is included in S.

| S | Distance to all other vertices | | | | | | |
|---|---|---|---|---|---|---|---|
| | X | $a$ | $b$ | $c$ | $d$ | $e$ | Y |
| X, $a$, $c$ | 0 | 2(X) | 6(X, $a$) | 3(X) | 7(X, $c$) | $\infty$ | $\infty$ |

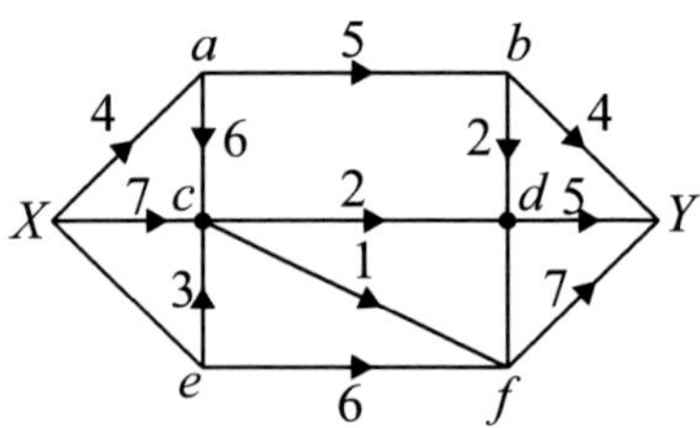

**Fig. 58(b)**

**Example 22:**  Find the shortest path between X and Y in the graph shown in Fig. 58(b) by using Dijkstra's Algorithm

***Solution***

**Step I.**  Include the vertex X in set S and determine all the direct paths from X to all other vertices without going through any other vertex.

| **S** | **Distance to all other vertices** | | | | | | | |
|---|---|---|---|---|---|---|---|---|
| | X | $a$ | $b$ | $c$ | $d$ | $e$ | $f$ | Y |
| X | 0 | 4(X) | $\infty$ | 7(X) | $\infty$ | 2(X) | $\infty$ | $\infty$ |

**Step II.**  Include the vertex in S that is nearest to X and determine shortest paths to all the vertices going through this vertex and update the values. The nearest vertex is '$e$' and is included in S.

| **S** | **Distance to all other vertices** | | | | | | | |
|---|---|---|---|---|---|---|---|---|
| | X | $a$ | $b$ | $c$ | $d$ | $e$ | $f$ | Y |
| X, $e$ | 0 | 4(X) | $\infty$ | 5(X, $e$) | $\infty$ | 2(X) | 8(X, $e$) | $\infty$ |

**Step III.**  The vertex, which is 2nd nearest to X is '$a$' and is included in S.

| **S** | **Distance to all other vertices** | | | | | | | |
|---|---|---|---|---|---|---|---|---|
| | X | $a$ | $b$ | $c$ | $d$ | $e$ | $f$ | Y |
| X, $e$, $a$ | 0 | 4(X) | 9(X, $a$) | 5(X, $e$) | $\infty$ | 2(X) | 8(X, $e$) | $\infty$ |

**Step IV.** The vertex, which is 3rd nearest to X is '*c*' and is included in S.

| S | Distance to all other vertices | | | | | | | |
|---|---|---|---|---|---|---|---|---|
| | X | *a* | *b* | *c* | *d* | *e* | *f* | Y |
| X, *e*, *a*, *c* | 0 | 4(X) | 9(X, *a*) | 5(X, *e*) | 7(X, *e*, *c*) | 2(X) | 6(X, *e*, *c*) | ∞ |

**Step V.** The vertex, which is 4th nearest to X '*f*' and is included in S.

| S | Distance to all other vertices | | | | | | | |
|---|---|---|---|---|---|---|---|---|
| | X | *a* | *b* | *c* | *d* | *e* | *f* | Y |
| X, *e*, *a*, *c*, *f* | 0 | 4(X) | 9(X, *a*) | 5(X, *e*) | 7(X, *e*, *c*) | 2(X) | 6(X, *e*, *c*) | 13(X, *e*, *c*, *f*) |

**Step VI.** The vertex, which is 5th nearest to X '*d*' and is included in S

| S | Distance to all other vertices | | | | | | | |
|---|---|---|---|---|---|---|---|---|
| | X | *a* | *b* | *c* | *d* | *e* | *f* | Y |
| X, *e*, *a*, *c*, *f* | 0 | 4(X) | 9(X, *a*) | 5(X, *e*) | 7(X, *e*, *c*) | 2(X) | 6(X, *e*, *c*) | 13(X, *e*, *c*, *f*) |

**Step VII.** The vertex, which is 6th nearest to X '*b*' and is included in S.

| S | Distance to all other vertices | | | | | | | |
|---|---|---|---|---|---|---|---|---|
| | X | *a* | *b* | *c* | *d* | *e* | *f* | Y |
| X, *e*, *a*, *c*, *f*, *d*, *b* | 0 | 4(X) | 9(X, *a*) | 5(X, *e*) | 7(X, *e*, *c*) | 2(X) | 6(X, *e*, *c*) | 12(X, *e*, *c*, *d*) |

Since, $n - 1$ vertices are included in S, so, we have found the shortest distance from X to all other vertices. Thus, the shortest distance between X and Y is 12 and the shortest path is X, *e*, *c*, *d*, Y.

## Exercise 10.4

1. Find the shortest path between X and Y in the graph shown in figure Ex. 1 to Ex. 3.

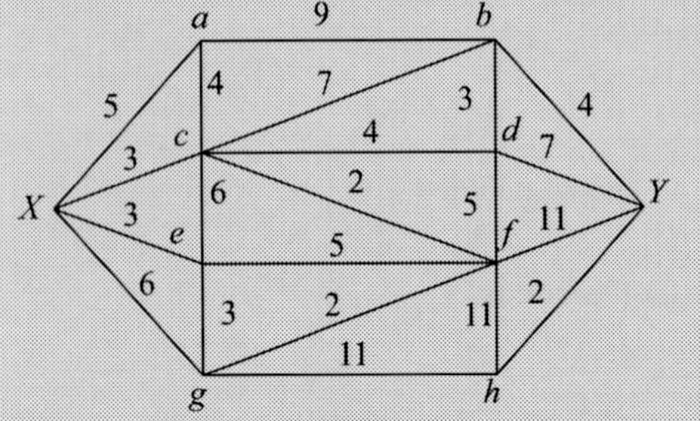

Fig. Ex. 1

Fig. Ex. 2

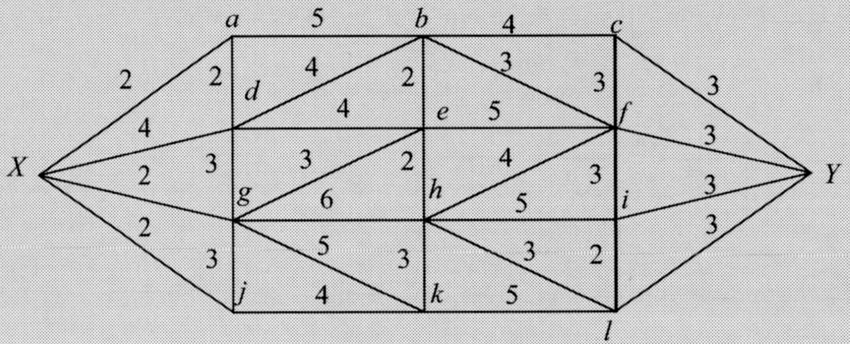

Fig. Ex. 3

## Answers to Selected Problems

1. (i) Shortest path for Fig. Ex. 1 is $XcdY$, total distance is 14.

   (ii) Shortest path for Fig. Ex. 2 is $XcabY$, total distance is 8.

   (iii) Shortest path for Fig. Ex. 3 is $XgefY$, total distance is 13.

# 11

# TREES

## 11.1 INTRODUCTION

Trees form one of the most widely used subclasses of graphs. Trees were used as long ago as 1857, when the English mathematician Arthur Cayley used them to count certain types of chemical compounds. Trees are particularly useful in computer science, where they are employed in a wide range of algorithms. For instance, trees are used to construct efficient algorithms for locating items in a list. They can be used in algorithms, such as Huffman coding, that construct efficient codes saving costs in data transmission and storage.

A connected (undirected) graph that contains no simple circuits or a graph with no cycles is called a tree. We can assign weights to the edges of a tree to model many problems. For example, using weighted trees we can develop algorithms to construct networks containing the least expensive set of telephone lines linking, different network nodes.

**Definition:** A *tree* is a connected undirected graph with no simple circuits
A tree is a finite set of one or more nodes such that:

   (i)  There is a node root.
   (ii) Remaining nodes are partitioned into $n \geq 0$ disjoint sets
        $T_1 \ldots T_n$ where each of these sets is a tree $T_1 \ldots T_n$ are called subtrees of the root.

The 'level' of a node is defined by initially letting the root be at level 1. If a node is at level l, then its children are at level l + 1.

## 11.2 PROPERTIES OF TREES

   1. There is a unique path between every pair of vertices in a tree T.
   2. The number of edges is one less than the number of vertices in a tree.
   3. There is no circuit.
   4. The depth or height of a tree is defined as the maximum level number in the tree.
   5. The degree of a node is the number of its subtrees. Obviously degree of terminal node is zero.

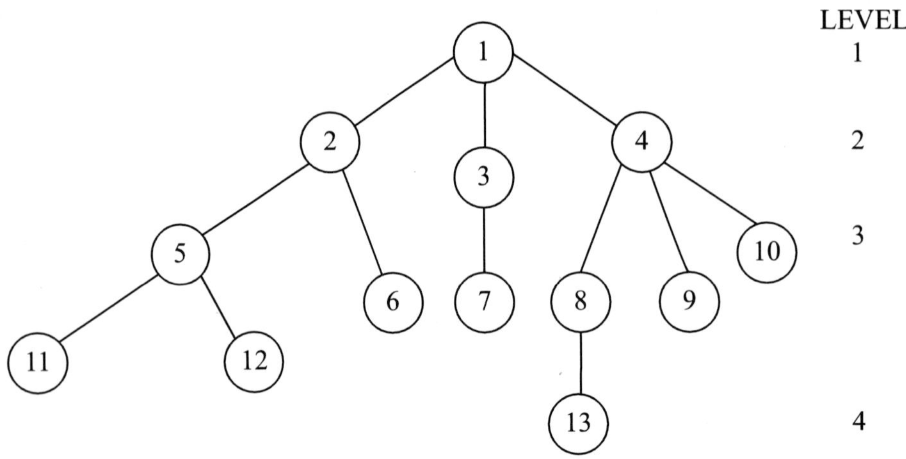



**Fig. 1**

**Theorem 1** Prove that there is a unique path between every pair of vertices in a tree.

*Proof*

Since, a tree is a connected graph; there is at least one path between every two vertices. However if there were two or more paths between a pair of vertices, there would be a circuit in the tree, which is not possible in a tree. Hence there is unique path between every pair of vertices in a tree.

**Theorem 2** Let G be a graph such that there is exactly one path between every pair of vertices. Prove that the graph G is a tree.

*Proof*

We know that a graph in which there is a path between every pair of vertices is connected. Moreover, if these paths are unique the graph cannot contain a circuit, since the existence of circuit implies the existence of two distinct paths between a certain pair of vertices.

Thus we conclude that a graph in which there is unique path between every pair of vertices is a tree.

**Rooted Tree:** A directed tree is called a rooted tree if there is exactly one vertex whose incoming degree is 0 and incoming degree of all other vertices is 1. The vertex with incoming degree 0 is called root of rooted tree. In a rooted tree, a vertex whose out degree is 0 is called a leaf or terminal node. For example tree shown in figure 1 is a rooted tree, with node 1 as the root node.

**Directed Tree:** A directed graph is said to be a directed tree if it becomes a tree when the direction of edges are ignored.

**Ordered Tree:** An ordered tree is a rooted tree with the edges incident from branch node labelled with integers 1, 2, ..., $i$, ...

**Theorem 3** Prove that the number of edges is one less than the number of vertices in a tree.

### *Proof*

Let $v$ = number of vertices and $e$ = number of edges in the tree T.
We will prove by method of induction

### **Basic Step**

Let $v = 1$, i.e. there is a single vertex, so there is no edge, i.e. $e = 1 - 1 = 0$
∴   The result is true for $n = 1$

**Fig. 2**

### **Induction Step**

Let, the result be true for $v = k$ where $k > 1$, i.e. number of vertices is $k$ and number of edges is $k - 1$. Now, we will prove it for $v = k + 1$. i.e. we will add one vertex to T. If this is connected to only one of the vertices $v_1, v_2, ..., v_k$, by means of an edge '$e$' then number of edges will become $k - 1 + 1 = k$. It cannot connect to two or more vertices, otherwise their would be a circuit.

Hence the result is true for $v = k + 1$.   i.e. $e = v - 1$, where $e$ = number of edges and $v$ = number of vertices.

**Theorem 4** Prove that a connected graph with $v$ vertices and $v - 1$ edges is a tree.

### *Proof*

Let $v$ = number of vertices and $e$ = number of edges in the Graph G
    $e = v - 1$,        let's assume that G contains a simple circuit C.

Let $c$ denote the number of vertices in C. Clearly number of edges in C is equal to $c$. Since G is connected every vertex of G which is not in C must be connected to vertices in C. Now each edge of G that is not in C can connect only one additional vertex to the vertices in C.

There are $v - c$ vertices that are not in C, so G must contain at least $v - c$ edges that are not in C. Thus we must have $e \geq c + (v - c) = v$, which is a contradiction. It follows that G doesn't contain any circuit and is, therefore, a tree.

**Example 1:** Which of the following graphs are trees?

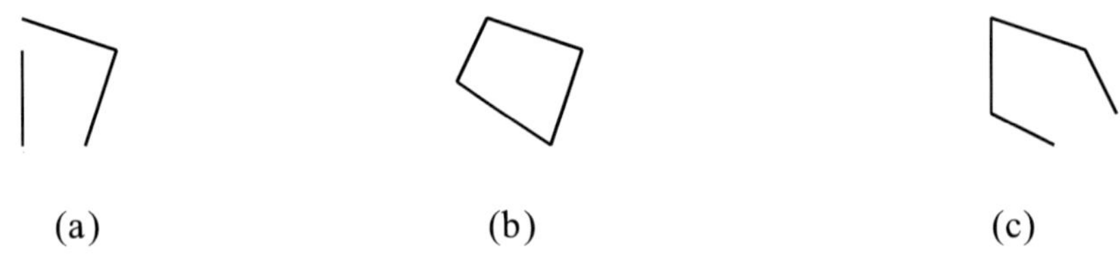

(a)      (b)      (c)

**Fig. 3**

*Solution*

(a) no, as the graph is not connected.
(b) no, as the graph contains a circuit.
(c) yes, as the graph is connected and doesn't contain any circuit.

## 11.3 BINARY TREE

A binary tree is a finite set of nodes which is either empty or consists of a root and two disjoint binary trees called left and right subtrees.

  (i) Maximum number of nodes on level L of a binary tree is $2^{L-1}$ $L \geq 1$
  (ii) Maximum number of nodes in binary tree of depth $k$ is $2^k - 1$, $k \geq 1$

**Complete Binary Tree:** A binary tree in which all levels, except possibly the last, have the same number of possible nodes as far left as possible.

**Full Binary Tree:** A full binary tree is defined as a rooted tree in which each interior vertex is of out degree 2 and all leaves are at same depth. A full binary tree of height $h$ has $2^h$ leaves and $2^{h+1} - 1$ vertices.

**Balanced Tree:** A tree is said to be balanced if all of its vertices are balanced. i.e. if its left and right subtrees do not have the difference of more than unity in their respective degrees.

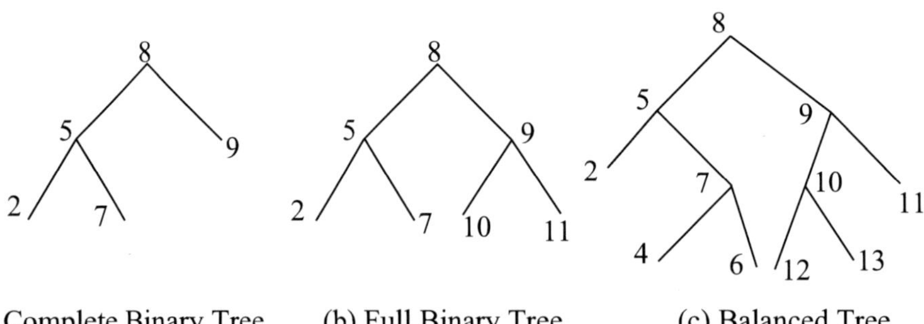

(a) Complete Binary Tree  (b) Full Binary Tree  (c) Balanced Tree

**Fig. 4**

**Theorem 5** Show that a regular binary tree has an odd number of vertices.

*Proof*

In a regular binary tree every node has 2 children. So at any level the total number of nodes is even. Hence, the total number of nodes in the tree is the root and the total number of children which is odd. Hence a regular binary tree has an odd number of vertices.

## 11.4  TRAVERSING BINARY TREES

We can use trees to store information in a computer. Therefore, some procedures are needed for accessing the information easily (or for visiting each vertex easily). Traversing a tree means visiting all nodes in tree. There are three methods to traverse the binary trees.

1.  Preorder Traversal
2.  Inorder Traversal
3.  Postorder Traversal

**1. Preorder Traversal:** In preorder traversal of a binary tree each subtree of a binary tree can be traversed as follows:

a.  Expand the root of subtree
b   Traverse the left subtree of the root in preorder
c.  Traverse the right subtree of the root in preorder

**2. Inorder Traversal:**  In inorder traversal of a binary tree each subtree of a binary tree can be traversed inorder as follows:

a.  Traverse the left subtree of the root in inorder
b.  Expand the root of subtree
c.  Traverse the right subtree of the root in inorder

**3. Postorder Traversal:**  In postorder traversal of a binary tree each subtree of a binary tree can be traversed as follows:

a.  Traverse the left subtree of the root in postorder
b.  Traverse the right subtree of the root in postorder
c.  Expand the root of subtree

Observe that each algorithm contains the same three steps, and that the left subtree of R is always traversed before the right subtree. The difference between the algorithms is the time at which the root R is processed. Specifically, in the "pre" algorithm, the root R is processed before the subtrees are traversed; in the "in" algorithm, the root R is processed between the traversals of the subtrees; and in the "post" algorithm, the root R is processed after the subtrees are traversed.

**Example 2:** Consider the binary tree in Fig. 5. Observe that A is the root that its left subtree $L_T$ consists of nodes B, D, and E, and its right subtree $R_T$ consists of nodes C and F.

(a) The preorder traversal of T processes A, traverses $L_T$, and traverses $R_T$. However, the preorder traversal of $L_T$ processes the root B and then D, and E, and the preorder traversal of $R_T$ processes the root C and then F. Hence ABDECF is the preorder traversal of T.

(b) The inorder traversal of T traverses $L_T$, processes A, and traverses $R_T$. However, the inorder traversal of $L_T$, processes D, B, and then E and the inorder traversal of $R_T$ processes C and then F. Hence DBEACF is the inorder traversal of T.

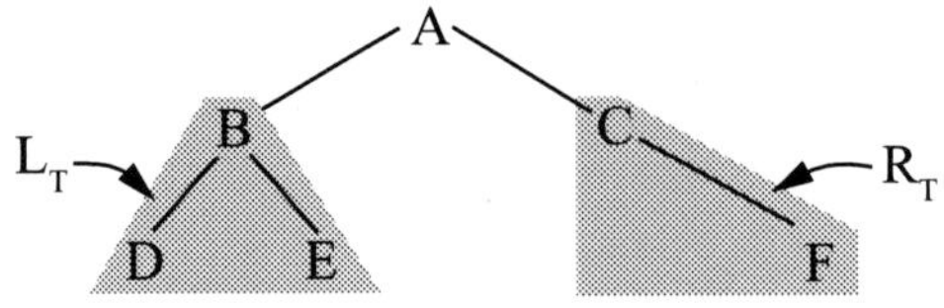

**Fig. 5**

(c) The postorder traversal of T traverses $L_T$, traverses $R_T$, and processes A. However, the postorder traversal of $L_T$ processes D, E, and then B, and the postorder traversal of $R_T$ processes F and then C. Accordingly, DEBFCA is the postorder traversal of T.

**Example 3:** Consider the binary tree T in Fig. 6.

(a) Find the depth $d$ of T.
(b) Traverse T using the preorder algorithm.
(c) Traverse T using the inorder algorithm.
(d) Traverse T using the postorder algorithm.
(e) Find the terminal nodes of T, and the order that they are traversed in (b), (c), and (d).

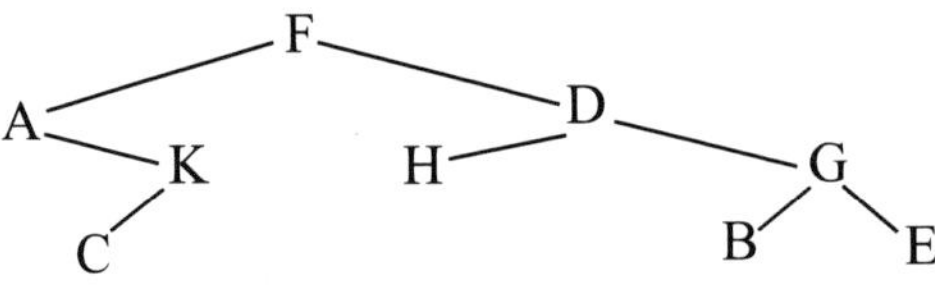

**Fig. 6**

*Solution*

(a) The depth $d$ is the number of nodes in a longest branch of T; hence $d = 4$.

(b) The preorder traversal of T is a recursive NLR algorithm, that is, it first processes a node N, then its left subtree L, and finally its right subtree R. Letting $[A_1 \ldots A_k]$ denote a subtree with nodes $A_i \ldots, A_k$, and the tree T is traversed as follows:

$$F\text{-}[A, K, C][D, H, G, B, E] \quad \text{or} \quad F\text{-}A\text{-}[K, C]\text{-}D \; [H][G, B, E]$$

or, finally,   F-A-K-C-D-H-G-B-E

(c) The inorder traversal of T is a recursive LNR algorithm, that is, first processes a left subtree L, then its node N, and finally its right subtree R. Thus T is traversed as follows:

$$[A, K, C] \; F\text{-}[D, H, G, B, E] \quad \text{or} \quad A\text{-}[K, C]\text{-}F\text{-}[H]\text{-}D\text{-}[G, B, E]$$

or, finally,   A, C, K, F, H, D, B, G, E

(d) The postorder traversal of T is a recursive LRN algorithm, that is, it first processes a left subtree L, then its right subtree R, and finally its node N. Thus T is traversed as follows:

$$[A, K, C][D, H, G, B, E]\text{-}F \quad \text{or} \quad [K, C]\text{-}A\text{-}[H][G, B, E]\text{-}D\text{-}F$$

or, finally,   C-K-A-H B-E-G-D-F

(e) The terminal nodes are the nodes without children. They are traversed in the same order in all three traversal algorithms: C, H, B, E.

**Example 4:** Following is the Preorder and Inorder traversal of a binary tree:

Preorder: ABDECF
Inorder: DBEACF
Construct the tree and explain the procedure.

*Solution*

The tree T is drawn from its root $r$ downward as follows:

(a) The root of T is obtained by choosing the first node in its preorder. Thus A is the root.

(b) The left child is obtained as follows:

First use inorder of T to find nodes in left subtree $T_1$ of A. Thus $T_1$ consists of nodes DBE which are to the left of A in the inorder. Then left child of A is obtained by choosing the first node in preorder of $T_1$ next to A in preorder T. Thus B is left child of A.

(c) Similarly right subtree $T_2$ of A consists of nodes C, F & C is root of $T_2$, i.e. C is right child of A.

Similarly the whole tree is constructed by using $T_1$, $T_2$ and so-on as roots

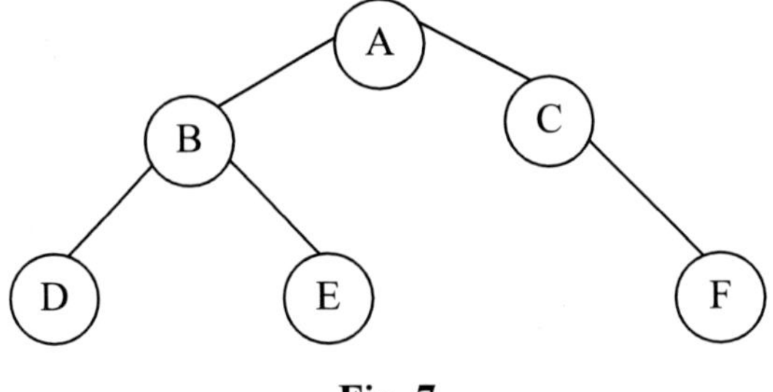

**Fig. 7**

## 11.5 BINARY SEARCH TREES

A Binary tree (T) in which information in nodes are represented in such a way that all information in its left subtree is less than or equal to information stored in the vertex, and all information stored in right subtree is greater than the information stored in the vertex. It is not difficult to see that the above property guarantees that the inorder traversal of T will yield a sorted listing of the elements of Tree (T).

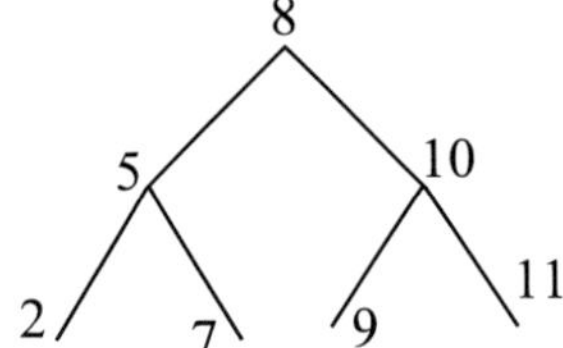

**Fig. 8** Binary Search Tree

## 11.6 SPANNING TREE

A **spanning tree** in a connected graph G is a subgraph which includes every vertex and is a tree. For instance figure shows the complete graph K5 and several possible spanning trees. Large and complex graphs may have very many spanning trees.

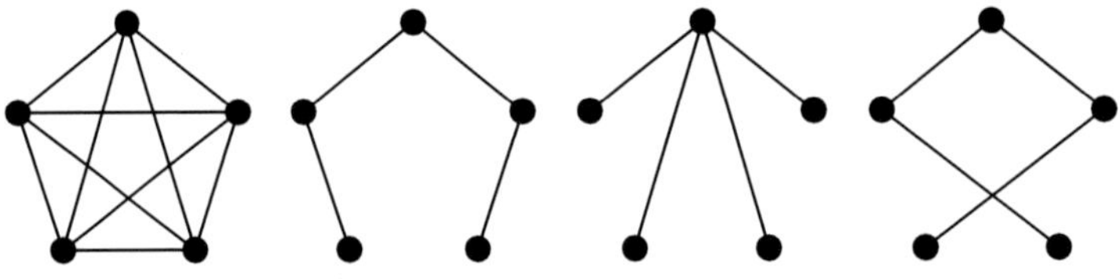

**Fig. 9**

We tackle the problem in stages. First of all we need the concept of a minimum spanning tree.

A **minimum spanning tree or minimum connector** is a spanning tree in a connected weighted graph which has minimum total weight.

**Figure 10** shows a number of possible spanning trees for the graph shown.

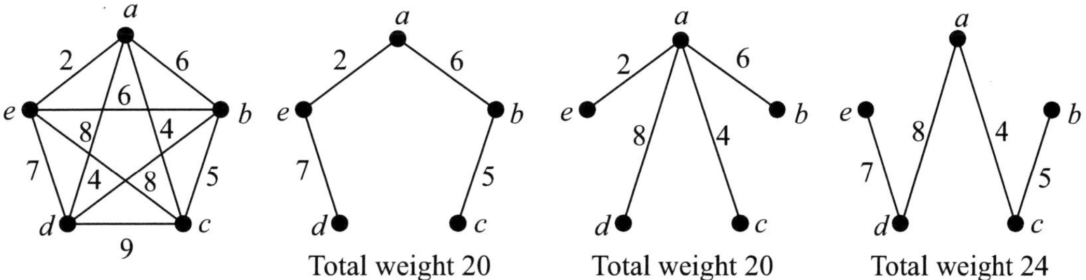



**Fig. 10**

There are two algorithms for constructing minimum spanning trees.

## Kruskal's Algorithm

Start with a finite set of vertices ($v$) for a weighted graph ($g$), each pair joined by a weighted edge.

1. List all the weights in ascending order.
2. Draw the vertices and weighted edge corresponding to the first weight in the list provided that, in so doing, no cycle is formed. Delete the weight from the list.
3. Repeat step 2 until all vertices are connected, then stop. The weighted graph obtained is a minimum connector, and the sum of weights on its edges is the total weight of the minimum connector.

**Example 5:** Construct a minimum spanning tree for the given weighted graph.

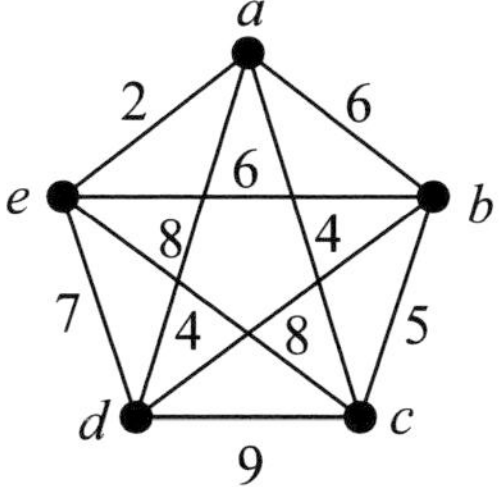

**Fig. 11**

### *Solution*

The given graph G has 5 vertices and 10 edges. We proceed with the arrangement of the edges in increasing order of weights. After arranging, the weights are

$$2\,(ae)\quad 4\,(ac)\quad 4\,(ce)\quad 5\,(bc)\quad 6\,(ab)\quad 6\,(be)\quad 7\,(de)\quad 8\,(ad)\quad 8\,(bd)\quad 9\,(cd)$$

So the steps of Kruskal's algorithm will give the following:

> choose *ae*
> choose *ac*
> reject *ce* (would form cycle *acea*)
> choose *bc*
> reject *ab* (would form cycle *acba*)
> reject *be* (would form cycle *acbea*)
> choose *de*.

All vertices now connected. The total weight is 18.

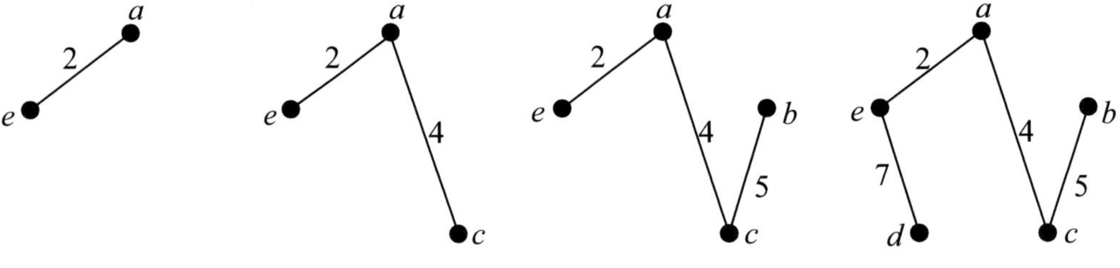

**Fig. 12**

Although, in the above example, the subgraph constructed was a connected subgraph at each stage this isn't necessarily the case. For instance suppose edge *bc* had weigh 3 instead of 5. Now the weights are

$$2\,(ae)\quad 3\,(bc)\quad 4\,(ac)\quad 4\,(ce)\quad 6\,(ab)\quad 6\,(be)\quad 7\,(de)\quad 8\,(ad)\quad 8\,(bd)\quad 9\,(cd)$$

so the steps of Kruskal's algorithm will give the following:

> choose *ae*
> choose *bc*
> choose *ac*
> reject *ce* (would form cycle)
> reject *ab* (would form cycle)
> reject *be* (would form cycle)
> choose *de*.

All vertices now connected. The total weight is 16.

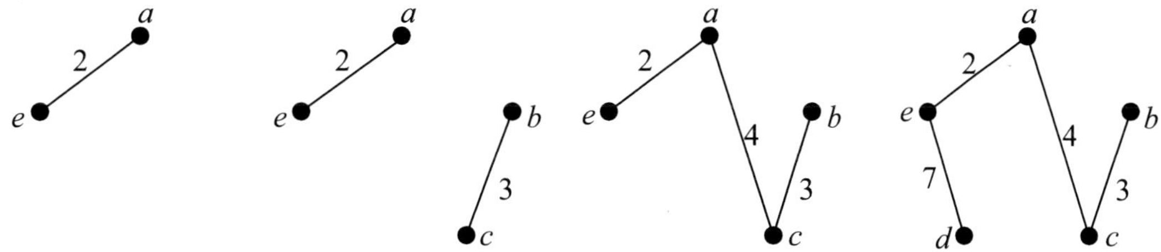

**Fig. 13**

## Prim's Algorithm

Start with a finite set of vertices, each pair joined by a weighted edge.

1.  Choose and draw any vertex
2.  Find the edge of least weight joining a drawn vertex to a vertex not currently drawn. Draw this weighted edge and the corresponding new vertex.

Repeat step 2 until all vertices are connected, then stop.

**Example 6:** To construct a minimum spanning tree for the graph of figure using Prim's algorithm.

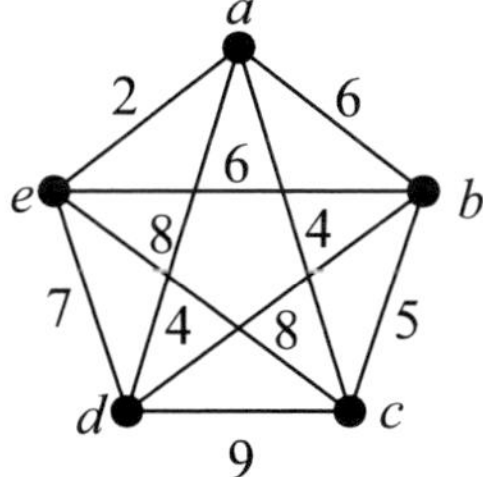

**Fig. 14**

### *Solution*

We proceed thus. Choose vertex $a$ to start, minimum weight edge is $ae$ (weight 2), minimum weight edge from $\{a, e\}$ to new vertex is $ac$ (weight 4), minimum weight edge from $\{a, c, e\}$ to new vertex is $cb$ (weight 5), minimum weight edge from $\{a, b, c, e\}$ to new vertex is $ed$ (weight 7), all vertices now connected. Total weight is 18.

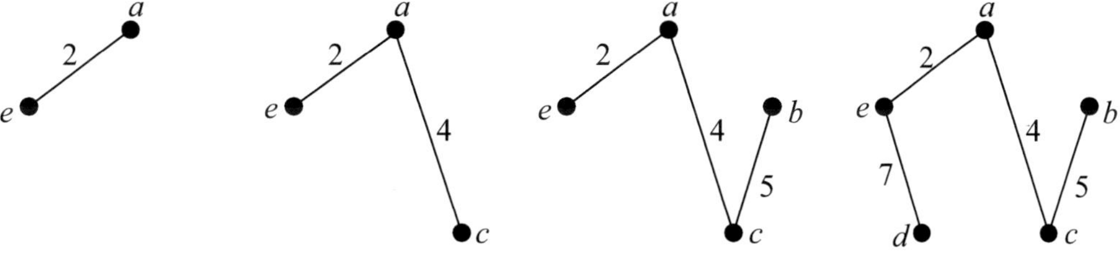

**Fig. 15**

Of course the choice of $a$ as starting vertex is arbitrary – let's see what happens if we start somewhere else. Choose vertex $b$ to start, minimum weight edge is $bc$ (weight 5), minimum weight edge from $\{b, c\}$ to new vertex is $ca$ or $ce$ (both weight 4), choose $ce$, minimum weight edge from $\{b, c, e\}$ to new vertex is $ea$ (weight 2), minimum weight edge from $\{a, b, c, e\}$ to new vertex is $ed$ (weight 7), all vertices now connected. Total weight is 18.

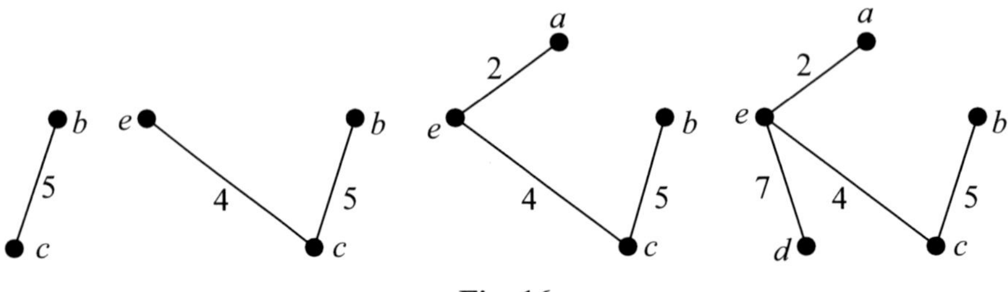

**Fig. 16**

In this case we see that we get a different minimum spanning tree (if we had chosen $ca$ instead of $ce$ at stage 2 we would have got the same minimum spanning tree as previously). This is because, in this graph, there is more than one minimum spanning tree.

### Exercise 11.1

1. Let $u$ and $v$ be any two vertices in a tree T and the edge $e$ between $u$ and $v$ is in T. Show that the graph $T - e$ is not a tree

2. Prove that a full binary tree of height $h$ has $2^h$ leaves and $2^{h+1} - 1$ vertices.

3. Show that the sum of degrees of the vertices of a tree with $n$ vertices is $2n - 2$.

4. Show the sequential order in which the vertices of tree in fig. are visited using preorder, inorder and postorder traversal.

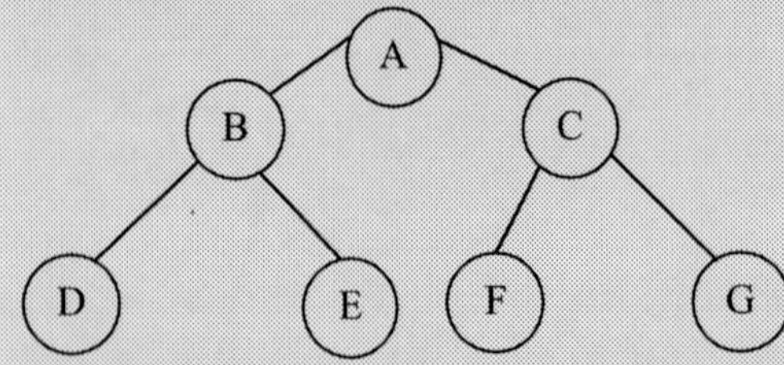

5. Consider the algebraic expression $E = (2x + y)(5a - b)^3$.

(a) Draw the tree T which corresponds to the expression E.

(b) Find the preorder of T.

6. Determine the minimum spanning tree for the graph shown below:

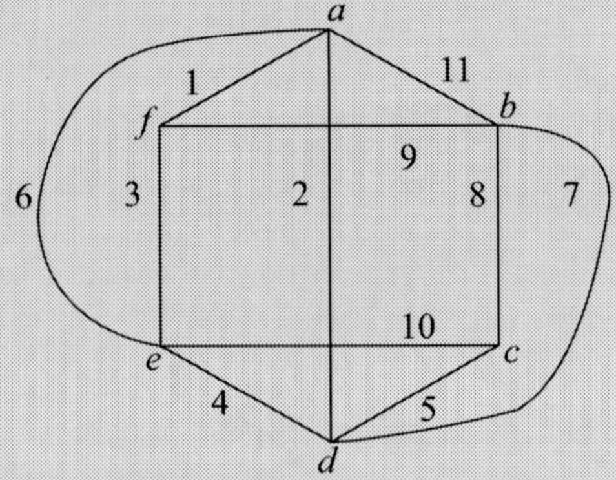

## Answers to Selected Problems

5. (a) Tree corresponding to expression E

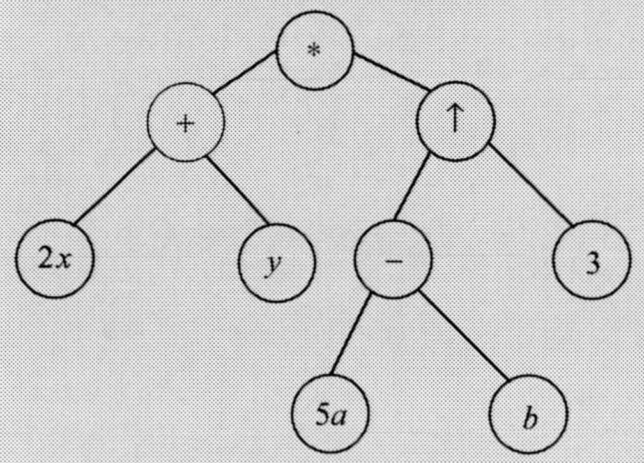

   (b) Preorder of tree T is given as

      $*, +, 2x, y, \uparrow, -, 5a, b, 3$

6. Minimum spanning tree of given graph is given as

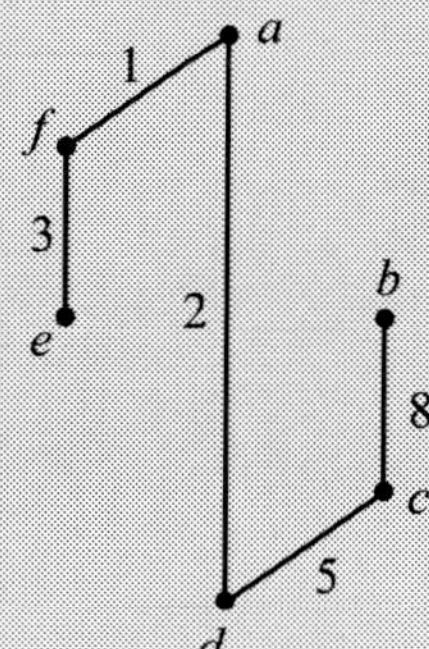

# Section 5

# 12

# Algebraic System

## 12.1 BINARY OPERATION (OR BINARY COMPOSITION)

Let A be a non-empty set. An operation on set A is the function $*$ from $A \times A$ into A. In such a case we usually write

$$a * b \text{ or sometimes } ab \text{ instead of } *(a, b).$$

The set A and the operation $*$ on A is denoted by $(A, *)$.

- An operation from $A \times A$ into A is generally called a *binary* operation.
- A *unary* operation is a function from A into A.
- For example, the absolute value $|n|$ of an integer $n$ is a *unary* operation on $\mathbf{Z}$.
- A ternary (3-ary) operation is a function from $A \times A \times A$ into A.
- More generally, an $n$-ary operation is a function from $A \times A \ldots \times A$ ($n$-times) into A.

For example, consider the set $\mathbf{Z}$ of integers.
Addition (+), subtraction (−) and multiplication (×) are operations on set $\mathbf{Z}$. However, division (/) is not an operation on set $\mathbf{Z}$ since division of two integers need not to be an integer. For example, 8/5 is not an integer.

**Example 1:**

Let $A = \{-1, 0, 1\}$.

Then subtraction is not an operation on A since $-1 - 1 = -2$ is not an element of A. On the other hand, multiplication is an operation.

**Example 2:**

Let $A = \{1, 2, 3, 4\}$. Suppose that the operation $*$ is defined by the formula
$$x * y = 3x + 2y,$$

where $x$ and $y$ are any elements of A. The operation $*$ can be expressed using table as below:

|     | $x*y$ | 1 | 2 | 3 | 4 |
|-----|-------|---|---|---|---|
|     | 1 | 5 | 7 | 9 | 11 |
| $x$ | 2 | 8 | 10 | 12 | 14 |
|     | 3 | 11 | 13 | 15 | 17 |
|     | 4 | 14 | 16 | 18 | 20 |

## 12.2 ALGEBRAIC SYSTEMS (OR ALGEBRAIC STRUCTURE)

A system which consists of a non-empty set A and one or more operations on that set is known as algebraic system and is usually denoted by notation such as (A, $*$, $\bullet$) where A is a set and $*$, $\bullet$ are operations on set A.

## 12.3 PROPERTIES OF BINARY OPERATIONS

The following are some properties of binary operations:

**(1) Closure Property:**

The operation $*$ on set A is said to be closed if

$$a * b \in A \qquad \forall\, a, b \in A$$

For example, the operations of addition and multiplication are closed operation on the set of positive integers (A) while subtraction is not a closed operation as $1-3 = -2 \notin A$

**(2) Commutative Property:**

The operation $*$ on set A is said to be commutative or satisfy commutative law if

$$a * b = b * a \qquad \forall\, a, b \in A$$

For example, the operation of addition is commutative on the set $\mathbf{Z}$ of integers, while subtraction is noncommutative as $2 - 4 \neq 4 - 2$.

**(3) Associative Property:**

The operation $*$ on set A is said to be associative or satisfy associative law if

$$(a * b) * c = a * (b * c) \qquad \forall\, a, b, c \in A.$$

For example, the operation of addition and multiplication are associative on the set $\mathbf{Z}$ of integers, while subtraction is nonassociative as $(7 - 3) - 2 = 2$ but $7 - (3 - 2) = 6$.

**(4) Identity Property:**

Consider the operation $*$ on set A. Then the operation $*$ has an identity property if there exists an element $e$ in A such that

$$a * e = e * a = a \qquad \forall\, a \in A.$$

Generally, an element $e$ is called $a$ left identity if $e * a = a$ $\forall\, a \in$ A or $a$ right identity if $a * e = a$ $\forall\, a \in$ A. The identity is always unique.

For example, the element 0 is identity element for the set A = {0, 1, 2, ...} under the operation of addition.

## (5) Inverse Property:

Consider the operation $*$ on set A with identity element $e$. Then the inverse of an element $a$ in A is an element $b$ such that

$$a * b = b * a = e$$

Note that if $b$ is the inverse of $a$, then $a$ is the inverse of $b$. Thus inverse is a symmetric relation. The element $b$ is said to be left inverse of $a$ if $b * a = e$. The element $b$ is said to be right inverse of $a$ if $a * b = e$.

For example, suppose A = {−1, 0, +1}

The identity element under the operation of addition is zero. So, the inverse of element − 1 is need not 1 and of element need not 1 is − 1.

## (6) Distributive Property:

Consider the non-empty set A with two binary operation $*$ and +. Then the operation $*$ distributes over + if

$$a * (b + c) = (a * b) + (a * c) \quad \text{[Left distributive]}$$
$$(b + c) * a = (b * a) + (c * a) \quad \text{[Right distributive]}$$

$\forall\ a, b, c \in$ A.

**Example 3:** Consider the binary operation $*$ on the set of rational numbers Q defined by, $a * b = ab / 4$ for all $a, b \in$ Q. Determine whether operation $*$ is (i) commutative (ii) associative.

***Solution***

(i) Let us assume that $a, b \in$ Q, then we have

$$a * b = ba/4 = b * a$$

Hence, the operation $*$ is commutative.

(ii) Let us assume that $a, b, c \in$ Q, then we have

$$(a * b) * c = (ab / 4) * c = ((ab / 4) \cdot c) / 4 = abc / 16$$
$$a * (b * c) = a * (bc / 4) = (a \cdot (bc / 4)) / 4 = abc / 16$$

Therefore, $(a * b) * c = a * (b * c)$.

Hence, the operation $*$ is associative.

**Example 4:** Consider the binary operation $*$ on $I_+$, the set of positive integers defined by, $a * b = ab / 4$. Determine the identity for the binary operation $*$, if exists.

### *Solution*

Let us assume that $e$ be $a$ positive integer, then

$$e * a = a, a \in I_+$$
$$ea / 4 = a \Rightarrow e = 4 \qquad\qquad\text{(i)}$$

Similarly, we have

$$a * e = a, a \in I_+$$
$$ae / 4 = a \quad \Rightarrow e = 4 \qquad\qquad\text{(ii)}$$

From (i) and (ii), we have $e * a = a * e = a$
Therefore, 4 is the identity element.

## 12.4 GROUPOID

An algebraic structure (A, $*$) is called groupoid if it satisfies the Closure property i.e. $\forall\ a, b \in$ A, $a * b \in$ A

## 12.5 SEMIGROUP

Let (A, $*$) be an algebraic system where $*$ is a binary operation on set A. Then the system (A,$*$) is said to be semigroup if it satisfies the following two properties:

(1)  The operation $*$ is a closed operation.
(2)  The operation $*$ is an associative operation.

For example, let A be the set of all the positive integers $\{1, 2, 3, \ldots\}$ and $+$ be the addition operation over this set of positive integer. Since addition ($+$) is closed as well as associative operation on A, so, $\{A, *\}$ is semigroup.

**Example 5:**  Consider the algebraic system $\{(-1, 0, 1), *\}$ where $*$ is the ordinary multiplication operation. Determine whether $\{(-1, 0, 1), *\}$ is semigroup?

### *Solution*

**Closure property:**
The operation $*$ is closed on the given set as
$$-1 * -1 = 1, -1 * 0 = 0, -1 * 1 = -1, 0 * 0 = 0, 0 * 1 = 0, 1 * 1 = 1$$

**Associative property:**
The operation $*$ is associative as we have
$$a * (b * c) = (a * b) * c, \quad \text{for all } a, b, c$$
Since the algebraic system is closed and associative.
Hence, it is semigroup.

**Example 6:** Consider the algebraic system (A, $*$) where A = {2, 4, 6, ...} the set of all positive even integer and $*$ is the ordinary multiplication operation. Determine whether (A, $*$) is semigroup?

*Solution*

**Closure property:**

The operation $*$ is closed as multiplication of two positive even integers is always a positive even integer.

**Associative property:**

The operation $*$ is associative as we have

$$a * (b * c) = (a * b) * c, \qquad \text{for all } a, b, c \in A.$$

Since the algebraic system is closed and associative.
Hence, the given system (A, $*$) is semigroup.

**Example 7:** Let (A, $*$) be a semigroup. Show that for any $a, b, c \in A$, if $a * c = c * a$ and $b * c = c *$, then $(a * b) * c = c * (a * b)$.

*Solution*

L.H.S.
We have $(a * b) * c$
$\Rightarrow \quad a * (b * c) \qquad [* \text{ is associative as (A, } *) \text{ is semigroup}]$
$\Rightarrow \quad a * (c * b) \qquad [b * c = c * b]$
$\Rightarrow \quad (a * c) * b \qquad [* \text{ is associative}]$
$\Rightarrow \quad (c * a) * b \qquad [c * a = a * c]$
$\Rightarrow \quad c * (a * b) \qquad [* \text{ is associative}]$
$= \text{R. H. S.}$

Hence, $(a * b) * c = c * (a * b)$.

## 12.5.1 Subsemigroup

Let (A, $*$) be a semigroup where $*$ is the binary operation on A. Also, let B is a non-empty subset of A i.e. B $\subseteq$ A. Then the system (B, $*$) is a subsemigroup if the set B satisfies the properties of semigroup under the operation $*$ i.e. B is closed and associative under the operation $*$. Since the elements of B are also the elements of A, so, the associative law automatically holds for the elements of B. Therefore, B is a subsemigroup if and only if B is closed under the operation $*$.

For example, consider a semigroup (N, $\times$) where N is the set of all the natural numbers and let A be the set of all the odd positive integers (A $\subseteq$ N). Then (A, $\times$) is subsemigroup as set A is closed under multiplication.

On the other hand (A, $+$) is not a subsemigroup of (N, $+$) as A is not closed under addition since addition of two odd integers is always an even integer.

**Example 8:** Consider the algebraic system (N, +). Determine whether (A, +) is a subsemigroup of (N, +), where A = {2, 4, 6, …} the set of all positive even integer and + is the ordinary addition operation?

### *Solution*

**Closure property:**
The operation + is closed as addition of two positive even integers is always a positive even integer.

**Associative property:**
The operation + is associative as we have

$$a + (b + c) = (a + b) + c, \qquad \text{for all } a, b, c \in A.$$

Since the algebraic system is closed and associative.

Hence, the given system (A, +) is subsemigroup of (N, +).

**Example 9:** Consider the set Q of rational numbers and let $*$ be the operation on Q defined by $a * b = a + b + ab$

   (i) Find $2 * 3$ and $3 * (-5)$
   (ii) Is (Q, $*$) a semigroup? Is it commutative?
   (iii) Find the identity element for $*$.
   (iv) Do any element in Q have an inverse?

### *Solution*

(i) $2 * 3 = 2 + 3 + 2 * 3 = 11$ and $3 * (-5) = 3 + (-5) + 3 * (-5) = -17$
(ii) We have

$$\begin{aligned}
(a * b) * c &= (a + b + ab) * c \\
&= (a + b + ab) + c + (a + b + ab) \cdot c \\
&= a + b + ab + c + ac + bc + abc \\
&= a + b + c + ab + ac + bc + abc \\
a * (b * c) &= a * (b + c + bc) \\
&= a + (b + c + bc) + a \cdot (b + c + bc) \\
&= a + b + c + bc + ab + ac + abc \\
&= a + b + c + ab + ac + bc + abc
\end{aligned}$$

Hence, $*$ is associative and (Q, $*$) is semigroup.
   Also, we have

$$a * b = a + b + ab = b * a$$

Hence, (Q, $*$) is a commutative group.
(iii) Computation of identity element is as follows:

$$a * e = a \text{ for all } a \in Q$$
$$a + e + ae = a \quad \Rightarrow \quad e(1 + a) = 0 \quad \Rightarrow \quad e = 0$$

Accordingly, 0 is the identity element.

(iv) Let $b$ be inverse of element $a \in$ Q.

Then, we have $a * b = 0$ as 0 is the identity element by part (iii)

$\Rightarrow \quad a + b + ab = 0$

$\Rightarrow \quad b(1 + a) = -a$

$\Rightarrow \quad b = -a/(1 + a)$

Thus, if $a \neq -1$, then the inverse element of $a$ is $-a / (1 + a)$.

### 12.5.2 Congruence relation

An equivalence relation R on the semigroup (A, $*$) is called a congruence relation if a R $a'$ and $b$ R $b' \Rightarrow (a * b)$ R $(a' * b')$.

**Example 10:** Let (I, $+$) be a semigroup and R is an equivalence relation on I defined by $aRb$ iff $a \equiv b$ (MOD 5). Determine if R is a congruence relation?

### *Solution*

The congruence relation

$\qquad a \equiv b$ (MOD 5).

is defined as 5 divides the difference $a - b$.

So, we can write

$\qquad a - b = 5m$ $\hfill$ (i)

Similarly, if we have

$\qquad c \equiv d$ (MOD 5).

Then, it is defined as 5 divide the difference $c - d$.

And we can write

$\qquad c - d = 5n$ $\hfill$ (ii)

where $m$ and $n$ are some integers of I.

Adding (i) and (ii), we have

$\qquad (a - b) + (c - d) = 5m + 5n$

$\qquad$ or

$\qquad (a + c) - (b + d) = 5m + 5n$

$\qquad$ or

$\qquad (a + c) \equiv (b + d)(\text{MOD } 5)$

Hence, the relation R is congruence relation.

## 12.6 MONOID

Let (A, $*$) be an algebraic system where $*$ is a binary operation on set A. Then the system (A, $*$) is said to be monoid if it satisfies the following properties:

(1) The operation $*$ is a closed operation.

(2) The operation $*$ is an associative operation.

(3) There exists an identity element $e$ w.r.t. operation $*$, i.e. $e * a = a * e$

For example, consider an algebraic system $(A, \times)$ where A is the set of all the positive integers and $\times$ is the multiplication operation. Then $(A, \times)$ is monoid since A is closed, associative and have an identity element 1 w.r.t operation $\times$.

**Example 11:** Consider the algebraic system $(N, +)$, where $N = \{0, 1, 2, \ldots\}$ and $+$ is the ordinary addition operation. Determine whether $(N, +)$ is a monoid?

### *Solution*

**Closure property:**

The operation $+$ is closed as addition of two natural numbers is always a natural number.

**Associative property:**

The operation $+$ is associative as we have

$$a + (b + c) = (a + b) + c, \text{ for all } a, b, c \in N.$$

**Identity element:**

The identity element here is zero.
Since the algebraic system is closed, associative and have the identity element.
Hence, the given system $(N, +)$ is a monoid.

## 12.6.1 Submonoid

Let $(A, *)$ be a monoid where $*$ is the binary operation on A. Also, let B is a non-empty subset of A i.e. $B \subseteq A$. Then the system $(B, *)$ is a submonoid if the set B satisfies the properties of monoid under the operation $*$ as given below:

(1) The operation $*$ is a closed operation.
(2) The operation $*$ is an associative operation.
(3) There exists an identity element w.r.t operation $*$.

Since the elements of B are also the elements of A, so, the associative law automatically holds for the elements of B. Therefore, B is a submonoid if and only if B is closed under the operation $*$ and have an identity element w.r.t. operation $*$.

For example, consider a monoid $(I, \times)$ where I is the set of all the integers and $\times$ is the multiplication operation. Then $(A, \times)$ is submonoid where A is the set of all the positive integers.

**Example 12:** Determine whether $(N, +)$ and $(N, *)$ are semigroup or monoid or both where N is the set of positive integers and $+$ and $*$ are ordinary addition and multiplication operation.

### *Solution*

**Closure Property:**
Since addition and multiplication of two positive integers is always a positive integer. So, $+$ and $*$ are closed operation on set N.

**Associative Property:**
The operation $+$ and $*$ are associative operation as we have

$$a * (b * c) = (a * b) * c$$
and
$$a + (b + c) = (a + b) + c$$

Since, addition and multiplication are closed and associative on set N.
Hence, (N, +) and (N, *) are semigroup.

### Identity Property:

There is no identity element with respect to addition operation.
Hence, (N, +) is not a monoid.
The identity element w.r.t multiplication operation is one.
Hence, (N, *) is a monoid.

## 12.7 GROUP

Let (A, *) be an algebraic system where * is a binary operation on set A. Then the system (A, *) is said to be group if it satisfies the following properties:

(1) The operation * is a closed operation.
(2) The operation * is an associative operation.
(3) There exists an identity element w.r.t. operation *.
(4) Every element in A has an inverse i.e. there exists an element $a^{-1} \in A$ $a \, \forall \, A$ such that

$$a^{-1} * a = a * a^{-1} = e$$

For example, consider an algebraic system (Q, ×) where Q is the set of rational numbers and × is the multiplication operation. Then (Q, ×) is the group with an identity element 1 w.r.t. operation × and $-2$ and $-1/2$ are the inverses since

$$(-2) \times (-1/2) = (-1/2) \times (-2) = 1.$$

**Example 13:** Let M be the set of $2 \times 2$ matrices with rational entries under the operation of matrix multiplication. Determine if M is a group?

### *Solution*

Since the determinant of a given matrix may be zero, so the inverses don't always exist. Hence, M is not a group.

**Example 14:** Let G be the subset of $2 \times 2$ matrices with non-zero determinant and with rational entries under the operation of matrix multiplication. Determine if G is a group?

### *Solution*

G is a group under matrix multiplication and the identity element is given as

$$I = \begin{bmatrix} 1 & 0 \\ 0 & 1 \end{bmatrix} \text{ and the inverse of } A = \begin{bmatrix} a & b \\ c & d \end{bmatrix} \text{ is } A^{-1} = \begin{bmatrix} \dfrac{d}{|A|} & \dfrac{-b}{|A|} \\ \dfrac{c}{|A|} & \dfrac{a}{|A|} \end{bmatrix}$$

### 12.7.1 Subgroup

Let $(A, *)$ be a group where $*$ is the binary operation on A. Also, let B is a non-empty subset of A i.e. $B \subseteq A$. Then the system $(B, *)$ is a subgroup of $(A, *)$ if the set B satisfies the properties of group under the operation $*$ as given below:

    (1) The operation $*$ is closed operation.
    (2) The operation $*$ is associative operation.
    (3) The identity element of $(A, *)$ must also belong to $(B, *)$.
    (4) Every element in B has an inverse i.e. there exists an element $a^{-1} \in B$ $a \forall B$ such that

$$a^{-1} * a = a * a^{-1} = e$$

For example, consider a group $(I, +)$ where I is the set of all the integers and $+$ is the addition operation. Then $(A, +)$ is a subgroup where A is the set of all the even integers including zero.

### 12.7.2 Abelian Group

Let $(A, *)$ be an algebraic system where $*$ is a binary operation on set A. Then the system $(A, *)$ is said to be an abelian group or commutative group if it satisfies all the properties of a group and an additional property as given below:

    (a) The operation $*$ is commutative i.e.

$$x * y = y * x \quad \forall\, x, \quad y \in A$$

For example, consider an algebraic system $(I, +)$ where I is the set of all the integers and $+$ is the addition operation. Then system $(I, +)$ is an abelian group as it satisfies all the properties of an abelian group.

**Example 15:** Consider the algebraic system $(G, *)$ where G is the set of all the real numbers and $*$ is the binary operation is defined by

$$a * b = ab\,/\,2$$

Determine if $(G, *)$ is an abelian group?

*Solution*

**Closure Property:**

The set G is closed under the operation $*$ as we have $a * b = ab\,/\,2$ which is a real number, for all $a, b \in G$.

**Associative Property:**

The operation $*$ is associative as we have

$$(a * b) * c = (ab\,/\,2) * c = (ab\,/\,2 \cdot c)/2 = abc\,/\,4$$
$$a * (b * c) = a * (bc\,/\,2) = (a \cdot bc\,/\,2)/2 = abc\,/\,4$$

Hence, $a * (b * c) = (a * b) * c$

**Identity Element:**
Let us assume that $e$ is a positive real number.
Then $e * a = a$ for all $a \in G$
$$\therefore \quad ea / 2 = a \Rightarrow e = 2$$
Similarly, $a * e = a$ for all $a \in G$
$$\therefore \quad ae / 2 = a \Rightarrow e = 2$$
Hence, the identity element is 2.

**Inverse element:**
Let us assume that $b \in G$ be the inverse of $a \in G$.
Then, we have

$$a * b = e \text{ where } e \text{ is the identity element}$$
$$\therefore \quad a * b = 2 \text{ here 2 is the identity element}$$
$$\Rightarrow \quad ab / 2 = 2$$
$$\Rightarrow \quad b = 4 / a$$
$$\Rightarrow \quad \text{Similarly,} \quad b * a = 2$$
$$\Rightarrow \quad ba / 2 = 2$$
$$\Rightarrow \quad b = 4 / a$$
So, the inverse of element $a$ is $4 / a$.

**Commutative:**

The operation $*$ is commutative as

$$A * b = ba/2 = b * a$$

As the algebraic system satisfies all the properties of an abelian group.
Hence, the system $(G, *)$ is an abelian group.


### 12.7.3 Finite and Infinite Group

A group $(A, *)$ is said to be a finite group if A is a finite set.
For example, the group $A = \{1, 2, 3, 4, 5, 6, 7, 8\}$ under multiplication modulo 9 is a finite group as the set A is finite.
A group $(A, *)$ is said to be infinite group if A is an infinite set.
For example, the group $(I, +)$ is an infinite group as the set I is infinite.


**Example 16:** State which of the following sets are closed under the operation of multiplication

(a) $A = \{-1, 0, +1\}$        (b) $B = \{X: x \text{ is odd}\}$
(c) $C = \{x: x \text{ is prime}\}$      (d) $D = \{1, 3\}$

***Solution***

(a) We have:
$-1 \cdot -1 = 1, -1 \cdot 0 = 0, -1 \cdot +1 = -1, 0 \cdot 0 = 0, 0 \cdot +1 = 0, +1 \cdot +1 = 1$
Hence, A is closed under the operation of multiplication.

  (b) The product of odd number is odd. Hence, B is closed.

  (c) Note that 2 and 3 are prime but $2 \cdot 3 = 6$ is not a prime.
     Hence, C is not closed.

  (d) We have $3 \cdot 3 = 9$, which doesn't belong to D.
     Hence, D is not closed.

## 12.8 COSETS

Let $(G, *)$ be a group where $*$ is binary operation on G and let H is a subset of i.e. $H \subseteq G$. Then left cossets $a * H$ of H is defined as the set of elements given by:

$$a * H = \{a * h \,/\, h \in H\} \text{ where } a \text{ is any element of G.}$$

Similarly, the right cossets $H * a$ of H is defined as the set of elements given by:

$$H * a = \{h * a \,/\, h \in H\} \text{ where } a \text{ is any element of G.}$$

**Example 17:** Let $(I, +)$ be a group where I is the set of all the integers and $+$ is the ordinary addition operation and let $H = \{\ldots, -6, -3, 0, +3, +6, \ldots\}$ be a subgroup which consists of multiples of 3. Determine all the left cossets of H in I.

### *Solution*

The left cossets of H in I are as follows:

$$0 + H = \{\ldots, -6, -3, 0, +3, +6, \ldots\} = H$$
$$1 + H = \{\ldots, -5, -2, +1, +4, +7, \ldots\}$$
$$2 + H = \{\ldots, -4, -1, +2, +5, +8, \ldots\}$$

There are no other distinct left cossets because any other left cossets coincide with the cossets given above.

## 12.9 NORMAL SUBGROUPS

Let $(G, *)$ be a group and $(H, *)$ be a subgroup of G. Then the subgroup $(H, *)$ is called a normal subgroup if

$$aH = Ha \qquad \text{for all } a \in G$$

  i.e. if the right and left cosets coincide.
  Note that every subgroup of an abelian group is normal.

## 12.10 ORDER OF A GROUP

The order of a group G is defined as the number of elements in the group G. It is denoted by $|G|$.

  A group of order 1 has only one element i.e. $\{(e), *\}$.

A group of order 2 has two elements i.e. $\{(e, x), *\}$ i.e. one identity element and one some other element.

**Lagrange's Theorem:** Let H be a subgroup of a finite group G. Then the order of H divides the order of G.

### *Proof*

One need to show that the right cosets of H in G, called the index of H in G, is equal to the number of left cosets of H in G and both numbers are equal to $|G|$ divided by $|H|$.

## 12.11 ISOMORPHISM

Let $(A, *)$ and $(B, \times)$ be two algebraic systems where $*$ and $\times$ are two binary operations on set A and B respectively. Then, we can say that the algebraic system $(B, \times)$ is said to be *isomorphic* to the algebraic system $(A, *)$ if we can obtain $(B, \times)$ from $(A, *)$ by renaming the elements and/or the operation in $(A, *)$. In other words, we can say that the system $(B, \times)$ is isomorphic to the system $(A, *)$ if there exists a one-to-one function $f$ from A to B such that

$$f(a * b) = f(a) \times f(b)$$

for all $a, b \in A$.

The function $f$ is called an *isomorphism* from $(A, *)$ to $(B, \times)$ and $(B, \times)$ is called an *isomorphic image* of A.

For example, consider the algebraic system as shown in fig. 1 below:

| $*$ | $a$ | $b$ |
|---|---|---|
| $a$ | $a$ | $b$ |
| $b$ | $b$ | $a$ |

**Fig. 1**

| $\times$ | $\alpha$ | $\beta$ |
|---|---|---|
| $\alpha$ | $\alpha$ | $\beta$ |
| $\beta$ | $\beta$ | $\alpha$ |

**Fig. 2**

Then by simply changing $a$, $b$ with $\alpha$, $\beta$ and $*$ to $\times$, we obtain the algebraic system $(B, \times)$ as in Fig. 2. Clearly, two systems $(A, *)$ and $(B, \times)$ are "essentially the same" i.e. are isomorphic to each other and the function $f$ is define by

$$f(a) = \alpha$$
$$f(b) = \beta$$

The function $f$ is called an isomorphism form the algebraic system $(A, *)$.

Example, the algebraic system $(C, +)$ and $(D, \times)$ as shown in Fig. 3 and Fig. 4 below are all isomorphic to the algebraic system $(A, *)$ in Fig. 1.

| + | Even | Odd |
|---|---|---|
| Even | Even | Odd |
| Odd | Odd | Even |

**Fig. 3**

| × | 0 | 180 |
|---|---|---|
| Even | 0 | 180 |
| Odd | 180 | 0 |

**Fig. 4**

The system (C, +) corresponds to addition of even and odd numbers and the system (D, ×) corresponds to the rotation of geometric figures in the plane by 0 and 180 degree.

## 12.12 AUTOMORPHISM

An isomorphism from an algebraic system (A, *) to (A, *) is called *automorphism* on (A, *). For example, the function $f$ such that

$$f(a) = b$$
$$f(b) = a$$

is an automorphism on the algebraic system (A, *).

A physical interpretation of an automorphism on the algebraic system is the way in which the elements in the system interchange their roles.

## 12.13 HOMOMORPHISM

Let (A, *) and (B, ·) be two groups where * and · are binary operations. Then, $f$ be a function from A onto B such that

$$f : A \rightarrow B$$

$$f(a * b) = f(a) \cdot f(b)$$

for every $a, b \in A$. The function $f$ is called *homomorphism* from (A, *) to (B, ·) and (B, ·) is called *homomorphic image* of (A, *).

For example, consider the algebraic system (I, ·), where I is the set of integers and · (juxtaposition) is the ordinary multiplication operation of integers. Suppose that we want to make distinction between positive integers, negative integers and zero. Then the function $f$ defined by

$$f(n) = \begin{cases} \text{Positive} & \text{if } n \text{ is positive integer} \\ \text{Negative} & \text{if } n \text{ is negative integer} \\ \text{Zero} & \text{if } n = 0 \end{cases}$$

is a homomorphism from (I, ·) to (B, ⊗) where the system (B, ⊗) is as given below:

| ⊗ | Positive | Negative | Zero |
|---|---|---|---|
| Positive | Positive | Negative | Zero |
| Negative | Negative | Positive | Zero |
| Zero | Zero | Zero | Zero |

**Example 18:** Consider the group G = {1, 2, 3, 4, 5, 6} under the multiplication modulo 7.

(a) Find the multiplication table of G.
(b) Find $2^{-1}$, $3^{-1}$ and $6^{-1}$
(c) Find the order and subgroups generated by 2 and 3.
(d) Is G cyclic?

### Solution

(a) To find $a * b$ in G, find the remainder when the product $ab$ is divided by 7.
For example, $4 \cdot 5 = 20$, which yields a remainder of 6 when divided by 7.
Hence, $4 * 5 = 6$ in G.

The multiplication table of G is as below:

| * | 1 | 2 | 3 | 4 | 5 | 6 |
|---|---|---|---|---|---|---|
| 1 | 1 | 2 | 3 | 4 | 5 | 6 |
| 2 | 2 | 4 | 6 | 1 | 3 | 5 |
| 3 | 3 | 6 | 2 | 5 | 1 | 4 |
| 4 | 4 | 1 | 5 | 2 | 6 | 3 |
| 5 | 5 | 3 | 1 | 6 | 4 | 2 |
| 6 | 6 | 5 | 4 | 3 | 2 | 1 |

(b) The identity element of G is 1.
If $b$ is the inverse element of $a$, then we have $a * b = 1$
Hence, $2^{-1} = 4$, $3^{-1} = 5$, $6^{-1} = 6$
(c) We have $2^1 = 2$, $2^2 = 4$ and $2^3 = 1$. Hence $|2| = 3$ and $gp(2) = \{1, 2, 4\}$.
Similarly, $3^1 = 3$, $3^2 = 2$, $3^3 = 6$, $3^4 = 4$, $3^5 = 5$ and $3^6 = 1$.
Hence, $|3| = 6$ and $gp(3) = \{1, 2, 3, 4, 5, 6\} = G$.
G is cyclic since $G = gp(3)$

## 12.14 RINGS

We have so far studied algebraic system with only one binary operation. Now we will study briefly several classes of algebraic system with two binary operations. Under this class come rings, integral domain and fields.

Let (A, +, $*$) be an algebraic system where + and $*$ are two binary operation on set A. Then the system (A, +, $*$) is called *ring* if

1. (A, +) is an abelian group.
2. (A, $*$) is semigroup.
3. The operation $*$ is distributive over the operation +

   i.e. $a * (b + c) = a * b + a * c$   and   $(b + c) * a = b * a + c * a$.

For example, consider the algebraic system $(Z_n, \oplus, \otimes)$ where $Z_n$ is defined as the set of integers $\{0, 1, 2, 3, ..., n-1)$.

Let $\oplus$ be a binary operation on $Z_n$ defined by

$$a \oplus b = \begin{vmatrix} a + b & \text{if } a + b < n \\ a + b - n & \text{if } a + b >= n \end{vmatrix}$$

Let $\otimes$ be a binary operation on $Z_n$ such that
$a \otimes b = ab \bmod n$.
It can be shown that $(Z_n, \oplus)$ is an abelian group.
Furthermore, $(Z_n, \otimes)$ is a semigroup as $\otimes$ is closed and associative operation.
Also, $\otimes$ distributes over $\oplus$. So, the system $(Z_n, \oplus, \otimes)$ is a ring.

The identity of the abelian group (A, +) is referred to as *additive identity* and will $b$ denoted by 0. The inverse of an element $a$ of the group (A, +) is referred to as *additive inverse* of $a$ and will be denoted by $-a$.

### 12.14.1 Types of Rings

The followings are the different types of rings:

1. **Commutative ring:**
   A ring (R, +, $\cdot$) is called a commutative ring if it holds the commutative law under the operation of multiplication i.e.

   $$a \cdot b = b \cdot a \qquad \text{for all } a, b \in \text{R}.$$

   For example, the system (E, +, $\cdot$) is a commutative ring where E is the set of even integers and + and . are ordinary operation of integers.
2. **Ring with unity:**
   A ring (R, +, $\cdot$) is called *a ring with unity* if it has a multiplicative identity i.e.
   $$a \cdot e = e \cdot a = a \qquad \text{for all } a \in \text{R}$$
3. **Ring with zero divisors:**
   A ring (R, +, $\cdot$) is called *a ring with zero divisors* if there exists non-zero element $a, b \in$ R such that $ab = 0$. In such case $a$ and $b$ are called ***zero divisors***.

### 12.14.2 Subrings

A subset S of R is a *subring* of R if S itself is a ring under the operation in R.
Note that S is a subring of R if

1. $0 \in S$
2. for any $a, b \in S$, we have $a - b \in S$ and $ab \in S$.

For example, the system $(I, +, \cdot)$ of integers is a subring of ring $(R, +, \cdot)$ of real numbers.

### 12.14.3 Special types of Rings

There are two special types of rings: integral domain and fields.

### 12.15 INTEGRAL DOMAIN

Let $(A, +, *)$ be an algebraic system where $+$ and $*$ are two binary operations on set A. Then the system $(A, +, *)$ is called *integral domain* if

1. $(A, +)$ is an abelian group.
2. The operation $*$ is commutative. Furthermore, if $c \neq 0$ and $c * a = c * b$, then $a = b$, where 0 denotes the additive identity.
3. The operation $*$ is distributive over the operation $+$

$$\text{i.e.} \quad a * (b + c) = a * b + a * c \quad \text{and} \quad (b + c) * a = b * a + c * a$$

In other words, *a* commutative ring R is an *integral domain* if R has no zero divisors i.e. if $ab = 0$ implies that $a = 0$ or $b = 0$.

For example, consider the system $(I, +, *)$ where I is the set of all integers and $+$ and $*$ are ordinary addition and multiplication operations of integers.

We note that $(I, +)$ is an abelian group with 0 being the identity element and $-n$ being the inverse of *n* for any integer *n*.

We also note that the operation $*$ is commutative.

Furthermore, for any non-zero integer $c$, $c * a = c * b$ implies that $a = b$.

Also, the operation $*$ is distributive over $+$. So, the system $(I, +, *)$ is an integral domain.

### 12.16 FIELDS

Let $(A, +, *)$ be an algebraic system where $+$ and $*$ are two binary operation on set A. Then the system $(A, +, *)$ is called *field* if

1. $(A, +)$ is an abelian group.
2. Every non-zero element of A with identity element one (not equal to zero) has a multiplicative inverse.
3. The operation $*$ is distributive over the operation $+$

$$\text{i.e.} \quad a * (b + c) = a * b + a * c \quad \text{and} \quad (b + c) * a = b * a + c * a$$

Note that a field is necessarily an integral domain, for if $ab = 0$ and $a \neq 0$, then
$$b = 1 \cdot b = a^{-1} ab = a^{-1} \cdot 0 = 0$$
For example, the system $(Q, +, *)$ where $Q$ is the set of all rational numbers and $+$ and $*$ are ordinary addition and multiplication operations of rational numbers.

Other example of fields is $(R, +, *)$, where $R$ is the set of all real numbers and $+$ and $*$ are ordinary addition and multiplication operations of real numbers.

Another example of fields is $(C, +, *)$, where $C$ is the set of all complex numbers and $+$ and $*$ are ordinary addition and multiplication operations of complex numbers.

**Example 19:**  The set $Z_n = \{0, 1, 2, 3, \ldots, n-1\}$ under the operation of addition and multiplication modulo $n$ is a ring. It is called *ring of integers modulo n.*

If $n$ is a prime, then $Z_n$ is a field.

If $n$ is not a prime, then $Z_n$ has zero divisors. For example, as in the ring $Z_6$
$2 \cdot 3 = 0$ but $2 \neq 0 \pmod 6$ and $3 \neq 0 \pmod 6$

**Example 20:**  Let M denote the set of $2 \times 2$ matrices with the integer entries. Then M is a noncommutative ring with zero divisors under the operation of matrix addition and matrix multiplication. M does have an identity element, the identity matrix.

**Example 21:**  In an integral domain show that if $ab = ac$ with $a \neq 0$, then $b = c$.

### Solution

Since $ab = ac$, we have
$$ab - ac = 0$$
$$\Rightarrow \quad a(b - c) = 0$$
Since, $a \neq 0$, we must have $b - c = 0$
Since, D has no zero divisors, hence $b = c$.

### Exercise 12.1

1. Let $(S, *)$ be a commutative semigroup. Show that if $x * x = x$ and $y * y = y$, then $(x * y) * (x * y) = x * y$.

2. Consider the set N of positive integers and let $*$ denote the operation of least common multiple (l cm) on N.

   (i) Find $2 * 3$, $9 * 12$ and $3 * 5$
   (ii) Is $(N, *)$ a semigroup? Is it commutative?
   (iii) Find the identity element for $*$.
   (iv) Which elements in N have inverses, if any?

3. Consider the set Q of rational numbers and let $*$ be the operation on Q defined by
   $$a * b = a + b + 3ab$$

   (i) Find $3 * 7$ and $2 * (-5)$
   (ii) Is $(Q, *)$ a semigroup? Is it commutative?
   (iii) Find the identity element for $*$.
   (iv) Does any element in Q have an inverse?

4. Let $A = \{\ldots, -8, -4, 0, +4, +8, \ldots\}$, i.e. multiples of 4. Is A closed under

   (i) addition

    (ii) multiplication

    (iii) subtraction

    (iv) division (except by 0)

5. Consider the group G = {1, 2, 4, 7, 8, 11, 13, 14} under the multiplication modulo 15.

    (i) Find the multiplication table of G.

    (ii) Find $2^{-1}$, $7^{-1}$ and $11^{-1}$

    (iii) Find the order and subgroups generated by 2, 7 and 11.

    (iv) Is G cyclic?

6. Consider the group G = {1, 5, 7, 11, 13, 17} under the multiplication modulo 18.

    (i) Find the multiplication table of G.

    (ii) Find $5^{-1}$, $7^{-1}$ and $17^{-1}$

    (iii) Find the order and subgroups generated by 5 and 13.

    (iv) Is G cyclic?

7. Let $Z_n$ denote the set of integers {0, 1, 2, 3, …}. Let $\otimes$ be binary operation on $Z_n$ such that $a \otimes b$ = the remainder of $ab$ divided by $n$:

    (i) Construct the table for the operation $\otimes$ for $n = 4$

    (ii) Show that $(Z_n, \otimes)$ is a semigroup for any $n$.

**Answers to Selected Problems**

2. (i) 6, 36, 15

    (ii) (N, ∗) is a semigroup and is commutative

    (iii) 1  (iv) 1

5. (i) Multiplication table of 6

| ∗ | 1 | 2 | 4 | 7 | 8 | 11 | 13 | 14 |
|---|---|---|---|---|---|----|----|----|
| 1 | 01 | 2 | 4 | 7 | 8 | 11 | 13 | 14 |
| 2 | 2 | 4 | 8 | 14 | 01 | 07 | 11 | 13 |
| 4 | 4 | 8 | 01 | 13 | 2 | 14 | 07 | 11 |
| 7 | 7 | 14 | 13 | 04 | 11 | 02 | 01 | 08 |
| 8 | 8 | 01 | 02 | 11 | 04 | 02 | 14 | 07 |
| 11 | 11 | 07 | 14 | 02 | 13 | 01 | 08 | 04 |
| 13 | 13 | 11 | 07 | 01 | 14 | 08 | 04 | 02 |
| 14 | 14 | 13 | 09 | 08 | 07 | 04 | 02 | 01 |

    (ii) $2^{-1} = 8$; $7^{-1} = 13$; $11^{-1} = 11$;

    (iv) G is cyclic group.

7. (i)

| ∗ | 0 | 1 | 2 | 3 | 4 |
|---|---|---|---|---|---|
| 0 | 0 | 0 | 0 | 0 | 0 |
| 1 | 0 | 1 | 2 | 3 | 0 |
| 2 | 0 | 2 | 0 | 2 | 0 |
| 3 | 0 | 3 | 2 | 1 | 0 |
| 4 | 0 | 0 | 0 | 0 | 0 |

# 13

# Finite State Machine

## 13.1 INTRODUCTION

A Finite State Machine (FSM) is a model of behavior using states and state transitions. A transition is a state change triggered by an input event, i.e. transitions map some state-event pairs to other states. As indicated in the name, the set of states should be finite. Also, it is assumed that there is a finite set of distinct input events or their categories (types, classes). Subsequently, the set of transitions is finite as well.

Given a string of zeroes and ones, we want to recognize when the input contains the substring 001 but does not start with 11. We are going to design a simple theoretical "machine" that will be able to solve pattern recognition problem like this one.

Finite automata are primarily used in parsing for recognizing languages. Input strings belonging to a given language should turn an automaton to final states and all other input strings should turn this automaton to states that are not final.

If the output function depends on a state and input symbol, then it is called a Mealy automaton. If the output function depends only on a state, then it called is a Moore automaton. In this chapter we will restrict our discussion to finite state machines.

## 13.2 AUTOMATON

Automaton is defined as a system where energy, materials and information are transformed, transmitted and used for performing some functions without direct participation of man. e.g., automatic washing machine, automatic photo printing machine.

### 13.2.1 The Characteristics of Automaton

1. **Input:** At each of the discrete instants of time $t_1$, $t_2$, ... $t_m$ the input values $l_1$, $l_2$, ... $l_p$ each of which can take a finite number of fixed values from the input alphabet $\Sigma$.

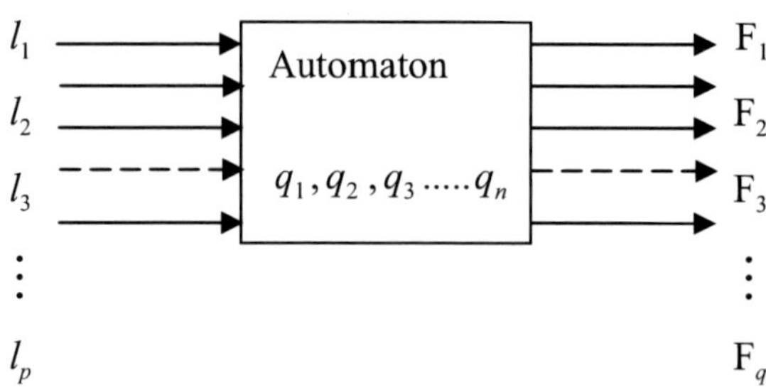


**Fig. 13.1**

2. **Output:** $F_1$, $F_2$, ... $F_q$ are the outputs of the model, each of which can take a finite number of fixed values from an output F.
3. **States:** At any instant of time the automaton can be in one of the states $q_1$, $q_2$, $q_3$, ... $q_n$.
4. **State relation:** The next state of an automaton at any instant of time is determined by the present state and the present input.
5. **Output relation:** The output is related to either state only or to both the input and the state.

### 13.2.2 Types of Automaton

(a)  Deterministic Automaton
(b)  Non-deterministic Automaton

A deterministic automaton is one in which each move is uniquely determined by the current configuration. If the internal state, input and contents of the storage are known, it is possible to predict the future behavior of the automaton. This is said to be deterministic automaton otherwise it is nondeterministic automaton.

### 13.2.3 Finite State machine (FSM)

**A finite state machine or deterministic finite automaton (DFA):**

M is an abstract model of a very simple computer (with limited memory).

$$M = (Q, \Sigma, \delta, q_0, F)$$

It has the following components:

- A finite non-empty set Q of states.
- A finite non-empty set $\Sigma$ of input symbols.
- An initial state $q_0 \in Q$
- A Finite set F of Output symbols
- A transition function $\delta: Q \times \Sigma \to Q$

The machine starts in the initial state. It then receives input, one symbol at a time. Upon receiving each input symbol, the transition function determines the next state of the machine. If it was in state $q$ and it receives input symbol $a$, then $\delta$ $(q, a)$ tells us the next state of the machine. Furthermore at each state the machine produces an output letter according to the output function.

### 13.2.4 Transition system

A transition system is a finite directed labelled graph in which each node represents a state and the directed edges indicate the transition of the state and the edges are labelled with input / output.

A transition system is a 5-tuple: $(Q, \Sigma, \delta, q_0, F)$

## 13.3 TRANSITION DIAGRAMS

A transition diagram is a finite directed graph in which each vertex represents a state and directed edges indicates the transition from one state to another. Edges are labelled with input / output in the representation the initial state is represented by a circle with an arrow towards it the final state by two concentric circle and the other intermediate states are represented by a circle.

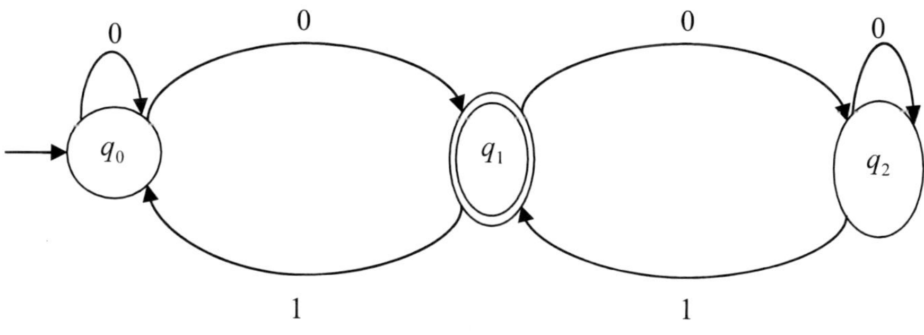

**Fig. 13.2**

Here in this figure $q_0$ is the initial state, $q_1$ is the final state, and $q_2$ is the intermediate state.

## 13.4 TRANSITION TABLE

Transition table is the tabular representation of transition diagram. Transition table is also known as transition function table or state table. In this representation initial state is represented by an arrow towards it and final state by a circle.

**Example 1:** Find the transition diagram of the finite state automation

$M = (Q, \Sigma, \delta, s_0, O)$, where $\Sigma = \{0, 1\}$, $Q = \{s_0, s_1, s_2\}$, $F = \{s_2\}$, $s_0$ is the initial state, and the transition function $\delta$ given by the table

**Table 13.1**

| $\Sigma$ | $\delta$ | |
|---|---|---|
| $Q$ | 0 | 1 |
| $s_0$ | $s_1$ | $s_0$ |
| $s_1$ | $s_2$ | $s_0$ |
| $\widehat{s_2}$ | $s_2$ | $s_0$ |

Also redraw the transition diagram of this FSA as a transition diagram of a finite state machine.

### *Solution*

The transition diagrams of this finite state automaton is

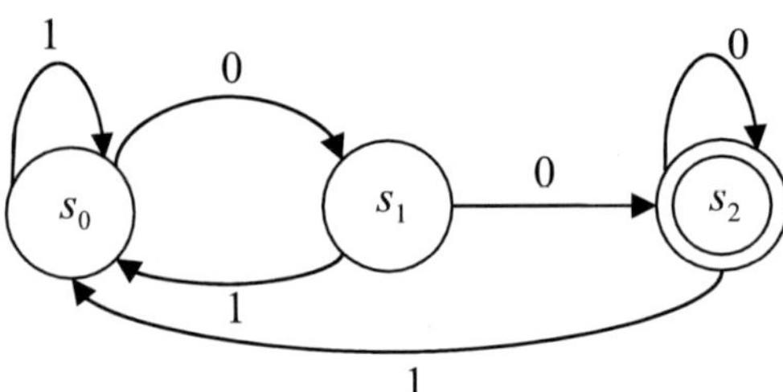


**Fig. 13.3**

## 13.5 ACCEPTABILITY OF A STRING BY A FINITE AUTOMATON

A string $x$ is accepted by a finite automaton $M = (Q, \Sigma, \delta, s_0, F)$ if $\delta(q_0, x) = q$ for some $q \in F$. This is basically the acceptability of a string by the final state. A Final state is also called accepting state.

Now a finite state automaton is a finite state machine with outputs 0 and 1. The incoming edges in an accepting state has output 1 and incoming edges in a non-accepting state has output 0. Thus the given finite state automaton is a finite state machine whose transition diagram is given below:

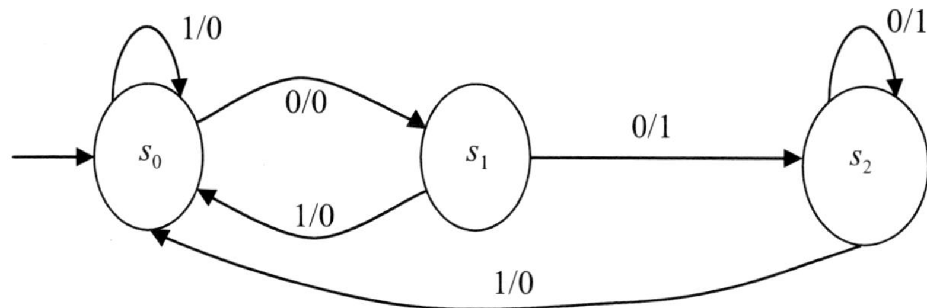

**Fig. 13.4**

**Example 2:** The finite state machine (1) described below (1) will accept all strings of zeroes and ones that contain at least two ones. Instead of describing Q, $\Sigma$, $q_0$, $\delta[Q, \Sigma, \delta, q_0, F]$ and A, we almost always represent a finite state machine with a diagram like the following:

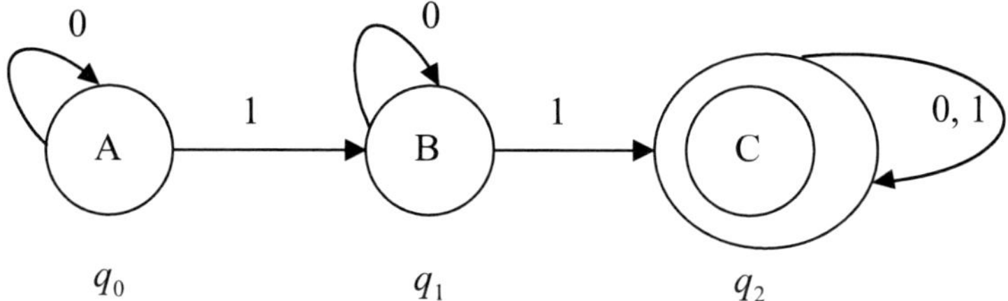

**Fig. 13.5**

The diagram above is a nice graphical way of representing all the information that would be in the official, format description above are the following:

- The set of states is {A, B, C}, {$q_0$, $q_1$, $q_2$}.
- The set of input symbols is {0, 1}.
- The initial state is A. (This is indicated by the unlabelled arrow going into $q_0$).
- The transition function $\delta$ can be described by the following transition table:

|  |  | **0** | **1** |
|---|---|---|---|
| $q_0$ | A | A | B |
| $q_1$ | B | B | C |
| $q_2$ | C | C | C |

The top row tells us that, if we are in state $q_0$, an input of 0 causes us to stay in state $q_0$, while an input of 1 cause us to move to state $q_1$.

- The set of accepting states (indicated by a double circle) is {$q_2$}. If the sequence of symbols in an input string will take us to state $q_2$, then that string is accepted by the machine;

otherwise, it is not accepted. Let us consider the example input string 1011000 and trace how the finite state machine would process it.

- The machine starts in state $q_0$.
- The first input symbol is 1. When we are in state $q_0$ and we see an input of 1, the diagram (or the transition table) tells us to move to state $q_1$.
- The second input symbols are 0. When we are in state $q_1$ and we see an input of 0, we stay is state $q_1$.
- The third input symbol is 1 this causes us to move to state $q_2$. Note that, in this example, once we are in state $q_2$, there is no way to leave. We will always stay in state $q_2$, no matter how much additional input we see. Therefore, in this case, the rest of the input (1000) does not really matter. We know that will end up staying in state $q_2$ and that the string will be accepted.
- This is the right decision. The string does not contain two or more ones and so we should accept it.

Let us trace the second example through our machine: 0010.

- The machine starts in state $q_0$.
- The first input symbol is 0. When we are in state $q_0$ and we see an input of 0, the diagram (or transition table) tells us to state in stay $q_0$.
- This happens again upon seeing the second 0 in the input.
- The third input symbol is 1. This causes us to move to state $q_1$.
- Finally the fourth input symbol is 0 and we stay in state $q_1$. Because the input is now finished and because we didn't end up in an accepting state, this string is not accepted. Also as the string doesn't have two or more ones and so it should not be accepted.

**Example 3:**  Design a finite state automaton that accepts those strings over $\{0, 1\}$ such that the number of zeros is divisible by 3.

### *Solution*

Let $M = (Q, \Sigma, \delta, s_0, F)$ be the required finite state automaton, where $\Sigma = \{0,1\}$.

Let $s_0$ be the initial state of M. Thus $s_0$ is the state of M after zero number of 0 has been input. Since zero is divisible by 3, $s_0$ must be an accepting state. Now $s_1$ be a state of M after one 0 has been input and $s_2$ be the state, where two 0's have been input. We observe that

- From the state $s_0$, three 0's are needed, so that total number of zeros are divisible by 3.
- From the state $s_1$, two 0's are needed, so that total number of zeros are divisible by 3.
- From the state $s_2$, one 0 is needed, so that total number of zeros are divisible by 3.

Further after three 0's are input, three more 0's are needed to get a new total divisible by 3. In general if $k \geq 0$ be integers and $(3k)$ 0's are input, then 3 more 0's are needed to get a new total divisible by 3.

Thus we will have a transition diagram of M to be of the type shown below.

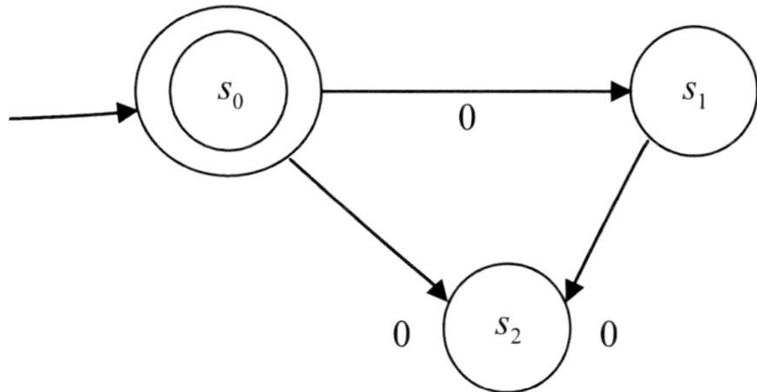

**Fig. 13.6**

Further if M is in any state and 1 is input, the total number of zeros in the input string remains unchanged. Thus there is a loop at each state labelled 1. Hence the transition diagram of FSA is

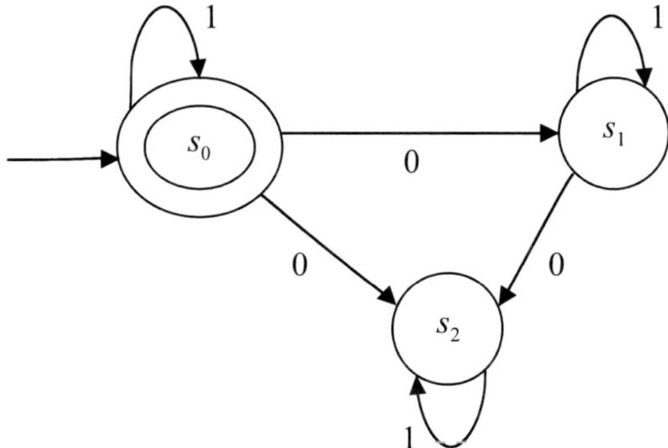

**Fig. 13.7**

Thus we have $\Sigma = \{0, 1\}$, $Q = \{s_0, s_1, s_2\}$, $F = \{s_0\}$, $s_0$ as the initial state and the next state function $\delta$ defined by

$$\delta(s_0, 0) = s_1, \quad \delta(s_0, 1) = s_0, \quad \delta(s_1, 0) = s_2, \quad \delta(s_1, 1) = s_1, \quad \delta(s_2, 0) = s_3, \quad \delta(s_2, 1) = s_2$$

**Example 4:** Let the alphabet set be $\{a, b\}$, design a FMS which accepts any string such that $a$ occurs on all odd number of position in the string. The position starts from 1. Such that $\{ababab \ldots\}$.

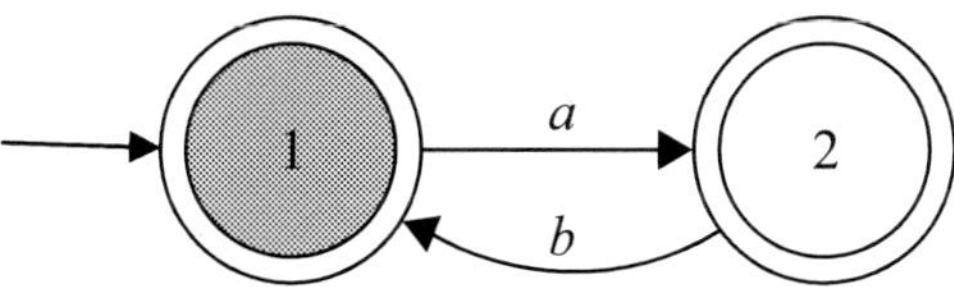

**Fig. 13.8**

**Example 5:** Design a FSM to accept any string such that in the string, if a occurs, it is always in a group of three. For instance, $\{baaabbbbbbaaabaaabb\}$.

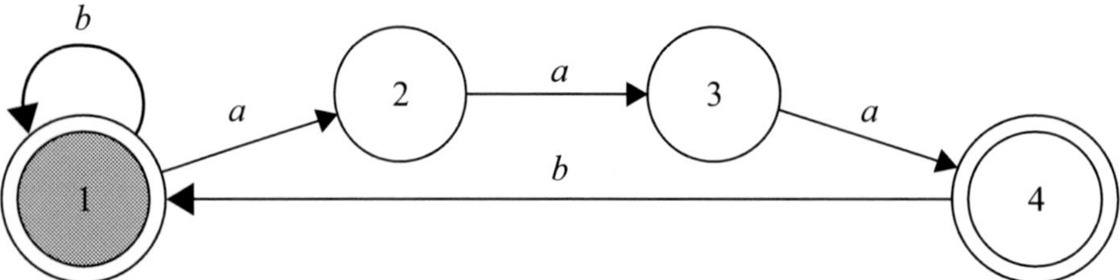

**Fig. 13.9**

**Example 6:** Design a finite state machine M which accepts the language $L(M) = \{w \in (a, b)^*\}$: $w$ does not contain three consecutive $b$'s.

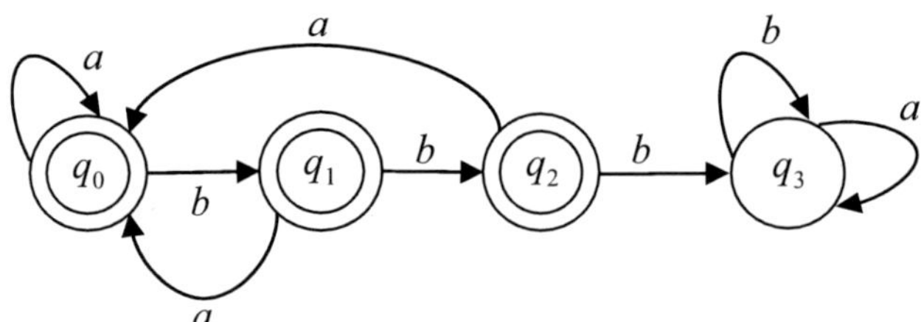

**Fig. 13.10**

### Solution

Here $q_0$, $q_1$, and $q_2$ are final states. Therefore any input string not containing three consecutive $b$'s will be accepted. If we get three consecutive $b$'s then the $q_3$ state is reached, which is not the final state hence M will remain in this state irrespective of any other symbol in the rest of the string. This kind of state is called dead state.

**Example 7:** Design a DFA, the language recognised by the automaton being $L = \{a^n b: n \geq 0\}$.

### Solution

In this example the string could be $b$, $ab$, $a^2b$, $a^3b$, ... Thus the DFA accepts all strings consisting of an arbitrary number of $a$'s followed by a single $b$.

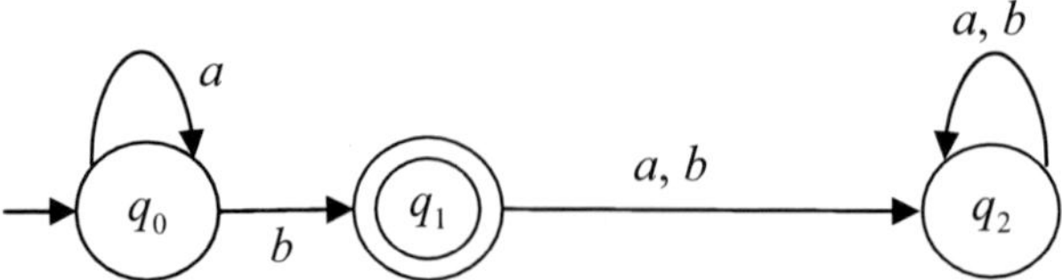

**Fig. 13.11**

**Example 8:** Determine the DFA with the

(a) Set of integers.
(b) Set of signed integers.
(c) Set of strings beginning with '*a*' and ending with '*b*'.
(d) Set of strings having '*aba*' as subword.

***Solution***

(a) Set of integers

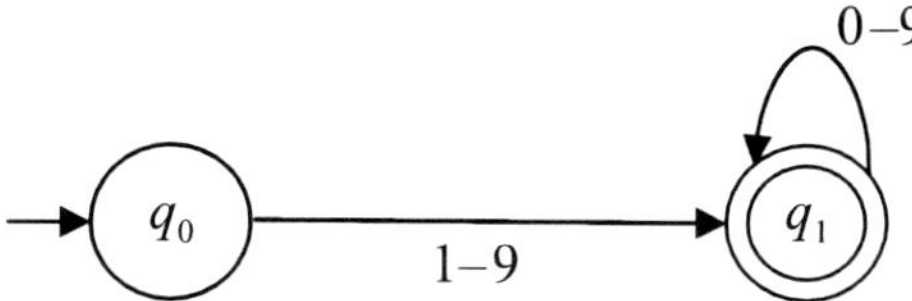

**Fig. 13.12**

(b) Set of signed integers

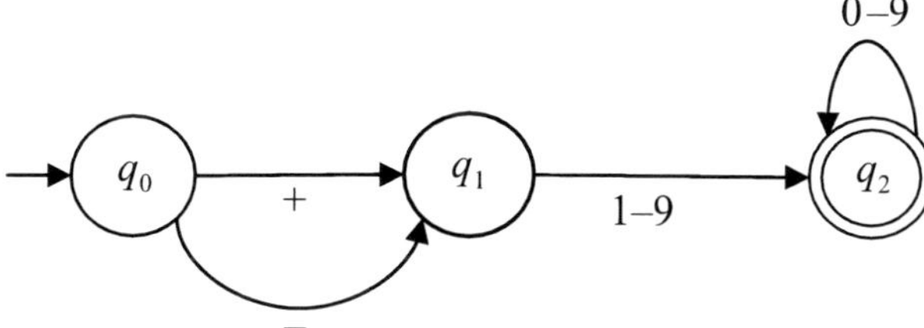

**Fig. 13.13**

(c) Set of strings beginning with '*a*' and ending with '*b*'

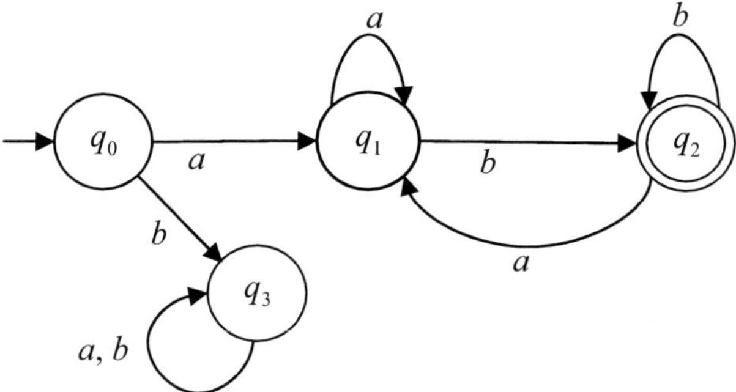

**Fig. 13.14**

$q_3$ is the dead state.

(d) Set of strings having '*aba*' as subword.

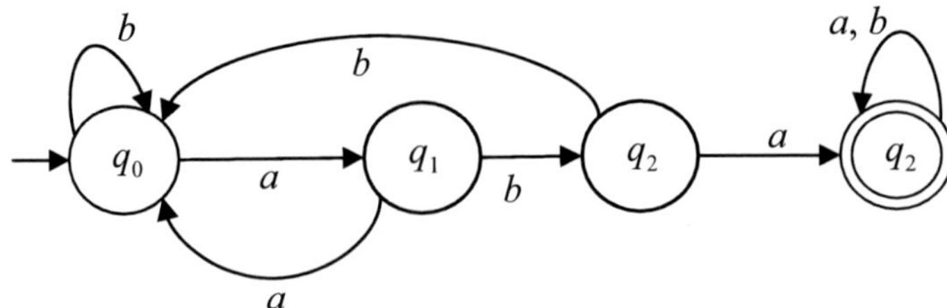

**Fig. 13.15**

**Example 9:** Construct a DFA accepting all string over $\{0, 1\}$.

(a) Having odd number of 0's.
(b) Having even number of 0's and even number of 1's.
(c) Even number of 0's.

***Solution***

(a) Having odd number of 0's.

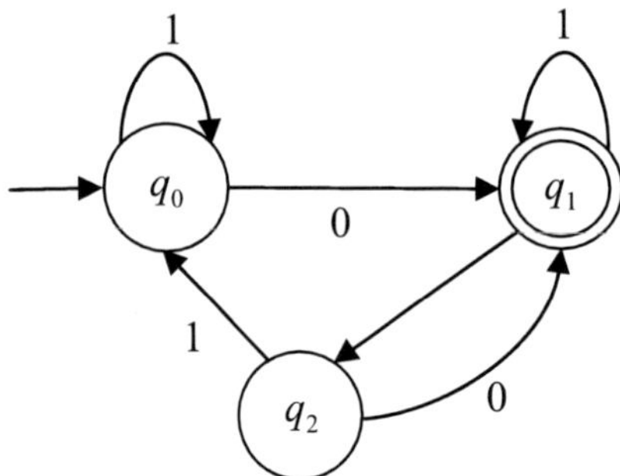

**Fig. 13.16**

(b) Having even number of 0's and even number of 1's.

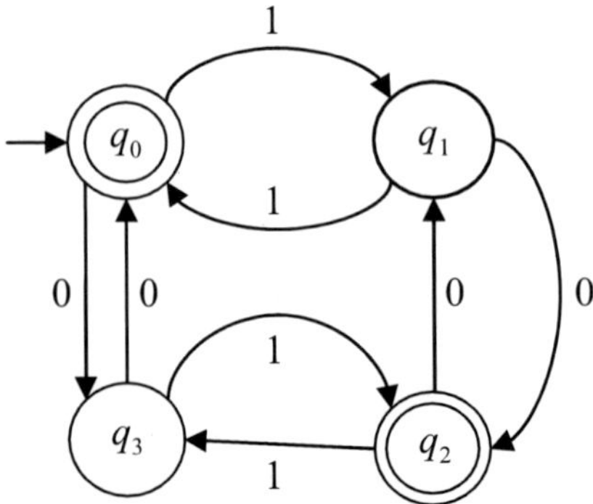

**Fig. 13.17**

(c) even number of 0's.

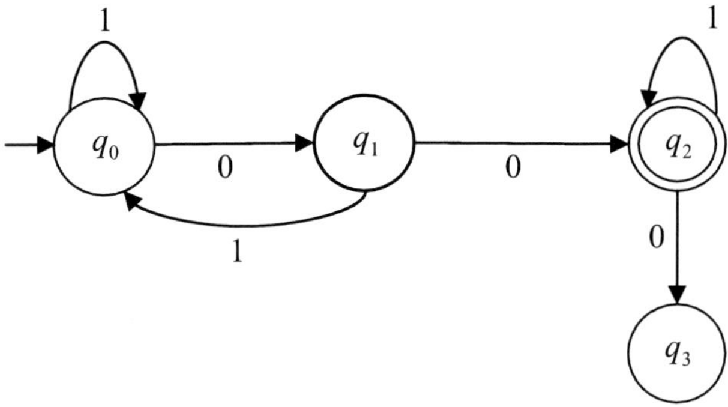

**Fig. 13.18**

**Example 10:** Consider the finite state machine whose transition function $\delta$ is given by Table 13.2. Here, $Q = \{q_0, q_1, q_2, q_3\}$, $\Sigma = \{0, 1\}$, $F = \{q_0\}$. Give the entire sequence of states for the input string 110001.

**Table 13.2**

| State | Input 0 | 1 |
|---|---|---|
| $q_0$ | $q_2$ | $q_1$ |
| $q_1$ | $q_3$ | $q_0$ |
| $q_2$ | $q_0$ | $q_3$ |
| $q_3$ | $q_1$ | $q_2$ |

*Solution*

$$\delta(q_0, 110101) = \delta(q_1, 10101)$$
$$= \delta(q_0, 0101)$$
$$= \delta(q_2, 101)$$
$$= \delta(q_3, 01)$$
$$= \delta(q_1, 1)$$
$$= \delta(q_0, \phi)$$
$$= q_0$$

hence the sequence is given by $q_0, q_1, q_0, q_2, q_3, q_1, q_0$.

## 13.6 NON-DETERMINISTIC FINITE STATE AUTOMATON (NFA)

In order to explain the concept of NFA using the transition diagram 13.2. Here from diagram if the automaton is in a state $\{q_0\}$ then the next state will either be $\{q_0\}$ or $\{q_1\}$. Thus some moves of the automaton cannot be determined uniquely by the input symbol and the present state. Such automatons are called NFA.

**A nondeterministic finite state automaton (NFA) is a 5-tuple** $M = (Q, \Sigma, \delta, q_0, F)$ where

- A finite nonempty set Q of states.
- A finite nonempty set $\Sigma$ of input symbols.
- An initial state $q_0 \in Q$
- A finite set $F \subset Q$ of Output symbols
- A transition function $\delta: Q \times \Sigma \to 2^Q$

We now consider modifying the finite automaton model to allow zero, one or more transitions from a state on the same input symbol. The new model is called a non-deterministic finite automaton. Any set accepted by a NFA can also be accepted by a deterministic finite automaton (DFA). The concept of NFA plays a central role in both the theory of languages and the theory of computation. A transition diagram for a NFA is shown in Fig. 13.19. Observe that there are two edges labelled 0 out of state $s_0$, one going back to state $s_0$ and one going to state $s_3$.

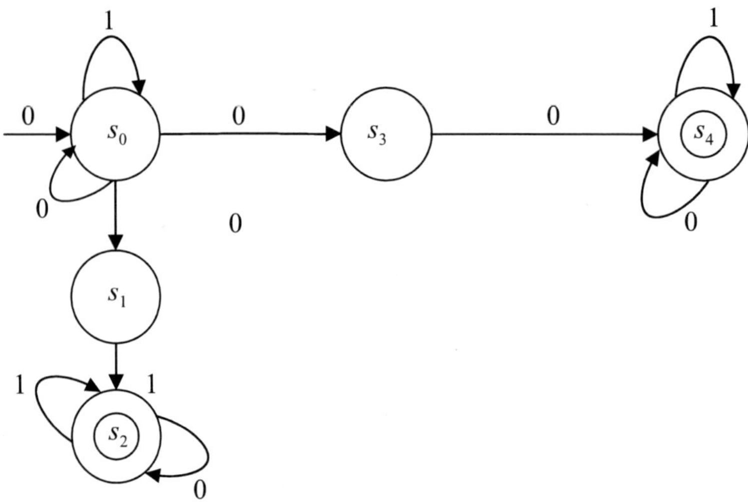

**Fig. 13.19**

**Difference between NFA and DFA**

1. In DFA, for a given state, on a given input we reach to deterministic and unique state. On the other hand, in NFA we may lead to more than one state for a given input.
2. The DFA is a subset of NFA.

3. In NFA the range δ is the power set of $2^Q$, so that its value is not a single element of Q but a subset of it. This subset defines the set of all possible states that can be reached by transition. But in DFA, it is not possible.

4. In NFA $\delta(q_0, n) = q$ it can change state without consumption of any symbol but it is not possible in DFA.

**Example 11:** Find an NFA with four states for $L = \{a^n: n \geq 0\} \cup \{b^n a: n \geq 1\}$.

***Solution***

In this language two cases arise:

(i) $a^n: n \geq 0$
(ii) $b^n a: n \geq 1$

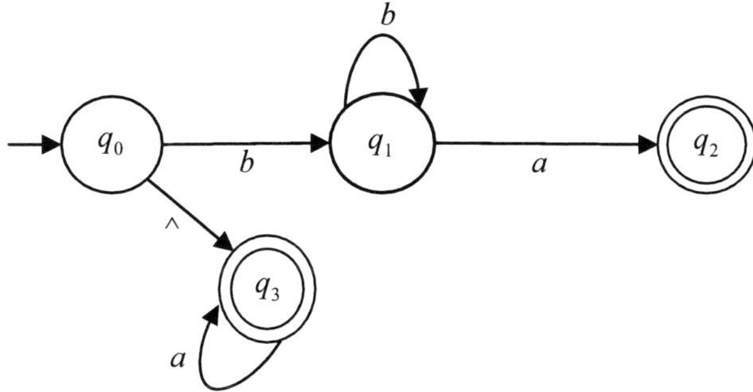

**Fig. 13.20**

**Example 12:** Design an NFA with three states that accepts the language $\{ab, abc\}$.

***Solution***

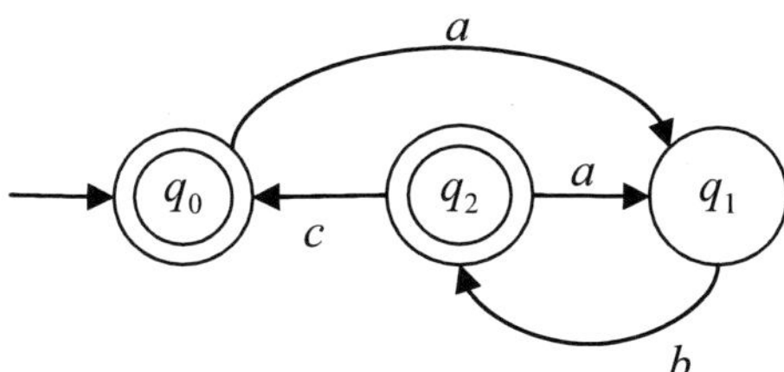

**Fig. 13.21**

## 13.7 EQUIVALENCE OF FINITE STATE MACHINES

Two finite state machines are said to be equivalent if, starting from their respective initial states, they will produce the same output sequence when they are given same input sequence, i.e. the two machines shave same terminal behaviours even though their internal structure might be different and may contain different number of states.

**Example 13:** Consider the two finite state machines given in Fig. (a) and Fig. (b). These two FSM are equivalent.

|  | Input | | |
|---|---|---|---|
| **State** | **1** | **2** | **Output** |
| A | B | C | 0 |
| B | F | D | 0 |
| C | G | E | 0 |
| D | H | B | 0 |
| E | B | F | 1 |
| F | D | H | 0 |
| G | E | B | 0 |
| H | B | C | 1 |

**Fig. (a)**

|  | Input | | |
|---|---|---|---|
| **State** | **1** | **2** | **Output** |
| A | B | C | 0 |
| B | C | D | 0 |
| C | D | E | 0 |
| D | E | B | 0 |
| E | B | C | 1 |

**Fig. (b)**

Consider the input sequence 1122212212, both machines will produce output sequence 0001010001. We may observe that for any input sequence we will get same output sequence by these two machines.

It uses in minimization of finite automaton.

## 13.8 TYPES OF FINITE STATE MACHINES

Many different finite state machines have been developed to model computing machines. If the outputs correspond to transitions between states, the finite state machine is called **Mealy machine**, since they were introduced by G.H. Mealy. If the output is determined only by the state, finite state machine is called **Moore Machine**, since they were introduced by E.F. Moore.

In the previous section the finite automata we consider have binary outputs i.e. either they accept the string or they do not accept the string. This acceptability was decided on the basis of reachability of final state by the initial state. Now we remove the restriction and consider the model where the outputs can be chosen from some other alphabet. The value of the output function $Z(t)$ in the most general case is a function $f$ the present state $q(t)$ and the present input $x(t)$.

$Z(t) = \lambda\{q(t), x(t)\}$, where $\lambda$ is called the output function. This generalised model is usually called the Mealy **machine**. If the output function $Z(t)$ depends only on the present state and is independent of the current input, the output function may be written as

$Z(t) = \lambda \{q(t)\}$, This restricted model is called the **Moore Machine**.

A **Moore Machine** is a 6-tuple $(Q, \Sigma, \Delta, \delta, \lambda, q_0)$, where

(i) Q is a finite set of states.
(ii) $\Sigma$ is the input alphabet.
(iii) $\Delta$ is the output alphabet.
(iv) $\delta$ is the transition function $\Sigma \times Q$ into Q
(v) $\lambda$ is the output function mapping Q into $\Delta$
(vi) $q_0$ is the initial state.

A **Mealy Machine** is a 6-tuple $(Q, \Sigma, \Delta, \delta, \lambda, q_0)$, where all the symbols except $\lambda$ have same meaning as in **Moore Machine**. $\lambda$ is the output function mapping $\Sigma \times Q$ into $\Delta$.

**Example 14:** Describe the Moore Machine given by the figure

| | Input | | |
| State | 1 | 2 | Output |
| --- | --- | --- | --- |
| A | B | C | 0 |
| B | C | D | 0 |
| C | D | E | 0 |
| D | E | B | 0 |
| E | B | C | 1 |

*Solution*

Here, the set of states Q is {A, B, C, D, E}, the set of input symbols $\Sigma$ is {1, 2}, and set of output letters O is {0, 1}.

The double arrow pointing to A denotes that A is initial state $q_0$.

**Example of Moore Machine:**

| | Next state $\delta$ | | Output |
| Present State | $a = 0$ | $a = 1$ | $\lambda$ |
| --- | --- | --- | --- |
| $q_0$ | $q_3$ | $q_1$ | 0 |
| $q_1$ | $q_1$ | $q_2$ | 1 |
| $q_2$ | $q_2$ | $q_3$ | 0 |
| $q_3$ | $q_3$ | $q_0$ | 0 |

For the input string 0111, the transition of states is given by $q_0 \rightarrow q_3 \rightarrow q_0 \rightarrow q_1 \rightarrow q_2$. The output string is 00010. For the input string $\wedge$, the output is $\lambda(q_0) = 0$.

**Example of Mealy Machine:**

|  | **Next state** | | | |
| --- | --- | --- | --- | --- |
|  | **a = 0** | | **a = 1** | |
| **Present State** | **state** | **output** | **state** | **output** |
| $q_1$ | $q_3$ | 0 | $q_2$ | 0 |
| $q_2$ | $q_1$ | 1 | $q_4$ | 0 |
| $q_3$ | $q_2$ | 1 | $q_1$ | 1 |
| $q_4$ | $q_4$ | 1 | $q_3$ | 0 |

For the input string 0011, the transition of states is given by $q_1 \to q_3 \to q_2 \to q_4 \to q_3$. The output string is 0100. In the case of Mealy Machine, we get an output only on the application of an input symbol. So for the input string $\wedge$, the output is only $\wedge$.

**Example 15:** Design a Moore machine to generate 1's complement of given binary number.

***Solution***

As we know, to generate 1's complement of given binary number the simple logic is that if input is 0 then output will be 1 and if the input is 1 then the output will be 0.
Transition table is given by:

|  | **Next State $\delta$** | | **Output** |
| --- | --- | --- | --- |
| **Present State** | **a = 0** | **a = 1** | **$\lambda$** |
| $q_0$ | $q_1$ | $q_2$ | 0 |
| $q_1$ | $q_1$ | $q_2$ | 1 |
| $q_2$ | $q_2$ | $q_2$ | 0 |

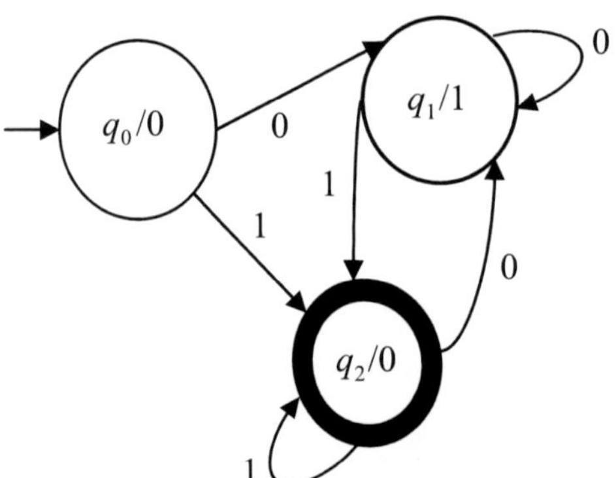


The Moore machine M = (Q, $\Sigma$, $\Delta$, $\delta$, $\lambda$, $q_0$), where Q = $\{q_0, q_1, q_2\}$, $\Sigma$ = (0, 1) and $\Delta$ = $\{0, 1\}$.

**Example 16:**  Design a Mealy machine which will increment the given binary number by 1.

*Solution*

In this we will read the binary number from LSB one at a time

e.g.

|     |     |     |     |     |     |
|-----|-----|-----|-----|-----|-----|
| 1   | 1   | 0   | 0   | 1   | 1   |
| 1   | 1   | 0   | 1   | 0   | 0   |

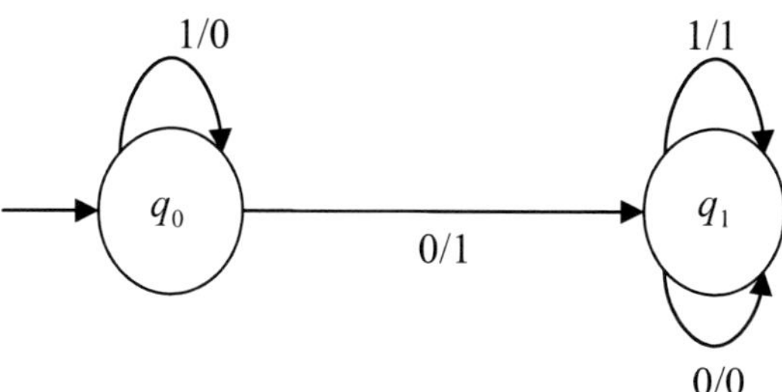

**Example 17:**  Design a Mealy machine to find 2's complement of a given binary number.

*Solution*

To find 2's complement of a binary number we assume that input is read from LSB to MSB. We will keep the binary number as it is until we read first1. Keep this 1 as it is and then change the remaining 1's by 0's and 0's by 1's.

Thus, 2's complement of 10011 is 01101.

The required Mealy machine will be

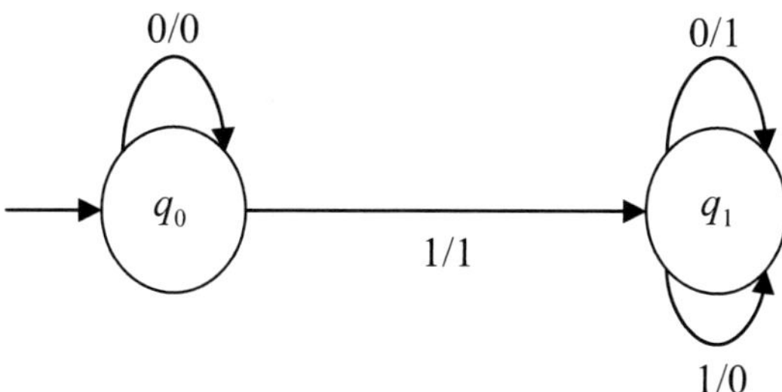

**Example 18:**  Construct a Moore machine to determine residue mod 3 for binary number.

*Solution*

This is also called remainder 3 tester. In this machine we will get remainder 0, remainder 1 and remainder 2.

To interpret the given binary number in its decimal value, we consider $n$ as a number. If 0 is written after $n$ then its value becomes $2n$ and if 1 is written after $n$ then its value becomes $2n + 1$.

For example, if $n = 0$ then its decimal value is 0 then

$$01 = 2n + 1 = 2 \times 0 + 1 = 1$$
$$011 = 2n + 1 = 2 \times 1 + 1 = 3$$
$$0111 = 2n + 1 = 2 \times 3 + 1 = 7$$

If $n = 1$ then its decimal value is 1, then
$$10 = 2n = 2 \times 1 = 2$$
$$101 = 2n + 1 = 2 \times 2 + 1 = 4 + 1 = 5$$
$$1010 = 2n = 2 \times 5 = 10$$

With the help of this logic, we can construct Moore machine as follows:

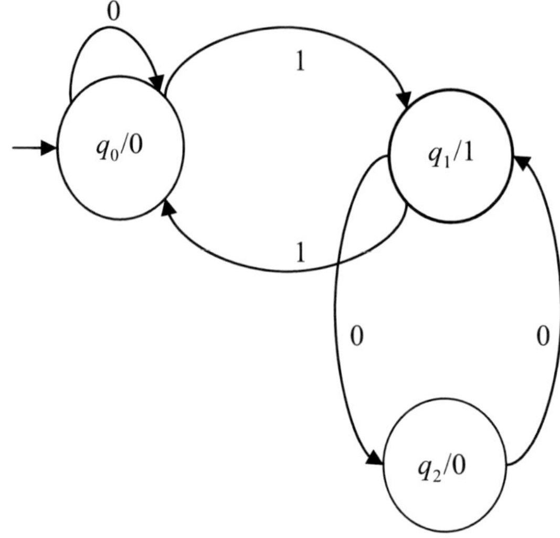


Here $q_0$ is the start state and is considered as remainder 0 state, $q_1$ is considered as remainder 1 state and $q_2$ is regarded as remainder 2 state.

**Example 19:** Design a Moore machine for a binary input sequence such that if it has a substring 101 the machine output A if input has substring 110 it outputs B otherwise it output C.

### *Solution*

First we design a machine in which if we get 101 the output will be A and if we recognize 110 the output will be B and for other strings the output will be C.

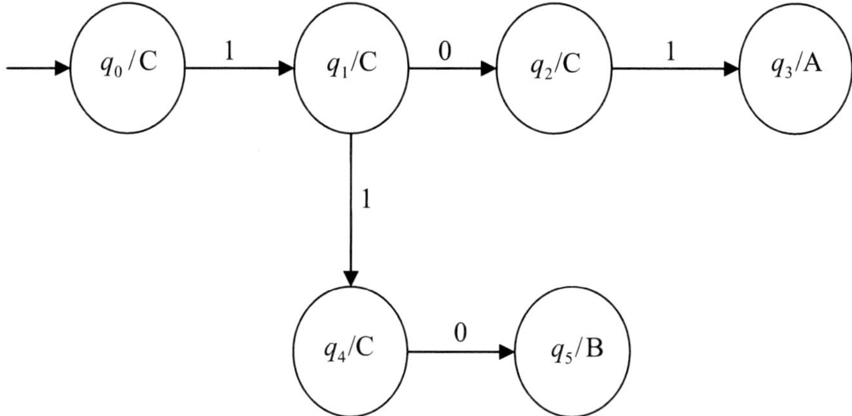

Now we will insert the possibilities of 1's and 0's for each state then it becomes

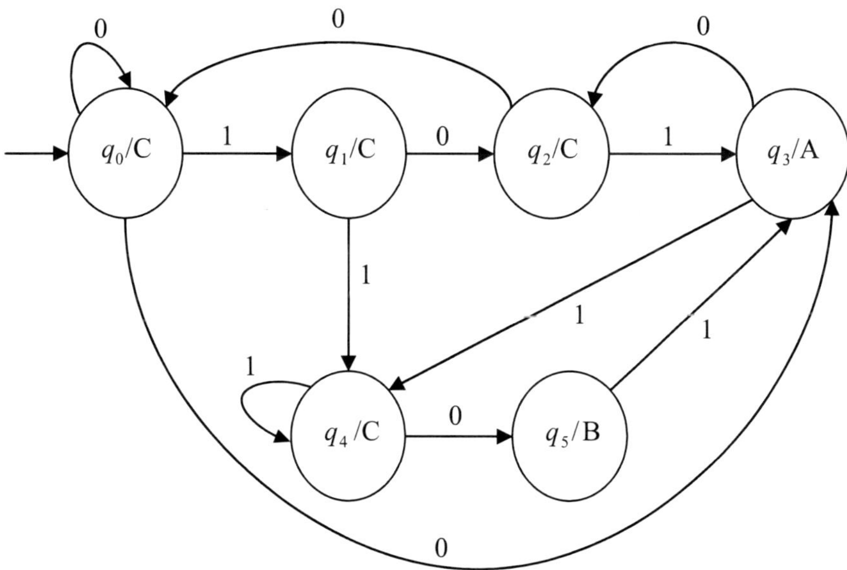

## 13.9 MINIMIZATION OF FSM

Occasionally, we may find that two finite state machines perform the same tasks, i.e. give the same output strings for the same input string, but contain different numbers of states. The question arises that some of the states are "essentially redundant". We may wish to find a process to reduce the number of states. In other words, we would like to simplify the finite state machine. The answer to this question is given by the Minimization process.

### 13.9.1 Equivalent States

In a finite state machine, two states $s_1$ and $s_2$ are said to be equivalent if for any input sequence the machine will produce the same output sequence whether it starts in $s_1$ or $s_2$.

How to determine whether two states are equivalent?

- 0-equivalent: Two states are said to be 0-equivalent if they have the same output.
- 1-equivalent: Two states are said to be 1-equivalent if they have the same output and if, for every input letter, their successors are 0-equivalent.
- $k$-equivalent: Two states are said to be $k$-equivalent if they have the same output and if, for every input letter, their successors are $(k-1)$-equivalent.

Two states are equivalent if they are $k$-equivalent for all $k$. $k$-equivalent is a equivalent relation, and forms a partition of the set of states.

### PROPOSITIONS

(a) For every $k \in N$, the relation $\cong k$ is an equivalence relation on S.
(b) Suppose that $s_i, s_j \in S$, $k \in N$ and $k \geq 2$, and that $s_i \cong {}_k s_j$, then $s_i \cong {}_{k-1} s_j$. In other words, $k$-equivalence implies $(k-1)$-equivalence.
(c) Suppose that $s_i, s_j \in S$ and $k \in N$, then $s_i \cong {}_{k+1} s_j$ if and only if $s_i \cong {}_k s_j$, and $v(s_i, x) \cong {}_k v(s_j, x)$ for every $x \in I$.

### The Minimization Process

(1) Start with $k = 1$. By examining the rows of the transition table, determine 1-equivalent states. Denote by $P_1$ the set of 1-equivalence classes of S (induced by $\cong 1$).
(2) Let $P_k$ denote the set of $k$-equivalence classes of S (induced by $\cong k$). In view of proposition (b), we now examine all the states in each $k$-equivalence class of S, and use proposition (c) to determine $P_{k+1}$, the set of all $(k+1)$-equivalence classes of S (induced by $\cong k + 1$).
(3) If $P_{k+1} \neq P_k$, then increase k by 1 and repeat (2).
(4) If $P_{k+1} = P_k$ then the process is complete. We select one state from each equivalence class.

**Example 20:** Suppose that $I = O = \{0, 1\}$. Consider a finite state machine with the transition table below:

|       | $w$ | | $v$ | |
|-------|---|---|---|---|
|       | 0 | 1 | 0 | 1 |
| $s_1$ | 0 | 1 | $s_4$ | $s_3$ |
| $s_2$ | 1 | 0 | $s_5$ | $s_2$ |
| $s_3$ | 0 | 0 | $s_2$ | $s_4$ |
| $s_4$ | 0 | 0 | $s_5$ | $s_3$ |
| $s_5$ | 1 | 0 | $s_2$ | $s_5$ |
| $s_6$ | 1 | 0 | $s_1$ | $s_6$ |

For the time being, we do not indicate which state is the starting state. Clearly

$$P_1 = \{\{s_1\}, \{s_2, s_5, s_6\}, \{s_3, s_4\}\}.$$

We now examine the 1-equivalence classes $\{s_2, s_5, s_6\}$ and $\{s_3, s_4\}$ separately to determine the possibility of 2-equivalence. Note at this point that two states from different 1-equivalence classes cannot be 2- equivalent, in view of Proposition (b). Consider first $\{s_2, s_5, s_6\}$. Note that

$$v(s_2, 0) = s_5 \cong_1 s_2 = v(s_5, 0) \quad \text{and} \quad v(s_2, 1) = s_2 \cong_1 s_5 = v(s_5, 1)$$

It follows from Proposition (c) that $s_2 \cong_2 s_5$. On the other hand,

$$v(s_2, 0) = s_5 \not\cong s_1 = v(s_6, 0)$$

It follows that $s_2 \not\cong_2 s_6$, and hence $s_5 \not\cong_2 s_6$ also. Consider now $\{s_3, s_4\}$. Note that

$$v(s_3, 0) = s_2 \cong s_5 = v(s_4, 0) \quad \text{and} \quad v(s_3, 1) = s_4 \cong s_3 = v(s_4, 1)$$

It follows that $s_3 \cong s_4$. Hence

$$P_2 = \{\{s_1\}, \{s_2, s_5\}, \{s_3, s_4\}, \{s_6\}\}.$$

We now examine the 2-equivalence classes $\{s_2, s_5\}$ and $\{s_3, s_4\}$ separately to determine the possibility of 3-equivalence. Consider first $\{s_2, s_5\}$. Note that

$$v(s_2, 0) = s_5 \cong_2 s_2 = v(s_5, 0) \quad \text{and} \quad v(s_2, 1) = s_2 \cong_2 s_5 = v(s_5, 1)$$

It follows from Proposition (c) that $s_2 \cong_3 s_5$. On the other hand,

$$v(s_3, 0) = s_2 \cong_2 s_5 = v(s_4, 0) \quad \text{and} \quad v(s_3, 1) = s_4 \cong_2 s_3 = v(s_4, 1)$$

It follows that $s_3 \cong_3 s_4$. Hence

$$P_3 = \{\{s_1\}, \{s_2, s_5\}, \{s_3, s_4\}, \{s_6\}\}$$

Clearly $P_2 = P_3$. It follows that the process ends. We now choose one state from each equivalence class to obtain the following simplified transition table:

|       | $w$ | | $v$ | |
|-------|-----|---|-----|---|
|       | 0 | 1 | 0 | 1 |
| $s_1$ | 0 | 1 | $s_3$ | $s_3$ |
| $s_2$ | 1 | 0 | $s_2$ | $s_2$ |
| $s_3$ | 0 | 0 | $s_2$ | $s_3$ |
| $s_6$ | 1 | 0 | $s_1$ | $s_6$ |

Next, we repeat the entire process in a more efficient form. We observe easily from the transition table that

$$P_1 = \{\{s_1\}, \{s_2, s_5, s_6\}, \{s_3, s_4\}\}$$

Let us label these three 1-equivalence classes by A, B, C respectively. We now attach an extra column to the right hand side of the transition table where each state has been replaced by the 1-equivalence class it belongs to. Next, we repeat the information on the next-state function, again where each state has been replaced by the 1-equivalence class it belongs to.

| | $w$ | | $v$ | | |
|---|---|---|---|---|---|
| | 0 | 1 | 0 | 1 | $\cong_1$ |
| $s_1$ | 0 | 1 | $s_4$ | $s_3$ | A |
| $s_2$ | 1 | 0 | $s_5$ | $s_2$ | B |
| $s_3$ | 0 | 0 | $s_2$ | $s_4$ | C |
| $s_4$ | 0 | 0 | $s_5$ | $s_3$ | C |
| $s_5$ | 1 | 0 | $s_2$ | $s_5$ | B |
| $s_6$ | 1 | 0 | $s_1$ | $s_6$ | B |

| | $w$ | | $v$ | | | $v$ | |
|---|---|---|---|---|---|---|---|
| | 0 | 1 | 0 | 1 | $\cong_1$ | 0 | 1 |
| $s_1$ | 0 | 1 | $s_4$ | $s_3$ | A | C | C |
| $s_2$ | 1 | 0 | $s_5$ | $s_2$ | B | B | B |
| $s_3$ | 0 | 0 | $s_2$ | $s_4$ | C | B | C |
| $s_4$ | 0 | 0 | $s_5$ | $s_3$ | C | B | C |
| $s_5$ | 1 | 0 | $s_2$ | $s_5$ | B | B | B |
| $s_6$ | 1 | 0 | $s_1$ | $s_6$ | B | A | B |

Recall that to get 2-equivalence, two states must be 1-equivalent, and their next states must also be 1-equivalent. In other words, if we inspect the new columns of our table, two states are 2-equivalent precisely when the patterns are the same. Indeed,

$$P_2 = \{\{s_1\}, \{s_2, s_5\}, \{s_3, s_4\}, \{s_6\}\}$$

Let us label these four 2-equivalence classes by A, B, C, D respectively. We now attach an extra column to the right hand side of the transition table where each state has been replaced by the 2-equivalence class it belongs to.

| | $w$ | | $v$ | | | $v$ | | |
|---|---|---|---|---|---|---|---|---|
| | 0 | 1 | 0 | 1 | $\cong_1$ | 0 | 1 | $\cong_2$ |
| $s_1$ | 0 | 1 | $s_4$ | $s_3$ | A | C | C | A |
| $s_2$ | 1 | 0 | $s_5$ | $s_2$ | B | B | B | B |
| $s_3$ | 0 | 0 | $s_2$ | $s_4$ | C | B | C | C |
| $s_4$ | 0 | 0 | $s_5$ | $s_3$ | C | B | C | C |
| $s_5$ | 1 | 0 | $s_2$ | $s_5$ | B | B | B | B |
| $s_6$ | 1 | 0 | $s_1$ | $s_6$ | B | A | B | D |

Next, we repeat the information on the next-state function, again where each state has been replaced by the 2-equivalence class it belongs to.

| | $w$ | | $v$ | | | $v$ | | | $v$ | |
|---|---|---|---|---|---|---|---|---|---|---|
| | 0 | 1 | 0 | 1 | $\cong_1$ | 0 | 1 | $\cong_2$ | 0 | 1 |
| $s_1$ | 0 | 1 | $s_4$ | $s_3$ | A | C | C | A | C | C |
| $s_2$ | 1 | 0 | $s_5$ | $s_2$ | B | B | B | B | B | B |
| $s_3$ | 0 | 0 | $s_2$ | $s_4$ | C | B | C | C | B | C |
| $s_4$ | 0 | 0 | $s_5$ | $s_3$ | C | B | C | C | B | C |
| $s_5$ | 1 | 0 | $s_2$ | $s_5$ | B | B | B | B | B | B |
| $s_6$ | 1 | 0 | $s_1$ | $s_6$ | B | A | B | D | A | D |

Recall that to get 3-equivalence, two states must be 2-equivalent, and their next states must also be 2-equivalent. In other words, if we inspect the new columns of our table, two states are 3-equivalent precisely when the patterns are the same. Indeed,

$$P_3 = \{\{s_1\}, \{s_2, s_5\}, \{s_3, s_4\}, \{s_6\}\}$$

Let us label these four 3-equivalence classes by A, B, C, D respectively. We now attach an extra column to the right hand side of the transition table where each state has been replaced by the 3-equivalence class it belongs to.

| | $w$ | | $v$ | | | $v$ | | | $v$ | | |
|---|---|---|---|---|---|---|---|---|---|---|---|
| | 0 | 1 | 0 | 1 | $\cong_1$ | 0 | 1 | $\cong_2$ | 0 | 1 | $\cong_3$ |
| $s_1$ | 0 | 1 | $s_4$ | $s_3$ | A | C | C | A | C | C | A |
| $s_2$ | 1 | 0 | $s_5$ | $s_2$ | B | B | B | B | B | B | B |
| $s_3$ | 0 | 0 | $s_2$ | $s_4$ | C | B | C | C | B | C | C |
| $s_4$ | 0 | 0 | $s_5$ | $s_3$ | C | B | C | C | B | C | C |
| $s_5$ | 1 | 0 | $s_2$ | $s_5$ | B | B | B | B | B | B | B |
| $s_6$ | 1 | 0 | $s_1$ | $s_6$ | B | A | B | D | A | D | D |

It follows that the process ends. We now choose one state from each equivalence class to obtain the simplified transition table as earlier.

**Unreachable States**

There may be a further step following minimization, as it may not be possible to reach some of the states of a finite state machine. Such states can be eliminated without affecting the finite state machine. We shall illustrate this point by considering an example.

**Example 21:** Note that in Example 5, we have not included any information on the starting state. If $s_1$ is the starting state, then the original finite state machine can be described by the following state diagram:

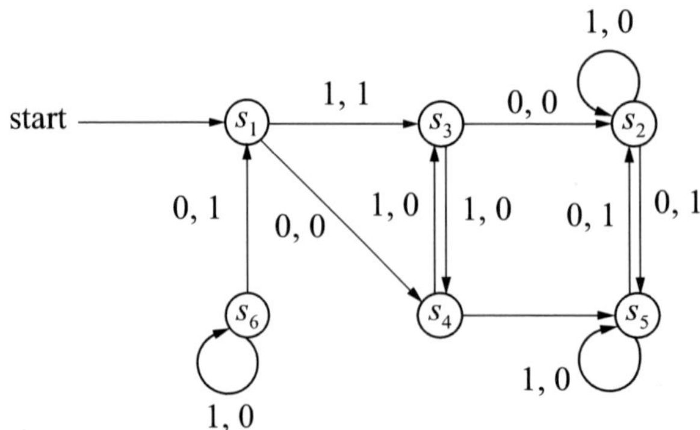

After minimization, the finite state machine can be described by the following state diagram:

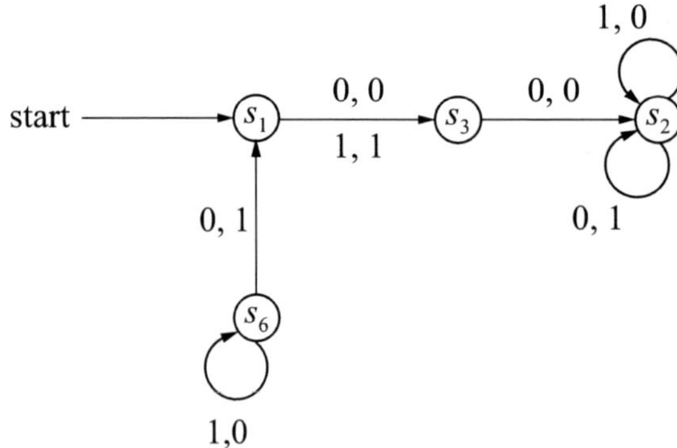

It is clear that in both cases, if we start at state $s_1$, then it is impossible to reach state $s_6$. It follows that state $s_6$ is essentially redundant, and we may omit it. As a result, we obtain a finite state machine described by the transition table

|       | $w$ | | $v$ | |
|-------|-----|-----|-------|-------|
|       | **0** | **1** | **0** | **1** |
| $s_1$ | 0 | 1 | $s_3$ | $s_3$ |
| $s_2$ | 1 | 0 | $s_2$ | $s_2$ |
| $s_6$ | 0 | 0 | $s_2$ | $s_3$ |

or the following state diagram:

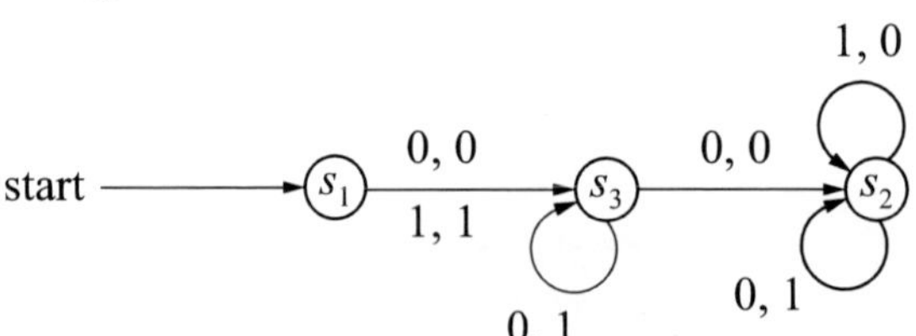

**States** such as $s_6$ in Example 21 are said to be unreachable from the starting state, and can therefore be removed without affecting the finite state machine, provided that we do not alter the starting state. Such unreachable states may be removed prior to the Minimization process, or afterwards if they still remain at the end of the process. However, if we design our finite state machines with care, then such states should normally not arise.

**Limitation of Machines**: There is no finite state machine that can perform binary multiplication.

### Exercise 13.1

1. Let A = {$a, b$}. Construct an automaton M which will accept those words from A which begin with a followed by (zero or more) $b$'s.

2. Suppose L is a language over A which is accepted by the automaton M = (A, S, Y, $s_0$, F). Find an automaton N which accepts L', that is, those words from A which do not belong to L.

3. Let M be the finite state machine with state table appearing in Fig.

   (a) Find the input set A, the state set S, the output set Z. and the initial state.

   (b) Draw the state diagram D = D(M) of M.

   (c) Suppose $w = aababaabbab$ is an input word (string). Find the corresponding output word $v$.

| F | $a$ | $b$ |
|---|-----|-----|
| $s_0$ | $s_1, x$ | $s_2, y$ |
| $s_1$ | $s_3, y$ | $s_1, z$ |
| $s_2$ | $s_1, z$ | $s_0, x$ |
| $s_3$ | $s_0, z$ | $s_2, x$ |

4. Construct a non-deterministic finite automaton accepting {$ab, ba$}, and use it to find a deterministic automaton accepting the same set.

5. Construct a non-deterministic finite automaton accepting the set of all strings over {$a, b$} in $aba$. Use it to construct DFA accepting the same set of strings.

6. Design a Mealy machine which calculates residue mode 4 for each binary string treated as binary integer.

7. Give Moore machine for $\Sigma = \{0, 1, 2\}$ print the residue MOD 5 of input treated as a ternary number.

8. Construct the minimum state automaton for the transition diagram given below.

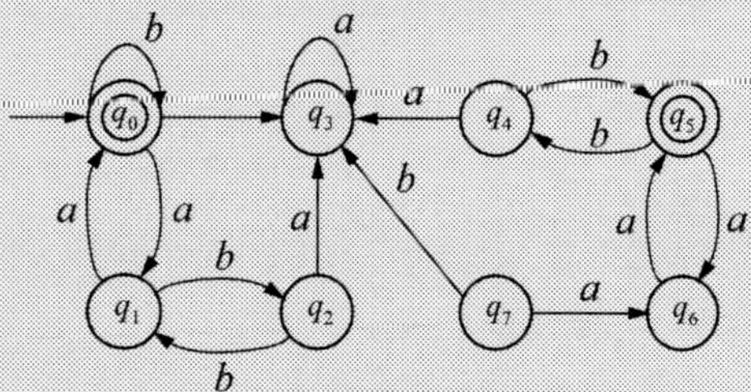

**Answers to Selected Problems**

1.

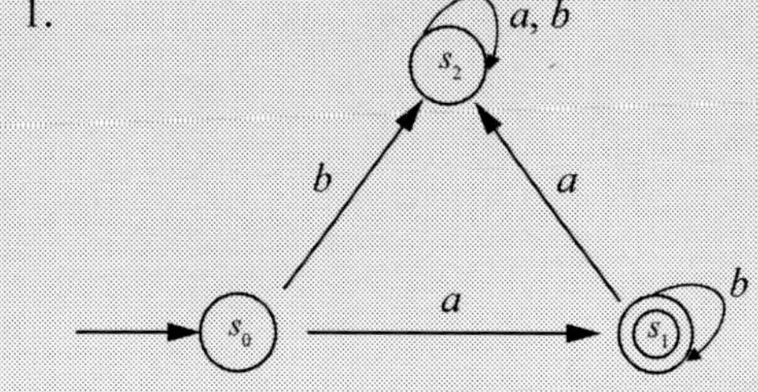

2. Interchange accepting and rejecting states in M to obtain N. Now $w$ will be accepted in the new machine N iff $w$ is rejected in M i.e., iff $w$ belongs to L.

3. (b)

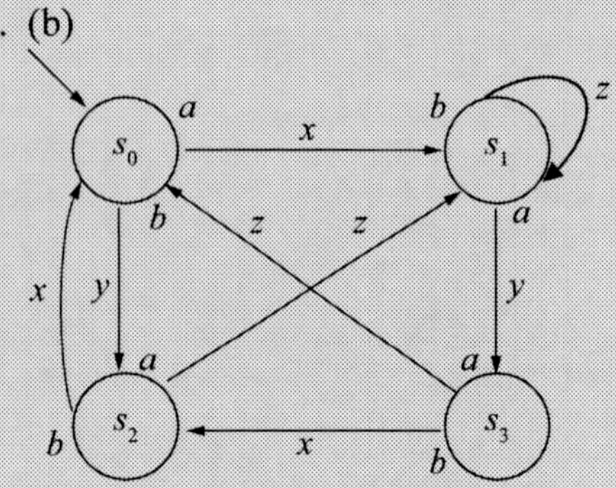



(c) Output word $v = xyxzzyzyxxz$.

# Index